Membrane Transporter Diseases

Membrane Transporter Diseases

Edited by

Stefan Bröer

Australian National University
Canberra, Australia

and

Carsten A. Wagner

University of Zurich
Zurich, Switzerland

Kluwer Academic/Plenum Publishers
New York, Boston, Dordrecht, London, Moscow

Library of Congress Cataloging-in-Publication Data

Membrane transporter diseases/edited by Stefan Bröer and Carsten A. Wagner.
     p.   cm.
   Includes bibliographical references and index.
   ISBN 0-306-47883-8
      1. Carrier proteins—Pathophysiology.   I. Bröer, Stefan.   II. Wagner, Carsten A.

   RB113.M467 2004
   616.3'9042—dc22

                                                                          2003054471

ISBN 0-306-47883-8

©2003 Kluwer Academic/Plenum Publishers, New York
233 Spring Street, New York, New York 10013

http://www.wkap.com

10   9   8   7   6   5   4   3   2   1

A C.I.P. record for this book is available from the Library of Congress

Permissions for books published in Europe: permissions@wkap.nl
Permissions for books published in the United States of America: permissions@wkap.com

Printed in the United States of America

Preface

Every cell and organism faces the problem of generating a confined space in which metabolic and anabolic reactions take place and at the same time allowing entry and exit of metabolites, ions, proteins, and signals across its border. Evolution has solved the problem by generating lipid membranes that contain transporters, ion channels, and receptors. In eukaryotic cells, this problem is exacerbated by the presence of multiple organelles, which are confined spaces in their own right. Even the lipid membrane consists of two relatively separate spaces, made up of the two leaflets of the lipid bilayer. The importance of traffic and signaling across membranes is reflected by the estimate that 20% of all genes in the human genome encode membrane proteins. A failure of any of these proteins may have dramatic consequences for cell function. In recent years much attention has been paid to diseases resulting from nonfunctional ion channels ("channelopathies"). Not surprisingly, many of these diseases affect the excitability of cells. Transporter diseases (perhaps coined "carrier

diseases") are more related to metabolic diseases. Transporters are frequently found at the beginning or the end of metabolic pathways and as a result can have similar effects to a missing enzyme. The pathological effect is often the result of an overload or lack of a certain metabolite or ion. Thus, "carrier diseases" are quite different from "channelopathies" and, in our view, deserve to be treated in a comprehensive monograph.

The idea of this book was born while both editors were working together in S.B.'s laboratory during a 3 months leave of absence of C.A.W. It may have been the elegant design of the native Australian Joey-transporters (photograph: from left to right C.A.W., Kangaroo with Joey, S.B.) that finally convinced us to contact authors for the individual chapters of the book.

Transporters are such a diverse class of proteins that we decided to write introductory chapters to functionally related groups of transporters. These are aimed at more inexperienced readers. Rather than providing a complete overview of transport functions, the introductions are specifically tailored to aid in the understanding of the individual expert chapters.

Each chapter is concluded by a summary, and most chapters also contain an overview of the clinical features of a particular transporter disease.

The editors are indebted to the many people who helped us in the preparation of the book. First of all, we would like to thank the authors of the individual chapters for their contributions. In addition, many friends and colleagues helped us with specific parts of the book. Chapters have been reviewed and proofread by a number of people, and we would especially like to thank Kiaran Kirk, Susan Howitt, Juleen Kavanaugh, Alice Kingsland, and Pauline Junankar, and also Heini Murer, Jürg Biber, and Francois Verrey for many critical discussions. S.B. would like to thank Angelika for her understanding and patience during the final stages of the book, and Karl-Ernst Bröer for valuable discussion about the term "carrier diseases."

Finally, we would like to thank Kluwer Academic Publishers and their staff for the opportunity to publish this book.

Canberra and Zürich, May 2003

Contents

List of contributors

Jinhi Ahn

Seth Alper
Harvard Medical School
Molecular Medicine and Renal Units
RW763 Beth Israel Deaconess Medical Center
East Campus, 330 Brookline Avenue
Boston, MA 02215
USA
e-mail: salper@caregroup.harvard.edu

Cristina Amat di San Filippo

Silke Beismann-Driemeyer

Joan Bertran

Stefan Bröer
School of Biochemistry and
    Molecular Biology
Faculty of Science
Australian National University
Canberra ACT 0200
Australia
e-mail: stefan.broeer@anu.edu.au

Josep Chillarón

Janice Yang Chou
Section of Cellular Differentiation
HDB, NICHD, NIH
Building 10, Room 9S241
National Institutes of Health
Bethesda
MD 20892-1830
USA
e-mail: chou@helix.nih.gov

Judith C. Fleming

Marc Foretz

Laura M. Garrick

Michael D. Garrick
State University of New York
140 Farber Hall, Buffalo
NY 14214-3000
USA
e-mail: mgarrick@buffalo.edu

Giuseppe Inesi
University of Maryland
School of Medicine
108 N. Greene Street
Baltimore
MD 21201-1503
USA
e-mail: ginesi@umaryland.edu

Wolfgang E. Kaminski

Keiko Kobayashi
Department of Biochemistry
Faculty of Medicine
Kagoshima University
8-35-1 Sakuragaoka
Kagoshima 890-8520
Japan
e-mail: dodoko12@m.kufm.kagoshima-u.
ac.jp

Klaus Peter Lesch
Department of Psychiatry and Psychotherapy
University of Würzburg

Füchsleinstrasse 15
97080 Würzburg
Germany
e-mail: kplesch@mail.uni-wuerzburg.de

Nicola Longo
Division of Medical Genetics
Department of Pediatrics
University of Utah
2C412 SOM
50 North Medical Drive
Salt Lake City UT 84103
USA
e-mail: Nicola.Longo@hsc.utah.edu

Grazia M. S. Mancini

Brian C. Mansfield

Daniel Markovich
School of Biomedical Sciences
Department of Physiology and
    Pharmacology
University of Queensland
Brisbane, QLD 4072
Australia
e-mail: d.markovich@uq.edu.au

Robert S. Molday
Biochemistry and Molecular Biology
Faculty of Medicine
University of British Columbia
2146 Health Sciences Mall
Vancouver, BC V6T1Z3
Canada
e-mail: molday@interchange.ubc.ca

Ellis J.Neufeld
Children's Hospital
Division of Hematology
300 Longwood Avenue
Boston
MA 02115
USA
e-mail: ellis.neufeld@tch.harvard.edu

Manuel Palacín
Dept. Bioquimica i Biologia Molecular

Universidad Barcelona
Avenida Diagonal 645
08028 Barcelona
Spain
e-mail: mpalacin@porthos.bio.ub.es

Marzia Pasquali

Antonello Pietrangelo
Unit for the Study of Disorders of
    Iron Metabolism
Department of Internal Medicine
University of Modena and Reggio
    Emilia
Via del Pozzo 71, 41100 Modena
Italy
e-mail: pietrangelo.antonello@unimo.it

Rajini Rao
Johns Hopkins University School of
Medicine
725 N. Wolfe Street
Baltimore, MD 21205
USA
e-mail: rrao@jhmi.edu

Michael F. Romero
Department of Physiology and
    Biophysics
Case Western Reserve University
School of Medicine
2119 Abington Road
Cleveland
OH 44106-4970
USA
e-mail: mfr2@po.cwru.edu

Takeyori Saheki

Gerd Schmitz
Institute for Clinical Chemistry and
    Laboratory Medicine
University Hospital Regensburg
Franz-Josef-Strauss-Allee 11
D-93053 Regensburg
Germany
e-mail: gerd.schmitz@klinik.
uni-regensburg.de

Robert Tampé
Institute of Biochemistry
Biocenter Frankfurt
Goethe-University, Frankfurt
Marie-Curie-Strasse 9
D-60439 Frankfurt
Germany
e-mail: tampe@em.uni-frankfurt.de

Bernard Thorens
Institute of Pharmacology and Toxicology
27 rue du Bugnon
CH 1005 Lausanne
Switzerland
e-mail: Bernard.Thorens@ipharm.unil.ch

Takeshi Uchiumi

Frans W. Verheijen
Department of Clinical Genetics
Erasmus MC

PO Box 1738
3000 DR Rotterdam
The Netherlands
e-mail: f.verheijen@erasmusmc.nl

Morimasa Wada
Department of Medical Biochemistry
Graduate School of Medical Sciences
Kyushu University
Fukuoka 812-8582
Japan
e-mail: wada@biochem1.med.kyushu-u.
ac.jp

Carsten Wagner
University of Zürich
Institute of Physiology
Winterthurerstrasse 190
Zürich 8057
Switzerland
e-mail: Wagnerca@physiol.unizh.ch

Membrane Transporter Diseases

STEFAN BRÖER* AND CARSTEN A. WAGNER**

Introduction to membrane transport

WHY DO WE NEED TRANSPORTERS?

Cells and cell compartments such as mito-chondria, endosomes, or lysosomes are surrounded by a membrane formed by lipid bilayers. This compartmentalization of organs and cells is a prerequisite for the organization of complex metabolic pathways, but also requires the shuttling of metabolites between compartments as well as the expression of distinct sets of enzymes, and the supply and removal of their substrates and products. Most substrates, however, are not lipophilic and cannot move freely through the lipid bilayers, which is otherwise necessary to ensure the specificity and local restriction of enzymatic reactions and to prevent the free distribution of potentially toxic or valuable substrates. Thus most substrates have to be transported across membranes. In general, two modes of crossing a membrane can be distinguished on the basis of certain characteristics: diffusion and carrier-mediated transport. Diffusion describes the movement of ions or substrates from a place of high concentration to a place of lower concentration along its concentration gradient.

As summarized in Fick's law of diffusion, the rate of diffusion depends on the initial concentration gradient, the distance between the two sites (or in case of membranes also their thickness and composition), and the temperature. Ultimately, diffusion results in the equilibration of concentrations and will never allow accumulation of a substrate. Diffusion has no specificity for its substrate, is not saturable, and does not require specific molecular structures. In contrast, carrier-mediated transport is mediated by proteins embedded into the lipid bilayer, shows specificity for its substrates, and can be saturated. Specificity implies the selective transport of substrates as well as the possibility to inhibit transport by transported or nontransported substrate analogs. Facilitated diffusion thus mediates a substrate specific equilibration of concentrations across distinct membranes (expressing the specific transporter). The rate of substrate translocation across the respective membrane is higher in facilitated diffusion than in simple diffusion. The direction of transport is primarily determined by the concentration of the respective substrate on either side of the membrane. However, the coupling of transport processes to enzymatic reactions may drive vectorial transport, as exemplified for glucose transport through the facilitative glucose transporters of the GLUT family (SLC2; see also Chapter 12) where intracellular glycolysis reduces glucose

* School of Biochemistry and Molecular Biology, Faculty of Science, Australian National University, Canberra ACT 0200, Australia

** University of Zürich, Institute of Physiology, Winterthurerstrasse 190, Zürich 8057, Switzerland

concentrations, forcing glucose transport inwards. In addition, many transport processes also depend on an energy source (see below) and are thus able to accumulate substrates against a substrate gradient. Transport systems are usually expressed at sites where the efficient and possibly complete transport of substrates from one compartment to another is required. Classic examples are found in the transporter families mediating the vectorial transport of substrates like glucose, amino acids, or phosphate across the epithelial cells lining the intestine or the renal nephron. In this sense, transporters are essential gatekeepers and helpers for substrate trafficking across membranes and are involved to ensure the specificity of cell or cell organelle function by providing means to regulate the exchange of substrates between different organ and cell compartments.

HOW DO TRANSPORTERS WORK AND WHAT DISCRIMINATES THEM FROM CHANNELS?

Transporters and channels are membrane proteins that mediate the traffic of metabolites and ions across biological membranes. Both types of proteins can adopt two different conformations (Figure 1). The two conformations of channels are "closed" and "open." Substrates (usually ions or small molecules) cannot pass through the channel in the "closed" conformation, whereas in the "open" conformation flux can occur in both directions. Movement of molecules through the open channel is governed by the electrochemical gradient of the transported ion/substrate itself. The two conformations of transporters are "outward-facing" and "inward-facing" (Figure 1). These two conformations are fundamentally different from those adopted by ion channels. Apart from a few debatable exceptions (see below), transporters never adopt an open conformation that allows diffusion-like movement of its substrate(s) along the electrochemical gradient. Transporters are

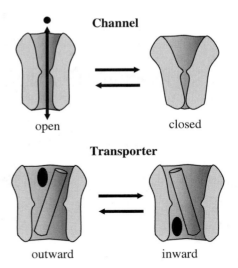

Figure 1 Comparison of conformational changes occurring in transporters and channels. Substrates are depicted as black circles or ovals.

more similar to enzymes than to channels and can be viewed as enzymes catalyzing a vectorial reaction. Substrate and product of the reaction are identical but they are placed in topologically different compartments. To illustrate the similarity to enzymes, transport catalysis can be compared with the reaction carried out by topoisomerases – enzymes that introduce or release superhelical twists in DNA molecules (Figure 2). Substrate and product of the reaction are chemically identical, but the product is topologically different from the substrate because the DNA was moved across another DNA strand. Acting like enzymes, transporters mediate the movement of substrates across membranes as part of a catalytic cycle in which the transport protein adopts two different conformations. Each conformation can bind only a very limited number of substrate molecules (perhaps up to five), and only the bound substrates are translocated across the membrane during (or after) the change from the outward- to the inward-facing conformation. An open channel, by contrast, can pass an indefinite number of molecules as long as it is open. Moreover, once open, movement of substrates through

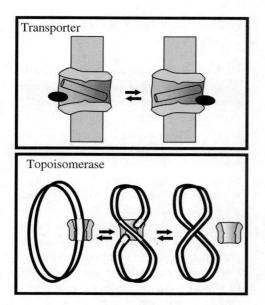

Figure 2 Topological reactions carried out by enzymes and transporters. Substrates and products of topological reactions are chemically identical but are topologically different. Transporter substrates are moved between compartments; topoisomerases induce twists in DNA molecules.

the channel occurs without further significant conformational changes (small dynamic movements of pore-lining residues are also required for permeation of ions through open channels, but the overall conformation of the channel remains "open").

Several lines of experimental evidence firmly support the two-conformation model of transporters.

1. Some transport inhibitors preferentially bind either to the inward- or the outward-facing conformation but not to both (Lauquin and Vignais 1976).
2. Fluorophores attached to certain residues or fluorescing tryptophan residues in the transporter molecule change fluorescence intensity in the absence and presence of substrate or cosubstrate of the transporter (Peerce 1988), indicating movement of these residues.

3. Cleavage of exposed loops by proteases changes in the absence and presence of transporter substrates (Gibbs *et al.* 1988).
4. The ability to modify cysteine residues in some parts of a transporter (or cysteine residues introduced by site-directed mutagenesis) changes in the absence or presence of substrate/cosubstrate even though the substrate does not directly block access of the reagent to the residue (Golovanevsky and Kanner 1999).

Transporters also provide high substrate specificity because weak interactions with substrate atoms occur at several places. Channels, by comparison, mainly discriminate by size and charge but do not discriminate between functional groups or between different molecule shapes.

Another feature that discriminates transporters from channels is their ability to couple the movement of one substrate to the movement of another substrate (or cosubstrate). The full conformational change does not occur until all substrates are bound to the transporter. This allows accumulation of one substrate at the expense of the energy provided by the electrochemical gradient of another substrate.

There are two principally different modes of energizing transport processes, namely *primary active* transport and *secondary active* transport (Figure 3). Primary active transporters use a chemical reaction to induce the conformational change of the transporter that is required for transport. The most common principle is the use of energy provided by the hydrolysis of adenosine triphosphate (ATP). ATP is used, for example, by ABC transporters (see Chapters 17–21) to move lipophilic substances across membranes. P-type ATPases are structurally different from ABC transporters but also use the free energy of ATP hydrolysis to move ions across the membrane (see Chapters 5 and 24). Primary active transporters can be found in physiological processes that require either a high accumulation of the substrate (e.g., Ca^{2+}) or large

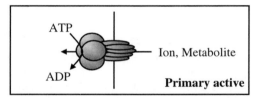

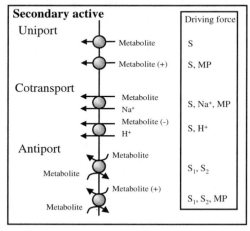

Figure 3 Energization of transport processes. Primary active transport is driven by chemical reactions such as ATP hydrolysis. Secondary active transport is energized by electrochemical gradients of ions and substrates. Driving forces of a particular transport are indicated as follows: S, substrate gradient; MP, membrane potential; Na^+, sodium gradient; H^+, proton gradient.

amounts of energy such as the removal of phospholipids from one leaflet of the membrane followed by its extrusion from the cell. In line with the two-conformation model, the free energy of ATP hydrolysis is used to induce a conformational change in the transport protein, which has recently been visualized in the crystal structure of the Ca^{2+}-ATPase (see below and Chapter 24).

Most transport processes, so-called secondary active transport processes, are driven by the electrochemical gradient(s) of one or several substrates and possibly cosubstrates. Three different mechanisms are recognized, but some transporters may actually obey a mixture of these modes such as the glutamate transporters which translocate glutamate

together with three Na^+-ions and a proton and also in exchange for K^+ (Figure 3).

1. *Uniporters* dissipate the (electro)chemical gradient of the single substrate of the transporter. The substrate can be a neutral or may carry a charge. Such a transport process is also known as facilitated diffusion. To avoid any confusion with the mechanistically different diffusion of ions through an open channel, the term uniport is preferred. Uniporters, for example, mediate uptake and release of glucose in mammalian cells (see Chapter 12). Uniporters for inorganic ions are extremely rare and are usually replaced by ion channels.

2. *Cotransporters* catalyze the concomitant translocation of two or more substrates in the same direction. These can be ions, as for example in the case of the Na/K/2Cl cotransporter (see Chapter 2), or a mixture of ions and metabolites, as for example in the case of the Na^+/thiamine cotransporter (see Chapter 16) or the Na/Cl/serotonin cotransporter (see Chapter 23). The gradient of the cosubstrate is held constant by the cell and hence drives the transport of the substrate. For example, large inward-directed Na^+ gradient is maintained by the Na^+, K^+-ATPase of mammalian cells, which allows the accumulation of amino acids, vitamins, etc. by Na^+ cotransport. Cotransporters may or may not translocate net charges. In the case of the Na^+/thiamine cotransporter, both the Na^+ gradient and the membrane potential (inside negative) will aid in the transport of this vitamin into the cell.

3. *Antiporters* (or exchangers) exchange one substrate against another substrate. Translocation takes place only when substrates are present on both sides of the membrane. As a rule, antiporters can carry out both homologous and heterologous exchange. The aspartate–glutamate exchanger catalyses the obligatory antiport of aspartate against glutamate in the inner mitochondrial membrane (see Chapter 10). Charges of

substrates may differ. As a result some antiport processes are additionally driven by the membrane potential and not only by substrate gradients. This principle is used for example to aid in the uptake of cationic amino acid in the kidney and intestine, which occurs in exchange for neutral amino acids (see Chapter 14).

The features described above clearly discriminate transporters from channels. In contrast with this consensus view of transporter function by a two-conformation-never-open model, channel-like functions have been observed in some transporters. Glutamate transporters, for example, display substrate-induced chloride currents (Fairman and Amara 1999). However, in this case the substrate of the transporter is different from the substrate of the associated ion conductance. Glutamate itself is still transported by an enzyme-like catalytic mechanism. Moreover, it has not yet been resolved whether chloride ions pass through the same pathway as the substrate glutamate or diffuse through a separate pore. Transporters for monoamines (serotonin, epinephrine, and norepinephrine; see Chapter 23), as another example, have been reported to mediate fast diffusion-like transport of their substrates in a channel-like fashion (Galli et al. 1996). These transporters usually couple the movement of a monoamine with the movement of one or two Na^+ ions and one Cl^- ion. However, the channel-like transport mode may also be interpreted as a uniport of the substrate, which is not accompanied by the cotransport of ions under circumstances that remain to be defined (Nelson et al. 2002). As a result, substrate movement would still occur during a reaction cycle involving two conformations but without the concomitant translocation of cosubstrates.

Another notable difference between transporters and channels is the velocity of substrate movement. Movement of ions through a channel involves interactions with side groups of amino acid residues lining the pore, but may in some cases come close to the limit set by diffusion. This is orders of magnitude faster than the velocity provided by the reaction cycle of a transporter involving conformational changes and multiple interactions with the substrate. Conformational changes of transporters occur at rates <200 times per second, and each turnover moves one or a few substrates. During the same period of time channels can pass several thousand ions. As a result, electrochemical equilibrium of ions can be reached in milliseconds, as exemplified by the time course of an action potential, whereas electrochemical equilibrium of transported substrates may only be reached after several minutes.

In summary, transporters and channels are fundamentally different mediators of transmembrane movement, which is also reflected by their structural design (see below). Channels usually mediate fast transport, whereas slow transport, particularly when it involves the expense of larger amounts of energy, is mediated by transporters. Not surprisingly, mutations of ion channels frequently affect excitability of cells (Ashcroft 2000). Mutations of transporters can affect a variety of biological functions, but usually on a slower scale. As a result they resemble metabolic diseases.

WHAT DO TRANSPORTERS LOOK LIKE?

For many years only limited information about the structure and topology of transporters was available. The most conspicuous feature of the primary sequence of transporters is repeated stretches of 20–25 hydrophobic amino acids. These have been interpreted as regions of the transporter molecule that span the lipid bilayer. Hydrophobicity analysis has been developed to depict this characteristic feature of transport proteins, pioneered by the algorithm of Kyte and Doolittle (1982). For hydrophobicity analysis a score is given to each amino acid. The score is based on the solubility of amino acids in aqueous and apolar solutions. Hydrophobic amino acids show a higher solubility than hydrophilic amino acids in apolar solutions. Using these scores, the average hydrophobicity is calculated for the central amino acid of a stretch of 10–20 amino acids. The window is moved one amino acid further and the average hydrophobicity is calculated again. The hydrophobicity plot is generated by moving the window along the complete primary sequence of the protein (Figures 4 and 5). Long stretches of hydrophobic amino acids will appear as peaks in such a plot. As a starting point it is assumed that each peak passing a certain threshold represents a transmembrane region. About 20 amino acids are required to span the lipid bilayer in an α-helical conformation. Thus it is generally assumed that α-helices are the main secondary structural elements of transporters and that they are connected by relatively short loops. For membrane proteins with large loops between the putative membrane spanning regions, such as ion channels, a large window is recommended for the calculation of the hydropathy plot (e.g., 20 amino acids).

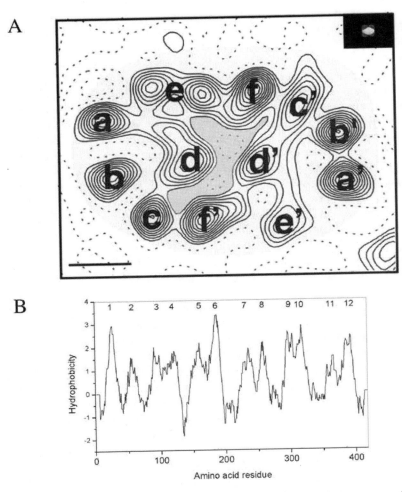

Figure 4 Comparison of (A) the projection structure of the *Oxalobacter formigenes* oxalate–formiate exchanger at 0.8 nm resolution reprinted with permission from (Heymann *et al.* 2001) with (B) the hydropathy plot (Kyte–Doolittle method, window = 12) of its amino acid sequence (for details see text).

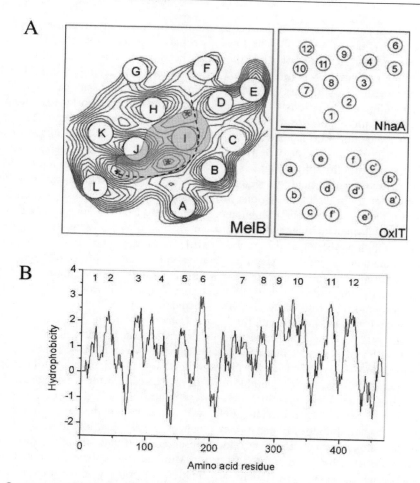

Figure 5 Comparison of (A) the projection structure of the *Escherichia coli* melibiose permease at 0.8 nm resolution reprinted with permission from (Hacksell *et al.* 2002) with (B) the hydropathy plot (Kyte–Doolittle method, window = 12) of its amino acid sequence (for details see text). In panel (A) the helical arrangements of the oxalate–formate exchanger and the *E. coli* Na$^+$/H$^+$ exchanger are shown as small panels on the right.

For membrane proteins with short loops between the membrane spanning regions, such as transporters, a smaller window of 11–12 amino acids is recommended. Using larger windows may result in the fusion of two putative transmembrane regions that are connected by only a short loop.

Although a number of assumptions are made in this type of analysis it has proven to be remarkably successful. Topology analysis of bacterial membrane transport proteins has confirmed that long stretches of amino acids do

indeed span the membrane (Manoil 1991). Most transport proteins have 10–14 transmembrane regions. However, some families, such as the mitochondrial transporter family (SLC25; see Chapter 10), have only about six transmembrane spanning regions. In such cases it is assumed that the transporter is a dimer. Owing to recent progress in the crystallization of membrane proteins, the reliability of hydropathy plots and related methods of membrane topology determination can now be assessed more rigorously. Two examples serve to illustrate the capacity and limitations of hydropathy plots (Figures 4 and 5). The low resolution projection structure of both the oxalate–formate exchanger (Heymann *et al.* 2001) and the melibiose permease (Hacksell *et al.* 2002) indicate 12 regions of higher electron density which span the membrane. In the case of the oxalate–formate exchanger, 12 transmembrane regions are also well predicted by the hydropathy plot (Figure 4). However, in the case of the melibiose permease, assignment of transmembrane regions in the hydropathy plot is much more ambiguous (Figure 5). Thus it appears that hydropathy plots provide a rough prediction of the number of transmembrane helices, but that experimental methods are required for a more precise determination of the topology. Crystallization of membrane proteins (currently more than 30; see http://blanco.biomol.uci.edu/Membrane_Proteins_xtal.html) has also confirmed that, with the exception of bacterial outer membrane proteins, the α-helix is indeed the predominant secondary structure of membrane spanning regions in membrane proteins.

The high-resolution three-dimensional structure of the bacterial multidrug transporter AcrB is the first atomic structure of a secondary active transporter (Murakami *et al.* 2002). Figure 6 shows the transmembrane domain of AcrB from the side (panel A) and from the cytosol (panel B), as generated by Cn3D software (Hogue 1997). The whole protein is much larger, but for simplicity only the transmembrane region is depicted. It demonstrates a number of principles of transporter architecture. The transmembrane domain is constituted by 12 transmembrane helices, which are also predicted by hydrophobicity analysis (not shown). All transmembrane regions have α-helical conformation. Helices are tilted with respect to the plane of the membrane and may have different lengths. An outer ring of helices is grouped around two centrally located helices. Many transporters also show a significant degree of symmetry between the amino-terminal half and the carboxy-terminal half, which is often detectable as a low but significant sequence homology between the two halves of the transporter. This symmetry can be observed in the structure of several transporters – for example, in Figure 4 where symmetric helices in each half of the oxalate/formate antiporter are indicated by unprimed and primed letters. The membrane domain of AcrB also shows mirror image symmetry (Figure 6B). Particularly in the case of antiporters, a symmetric structural design fits well with the symmetric transport function. In contrast, the melibiose permease, which carries out a Na^+ cotransport mechanism, does not display obvious symmetry (Figure 5). It is still not understood how secondary active transporters can adopt two different conformations. Studies of the mammalian glucose transporter SGLT1 and the bacterial lactose permease indicate that only a limited number of transmembrane helices may tilt and rotate during the translocation of the substrate (Wu *et al.* 1998; Wright 2001). Thus it appears possible that a more rigid ring of outer helices forms the scaffold for a movement of one or a few centrally located helices. In agreement with such a notion, the design of a ring of outer helices and two centrally located helices can be observed in all three structures discussed here.

More detail is known about the two conformations of primary active transporters, which in the case of P-type ATPases are called E1 and E2. The Ca^{2+}-ATPase (see Chapter 24) has been crystallized in the Ca^{2+}-free state (E2) (Toyoshima and Nomura 2002) and in the Ca^{2+}-bound state (E1) (Toyoshima *et al.* 2000) thus allowing a comparison between the different conformations. Figure 7 shows the transmembrane domain in the E1 conformation (Figure 7A) and in the E2 conformation (Figure 7B). The large cytosolic domain of the Ca^{2+}-ATPase was omitted. The Ca^{2+}-free conformation was induced by addition of the Ca^{2+}-ATPase inhibitor thapsigargin. The two Ca^{2+} ions in the E1 state interact with residues in helices 4, 5, 6, and 8 (Figure 7A). A change to the Ca^{2+}-free conformation involves a complex movement of several helices. Figures 7A and 7B were oriented so that helices 7 and 10,

A

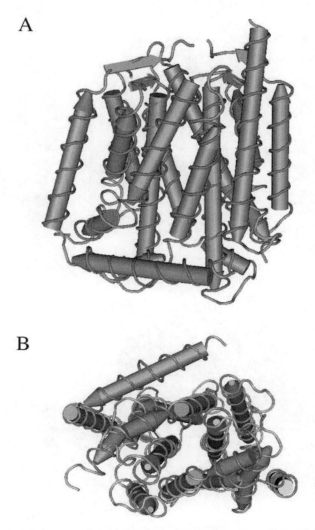

B

Figure 6 High-resolution structure (secondary structure elements shown) of the transmembrane domain of the *E. coli* multiple drug efflux transporter AcrB (Murakami *et al.* 2002). For clarity, periplasmic parts of the protein are not depicted. The picture was generated using Cn3D software of the NCBI database. (A) Membrane domain of AcrB viewed perpendicular to the plane of the membrane. (B) Same domain viewed from the cytosol.

which move very little between the two conformations, have roughly the same orientation. Major movements occur in helix 2, which tilts away from the core of the transmembrane domain. This has an impact on helices 3, 4, and 5, which move into the space cleared by helix 2. This movement causes a dramatic drop of affinity for the two Ca^{2+} ions, forcing them to move out of the transmembrane region. The example of the Ca^{2+}-ATPase illustrates the extent of helical movement that is required to move a substrate across the membrane. However, it also demonstrates that the helix packing and overall topology of transporters is not fundamentally different in the two conformations.

A

B

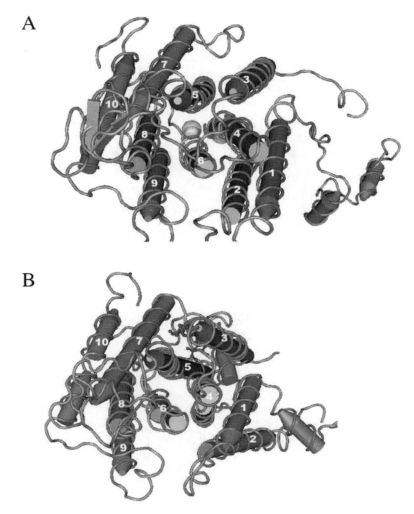

Figure 7 High-resolution structure of the transmembrane domain of the Ca^{2+}-ATPase. (A) Structure of the transmembrane domain of the Ca^{2+}-bound form of the Ca^{2+}-ATPase. The two Ca^{2+} ions are shown as balls in the center of the structure (protein database file 1EUL, graphic generated with Cn3D software). (B) Structure of the transmembrane domain of the Ca^{2+}-free form of the Ca^{2+}-ATPase (Protein database file 1IWO, graphic generated with Cn3D software).

HOW ARE TRANSPORTERS STUDIED AND IDENTIFIED?

The basic and traditional approach to the study of transport is the use of radiolabeled compounds. These are used as tracers to follow the uptake of metabolites and ions into tissues, cultured and isolated cells, and preparations of cellular compartments. The basic design of such experiments is to incubate the preparation with radiolabeled compounds for a period of time, usually a few minutes. Subsequently, transport of the labeled component is interrupted by removal of the extracellular solution and replacement with an identical solution that does not contain labeled compounds. It is usually important to remove the extracellular solution as completely as possible because it contains considerably more tracer than the specimen, which took up the compound. This is usually achieved by repetitive washing. As an alternative, a second labeled compound can be used which does not penetrate the membrane of the specimen. The amount of contamination can subsequently be calculated from the remaining amount of the second impermeable label.

Many variations of the basic protocol have been developed. Transport with cultured adherent cells can be performed in dishes, where changes of the extracellular solution are easy to perform. In tissues, in situ perfusion is used. In this method the radiolabeled compound is delivered to cells through their physiological route. For example, a solution containing radiolabeled compounds can be infused into a section of a nephron or into blood vessels by micropipettes. Samples are removed at a distance from the infusion site and the amount of radioactivity is compared with that of the infusion solution. To correct for nonspecific absorption a second radiolabeled or fluorescent compound is included, which is known to be inert to the cells of the tissue.

Changes of the intracellular ion concentration, as they occur during transport, can be monitored by additional techniques. Fluorescent dyes have been developed that change fluorescence intensity in the absence or presence of an ion. BCECF (2',7'-bis(2-carboxyethyl)-5(6)-carboxyfluorescein), for example, strongly fluoresces at pH 9 but only weakly at pH 5. As a result, fluorescence will decrease when H^+ ions are transported into a cell. Fluorescent dyes require calibration to quantify the transport of ions. For a limited number of transporters, fluorescent substrates analogs are available. This allows substrate transport to be monitored by changes of fluorescence in the cytosol of the cell type under study.

Transport of ions, such as H^+, Ca^{2+}, and Na^+, can also be monitored by ion-sensitive microelectrodes. Similar to the well-known pH electrode, coating of glass microcapillaries with certain ionophores can render them permeable only to a single ion species. The microcapillaries are impaled into cells. Changes of the intracellular ion concentration result in changes of the diffusion potential across the tip of the ion-sensitive electrode, which can be recorded with sensitive equipment.

Transport of substrates and ions, when accompanied by flux of charges, can generally be monitored by recording the corresponding currents. Influx of positive charges (or efflux of negative charges) causes cell depolarization; the opposite movement results in hyperpolarization of the cell. To measure these currents, the membrane potential is kept constant by injection of a current of opposite polarity into the cell through a microelectrode. As a result the injected current has the same size but opposite polarity to the current generated by the transport process. This approach is mainly used to characterize cloned transporters in heterologous expression systems.

Subcellular preparations, such as mitochondria or lysosomes, can be isolated for transport experiments. In contrast with experiments with tissue and cultured cells, transport experiments with subcellular preparations have to be performed in suspension. To separate the extracellular solution from the particle preparation, centrifugation or filtration is used. Silicone oil centrifugation, used to separate mitochondria and other organelles from their surrounding solution, makes use of liquid layers of different density. The first layer in a small centrifuge tube is a small volume of strong acid. On top of this layer, a silicone oil layer of intermediate density is placed. The buoyant density of the oil layer is adjusted to be higher than that of the extracellular solution (usually very close to 1.00) but lower than that of the subcellular particles (closer to 1.1). The particle suspension is placed on top of the oil, followed by a brief centrifugation at high speed. The centrifugation forces the subcellular particles through the oil because of their higher density. The extracellular solution stays on top of the oil.

The methods described above are used to investigate transporters in their natural environment. In many cases transport properties can be masked in these systems due to rapid metabolism or binding of substrates to proteins. Experimental systems of reduced complexity have been developed to avoid these complications. Membrane vesicles can be prepared by mechanical lysis of cells, which releases the cytoplasmic contents. The ruptured membranes usually reseal and form vesicles, which can be used to study transport. Cell lysis can be performed in solutions of different composition, which allows preloading of vesicles with transport substrates, different ions, etc. Vesicles do not have a membrane potential but can be energized using K^+ ionophores (valinomycin) in combination with preloading of KCl. Addition of the K^+ ionophores to KCl-loaded vesicles will cause an efflux of K^+, which is not balanced by a concomitant efflux of anions. This creates an inside negative membrane potential, which can drive the uptake of substrates for example by Na^+-metabolite cotransporters. Filtration is used to separate the extracellular medium from vesicles after the transport experiment.

A further reduction of biological complexity is achieved through the use of reconstituted systems. Membrane proteins are solubilized by treatment of membranes with mild detergents. The choice of the detergent is critical and difficult. Detergents should dissolve the lipid bilayer but at the same time should protect the secondary structure of the membrane protein. The secondary structure is often stabilized by hydrophobic interactions which are sensitive to detergents. After successful solubilization, membrane proteins are mixed with purified membrane lipids and the detergent is removed by dialysis, dilution, or column chromatography. As a result, artificial vesicles will be formed that contain incorporated membrane proteins. The orientation of a protein in the artificial membrane cannot be predicted. Therefore transport experiments might reveal a mixed picture of the properties of the cytosolic and extracellular face of the transporter.

Only very few transporters have been purified to homogeneity, which would allow their identification by protein sequencing or the generation of antibodies. Membrane proteins are inherently difficult to purify because most parts are well embedded in the membrane. Even in the solubilized state, which is a prerequisite for purification, most parts are masked by detergents. As a result, different membrane proteins display similar physical properties during purification and therefore are difficult to separate.

The molecular identification of mammalian transport proteins lagged behind other proteins for many years as a result of these problems. Arguably, the most significant breakthrough in the molecular identification of mammalian membrane transporters was the development of expression cloning for this group of proteins. Expression cloning requires only the transport activity as a signal. Several variants of expression cloning have been used, but protocols using *Xenopus laevis* oocytes have been proven to be most successful.

The use of *Xenopus* oocytes as an in vitro expression system was initiated in 1971 by the observation of John B. Gurdon that mRNA injected into *Xenopus* oocytes was translated into protein (Gurdon *et al.* 1971). However, it was not until 1987 that *Xenopus* oocytes were used to express transport proteins as pioneered by Hediger *et al.* (1987). Because of the large amount of stored protein, oocytes do not depend on extracellular resources for nutrition. As a result only limited numbers of endogenous membrane transporters are expressed. By contrast, large amounts of transporter proteins are synthesized following injection of mRNA. Transport experiments can be performed with single oocytes, owing to the large size of these cells. Oocytes are incubated with labeled compounds and the incorporated label is determined after removal of the extracellular medium. These properties have been exploited to isolate cDNAs encoding a multitude of membrane transport proteins (Figure 8).

Initially, a cDNA library is prepared from mRNA isolated from a tissue or cell line that is

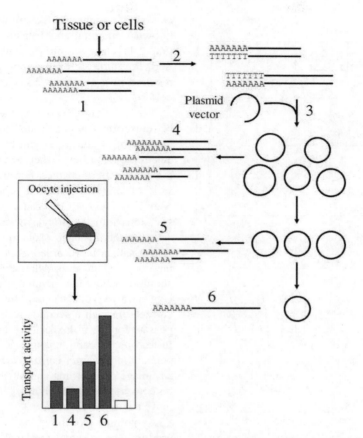

Figure 8 Expression cloning using *Xenopus laevis* oocytes. (1) mRNA is isolated from tissue or cells. (2) Using reverse transcriptase, mRNA is converted into cDNA. The cDNA strands are cloned into plasmid vectors (3), which are used to transform bacteria. The bacteria are plated out and form the cDNA library. Plasmid isolated from pools of the library is used to generate artificial RNA by in vitro transcription (4). The RNA is injected into oocytes and transport activity of oocytes is determined after a few days of expression. If an mRNA pool induces an increased transport activity, it is subfractionated and the procedure is repeated (5). Eventually, a single clone is isolated that induces a strong transport activity in oocytes (6). The transporter is then identified by sequencing of the cDNA.

known to express high levels of a certain transporter. Usually the relevant mRNA is enriched by a size selection procedure. The cDNA library, which may contain fewer than 100,000 different clones, is then plated out in portions about 200 clones. cDNA collections of these library portions are isolated and used for in vitro transcription into pools of RNA. The RNA pools are injected separately into oocytes (Figure 8). If one of the RNA pools contains the mRNA for the desired transporter, the oocyte translates it and incorporates the resulting protein into the membrane among other proteins. This may be detectable as a small increase of the transport activity over background. Subsequently, the initial positive cDNA fraction is subdivided into smaller portions, the RNA derived from which is injected into oocytes again. Since the RNA of interest becomes less diluted by contaminating RNA, an increase of the transport signal is expected. Following a lead signal in a portion of a cDNA

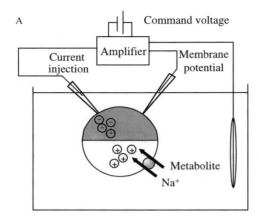

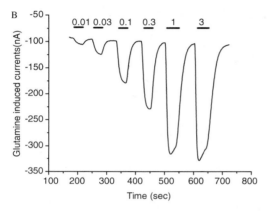

Figure 9 Characterization of transporters by voltage–clamp recordings. (A) Two electrodes are impaled into an oocyte that expresses an electrogenic transporter. One electrode records the membrane potential of the oocyte. The membrane potential is compared with a command voltage generated by an amplifier. If the membrane potential deviates from the command voltage, negative or positive charges are "injected" into the oocyte until the membrane potential matches the command voltage. The "injected" current equals the current generated by the expressed transporter. (B) Example of currents generated by Na^+–glutamine cotransport in oocytes expressing the amino acid transporter ATA1 (SLC38A1).

library may result in the identification of a single cDNA after several rounds of subfractionation. The method has been remarkably successful and has resulted in the initial cloning of members in most of the 40 currently known mammalian transporter families (Table 1). Once an initial member was identified, homology cloning approaches usually resulted in a quick extension of each transporter family.

The benefit of *Xenopus* oocytes for the study of transporters is not limited to cloning. The movement of substrates can be detected with electronic amplifiers, when accompanied by a net movement of charges. For example, expression of a Na^+–glutamine cotransporter in oocytes (ATA1, SLC38) causes a large flow of Na^+ ions into the oocyte when glutamine is added to the incubation solution. The influx of Na^+ ions can be recorded as depolarization of the oocyte (Figure 9). Alternatively, electronic amplifiers allow the membrane potential to be kept constant (voltage clamp) by the injection of currents through a second electrode that compensate for the influx of Na^+ ions (Figure 9). Instead of recording changes of the membrane potential, transporter currents are recorded. The advantage of recording currents is that they can be converted into the number of charges that move across the membrane.

As pointed out above, oocytes can also be used in experiments using labeled compounds. The extracellular medium can easily be removed from oocytes to stop the transport reaction. The low background activity makes oocytes an expression system superior to any other expression system for mammalian transporters.

HOW ARE TRANSPORTERS LINKED TO A CERTAIN DISEASE?

Some inherited diseases arise as a result of mutations in a single gene. Humans have 22 pairs of non-sex-specific chromosomes (autosomes) and a single pair of sex chromosomes (X and Y). Each child receives one set of chromosomes from each parent and consequently has two copies of each gene. A mutated gene is passed to the next generation following Mendelian rules. There are four basic patterns of inheritance of simple disorders: autosomal dominant, autosomal recessive, X-linked dominant, and X-linked recessive. The majority of mutations display recessive inheritance. This pattern can be interpreted to the effect that one intact copy of a gene is sufficient to carry out its physiological function and therefore the diseases will not develop in a heterozygous genotype. There are very few dominant mutations. For transporter gene defects, a dominant inheritance can be envisaged, for example, when the encoded protein is part of a homo-multimeric protein complex that is inactivated by a single defective subunit. The transmission of a disorder through a pedigree by itself provides information about the basic pattern of inheritance but no further information on the localization of the gene (unless it is located on the sex chromosomes). However, the localization of a gene on a chromosome may be deduced from linkage analysis (Figure 10) provided that certain requirements are met (see Risch 2000). Linkage refers to the fact that genes that lie on the same chromosome are not independently sorted into germ cells but are linked to each other. Nevertheless, linkage on a chromosome may not be complete. During meiosis crossover occurs between the two homologous chromosomes. On average one or two crossovers occur on each chromosome per meiosis. Genes that are far apart from each other on a chromosome may not appear to be linked. This deviation from complete linkage allows relative mapping of genes on a chromosome. Chromosomal maps rely on the positioning of highly polymorphic markers whose position is well documented using a number of different techniques. These markers are rarely internal to the coding sequence of the genes (because variation in sequence may lead to variation in the coding of amino acids and hence protein structure and function). If a marker sits close to a mutated gene, it will cosegregate with the disease gene in all meioses and therefore will appear linked with the mutated gene in all crosses. If the marker sits further away from the mutated gene, it might be separated from the disease-causing gene by a recombination event. Thus a child

might receive the marker that sits close to the defect gene but an intact copy of the gene, or vice versa (Figure 10). Complete (or almost complete) linkage indicates that the disease gene is close to the marker. Linkage analysis is a sophisticated statistical technique that aims at the simplest level to count the number of recombinations observed in a pedigree and compare this with what would be expected in a family in which the disease is not inherited. Combining linkage analysis from several markers also indicates the relative position to the marker (toward the telomere or toward the centromere). Linkage studies use the

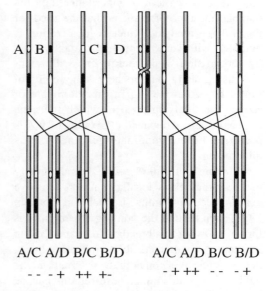

Figure 10 Linkage analysis reveals relative distances of genes from markers. Two crosses are shown. One with the normal order of alleles and the same cross in the case of a crossover during meiosis. The alleles of one marker are indicated as A, B, C, and D (short ovals and squares in black and white). The nonfunctional disease gene is shown in black (presence indicated as (–) below the chromosomes); the functional allele is given in white (+). Without the crossover the mutated gene is linked to marker alleles A and C. With crossover the mutated gene is linked to marker alleles B and C. The marker must be at a distance from the disease gene because it was separated during the crossover.

centimorgan (cM) as a unit; it is equal to 1% likelihood of recombination (50% likelihood of recombination is independent inheritance). Although there is no fixed ratio between a centimorgan and a certain length of a chromosome, it is roughly equivalent to 1,000,000 bp (1 Mbp). On average there are about 10 genes per Mbp but an individual gene stretches on average only about 14 kb on the chromosome. Generally, linkage analysis relies on the collection of large numbers of pedigrees in which the disease of interest is segregating. The ability to define the precise localization of the gene concerned will rely on a number of parameters including the number of pedigrees collected, the pattern of inheritance, and the number and type of markers that are used for the linkage analysis. Once a localization is obtained with some degree of certainty (met by certain statistical critera), fine-scale mapping using many markers across the region of interest can be used to further delineate or reduce the likely interval in which the gene lies. Normally a distance of less than 2 cM is preferred before candidate gene analysis is undertaken.

In some particular cases, homozygosity mapping may allow localization of a defective gene in recessive traits (Lander and Botstein 1987) even though the number of pedigrees is limiting. Owing to inbreeding (e.g., first-cousin marriages), an affected sibling may have inherited the same allele on both the paternal and the maternal chromosome. These regions can be identified because they have identical markers on both chromosomes, where two different markers would normally be expected. The extent of the homozygous region can be mapped by looking at several markers along the chromosome. For example, homozygosity mapping was used to identify the gene mutated in thiamine-responsive megaloblastic anemia (Chapter 16), where only a very limited number of families were available for mapping.

To narrow in further, gene analysis is required. Initially, protein-encoding regions are identified by reference to the databases of full-length genes, expressed sequence tags (ESTs), and genomic sequences. Thus genes may be identified by homology searches of known genes expressed in either humans or other organisms, or ESTs, or by computer predictions based on sequence analysis of genomic sequence. Several criteria can be applied to identify the disease-causing gene. For example, a gene responsible for the intestinal absorption of amino acids should be expressed in the intestine. Membrane transporter genes in general encode proteins with multiple membrane spanning regions, and these can be used to discriminate them from genes encoding soluble proteins. Final proofs rely primarily on the Mendelian pattern of inheritance, that is, co-inheritance of the mutation and disease in affected pedigrees as well as its absence in unaffected members of the pedigree. It may further involve functional characterization of the mutation, demonstrating its likely role in disease etiology. Final proof may also require in vivo studies in knockout and transgenic animal models.

HOW MANY TRANSPORTERS DO WE NEED?

Even though the human genome has been sequenced, the exact number of genes that it encodes is still unclear. Current estimates range from a lower end of 40,000 genes to something close to 80,000 genes. This demonstrates that recognition and annotation of genes by bioinformatic tools is far from perfect. Of the annotated proteins, 20% are thought to be membrane proteins (Lander et al. 2001). But how many of the membrane proteins are transporters, excluding channels? Molecular cloning and identification of membrane transporters gained momentum after the development of expression cloning for this class of proteins (see above). Subsequent to the identification of the Na^+–glucose cotransporter, a number of new families were uncovered using this method, followed by a rapid extension of those transporter families by homology cloning. These efforts resulted in the identification of now 40 solute carrier families (Unigene acronym SLC) that in total comprise slightly more than 200 members (Table 1). Another 60 members can be found in the family of ABC transporters (Unigene acronym ABC) (Table 2), and ion-motive ATPases constitute a family of about 40 members (Unigene acronym ATP) (Table 3). Despite these successes, it is clear that not all physiologically characterized transport process have a molecular correlate as yet,

Table 1 Solute carrier families

Family[a]	No.	Substrates	Comment
SLC1	7	Amino acids	Mostly glutamate
SLC2	13	Glucose (+)	Uniporter
SLC3	2	Amino acids	Heterodimers with SLC7
SLC4	11	Cl, HCO_3^-	Antiporter
SLC5	7	Carbohydrates, iodine, vitamins	Na^+ cotransporters
SLC6	16	Neurotransmitter (+)	Na^+ cotransporters
SLC7	12	Amino acids	Most heterodimers with SLC3
SLC8	3	Na^+/Ca^{2+}	Antiporter
SLC9	7	Na^+/H^+	Antiporter
SLC10	2	Bile acids	Na^+ cotransporter
SLC11	3	Metal$^{(2+)}$	H^+ cotransporter
SLC12	8	Na^+, K^+, Cl^-	Cotransporter
SLC13	4	Sulfate, dicarboxylates	Na^+ cotransporter
SLC14	2	Urea	Uniporter
SLC15	2	Peptides	H^+ cotransporter
SLC16	10	Monocarboxylates (+)	H^+ cotransporter
SLC17	7	Organic anions, glutamate	Vesicular transporters
SLC18	3	Neurotransmitter	Vesicular transporters
SLC19	3	Vitamins	Folate, Thiamine
SLC20	2	Phosphate	Na^+ cotransporter
SLC21	14	Organic anions (+)	Xenobiotics
SLC22	12	Organic cations (+)	Xenobiotics, amines
SLC23	2	Ascorbate	Na^+ cotransporter
SLC24	4	Na^+/Ca^{2+}, K^+	Antiporter
SLC25	22	Diverse	Mitochondrial carriers
SLC26	11	Anions	Antiporter
SLC27	6	Fatty acids	Transport and possibly metabolism
SLC28	3	Nucleosides	Na^+ cotransporters
SLC29	2	Nucleosides	Uniporter
SLC30	4	Zinc	Efflux
SLC31	2	Copper	Copper uptake
SLC32	1	GABA	Vesicular transporter
SLC33	1	Acetyl-CoA	putative Golgi acetyl-CoA transporter
SLC34	2	Phosphate	Epithelial transporter
SLC35	3	Nucleotide sugars	ER/Golgi
SLC36	2	Amino acids	Lysosomal + plasma membrane
SLC37	1	Glycerol-3P	Putative
SLC38	6	Amino acids	Na^+ cotransporters
SLC39	4	Zinc	Zinc uptake
SLC40	1	Iron	Iron export

[a] Solute carrier families are listed using the Unigene nomenclature of the NCBI database.
(+) additional substrates transported by some members.

Table 2   ABC transporter families

Family[a]	No.	Examples	Substrate
ABCA	12	ABCA1	Cholesterol
		ABCA4	Retinyl-phosphatidylethanolamine
		ABCA7	Phospholipids
ABCB	13	ABCB1 (multiple drug resistance protein 1 or P-glycoprotein)	Amphiphilic drugs
		ABCB11 (bile salt efflux pump, BSEP)	Bile salts
		TAP1/TAP2	Peptides
ABCC	14	ABCC1 (MRP1)	Drug conjugates
		ABCC2 (MRP2)	Drug conjugates
ABCD	4	ABCD1	Very-long-chain fatty acids
ABCE	1	No transporters	
ABCF	3	No transporters	
ABCG	6	ABCG5/8	Sterols

[a]ABC transporters are listed according to the Unigene nomenclature of the NCBI.

Table 3   ATPase families

Family[a]	Subunits	Substrate	Comment
ATP1	A, B	Na^+, K^+	Na^+, K^+-ATPase
ATP2	A, B, C	Ca^{2+}	Ca^{2+}-ATPase
ATP3		Mg^{2+}	
ATP4	A, B	H^+, K^+	Gastric H^+, K^+-ATPase
ATP5	A-S	H^+	Mitochondrial F_1F_0
ATP6	V_0, V_1	H^+	Lysosomal, vacuolar
ATP7	A, B	Cu^{2+}	Copper transport
ATP8	A, B	Aminophospholipids	Flippase
ATP9	A, B	Unknown	
ATP10	A, B	Unknown	
ATP11	A, B	Unknown	
ATP12	A, B	Unknown	

[a]ATPases are listed according to the unigene nomenclature of the NCBI. Only families with mammalian members are shown.

but it appears likely that the majority of transporters have already been identified. On this basis, the total number of transporters in the human genome might be estimated to be around 500, which is about 1% of all genes. It appears that most biological functions can be explained with the currently known three subgroups of transporters: solute carriers, ABC transporters, and ATPases. With respect to biological function, metabolite transporters can be grouped into four broad areas:

1. transporters of energy metabolites and building blocks (Chapters 9–16);
2. resorption of metabolites in epithelia (Chapters 9–16);
3. transporters involved in secretion of metabolites, drugs, and other compounds (Chapters 17–21);

Table 4 Inherited diseases of membrane transport

HUGO	Alternative name	Substrate	Associated disease	Reference
SLC				
2A1	GLUT-1	D-Glucose	De Vivo syndrome	Seidner *et al.* 1998
2A2	GLUT-2	D-Glucose	Hepatorenal glucogenosis with renal Fanconi syndrome (Fanconi–Bickel disease)	Santer *et al.* 1997 Chapter 12
3A1	rBAT, NBAT	Dibasic amino acids, cystine–neutral amino acids exchange	Cystinuria A	Calonge *et al.* 1994 Chapter 14
4A1	AE-1, band-3	Cl^-–bicarbonate exchange	Distal renal tubular acidosis, spherocytosis, Asian ovalocystosis	Bruce *et al.* 1997; Lima *et al.* 1997; Karet *et al.* 1998 Chapter 3
4A4	kNBC-1	Na^+–bicarbonate cotransport	Proximal renal tubular acidosis with blindness	Igarashi *et al.* 1999 Chapter 4
5A1	SGLT-1	Na^+–D-glucose cotransport	Glucose–galactose malabsorption	Martin *et al.* 1996
5A2	SGLT-2	Na^+–D-glucose cotransport	Autosomal recessive renal glucosuria	van den Heuvel *et al.* 2002
5A5	NIS	Na^+–iodide cotransport	Hypothyroidism	Fujiwara *et al.* 1997
7A7	y^+LAT-1	Dibasic amino acids–neutral amino acids + Na^+ exchange	Lysinuric protein intolerance	Borsani *et al.* 1999; Torrents *et al.* 1999 Chapter 14
7A9	$b^{0,+}$; AT	Dibasic amino acids, cystine–neutral amino acids exchange	Cystinuria B	Feliubadalo *et al.* 1999 Chapter 14
9A3	NHE-3	Na^+–H^+ exchange	Genetically not confirmed	Booth *et al.* 1985
10A2	NTCP2, ISBT	Na^+–bile acid	Primary bile acid malabsorption	Oelkers *et al.* 1997
11A3	Ferroportin FPN1	Iron	Hemochromatosis type 4	Njajou *et al.* 2001 Chapter 8
12A1	NKCC-2, BSC1	Na^+–K^+–$2Cl^-$ cotransport	Bartter's syndrome type I	Simon *et al.* 1996a
12A3	NCC, TSC	Na^+–Cl^- cotransport	Gitelman's syndrome	Simon *et al.* 1996b
17A5	Sialin	Lysosomal sialic acid	Sialic acid storage disease and sialuria, Salla disease	Verheijen *et al.* 1999 Chapter 15
19A2	THTR1	Thiamine	Megaloblastic anemia	Diaz *et al.* 1999; Fleming *et al.* 1999; Labay *et al.* 1999 Chapter 16

Table 4 Continued

HUGO	Alternative name	Substrate	Associated disease	Reference
22A5	OCTN-2	Na^+–carnitine cotransport	Primary carnitine deficiency	Nezu *et al.* 1999; Wang *et al.* 1999 Chapter 11
22A12	OAT4	Urate–anion exchange	Idiopathic renal hypouricemia	Enomoto *et al.* 2002
25A4	ANT1	ATP–ADP exchange	Progressive external ophthalmoplegia	Kaukonen *et al.* 2000
25A13	Citrin	Aspartate–glutamate exchange	Citrullinemia neonatal intrahepatic cholestasis	Kobayashi *et al.* 1999 Chapter 10
25A20	CACT	Carnitine–acylcarnitine	Carnitine/acylcarnitine translocase deficiency	Huizing *et al.* 1997
25A15	ORNT1	Ornithine–citrulline exchange	Hyperornithinemia–hyperammonemia–homocitrullinuria syndrome	Camacho *et al.* 1999
26A2	DTDST	Cl^-–sulfate exchange	Diastrophic dysplasia	Hastbacka *et al.* 1994; Superti-Furga *et al.* 1996 Chapter 6
26A3	DRA	Cl^-–bicarbonate exchange	Congenital diarrhea	Hoglund *et al.* 1996 Chapter 6
26A4	Pendrin	Cl^-–anion exchange (iodide, formate)	Pendred syndrome (hypothyroidism, deafness)	Everett *et al.* 1997; Scott *et al.* 1999 Chapter 6
34A1	NaPi-IIa	Na^+–phosphate	Functionally not confirmed	Prie *et al.* 2002
35		GDP-fucose	Leukocyte adhesion deficiency II	Luhn *et al.* 2001
CTNS		Cystine	Cystinosis	Town *et al.* 1998
G6PT1		Glucose-6-phosphate	Glycogen storage disease Ib and Ic	Hiraiwa *et al.* 1999 Chapter 13
ATP A2	SERCA2	Calcium	Darier–White disease	Sakuntabhai *et al.* 1999 Chapter 24
2C1		Calcium	Hailey–Hailey disease: benign familial pemphigus	Hu *et al.* 2000 Chapter 24
6V0A3	a3 V-H$^+$-ATPase	Protons	Malignant infantile osteopetrosis	Frattini *et al.* 2000
6V0A4	a4 V-H$^+$-ATPase	Protons	Distal renal tubular acidosis	Smith *et al.* 2000
6V1B1	B1 V-H$^+$-ATPase	Protons	Distal renal tubular acidosis with sensorineural deafness	Karet *et al.* 1999
7A		Copper	Menkes, Wilson, and occipital horn syndromes	Vulpe *et al.* 1993; Kaler *et al.* 1994

Table 4 Continued

HUGO	Alternative name	Substrate	Associated disease	Reference
7B		Copper	Menkes, Wilson, and occipital horn syndromes	Vulpe *et al.* 1993; Kaler *et al.* 1994
8B1		Aminophospholipid	Progessive familial intrahepatic cholestasis 1, Byler disease, benign recurrent intrahepatic cholestasis	Bull *et al.* 1998
ABC				
A1	ABC1	Phospholipid Cholesterol	Tangier's disease (HDL deficiency), hepatosplenomegaly, peripheral neuropathy	Bodzioch *et al.* 1999 Chapter 19
A4		Retinal, retinyl-phosphatidyl-ethanolamine	Stargardt disease type 1, retinitis pigmentosa, macular degeneration type 2, cone–rod dystrophy type 3	Allikmets *et al.* 1997; Martinez-Mir *et al.* 1998 Chapter 20
B2/B3	TAP1, TAP2	Peptides	Bare lymphocyte syndrome, tumor escape	de la Salle *et al.* 1994; Chen *et al.* 1996 Chapter 21
B7	ABC7, ATM1	Fe–S cluster	X-linked sideroblastic anemia with ataxia	Allikmets *et al.* 1998
B11	BSEP, SPGP	Bile acid	Progressive familial intrahepatic cholestasis	Strautnieks *et al.* 1998
C2	cMOAT	Organic anions	Dubin-Johnson syndrome with hyperbilirubinemia	Wada *et al.* 1998 Chapter 18
C3	Mrp3	Phosphatidylcholine	Type 3 progressive familial intrahepatic cholestasis (progressive familial intrahepatic cholestasis with elevated serum γ-glutamyltransferase)	de Vree *et al.* 1998
C6	Mrp6	Glutathione conjugates	Pseudoxanthoma elasticum	Bergen *et al.* 2000; Le Saux *et al.* 2000; Ringpfeil *et al.* 2000
D1	ALD	Very-long-chain fatty acids	Adrenoleukodystrophy	Mosser *et al.* 1993
G5/G8	Sterolin1	Plant sterols, cholesterol	Sitosterolemia	Lee *et al.* 2001; Lu *et al.* 2001

The table summarizes human inherited diseases caused by mutations in genes encoding for carriers or one of their subunits. The name assigned by HUGO is given as well as older alternative names, the main substrate, the disease caused by the mutations, and a reference describing first mutations in the respective gene. Carriers described in this book are referenced to the respective chapter.

4. metabolite transporters involved in signaling (Chapters 22–24).

Similarly, ion transporters can be assigned to four different biological functions:

5. ion transporters involved in resorption and secretion of ions in epithelia and in the homeostasis of intracellular ion concentration (Chapters 2–4);
6. transporters for biometals and mineral ions (Chapters 5–8);
7. ion transporters involved in signaling (Chapters 22–24);
8. Ion transporters maintaining driving force for other membrane transport processes (Na^+, K^+-ATPase, vacuolar H^+-ATPase).

The variety of functions carried out by transporters is reflected by the phenotypes of inherited membrane transporter diseases. Using linkage studies as described above and other approaches, about 50 diseases of membrane transport have been identified thus far (Table 4).

The chapters of this book are grouped according to physiological functions of transport proteins. An introduction to each group of chapters provides background information that facilitates understanding of the individual chapters. The individual chapters discuss selected examples of transporter diseases. Each chapter gives an overview of the physiological function of the transporter and how malfunction translates into the phenotype of the disease. Information about the clinical phenotype of each disease is outlined in a separate box in most chapters.

REFERENCES

Allikmets, R., Singh, N., Sun, H., Shroyer, N.F., Hutchinson, A., Chidambaram, A., et al. (1997). A Photoreceptor Cell-Specific ATP-Binding Transporter Gene (ABCR) is Mutated in Recessive Stargardt Macular Dystrophy. Nat. Genet., 15, 236–246.

Allikmets, R., Raskind, W.H., Hutchinson, A., Schueck, N.D., Dean, M., and Koeller, D.M. (1998). Mutation of a Putative Mitochondrial Iron Transporter Gene (ABC7) in X-Linked Sideroblastic Anemia and Ataxia (XLSA/A). Hum. Mol. Genet., 8, 743–749.

Ashcroft, F.M. (2000). Ion Channels and Disease. San Diego: Academic Press.

Bergen, A.A., Plomp, A.S., Schuurman, E.J., Terry, S., Breuning, M., Dauwerse, H., et al. (2000). Mutations in ABCC6 cause Pseudoxanthoma Elasticum. Nat. Genet., 25, 228–231.

Bodzioch, M., Orso, E., Klucken, J., Langmann, T., Bottcher, A., Diederich, W., et al. (1999). The Gene Encoding ATP-Binding Cassette Transporter 1 is Mutated in Tangier Disease. Nat. Genet., 22, 347–351.

Booth, I.W., Stange, G., Murer, H., Fenton, T.R., and Milla, P.J. (1985). Defective Jejunal Brush-Border Na^+/H^+ Exchange: A Cause of Congenital Secretory Diarrhoea. Lancet, 1, 1066–1069.

Borsani, G., Bassi, M.T., Sperandeo, M.P., De Grandi, A., Buoninconti, A., Riboni, M., et al. (1999). SLC7A7, Encoding a Putative Permease-Related Protein, is Mutated in Patients with Lysinuric Protein Intolerance. Nat. Genet., 21, 297–301.

Bruce, L.J., Cope, D.L., Jones, G.K., Schofield, A.E., Burley, M., Povey, S., et al. (1997). Familial Distal Renal Tubular Acidosis is Associated with Mutations in the Red Cell Anion Exchanger (Band 3, AE1) Gene. J. Clin. Invest., 100, 1693–1707.

Bull, L.N., van Eijk, M.J., Pawlikowska, L., DeYoung, J.A., Juijn, J.A., Liao, M., et al. (1998). A Gene Encoding a P-Type ATPase Mutated in Two Forms of Hereditary Cholestasis. Nat. Genet., 18, 219–224.

Calonge, M.J., Gasparini, P., Chillaron, J., Chillon, M., Gallucci, M., Rousaud, F., et al. (1994). Cystinuria Caused by Mutations in rBAT, a Gene Involved in the Transport of Cystine. Nat. Genet., 6, 420–425.

Camacho, J.A., Obie, C., Biery, B., Goodman, B.K., Hu, C.-A., Almashanu, S., et al. (1999). Hyperornithinaemia-hyperammonaemia-homocitrullinuria syndrome is caused by mutations in a gene encoding a mitochondrial ornithine transporter. Nat. Genet., 22, 151–158.

Chen, H.L., Gabrilovich, D., Tampe, R., Girgis, K.R., Nadaf, S., and Carbone, D.P. (1996). A functionally defective allele of TAP1 results in loss of MHC class I antigen presentation in a human lung cancer. Nat. Genet. 13, 210–213.

de la Salle, H., Hanau, D., Fricker, D., Urlacher, A., Kelly, A., Salamero, J., et al. (1994). Homozygous Human TAP Peptide Transporter Mutation in HLA Class I Deficiency. Science, 265, 237–241.

de Vree, J.M., Jacquemin, E., Sturm, E., Cresteil, D., Bosma, P.J., Aten, J., et al. (1998). Mutations in the MDR3 Gene Cause Progressive Familial Intrahepatic Cholestasis. Proc. Natl Acad. Sci. USA, 95, 282–287.

Diaz, G.A., Banikazemi, M., Oishi, K., Desnick, R.J., and Gelb, B.D. (1999). Mutations in a New Gene Encoding a Thiamine Transporter Cause Thiamine-Responsive Megaloblastic Anaemia Syndrome. Nat. Genet., 22, 309–312.

Enomoto, A., Kimura, H., Chairoungdua, A., Shigeta, Y., Jutabha, P., Cha, S.H., et al. (2002). Molecular

Identification of a Renal Urate Anion Exchanger That Regulates Blood Urate Levels. *Nature*, **417**, 447–452.

Everett, L.A., Glaser, B., Beck, J.C., Idol, J.R., Buchs, A., Heyman, M., et al. (1997). Pendred Syndrome is Caused by Mutations in a Putative Sulphate Transporter Gene (PDS). *Nat. Genet.*, **17**, 411–422.

Fairman, W.A. and Amara, S.G. (1999). Functional Diversity of Excitatory Amino Acid Transporters: Ion Channel and Transport Modes. *Am. J. Physiol.*, **277**, F481–F486.

Feliubadalo, L., Font, M., Purroy, J., Rousaud, F., Estivill, X., Nunes, V., et al. (1999). Non-Type I Cystinuria Caused by Mutations in SLC7A9, Encoding a Subunit (b$^{o,+}$AT) of rBAT. *Nat. Genet.*, **23**, 52–57.

Fleming, J.C., Tartaglini, E., Steinkamp, M.P., Schorderet, D.F., Cohen, N., and Neufeld, E.J. (1999). The Gene Mutated in Thiamine-Responsive Anaemia with Diabetes and Deafness (TRMA) Encodes a Functional Thiamine Transporter. *Nat. Genet.*, **22**, 305–308.

Frattini, A., Orchard, P.J., Sobacchi, C., Giliani, S., Abinun, M., Mattsson, J.P., et al. (2000). Defects in TCIRG1 Subunit of the Vacuolar Proton Pump Are Responsible for a Subset of Human Autosomal Recessive Osteoporosis. *Nat. Genet.*, **25**, 343–346.

Fujiwara, H., Tatsumi, K., Miki, K., Harada, T., Miyai, K., Takai, S., et al. (1997). Congenital Hypothyroidism Caused by a Mutation in the Na$^+$/I$^-$ symporter. *Nat. Genet.*, **16**, 124–125.

Galli, A., Blakely, R.D., and DeFelice, L.J. (1996). Norepinephrine Transporters Have Channel Modes of Conduction. *Proc. Natl Acad. Sci. USA*, **93**, 8671–8676.

Gibbs, A.F., Chapman, D., and Baldwin, S.A. (1988). Proteolytic Dissection as a Probe of Conformational Changes in the Human Erythrocyte Glucose Transport protein. *Biochem. J.*, **256**, 421–427.

Golovanevsky, V. and Kanner, B.I. (1999). The Reactivity of the Gamma-Aminobutyric Acid Transporter GAT-1 Toward Sulfhydryl Reagents Is Conformationally Sensitive. Identification of a Major Target Residue. *J. Biol. Chem.*, **274**, 23020–23026.

Gurdon, J.B., Lane, C.D., Woodland, H.R., and Marbaix, G. (1971). Use of Frog Eggs and Oocytes for the Study of Messenger RNA and its Translation in Living Cells. *Nature*, **233**, 177–182.

Hacksell, I., Rigaud, J.L., Purhonen, P., Pourcher, T., Hebert, H., and Leblanc, G. (2002). Projection Structure at 8 Å Resolution of the Melibiose Permease, an Na–Sugar Co-Transporter from *Escherichia coli*. *EMBO J.*, **21**, 3569–3574.

Hastbacka, J., de la Chapelle, A., Mahtani, M.M., Clines, G., Reeve-Daly, M.P., Daly, M., et al. (1994). The Diastrophic Dysplasia Gene Encodes a Novel Sulfate Transporter: Positional Cloning by Fine-Structure Linkage Disequilibrium Mapping. *Cell*, **78**, 1073–1087.

Hediger, M.A., Coady, M.J., Ikeda, T.S., and Wright, E.M. (1987). Expression Cloning and cDNA Sequencing of the Na$^+$/Glucose Co-Transporter. *Nature*, **330**, 379–381.

Heymann, J.A., Sarker, R., Hirai, T., Shi, D., Milne, J.L., Maloney, P.C., et al. (2001). Projection Structure and Molecular Architecture of OxlT, a Bacterial Membrane Transporter. *EMBO J.*, **20**, 4408–4413.

Hiraiwa, H., Pan, C.-J., Lin, B., Moses, S.W., and Chou, J.Y. (1999). Inactivation of the Glucose 6-Phosphate Transporter Causes Glycogen Storage Disease Type 1b. *J. Biol. Chem.*, **274**, 5532–5536.

Hoglund, P., Haila, S., Socha, J., Tomaszewski, L., Saarialho-Kere, U., Karjalainen-Lindsberg, M.L., et al. (1996). Mutations of the Down-Regulated in Adenoma (DRA) Gene Cause Congenital Chloride Diarrhoea. *Nat. Genet.*, **14**, 316–319.

Hogue, C.W. (1997). Cn3D: A New Generation of Three-Dimensional Molecular Structure Viewer. *Trends Biochem. Sci.*, **22**, 314–316.

Hu, Z., Bonifas, J.M., Beech, J., Bench, G., Shigihara, T., Ogawa, H., et al. (2000). Mutations in ATP2C1, Encoding a Calcium Pump, Cause Hailey–Hailey Disease. *Nat. Genet.*, **24**, 61–65.

Huizing, M., Iacobazzi, V., Ijlst, L., Savelkoul, P., Ruitenbeek, W., van den Heuvel, L., et al. (1997). Cloning of the Human Carnitine–Acylcarnitine Carrier cDNA and Identification of the Molecular Defect in a Patient. *Am. J. Hum. Genet.*, **61**, 1239–1245.

Igarashi, T., Inatomi, J., Sekine, T., Cha, S.H., Kanai, Y., Kunimi, M., et al. (1994). Mutations in SLC4A4 Cause Permanent Isolated Proximal Renal Tubular Acidosis with Ocular Abnormalities. *Nat. Genet.*, **23**, 264–266.

Kaler, S.G., Gallo, L.K., Proud, V.K., Percy, A.K., Mark, Y., Segal, N.A., et al. (1994). Occipital Horn Syndrome and a Mild Menkes Phenotype Associated with Splice Site Mutations at the MNK Locus. *Nat. Genet.*, **8**, 195–202.

Karet, F.E., Gainza, F.J., Gyory, A.Z., Unwin, R.J., Wrong, O., Tanner, M.J., et al. (1998). Mutations in the Chloride–Bicarbonate Exchanger Gene AE1 Cause Autosomal Dominant but not Autosomal Recessive Distal Renal Tubular Acidosis. *Proc. Natl Acad. Sci. USA*, **95**, 6337–6342.

Karet, F.E., Finberg, K.E., Nelson, R.D., Nayir, A., Mocan, H., Sanjad, S.A., et al. (1999). Mutations in the Gene Encoding B1 Subunit of H$^+$-ATPase Cause Renal Tubular Acidosis with Sensorineural Deafness. *Nat. Genet.*, **21**, 84–90.

Kaukonen, J., Juselius, J.K., Tiranti, V., Kyttala, A., Zeviani, M., Comi, G.P., et al. (2000). Role of Adenine Nucleotide Translocator 1 in mtDNA Maintenance. *Science*, **289**, 782–785.

Kobayashi, K., Sinasac, D.S., Iijima, M., Boright, A.P., Begum, L., Lee, J.R., et al. (1999). The Gene Mutated in Adult-Onset Type II Citrullinaemia Encodes a Putative Mitochondrial Carrier Protein. *Nat. Genet.*, **22**, 159–163.

Kyte, J. and Doolittle, R.F. (1982). A Simple Method for Displaying the Hydropathic Character of a Protein. *J. Mol. Biol.*, **157**, 105–132.

Labay, V., Raz, T., Baron, D., Mandel, H., Williams, H., Barrett, T., *et al.* (1999). Mutations in SLC19A2 Cause Thiamine-Responsive Megaloblastic Anaemia Associated with Diabetes Mellitus and Deafness. *Nat. Genet.*, **22**, 300–304.

Lander, E.S. and Botstein, D. (1987). Homozygosity Mapping: A Way to Map Human Recessive Traits with the DNA of Inbred Children. *Science*, **236**, 1567–1570.

Lander, E.S., Linton, L.M., Birren, B., Nusbaum, C., Zody, M.C., Baldwin, J., *et al.* (2001). Initial Sequencing and Analysis of the Human Genome. *Nature*, **409**, 860–921.

Lauquin, G.J. and Vignais, P.V. (1976). Interaction of (3H) Bongkrekic Acid with the Mitochondrial Adenine Nucleotide Translocator. *Biochemistry*, **15**, 2316–2322.

Lee, M.H., Lu, K., Hazard, S., Yu, H., Shulenin, S., Hidaka, H., *et al.* (2001). Identification of a Gene, ABCG5, Important in the Regulation of Dietary Cholesterol Absorption. *Nat. Genet.*, **27**, 79–83.

Le Saux, O., Urban, Z., Tschuch, C., Csiszar, K., Bacchelli, B., Quaglino, D., *et al.* (2000). Mutations in a Gene Encoding an ABC Transporter cause Pseudo-xanthoma Elasticum. *Nat. Genet.*, **25**, 223–227.

Lima, P.R., Gontijo, J.A., Lopes de Faria, J.B., Costa, F.F., and Saad, S.T. (1997). Band 3 Campinas: A Novel Splicing Mutation in the Band 3 Gene (AE1) Associated with Hereditary Spherocytosis, Hyperactivity of Na^+/Li^+ Countertransport and an Abnormal Renal Bicarbonate Handling. *Blood*, **90**, 2810–2818.

Lu, K., Lee, M.H., Hazard, S., Brooks-Wilson, A., Hidaka, H., Kojima, H., *et al.* (2001). Two Genes That Map to the STSL Locus Cause Sitosterolemia: Genomic Structure and Spectrum of Mutations Involving Sterolin-1 and Sterolin-2, Encoded by ABCG5 and ABCG8, Respectively. *Am. J. Hum. Genet.*, **69**, 278–290.

Luhn, K., Wild, M.K., Eckhardt, M., Gerardy-Schahn, R., and Vestweber, D. (2001). The Gene Defective in Leukocyte Adhesion Deficiency II Encodes a Putative GDP–Fucose Transporter. *Nat. Genet.*, **28**, 69–72.

Manoil, C. (1991). Analysis of Membrane Protein Topology Using Alkaline Phosphatase and Beta-Galactosidase Gene Fusions. *Methods Cell Biol.*, **34**, 61–75.

Martin, M.G., Turk, E., Lostao, M.P., Kerner, C., and Wright, E.M. (1996). Defects in $Na^+/Glucose$ Cotransporter (SGLT1) Trafficking and Function Cause Glucose–Galactose Malabsorption. *Nat. Genet.*, **12**, 216–220.

Martinez-Mir, A., Paloma, E., Allikmets, R., Ayuso, C., del Rio, T., Dean, M., *et al.* (1998). Retinitis Pigmentosa caused by a Homozygous Mutation in the Stargardt Disease Gene ABCR. *Nat. Genet.*, **18**, 11–12.

Mosser, J., Douar, A.M., Sarde, C.O., Kioschis, P., Feil, R., Moser, H., *et al.* (1993). Putative X-Linked Adrenoleukodystrophy Gene Shares Unexpected Homology with ABC Transporters. *Nature*, **361**, 726–730.

Murakami, S., Nakashima, R., Yamashita, E., and Yamaguchi, A. (2002). Crystal Structure of Bacterial Multidrug Efflux Transporter AcrB. *Nature*, **419**, 587–593.

Nelson, N., Sacher, A., and Nelson, H. (2002). The Significance of Molecular Slips in Transport Systems. *Nat. Rev. Mol. Cell. Biol.*, **3**, 876–881.

Nezu, J., Tamai, I., Oku, A., Ohashi, R., Yabuuchi, H., Hashimoto, N., *et al.* (1999). Primary Systemic Carnitine Deficiency is Caused by Mutations in a Gene Encoding Sodium Ion-Dependent Carnitine Transporter. *Nat. Genet.*, **21**, 91–94.

Njajou, O.T., Vaessen, N., Joosse, M., Berghuis, B., van Dongen, J.W.F., Breuning, M.H., *et al.* (2001). A Mutation in SLC11A3 Is Associated with Autosomal Dominant Hemochromatosis. *Nat. Genet.*, **28**, 213–214.

Oelkers, P., Kirby, L.C., Heubi, J.E., and Dawson, P.A. (1997). Primary Bile Acid Malabsorption Caused by Mutations in the Ileal Sodium-Dependent Bile Acid Transporter Gene (SLC10A2). *J. Clin. Invest.*, **99**, 1880–1887.

Peerce, B.E. (1990). Examination of Substrate-Induced Conformational Changes in the $Na^+/Glucose$ Cotransporter. *J. Biol. Chem.*, **265**, 1737–1741.

Prie, D., Huart, V., Bakouh, N., Planelles, G., Dellis, O., Gerard, B., *et al.* (2002). Nephrolithiasis and Osteoporosis Associated with Hypophosphatemia Caused by Mutations in the Type 2a Sodium–Phosphate Cotransporter. *N. Engl. J. Med.*, **347**, 983–991.

Ringpfeil, F., Lebwohl, M.G., Christiano, A.M., and Uitto, J. (2000). Pseudoxanthoma Elasticum: Mutations in the MRP6 Gene Encoding a Transmembrane ATP-Binding Cassette (ABC) Transporter. *Proc. Natl Acad. Sci. USA*, **97**, 6001–6006.

Risch, N.J. (2000). Searching for Genetic Determinants in the New Millennium. *Nature*, **405**, 847–856.

Sakuntabhai, A., Ruiz-Perez, V., Carter, S., Jacobsen, N., Burge, S., Monk, S., *et al.* (1999). Mutations in ATP2A2, Encoding a Ca^{2+} Pump, Cause Darier Disease. *Nat. Genet.*, **21**, 271–277.

Santer, R., Schneppenheim, R., Dombrowski, A., Gotze, H., Steinmann, B., and Schaub, J. (1997). Mutations in GLUT2, the Gene for the Liver-Type Glucose Transporter, in Patients with Fanconi–Bickel Syndrome. *Nat. Genet.*, **17**, 324–326.

Scott, D.A., Wang, R., Kreman, T.M., Sheffield, V.C., and Karniski, L.P. (1999). The Pendred Syndrome Gene Encodes a Chloride–Iodide Transport Protein. *Nat. Genet.*, **21**, 440–443.

Seidner, G., Alvarez, M.G., Yeh, J.I., O'Driscoll, K.R., Klepper, J., Stump, T.S., *et al.* (1998). GLUT-1 Deficiency Syndrome Caused by Haploinsufficiency of the Blood–Brain Barrier Hexose Carrier. *Nat. Genet.*, **18**, 188–191.

Simon, D.B., Karet, F.E., Hamdan, J.M., DiPietro, A., Sanjad, S.A., and Lifton, R.P. (1996a). Bartter's Syndrome, Hypokalaemic Alkalosis with Hypercalciuria, is caused by Mutations in the Na–K–2Cl Cotransporter NKCC2. *Nat. Genet.*, **13**, 183–188.

Simon, D.B., Nelson-Williams, C., Bia, M.J., Ellison, D., Karet, F.E., Molina, A.M., *et al.* (1996b). Gitelman's Variant of Bartter's Syndrome, Inherited Hypokalaemic Alkalosis, is caused by Mutations in the Thiazide-Sensitive Na–Cl Cotransporter. *Nat. Genet.*, **12**, 24–30.

Smith, A.N., Skaug, J., Choate, K.A., Nayir, A., Bakkaloglu, A., Ozen, S., *et al.* (2000). Mutations in ATP6N1B, Encoding a New Kidney Vacuolar Proton Pump 116-kD Subunit, cause Recessive Distal Renal Tubular Acidosis with Preserved Hearing. *Nat. Genet.*, **26**, 71–75.

Strautnieks, S.S., Bull, L.N., Knisely, A.S., Kocoshis, S.A., Dahl, N., Arnell, H., *et al.* (1998). A Gene Encoding a Liver-Specific ABC Transporter is Mutated in Progressive Familial Intrahepatic Cholestasis. *Nat. Genet.*, **20**, 233–238.

Superti-Furga, A., Hastbacka, J., Wilcox, W.R., Cohn, D.H., van der Harten, H.J., Rossi, A., *et al.* (1996). Achondrogenesis Type IB is caused by Mutations in the Diastrophic Dysplasia Sulphate Transporter Gene. *Nat. Genet.*, **12**, 100–102.

Torrents, D., Mykkanen, J., Pineda, M., Feliubadalo, L., Estevez, R., de Cid, R., *et al.* (1999). Identification of SLC7A7, Encoding γ^+LAT-1, as the Lysinuric Protein Intolerance Gene. *Nat. Genet.*, **21**, 293–296.

Town, M., Jean, G., Cherqui, S., Attard, M., Forestier, L., Whitmore, S.A., *et al.* (1998). A Novel Gene Encoding an Integral Membrane Protein is Mutated in Nephropathic Cystinosis. *Nat. Genet.*, **18**, 319–324.

Toyoshima, C. and Nomura, H. (2002). Structural Changes in the Calcium Pump Accompanying the Dissociation of Calcium. *Nature*, **418**, 605–611.

Toyoshima, C., Nakasako, M., Nomura, H., and Ogawa, H. (2000). Crystal Structure of the Calcium Pump of Sarcoplasmic Reticulum at 2.6 Å Resolution. *Nature*, **405**, 647–655.

van den Heuvel, L.P., Assink, K., Willemsen, M., and Monnens, L. (2002). Autosomal Recessive Renal Glucosuria Attributable to a Mutation in the Sodium Glucose Cotransporter (SGLT2). *Hum. Genet.*, **111**, 544–547.

Verheijen, F.W., Verbeek, E., Aula, N., Beerens, C.E.M.T., Havelaar, A.C., Joosse, M., *et al.* (1999). A New Gene, Encoding an Anion Transporter, is Mutated in Sialic Acid Storage Diseases. *Nat. Genet.*, **23**, 462–465.

Vulpe, C., Levinson, B., Whitney, S., Packman, S., and Gitschier, J. (1993). Isolation of a Candidate Gene for Menkes Disease and Evidence that it Encodes a Copper-Transporting ATPase. *Nat. Genet.*, **3**, 7–13.

Wada, M., Toh, S., Taniguchi, K., Nakamura, T., Uchiumi, T., Kohno, K., *et al.* (1998). Mutations in the Canilicular Multispecific Organic Anion Transporter (cMOAT) Gene, a Novel ABC Transporter, in Patients with Hyperbilirubinemia II/Dubin–Johnson Syndrome. *Hum. Mol. Genet.*, **7**, 203–207.

Wang, Y., Ye, J., Ganapathy, V., and Longo, N. (1999). Mutations in the Organic Cation/Carnitine Transporter OCTN2 in Primary Carnitine Deficiency. *Proc. Natl Acad. Sci. USA*, **96**, 2356–2360.

Wright, E.M. (2001). Renal Na(+)-Glucose Cotransporters. *Am. J. Physiol. Renal Physiol.*, **280**, F10–F18.

Wu, J., Hardy, D., and Kaback, H.R. (1998). Transmembrane Helix Tilting and Ligand-Induced Conformational Changes in the Lactose Permease Determined by Site-Directed Chemical Crosslinking In Situ. *J. Mol. Biol.*, **282**, 959–967.

RENAL TRANSPORT OF ELECTROLYTES AND ACID–BASE EQUIVALENTS

2

CARSTEN A. WAGNER*

Introduction

This part will focus on the carriers that are involved in the transport of electrolytes and acid–base equivalents, such Na^+, K^+, Cl^-, H^+, or HCO_3^-. Often both functions are intricately linked not only by directly coupling the transport of acid–base equivalents to that of electrolytes (i.e., Na^+/H^+ or Cl^-/HCO_3^- exchangers, Na^+/HCO_3^- cotransport) but also because both influence whole-cell function or body fluid composition and are often regulated by the same hormonal systems. Thus it is not surprising that disorders primarily affecting either electrolyte or acid–base transport often affect cellular or whole-body electrolyte and acid–base homeostasis. In contrast to the transport of acid–base equivalents, however, the transport of electrolytes is achieved by the concerted action of transporters and a variety of ion channels. Mutations in these ion channels have been identified and are also involved in inherited diseases affecting electrolyte transport. However, channelopathies have been the focus of numerous recent reviews and books, and thus will not be discussed in detail here.

As mentioned above, transport of electrolytes and acid–base equivalents is necessary for both normal cellular function and maintenance of a normal cellular environment, that is, the extracellular milieu or whole-body fluid composition. Thus ubiquitous isoforms of

many electrolyte and acid–base transporters such as the Na^+/H^+ (NHE1, SLC9A1) or Cl^-/HCO_3^- (AE2, SLC4A2) exchangers are expressed in most mammalian cells, whereas specialized isoforms of these transporters are mainly involved in the regulation of whole-body fluid composition. The kidneys play a central role in the regulation of both whole-body fluid composition and acid–base homeostasis. Thus the physiological role of electrolyte and acid–base equivalent transporters will be exemplified in the kidney and discussed against the background of their mutual influences and interactions with ion channel function.

In order to better understand transport mechanisms along the nephron and the interplay of different transporters in the overall reabsorption of solutes and electrolytes as well as in the excretion of toxic metabolites a brief description of the nephron and its function will be given (reviewed by Kriz and Kaissling 2000). The kidneys are organized in functional units, known as nephrons, of which approximately a million exist in each human kidney. Nephrons consist of a glomerulum where ultrafiltration of plasma occurs and a tubule lined by a single layer of epithelial cells where reabsorption and excretion of ions, solutes, and metabolic waste products take place. Several nephrons converge into collecting ducts, which in turn empty into the renal pelvis (Figure 1). Furthermore, a renal cortex and several regions in the medulla can be distinguished containing the glomerulum, proximal

*University of Zürich, Institute of Physiology, Winterthurerstrasse 190, Zürich 8057, Switzerland

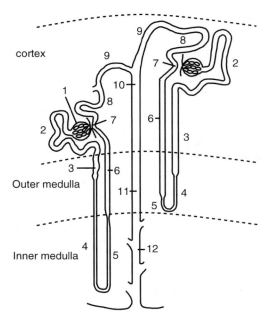

Figure 1 Schematic diagramm of nephron and collecting duct structure. Two nephrons are shown, a juxtamedullary nephron (left) and a nephron starting more superficially (right): (1) glomerulum; (2) convoluted part of the proximal tubule (S1, S2 segments); (3) straight part of the proximal tubule (S3 segment); (4) descending thin limb; (5) ascending thin limb; (6) thick ascending limb (TAL); (7) macula densa; (8) distal tubule; (9) connecting segment; (10) cortical collecting duct; (11) outer medullary collecting duct; (12) inner medullary collecting duct. Adapted from Kriz and Kaissling (2000).

tubule, part of the thick ascending limb of the loop of Henle, the distal tubule, and the cortical collecting duct (cortex), or the thin descending and ascending limb and the lower part of the thick ascending limb of the loop of Henle and the outer and inner medullary collecting duct (medulla). Several structurally and functionally distinct segments exist along the nephron as well as in the collecting duct. Ultrafiltration of plasma takes place in the glomerulum producing primary urine with a composition similar to plasma containing low molecular weight proteins, solutes (such as glucose, amino acids, bicarbonate, inorganic

phosphate), and electrolytes. In the early proximal tubule (also called the convoluted part or S1 and S2 segment) and the straight proximal tubule (S3 segment) the bulk of filtered water, solutes, and electrolytes are reabsorbed proportionally, producing an isotonic urine at the end of the S3 segment which is slightly acidic (pH 6.8) due to the quantitatively higher reabsorption of bicarbonate and the excretion of protons and acid equivalents (see below). Inherited forms of loss of reabsorptive transport activity in the proximal tubule may cause *glucosuria* due to mutations in the SGLT1 or SGLT2 (SLC5A1 or SLC5A2) transporters (Martin *et al.* 1996; Van den Heuvel *et al.* 2002), amino acidurias such as *Hartnups disease, cystinuria*, or *lysinuric protein intolerance* (see Chapter 14), primary carnitine deficiency (see Chapter 13), or *proximal renal tubular acidosis* due to loss of bicarbonate transport activity (see Chapter 4). However, the proximal tubule not only plays an important role in reabsorption but is also the major site in the kidney where excretion of waste products and detoxification occurs due to the activity of secretory transport systems (Russel *et al.* 2002). Several anionic substrates and drugs are excreted into urine in the proximal tubule, such as *p*-aminohippurate (PAH), urate, loop diuretics, salicylates, angiotensin-converting enzyme (ACE) inhibitors, penicillin, indomethacin, or probenecid involving a tertiary active transport system coupling the transport of dicarboxylates such α-ketoglutarate or citrate to the basolateral import and apical export of anionic substances. Whereas dicarboxylates are taken up into the cell from the urine through the NaDC1 transporter (SLC13A2) (one of the main substrates being divalent citrate) or from the blood through NaDC3 (SLC13A3) (one of the main substrates is α-ketoglutarate), the OAT transporters (organic anion transporter, SLC22 family) exchange these dicarboxylates for other anionic substrates (Burckhardt and Wolff 2000; Inui *et al.* 2000; Sweet *et al.* 2001). Very recently, mutations in the OAT4 (SLC22A12) transporter have been found in patients with *idiopathic renal hypouricemia* (i.e., increased urate concentrations in urine and urate kidney stones) demonstrating the importance of these transport pathways (Enomoto *et al.* 2002). Also cationic substrates and drugs (monoamines, creatinine, cimetidine, atropine, amiloride, amidated local anesthetics) can be transported by a related family of organic cation transporters (OCT transporters, SLC22 family) and are excreted into urine (Koepsell 1998). Also, transporters belonging to the ABC transporter family (see Chapters 17–21), such as MRP2 (ABCC2; see Chapter 18), are involved in detoxification, excreting substances as glucuronate, glutathione, or sulfate conjugates (Burckhardt and Wolff 2000; Inui *et al.* 2000).

The thin descending and ascending limb and thick ascending limb (TAL) of the loop of Henle play a major role in generating a hypertonic medulla necessary for the final concentration of urine along the collecting duct. Superficial nephrons starting in a glomerulum close to the organ surface have short loops, whereas juxtamedullary nephrons having a glomerulus near the cortex–outer medulla boundary have long loops of Henle reaching deep into the medulla. In the thin limb, urea secretion occurs through facilitative urea transporters (SLC14 family), whereas urea is strongly reabsorbed in the medullary collecting duct, contributing to the high osmolarity of the inner medulla (Sands 2003). In the thick ascending limb Na^+ and Cl^- are reabsorbed, whereas this segment is also characterized by its very low water permeability resulting in relatively diluted urine. In the following distal tubule, separated anatomically from the TAL by the macula densa, Na^+, Ca^{2+}, and Mg^{2+} transport takes place as well as bicarbonate reabsorption and H^+ excretion in the late or convoluted distal tubule. Defects in several components of the Na^+ and Cl^- reabsorbing machinery in the TAL and distal tubule lead to different types of *Bartter's syndrome* and *Gitelman's syndrome* (see below). The connecting tubule and cortical collecting duct serve the fine-tuning of water, Na^+, and acid–base transport. At least two major cell types are involved in this task: the principal cells (different subtypes are distinguished based on their localization and expression of transport systems) reabsorb Na^+ through the action of the aldosterone-regulated epithelial Na^+ channel (ENaC), secrete K^+ through the ROMK channel (renal outer medullary K^+ channel), and reabsorb water via the AQP-2 water channel which is under the hormonal control of the antidiuretic hormone (ADH, vasopressin). Even though only about 2% of the overall filtered Na^+ and water are reabsorbed in these segments, inherited or acquired loss or gain of function in these transport systems has dramatic consequences, as highlighted by severe forms of hypertension in Liddle's syndrome (gain of function mutation in ENaC subunits) or several forms of defective signaling through aldosterone (reviewed by Lifton *et al.* 2001). Dysfunction of the AQP-2 channel or the vasopressin receptor cause renal diabetes insipidus where patients loose 10–15 L of water daily (Agre *et al.* 2002). Intercalated cells, the second cell type, are the main site of acid–base transport (Wagner 2002). There are at least two main subtypes of intercalated cells: the type A cell which expresses a vacuolar-type H^+-ATPase secreting protons, and non-type A intercalated cells which secrete bicarbonate. The activity of the intercalated cells is adapted to metabolism ensuring that acid or base equivalents can be secreted into urine depending on their generation by metabolism. Defects in intercalated cell function lead to several syndromes of distal renal tubular acidosis (see below). In the final part of the collecting duct the medullary collecting duct, a further reabsorption of Na^+ and water and secretion of protons take place. However, this contribution is relatively small.

ELECTROLYTE TRANSPORT

About 180 L of serum are filtered in the glomeruli, yielding a primary urine having the same composition of solutes and electrolytes as serum. About 99% of water, solutes, and electrolytes are subsequently reabsorbed along the nephron, and toxic substrates and acid are excreted, resulting in about 1.5–2 L of final urine. Obviously, only small differences in the amount of water and electrolytes excreted will have a major impact on total body fluid volume and electrolyte composition. The main electrolytes determining total body fluid volume are Na^+ and Cl^- where water reabsorption through water channels (aquaporin family) follows along the osmotic gradient established by the active transport of these electrolytes. There has been a great interest in identifying and understanding mechanisms involved in Na^+ reabsorption and their regulation as these mechanisms contribute to, if not control, extracellular fluid volume and thus blood pressure (Lifton et al. 2001). A combination of genetic and physiological research has made great progress in recent years in dissecting pathways of Na^+ transport and highlighting their physiological importance by finding mutations in many components of these pathways in patients with inherited forms of high or low blood pressure (Lifton et al. 2001).

The quantitatively largest amount of Na^+ is reabsorbed in the proximal tubule through the coupling of solute transport to Na^+ and Na^+/H^+ exchange (Figure 2). The basolaterally localized Na^+, K^+-ATPase energizes apical reabsorption of Na^+ and solutes by generating an electrochemical gradient for Na^+ which is used to transport solutes such as inorganic phosphate (through Na^+/phosphate cotransporters, SLC34 family), glucose (Na^+/glucose cotransporters SLC5A1 and SLC5A2), or amino acids (system B^0, not yet identified on a molecular level) against their concentration gradient. In addition, apically localized Na^+/H^+ exchangers exchange Na^+ against intracellular H^+ and thus reabsorb bicarbonate and Na^+ (see below for details). Several isoforms of Na^+/H^+ exchangers (NHE-2, NHE-3, and recently NHE-8) have been identified in the brush border membrane of the proximal tubule. Knockout experiments for NHE-2 and NHE-3 suggest that NHE-3 plays an major role as mice deficient in this isoform show

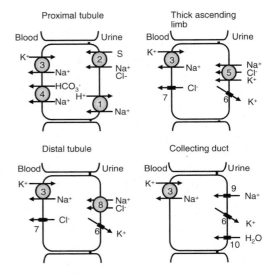

Figure 2 Na^+ transport in different cells along the nephron and collecting duct. In the *proximal tubule* apical Na^+ uptake from urine occurs mainly through Na^+-coupled solute reabsorption (2) and Na^+/H^+ exchange (1). The driving force for Na^+ entry is generated by the basolateral Na^+, K^+-ATPase (3). In addition, Na^+/bicarbonate cotransport releases Na^+ into blood (4). In the *thick ascending limb*, Na^+ is reabsorbed through the apical $Na^+/K^+/2Cl^-$ cotransporter NKCC2 (5), and released into blood by the basolateral Na^+, K^+-ATPase (3). In order to function, NKCC2 needs sufficient K^+ which is recycled across the apical membrane through the ROMK K^+ channel (6). The reabsorbed Cl^- leaves the cell on the basolateral side through Cl^--channels formed by ClC-Ka and ClC-Kb together with the β-subunit Barttin (7). Mutations in either NKCC2, ROMK, the ClC-Ka channel or Barttin lead to Bartter's syndrome. In the *distal tubule*, Na^+ reabsorption involves mainly the thiazide-sensitive Na^+/Cl^- cotransporter NCC (8) which is mutated in Gitelman's syndrome. In the *connecting segment and collecting duct*, the final tuning of Na^+ reabsorption occurs in aldosterone-sensitive principal cells expressing the epithelial Na^+ channel ENaC (9), the K^+-secretory ROMK channel (6), and the vasopressin-regulated water channel AQP-2 (10).

impaired Na^+ and water reabsorption in the proximal tubule (Woo et al. 2003). Reabsorption in the proximal tubule accounts for approximately 70% of totally filtered Na^+.

Most of the remaining sodium is reabsorbed in the TAL of Henle and the distal tubule (Figure 2). Distinct mechanisms involving two related transporters of the sodium/potassium/chloride cotransporter family SLC12 mediate the net reabsorption of Na^+ and Cl^-. In the TAL Na^+ and Cl^- are reabsorbed through the action of the apically localized loop diuretic (i.e., furosemide) sensitive $Na^+/K^+/2Cl^-$ cotransporter NKCC2/BSC1 (SLC12A1). In order to function, this transporter requires luminal K^+ which is recycled across the apical membrane through the ROMK channel after being reabsorbed. Na^+ is released into blood via the basolateral Na^+, K^+-ATPase, whereas Cl^- leaves the cell through basolateral heterodimeric ClC-Ka/Barttin and ClC-Kb/Barttin Cl^--channels. The water permeability of the TAL is very low and thus the urine will be diluted in this segment. As the reabsorption of Na^+ and Cl^- depends on the function of several transport pathways in the same cells of the TAL, mutations in each of these components alone can cause a clinically similar picture, which has been initially described as Bartter's syndrome. Subsequently, using genetics and a refined clinical characterization of patients suffering from Bartter's syndrome, it became clear that besides a shared pathophysiology distinct features can be used to discriminate between different genes/transporters affected. Currently four types of *Bartter's syndrome* are distinguished: type I due to mutations in the $Na^+/K^+/2Cl^-$ cotransporter NKCC2 (SLC12A1) (Simon *et al.* 1996a), type II due to mutations in the ROMK K^+ channel (Simon *et al.* 1996b), type III caused by mutations in the ClC-Kb Cl^- channel (Simon *et al.* 1997), and type IV (mutations in Cl^- channel β-subunit Barttin) (Birkenhager *et al.* 2001; Estevez *et al.* 2001). Type IV is associated with sensorineural deafness due to the importance of the Barttin subunit in the inner ear. This form of Bartter's syndrome is also the most severe in terms of metabolic consequences and renal failure. Types I and II are also termed antenatal Bartter's syndrome because of hydramnion (increased fetal

urinary secretion leading to an increase in hydramnic fluid), and are further characterized by hypokalemic metabolic alkalosis, Na^+ and subsequent water loss, and hypercalciuria. In type III, the onset of disease is somewhat later and less severe, but is associated with hypomagnesemia. Only type IV Bartter's syndrome shows inner-ear deafness.

In the distal tubule Na^+ reabsorption occurs via transport by the thiazidesensitive NCC/TSC Na^+/Cl^- cotransporter (SLC12A3) localized in the apical membrane (Figure 2). This transporter is defective in *Gitelman's syndrome*, a syndrome characterized by urinary loss of Na^+, hypokalemic metabolic alkalosis, loss of magnesium into urine, and increased calcium reabsorption (Simon *et al.* 1996c).

The importance of NCC for blood pressure regulation has also been highlighted by a large study on the impact of heterozygosity for NCC mutations on systemic blood pressure. Carriers with one inactivating mutation in NCC had significantly lower blood pressure (Cruz *et al.* 2001). Very recently, it has been suggested that NCC forms functional homodimers (De Jang *et al.* 2003) and it is conceivable that defects in one subunit can lead to inactivation of up to 75% of all homodimers, significantly reducing overall Na^+ absorption through this transporters.

The final fine-tuning of Na^+ reabsorption occurs in the connecting segment and cortical collecting duct through the action of the epithelial Na^+ channel (ENaC) (Verrey *et al.* 2000) (Figure 2). Gain- and loss-of-function mutations in all three subunits have been found in patients with high or low blood pressure, and this has been extensively reviewed elsewhere (Lifton *et al.* 2001).

It should also be mentioned here that an increasing number of inherited diseases are not caused by mutations in transport systems themselves but by mutations affecting associated proteins or regulatory pathways. Transport pathways for Na^+ are excellent examples of this. Several forms of inherited high or low blood pressure are caused by mutations underlying the renin–angiotensin–aldosterone regulatory pathway, which is the main hormonal regulatory system for Na^+ reabsorption (Lifton *et al.* 2001). More recently, mutations in the WNK kinase pathway (WNK1 and WNK4) have been identified leading to *pseudohypoaldosteronism type II*, featuring hypertension, increased renal salt reabsorption, and impaired K^+ and H^+ excretion (Wilson *et al.* 2001). The role of WNK kinases in the regulation of Na^+ transport is not fully understood, but it emerges that WNK kinases may be involved in the regulation of Na^+/Cl^- cotransporter NCC (SLC12A3) activity and that mutations may lead to disturbed retrieval and thus excessive Na^+ reabsorption (Wilson *et al.* 2003; Yang *et al.* 2003).

RENAL TRANSPORT OF ACID–BASE EQUIVALENTS

Whole-body acid–base homeostasis is controlled by two organs – the lungs and the kidneys. The major adaptation to changes in metabolic rates affecting acid–base homeostasis occurs through changes in renal transport of acid–base equivalents and generation of bicarbonate. Thus the kidney serves two main purposes in controlling acid–base homeostasis, reclaiming filtered bicarbonate and secreting acid–base equivalents produced by metabolism (Hamm and Alpern 2000).

As mentioned above, bicarbonate is freely filtered in the glomeruli and about 4500 mEq of bicarbonate have to be reabsorbed daily. The bulk of bicarbonate is reabsorbed in the proximal tubule (about 70–80%) involving several transport systems (Figure 2). Protons are secreted into the lumen via action of Na^+/H^+ exchangers and vacuolar H^+-ATPases. The main Na^+/H^+ exchanger isoforms involved in proton secretion are NHE-3 and another not yet identified isoform, potentially NHE-8. It has been estimated that about 40% of overall proton secretion in the proximal tubule may be mediated by vacuolar H^+-ATPases. The protons combine in the lumen with the filtered bicarbonate and under the influence of a carbonic anhydrase bound to the extracellular domain of the brush border (carbonic anhydrase IV), water and CO_2 are formed. CO_2 easily diffuses into the cell where it is rehydrated and bicarbonate and protons are formed again, catalysed by the cytosolic carbonic anhydrase isoform II. Bicarbonate is released into blood via the basolaterally expressed Na^+-dependent bicarbonate cotransporter NBC-1 (SLC4A4), whereas the protons formed are recycled across the apical membrane. Inherited mutations in the SLC4A4 basolateral bicarbonate transporter are associated with *proximal renal tubular acidosis* as reviewed here (Chapter 4).

About another 20% of the filtered bicarbonate is subsequently reabsorbed in the TAL and the distal tubule, involving similar mechanisms as in the proximal tubule. In the late distal tubule, the connecting segment and collecting duct, specialized cells—intercalated cells—are important in reabsorbing the remaining bicarbonate and also in generating new bicarbonate (Figure 3). Based on morphological and functional characteristics, at least two types of intercalated cells can be distinguished (Hamm and Alpern 2000; Wagner and Geibel 2002). Type A intercalated cells are proton secretory cells, characterized by the apical expression of vacuolar H^+-ATPases mediating proton export into urine. An intracellular carbonic anhydrase (isoform II) generates protons and bicarbonate. The newly generated bicarbonate is exchanged against chloride via the basolateral Cl^-/HCO_3^- exchanger AE-1 and thus secreted into blood. The structure, function, and inherited defects of AE-1 (SLC4A1) are discussed in Chapter 3. In contrast, type B intercalated cells express the vacuolar H^+-ATPase on the basolateral side and have an apical Cl^-/HCO_3^- exchanger; hence these cells mediate bicarbonate secretion into urine. The molecular identity of the apical Cl^-/HCO_3^- exchanger is not fully resolved. Two candidate proteins have been suggested, AE-4 and pendrin [the extrarenal function of pendrin (SLC26A4) is described in Chapter 8]. However, the apical expression of AE-4 in

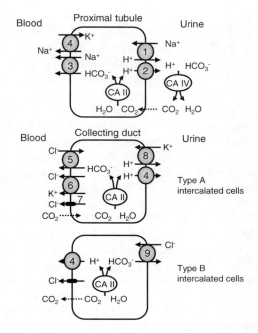

Figure 3 Main sites of acid–base transport along the nephron. In the *proximal tubule*, H^+ is secreted via Na^+/H^+ exchangers (1) and vacuolar H^+-ATPases (2), combines with filtered HCO_3^-, and forms water and CO_2 (under the influence of carbonic anhydrase CAIV). CO_2 diffuses into the cells and is rehydrated, and H^+ and HCO_3^- are generated (catalyzed by carbonic anhydrase CAII). H^+ is recycled whereas HCO_3^- is transported into blood via the basolateral Na^+/HCO_3^- cotransporter (3). The driving force for Na^+/H^+ exchange is maintained by the basolateral Na^+, K^+-ATPase (4) and the intracellular production of H^+. In the collecting duct two types of cells are involved in acid–base transport: H^+-secretory type A intercalated cells, and HCO_3^--secretory type B intercalated cells. Type A intercalated cells generate HCO_3^- catalyzed by CAII, which is transported into blood via the basolateral Cl^-/HCO_3^- exchanger AE-1 (5). The H^+ is secreted into urine by vacuolar H^+-ATPases (4). Cl^- is recycled across the basolateral membrane, possibly involving ClC-Ka/ Barttin Cl^- channels (7) and the K^+/Cl^- cotransporter KCC4 (6). In addition, an H^+, K^+-ATPase, which is mainly involved in reabsorbing K^+, is expressed apically (8). Type B intercalated cells excrete HCO_3^- through an apical Cl^-/HCO_3^- exchanger (9), possibly pendrin, and thus reabsorb Cl^-. A basolateral vacuolar H^+-ATPase (4) secretes H^+ into blood.

non-type-A intercalated cells seems to be species specific and suggests that this is not the main apical $Cl^--HCO_3^-$ exchanger. Pendrin (SLC26A4) is expressed on the apical membrane of non-type-A intercalated cells and is capable of mediating Cl^-/HCO_3^- exchange, and mice deficient in pendrin have a greatly reduced rate of bicarbonate secretion in the cortical collecting duct (Royaux *et al.* 2001). These results suggest that pendrin is either involved or directly mediates bicarbonate secretion in the cortical collecting duct. A third type of intercalated cell expressing vacuolar H^+-ATPases on the apical and basolateral membrane and having functional Cl^-/HCO_3^- exchange on both sides has been suggested by some authors (Wagner and Geibel 2002). Whether this is a distinct cell type or a substate of activity of type B intercalated cells remains controversial. In general, type B intercalated cells are only found in the connecting segment and cortical collecting duct, whereas type A intercalated cells are found throughout the last distal tubule to the initial segment of the inner medullary collecting duct. Depending on metabolism, dietary intake, or drugs, type A acid-secretory cells or type B bicarbonate secretory cells are active. Under normal conditions (human, mixed diet) type A intercalated cells predominate and generate about 70–80 mEq of new bicarbonate, which is needed to buffer the equivalent amount of protons generated by metabolism.

In addition to mutations in the AE-1 (SLC4A1) Cl^-/HCO_3^- exchanger, mutations in two subunits of the vacuolar H^+-ATPase have been identified in patients with *distal renal tubular acidosis* (Karet *et al.* 1999; Smith *et al.* 2000). Vacuolar H^+-ATPases are complex transport proteins with at least 13 subunits mediating the ATP-driven transport of protons across membranes in mammalians. Vacuolar H^+-ATPases are ubiquitously expressed and are found only in the intracellular membranes of most cells, such as lysosomes, endosomes, the Golgi or neurotransmitter-storing vesicles (renewed by Nishi and Forgac 2002). There, vacuolar H^+-ATPases are responsible for the acidification of these intracellular organelles, which is necessary for the proper function of enzymes in these compartments or for driving the transport of substrates across the organellar membrane (i.e., H^+-dependent uptake of neurotransmitters into storage vesicles in neurons). In some specialized cells, such as osteoclasts, macrophages, and some cells in the epididymis or inner ear and in the kidney, vacuolar H^+-ATPases secrete protons across the plasma membrane contributing to the regulation of systemic pH (i.e., in the kidney) or mediating cell-specific tasks such as dissolving bone matrix by osteoclasts. Organ and cell-specific isoforms exist for some of the 13 subunits of the vacuolar H^+-ATPase, and mutations in three subunits have been identified in inherited human diseases. Mutations in the a3 (ATP6VOA3) subunit were found in children suffering from the lethal *malignant infantile osteopetrosis disease*, where osteoclasts fail to form bone cavities necessary for bone marrow and erythropoiesis owing to the inactivation of the vacuolar H^+-ATPase (Frattini *et al*. 2000). Expression of the a3 subunit occurs almost exclusively in osteoclasts, limiting the disease to these cells. Inherited defects in the a4 isoform (ATP6V0A4) of the same subunit lead to *distal renal tubular acidosis* due to the almost exclusive expression of this isoform in the kidney (Smith *et al*. 2000). Also, mutations in the intercalated cell specific B1 isoform (ATP6V1B1) cause distal renal tubular acidosis combined with *sensorineural deafness* as this subunit is also expressed in specialized cells of the inner ear (Karet *et al*. 1999).

Renal excretion of acid equivalents not only involves the transport of protons and bicarbonate but also requires so-called titratable acids. If protons were only secreted and not buffered in urine in the collecting duct, the proton gradient between urine and cell interior would become so steep that the pumping efficiency of the vacuolar H^+-ATPase would be limited. In order to maintain proton secretion, titratable acids such as citrate, inorganic phosphate, and ammonia are used to buffer protons (Hamm and Alpern 2000). Citrate and phosphate

are filtered in the glomerulum and the non-reabsorbed portion of these substrates serves then as buffer. Ammonia, in contrast, is generated in the proximal tubule from glutamine while producing bicarbonate, and is then secreted into urine, reabsorbed in the TAL, and secreted again into urine across intercalated cells in the collecting duct. This complex mechanism involves several transport steps, most of which are not elucidated on a molecular level. Recently, a novel family of putative transport proteins has been identified that may mediate ammonia or ammonium transport. These proteins (RhCG and RhBG) belong to the family of blood-group determining proteins defining the rhesus antigen (Marini *et al*. 2000). Homologous proteins have been identified in yeast and plants and shown to be involved in ammonia/ ammonium transport, and the mammalian homologs are indeed able to complement yeast deficient in endogenous ammonia/ammonium transporters. The two mammalian putative ammonia transporters RhCG and RhBG are expressed in kidney on the apical and basolateral side of intercalated cells (Eladari *et al*. 2002; Quentin *et al*. 2003). However, it remains to be firmly established if these proteins are responsible for some of the ammonia/ammonium transport activities and what their specific role in this complex process may be.

Inherited or acquired disturbances of acid–base transport in the kidney may lead to several syndromes of metabolic acidosis due to renal tubular acidosis (RTA). The effects of inherited forms of renal tubular acidosis depend on the underlying defect as illustrated in Chapters 3 and 4. In general, different types of electrolyte disturbances (mainly involving potassium) and increased bone absorption to release phosphate and calcium carbonate in order to buffer protons are at the center of the associated pathophysiology. Release of calcium carbonate from bones leads to osteomalacia, growth retardation, and skeletal deformations in untreated children; the increased calcium load to the kidneys results in calcium-containing kidney stones. These phenotypes resolve almost completely under an adequate therapy with potassium citrate or bicarbonate. Four different types of renal tubular acidosis have been classified. The classic form, *type I RTA*, is *distal renal tubular acidosis* (dRTA) and is characterized by decreased proton secretion in the distal part of the nephron. Classically, urinary pH is not more acidic than pH 5.5 in the setting of an induced or spontaneous metabolic acidosis (i.e., arterial blood pH lower than 7.37). This type of dRTA is due to a defect in proton secretion or bicarbonate absorption in the collecting duct and mutations in AE-1 Cl^-/HCO_3^- exchanger (SLC4A1) and the a4 (ATP6V0A4) and B1 (ATP6V1B1) subunits of the vacuolar H^+-ATPase have been identified. *Type II RTA* is caused by a decrease in proximal tubular bicarbonate reabsorption (urinary bicarbonate wasting), thus termed pRTA, and mutations in the Na^+/bicarbonate cotransporter SLC4A4 have been found in patients. A mixed type of proximal and distal renal tubular acidosis (**type III**) can be caused by mutations in the gene encoding for carbonic anhydrase II, which is important for the generation of bicarbonate in both the proximal tubule and intercalated cells. However, the phenotype of these patients is dominated by cerebral calcifications and osteopetrosis due to the importance of the carbonic anhydrase II in brain and osteoclasts (Sly and Shah 2001). A fourth type of RTA is caused by aldosterone deficiency or resistance and is accompanied by hyperkalemia. Mutations in genes involved in aldosterone signaling or its targets (i.e., the epithelial Na^+ channel) have been associated (Hamm and Alpern 2000; Lifton *et al.* 2001).

REFERENCES

Agre, P., King, L.S., Yasui, M., Guggino, W.B., Ottersen, O.P., Fujiyoshi, Y., *et al.* (2002). Aquaporin Water Channels—From Atomic Structure to Clinical Medicine. *J. Physiol.,* **542**, 3–16.

Birkenhager, R., Otto, E., Schurmann, M.J., Vollmer, M., Ruf, E.M., Maier-Lutz, I., *et al.* (2001). Mutation of BSND Causes Bartter Syndrome with Sensorineural Deafness and Kidney Failure. *Nat. Genet.,* **29**, 310–314.

Burckhardt, G. and Wolff, N.A. (2000). Structure of Renal Organic Anion and Cation Transporters. *Am. J. Physiol. Renal Physiol.,* **278**, F853–866.

Cruz, D.N., Simon, D.B., Nelson-Williams, C., Farhi, A., Finberg, K., Burleson, L., *et al.* (2001). Mutations in the Na–Cl Cotransporter Reduce Blood Pressure in Humans. *Hypertension,* **37**, 1458–1464.

De Jong, J.C., Willems, P.H., Mooren, F.J., Van Den Heuvel, L.P., Knoers, N.V., and Bindels, R.J. (2003). The Structural Unit of the Thiazide-Sensitive Nacl Cotransporter (NCC) is a Homodimer. *J. Biol. Chem.,* **278**, 24302–24307.

Eladari, D., Cheval, L., Quentin, F., Bertrand, O., Mouro, I., Cherif-Zahar, B., *et al.* (2002). Expression of RhCG, a New Putative NH_3/NH_4^+ Transporter, Along the Rat Nephron. *J. Am. Soc. Nephrol.,* **13**, 1999–2008.

Enomoto, A., Kimura, H., Chairoungdua, A., Shigeta, Y., Jutabha, P., Cha, S.H., *et al.* (2002). Molecular Identification of a Renal Urate Anion Exchanger that Regulates Blood Urate Levels. *Nature,* **417**, 447–452.

Estevez, R., Boettger, T., Stein, V., Birkenhager, R., Otto, E., Hildebrandt, F., *et al.* (2001). Barttin is a Cl^- Channel Beta-Subunit Crucial for Renal Cl^- Reabsorption and Inner Ear K^+ Secretion. *Nature,* **414**, 558–561.

Frattini, A., Orchard, P.J., Sobacchi, C., Giliani, S., Abinun, M., Mattsson, J.P., *et al.* (2000). Defects in TCIRG1 subunit of the vacuolar proton pump are responsible for a subset of human autosomal recessive osteopetrosis. *Nat. Genet.,* **25**, 343–346.

Hamm, L.L. and Alpern, R.J. (2000). Cellular mechanisms of renal tubular acidification. In D.W. Seldin and G. Gebisch (eds), *The Kidney: Physiology and Pathophysiology* (3rd ed.), Philadelphia, PA: Lippincott–Williams & Wilkins, pp. 1935–1979.

Inui, K.I., Masuda, S. and Saito, H. (2000). Cellular and Molecular Aspects of Drug Transport in the Kidney. *Kidney Int.,* **58**, 944–958.

Karet, F.E., Finberg, K.E., Nelson, R.D., Nayir, A., Mocan, H., Sanjad, S.A., *et al.* (1999). Mutations in the Gene Encoding B1 Subunit of H^+-ATPase Cause Renal Tubular Acidosis with Sensorineural Deafness. *Nat. Genet.,* **21**, 84–90.

Koepsell, H. (1998). Organic Cation Transporters in Intestine, Kidney, Liver, and Brain. *Annu. Rev. Physiol.,* **60**, 243–266.

Kriz, W. and Kaissling, B. (2000). Structural organization of the mammalian kidney. In D. Seldin and G. Gebisch (eds), *The Kidney: Physiology and Pathopyhsiology* (3rd edn), Philadelphia, PA: Lippincott–Williams & Wilkins, pp. 587–654.

Lifton, R.P., Gharavi, A.G. and Geller, D.S. (2000). Molecular Mechanisms of Human Hypertension. *Cell,* **104**, 545–556.

Marini, A.M., Matassi, G., Raynal, V., Andre, B., Cartron, J.P., and Cherif-Zahar, B. The Human

Rhesus-Associated RhAG Protein and a Kidney Homologue Promote Ammonium Transport in Yeast. *Nat Genet.*, **26**, 341–344.

Martin, M.G., Turk, E., Lostao, M.P., Kerner, C., and Wright, E.M. (1996). Defects in Na$^+$/Glucose Cotransporter (SGLT1) Trafficking and Function Cause Glucose–Galactose Malabsorption. *Nat. Genet.*, **12**, 216–220.

Nishi, T. and Forgac, M. (2002). The vacuolar (H$^+$)-ATPases – Nature's Most Versatile Proton Pumps. *Nat. Rev. Mol. Cell. Biol.*, **3**, 94–103.

Quentin, F., Eladari, D., Cheval, L., Lopez, C., Goossens, D., Colin, Y., *et al.* (2003). RhBG and RhCG, the Putative Ammonia Transporters, Are Expressed in the Same Cells in the Distal Nephron. *J. Am. Soc. Nephrol.*, **14**, 545–554.

Royaux, I.E., Wall, S.M., Karniski, L.P., Everett, L.A., Suzuki, K., Knepper, M.A., *et al.* (2001). Pendrin, Encoded by the Pendred Syndrome Gene, Resides in the Apical Region of Renal Intercalated Cells and Mediates Bicarbonate Secretion. *Proc. Natl. Acad. Sci. USA*, **98**, 4221–4226.

Russel, F.G., Masereeuw, R., and van Aubel, R.A. Molecular Aspects of Renal Anionic Drug Transport. *Annu. Rev. Physiol.*, **64**, 563–594.

Sands, J.M. (2003). Mammalian Urea Transporters. *Annu. Rev. Physiol.*, **65**, 543–566.

Simon, D.B., Karet, F.E., Hamdan, J.M., DiPietro, A., Sanjad, S.A., and Lifton, R.P. (1996a). Bartter's Syndrome, Hypokalaemic Alkalosis with Hypercalciuria, is caused by Mutations in the Na–K–2Cl cotransporter NKCC2. *Nat. Genet.*, **13**, 183–188.

Simon, D.B., Karet, F.E., Rodriguez-Soriano, J., Hamdan, J.H., DiPietro, A., Trachtman, H., *et al.* (1996b). Genetic Heterogeneity of Bartter's Syndrome Revealed by Mutations in the K$^+$ Channel, ROMK. *Nat. Genet.*, **14**, 152–156.

Simon, D.B., Nelson-Williams, C., Bia, M.J., Ellison, D., Karet, F.E., Molina, A.M., *et al.* (1996c). Gitelman's Variant of Bartter's Syndrome, Inherited Hypokalaemic Alkalosis, is caused by Mutations in the Thiazide-Sensitive Na–Cl Cotransporter. *Nat. Genet.*, **12**, 24–30.

Simon, D.B., Bindra, R.S., Mansfield, T.A., Nelson-Williams, C., Mendonca, E., Stone, R., *et al.* (1997). Mutations in the Chloride Channel Gene, CLCNKB, Cause Bartter's Syndrome Type III. *Nat. Genet.*, **17**, 171–178.

Sly, W.S. and Shah, G.N. (2001). The Carbonic Anhydrase II Deficiency Syndrome: Osteopetrosis with Renal Tubular Acidosis and Cerebral Calcification. In C.R. Scriver, A.L. Baudet, W.S. Sly and D. Valle (eds), *The Metabolic and Molecular Bases of Inherited Disease*, New York: McGraw-Hill, pp. 5331–5343.

Smith, A.N., Skaug, J., Choate, K.A., Nayir, A., Bakkaloglu, A., Ozen, S., *et al.* (2000). Mutations in ATP6N1B, Encoding a New Kidney Vacuolar Proton Pump 116-kD Subunit, Cause Recessive Distal Renal Tubular Acidosis with Preserved Hearing. *Nat. Genet.*, **26**, 71–75.

Sweet, D.H., Bush, K.T., and Nigam, S.K. (2001). The Organic Anion Transporter Family: From Physiology to Ontogeny and the Clinic. *Am. J. Physiol. Renal Physiol.*, **281**, F197–F205.

van den Heuvel, L.P., Assink, K., Willemsen, M., and Monnens, L. (2002). Autosomal Recessive Renal Glucosuria Attributable to a Mutation in the Sodium Glucose Cotransporter (SGLT2). *Hum. Genet.*, **111**, 544–547.

Verrey, F., Hummler, E., Schild, L., and Rossier, B. (2000). Control of Na$^+$ Transport by Aldosterone. In D.W. Seldin and G. Griebisch (eds), *The Kidney: Physiology and Pathophysiology* (3rd edn), Philadelphia, PA: Lippincott–Williams & Wilkins, pp. 1441–1471.

Wagner, C.A. and Geibel, J.P. (2002). Acid–Base Transport in the Collecting Duct. *J. Nephrol.*, **Suppl. 5**, S112–S127.

Wilson, F.H., Disse-Nicodeme, S., Choate, K.A., Ishikawa, K., Nelson-Williams, C., Desitter, I., *et al.* (2001). Human Hypertension Caused by Mutations in WNK Kinases. *Science*, **293**, 1107–1112.

Wilson, F.H., Kahle, K.T., Sabath, E., Lalioti, M.D., Rapson, A.K., Hoover, R.S., *et al.* (2003). Molecular Pathogenesis of Inherited Hypertension with Hyperkalemia: The Na–Cl Cotransporter is Inhibited by Wild-Type but not Mutant WNK4. *Proc. Natl Acad. Sci. USA*, **100**, 680–684.

Woo, A.L., Noonan, W.T., Schultheis, P.J., Neumann, J.C., Manning, P.A., and Lorenz, J. Renal Function in NHE3-Deficient Mice with Transgenic Rescue of Small Intestinal Absorptive Defect. *Am. J. Physiol. Renal Physiol.*, **284**, F1190–F1198.

Yang, C.L., Angell, J., Mitchell, R. and Ellison, D.H. (2003). WNK Kinases Regulate Thiazide-Sensitive Na–Cl Cotransport. *J. Clin. Invest.*, **111**, 1039–1045.

3

SETH L. ALPER*

Diseases of mutations in the SLC4A1/AE1 (band 3) Cl⁻/HCO₃⁻ exchanger

INTRODUCTION

The *AE1/SLC4A1/EPB3* gene is the founding member of the *SLC4* bicarbonate transporter superfamily. The *SLC4* gene family includes nine known mammalian genes, each of which encodes multiple transcripts encoding variant polypeptides. The *SLC4A1* gene is one of three of these genes (*SLC4A1–3*) which have been shown to encode Na$^+$-independent, electroneutral Cl⁻/HCO₃⁻ exchanger polypeptides. (Although less thoroughly investigated, *SLC4A9* in at least some species has also been reported to express this activity.) The *SLC4A1* gene is unique among the human SLC4 Na$^+$-independent Cl⁻/HCO₃⁻ exchanger genes in its association with inherited human disease. The SLC4A1/AE1 polypeptide has recently been reviewed (Alper 2002; Knauf and Pal 2003), as has its roles in human disease (Shayakul and Alper 2000; Alper 2002; Kaset 2002; Steming and Casey 2002; Tannel 2002; Wrong *et al.* 2002; Jarolim 2003).

PHYSIOLOGICAL FUNCTION AND STRUCTURE OF AE1

The 911 amino acid (aa) AE1/band 3 polypeptide is the major intrinsic membrane protein of the erythrocyte. Erythroid AE1 (eAE1) comprises up to 25% of total membrane protein and up to 50% of membrane intrinsic protein. The AE1 protein is encoded by a 20 exon *AE1* gene on chromosome 17q21–22, with a TATA-less "erythroid-specific promoter" upstream of exon 1.

Kidney AE1 (kAE1) protein is localized in the basolateral membrane of type A intercalated cells of the renal collecting duct and lacks the 65 N-terminal amino acid residues (starting at Met 66). The kAE1 promoter is presumed to reside in intron 3 of the *AE1* gene. Both proteins mediate the tightly coupled electroneutral exchange of Cl⁻ for HCO₃⁻.

In the peripheral capillaries, prevailing anion concentration gradients in the CO₂-rich environment of the respiring tissues favor exchange of intracellular HCO₃⁻ (obtained by carbonic anhydrase mediated hydroxylation of CO₂ which has diffused into the red cell from the plasma) for extracellular Cl⁻. Conversely, in the CO₂-poor environment of the pulmonary capillaries, the prevailing anion concentration

* Harvard Medical School, Molecular Medicine and Renal Units, RW 763 Beth Israel Deaconess Medical Center, East Campus, 330 Brookline Avenue, Boston, MA 02215, USA

gradients favor exchange of intracellular Cl^- for extracellular HCO_3^- (rapidly converted by erythroid cytosolic carbonic anhydrase to CO_2, which can diffuse to the alveoli for exhalation). The combined processes serve to increase the total CO_2-carrying capacity of whole blood by severalfold and enhance CO_2 transfer from respiring tissues to the lungs. In the basolateral membrane of renal type A intercalated cells, the kidney isoform of AE1, kAE1, mediates exchange of intracellular HCO_3^- for extracellular Cl^-, contributing to collecting duct secretion of HCl (Alper 2002; Wrong et al. 2002; Knauf and Pal 2003). AE1 mRNA has been detected in other tissues such as heart (Kudryki et al. 1990) and colon (Rajendran et al. 2002), but polypeptide detection in those tissues has been controversial.

The AE1 polypeptide has three structural domains, a hydrophilic amino-terminal cytoplasmic domain of ~408 aa, a hydrophobic transmembrane domain predicted from its hydrophobicity profile to traverse the lipid bilayer with 12–14 transmembrane spans, and a 33 aa hydrophilic carboxy-terminal tail (Figure 1). In the membrane, AE1 is present as a homodimer or as an ankyrin-associated homo-tetramer (dimer of dimers). Crystals of aa 1–379 of the recombinant N-terminal cytoplasmic domain grown at pH 4.8 allowed detection of ordered structure at 2.6 Å resolution for the region encompassing aa 55–201 and aa 212–356 (Zhang et al. 2000). The unstructured N-terminus binds the glycolytic enzymes GAPDH, aldolase, and phosphofructokinase, as well as deoxyhemoglobin and

Figure 1 Mutations of the human AE1/SLC4A1 (Band 3) Cl^-/HCO_3^- exchanger associated with hereditary spherocytic anemia and altered erythroid shape. Mutations are shown as shaded amino acid residues, and include missense, nonsense, splicing, and deletion mutations. The Southeast Asian Ovalocytosis (SAO) mutation is Δ400–408. The secondary structure of the N-terminal cytoplasmic domain between residues 55 and 353 reflects the X-ray crystal structure of residues 55–353 (Bruce et al. 1994). The topography of the first transmembrane span and the residues deleted in SAO is from the model of Kanki et al. (2003). The topography of the remaining transmembrane spans reflects cysteine-scanning mutagenesis experiments of Casey and colleagues, most recently summarized by Zhu et al. (2003). Met 66 (arrow) marks the start of kidney AE1.

hemichrome. In the context of the AE1 homotetramer, this N-terminal segment also contributes, along with a loop consisting of aa 175–185, to a binding site for the Ank repeats of the β-spectrin-binding cytoskeletal protein Ankyrin-1 (Chang *et al.* 2003).

eAE1 in red cells also binds tightly to protein 4.2, which may link it to the Rh antigen/glycophorin B/CD47/Lutheran antigen complex (Bruce *et al.* 2003). AE1 binds less tightly to protein 4.1R, which links it to the glycophorin C/D/ p55 / spectrin/actin complex (Han *et al.* 2000). Protein 4.1 also brings AE1 into proximity with 4.1-bound calmodulin and ICln. Recombinant kAE1 (as expressed in renal intercalated cells) does not bind to Ank1 (Ding *et al.* 1994). The protein kanadaptin has been proposed to serve as a binding partner of kAE1 (Chen *et al.* 1998), but a later study has suggested that kanadaptin is a nuclear protein which does not colocalize with kAE1 (Hubner *et al.* 2002).

Removal of most of the N-terminal cytoplasmic domain does not diminish anion transport in either resealed red cell ghosts or recombinant expression systems. However, the region extending between the tryptic cleavage site at aa 361 and the cytoplasmic face of the plasma membrane (at ~aa 408) plays an important permissive role in maintaining transport competence, with aa 381–385 of particular importance (Kanki *et al.* 2003).

The topography of the complex polytopic eAE1 transmembrane domain has been extensively studied in erythrocytes by chemical labeling, proteolytic, and immunological methods. These have definitively mapped the first eight transmembrane domains. Both termini of the protein are cytoplasmically disposed. The transmembrane region of eAE1 has also been studied by antibody reactivity, with glycosylation insertion mutagenesis (Popov *et al.* 1999), and, most extensively, with cysteine accessibility scanning mutagenesis (Tang *et al.* 1998, 1999; Fujinaga *et al.* 1999). Application of these techniques has given rise to the evolving topographic model shown in Figure 1 (Zhu *et al.* 2003). Two-dimensional arrays of the native erythroid AE1 transmembrane domain have allowed electron microscopic tomographic imaging to a resolution of ~20 Å (Wang *et al.* 1994). Three-dimensional crystallization of native erythroid AE1 transmembrane domain has proven difficult (Lemieux *et al.* 2002). However, the structurally related YNL275W polypeptide of the yeast *Saccharomyces cerevisiae*, proposed to be a borate/ HCO_3^- exchanger (Takano *et al.* 2002), has been overexpressed and purified (Zhao and Reithmeier 2001), and may prove a better candidate for crystallization and structure determination.

Cysteine accessibility scanning mutagenesis and chemical modification studies have suggested the presence in the AE1 transmembrane domain of "vestibules" which penetrate deep into the plane of the lipid bilayer. Some of the residues lining these vestibules are accessible to labeling from the adjacent aqueous phase and likely contribute to the anion translocation pathway. Studies with the glutamate-reactive Woodword's reagent K (Jennings and Smith 1992; Jennings 1995), and the mutagenesis studies that followed (Sekler *et al.* 1995; Chernova *et al.* 1997) suggested that E681 in TM8 contributes to the permeability barrier within the AE1 polypeptide, and that H734 (human equivalent of murine H752) may interact with it (Muller-Berger *et al.* 1995). Many other residues have been defined which, when modified chemically or by mutagenesis, abolish or modify anion exchange (reviewed by Knauf and Pal 2003).

Carbonic anhydrase II (CA2) binds to the C-terminal cytoplasmic tail of AE1, with aa 886–890 comprising the core binding site (Vince and Reithmeier 2000). This interaction is important to Cl^-/HCO_3^- exchange activity of recombinant AE1 and other SLC4 polypeptides (Sterling *et al.* 2001). AE1 truncation mutants which exhibit loss of Cl^-/HCO_3^- exchange can still mediate wild-type Cl^-/Cl^- exchange, supporting the hypothesis of an important role of CA2 in metabolic channeling of HCO_3^- substrate to and from the transport site. AE1-bound CA2 can perform this function not only for its bound transporter polypeptide, but equally well for an adjacent CA2-nonbinding AE1 protomer within the AE1 oligomer (Dahl *et al.*, submitted). The extreme C-terminal tetrapeptide of AE1 has been suggested to function as a type II recognition motif for PDZ domain binding (Cowan *et al.* 2000), but validation of this suggestion in physiological conditions remains awaited.

SLC4A1/AE1 MUTATIONS IN HEREDITARY SPHEROCYTOSIS

Hereditary spherocytosis (HS) and the attendant hemolytic anemia results from membrane instability, leading to a decrease in erythroid surface-to-volume ratio (see Box: AE1-deficient Hereditary Spherocytosis). Most cases result from deficiency of extrinsic proteins of the erythroid cytoskeleton [β and (rarely) α spectrin, ankyrin-1, and (especially in Japan) protein 4.2] and arise from heterozygous mutations. Approximately one third of patients harbor heterozygous mutations in the *AE1* gene, associated with dominant inheritance. In general, the clinical presentation is milder in HS associated with mutant AE1 than in HS secondary to mutations in the erythroid extrinsic cytoskeletal proteins (Jarolim *et al.* 1996; Jarolim 2003).

HS mutations found throughout the coding region of the AE1 polypeptide include nonsense, frameshift, and missense mutations, as summarized in Table 1. mRNA products of the nonsense and frameshift mutant alleles are often severely reduced or absent in reticulocytes. In contrast, mRNA products of several point mutant alleles and of the Δ117–121 allele were present at levels equal to that of the wild-type allele. Spherocyte AE1 polypeptide content is usually ~60–80% of wild-type levels (Jarolim *et al.* 1996). Since the mutant polypeptide is rarely detectable, even when distinguishable from wild-type, this value represents compensatory increase in expression from the wild-type allele (or a decrease in AE1 degradation during erythropoiesis). Studies of recombinant spherocytosis missense mutant polypeptides in *Xenopus* oocytes (Hubner *et al.* 2002) and in HEK 293 cells (Quilty and Reithmeier 2000) demonstrated failure of mutant polypeptides to reach the medial Golgi compartment of the secretory pathway.

AE1-DEFICIENT HEREDITARY SPHEROCYTOSIS (HS, MIM 109270)

Hereditary spherocytic anemia (HS) is characterized by fragile and unstable erythrocyte membranes. These erythrocytes loose surface membrane in vitro by shedding submicron vesicles into the bathing medium. This process, which also occurs in the circulation, results in conversion of the normal erythroid biconcave disk into a spherocyte. Erythrocyte surface membrane is also internalized. The blood smear is characterized by microcytic cells lacking central pallor, occasional stomatocytes, and, uniquely in AE1-deficient HS, pincered cells which disappear after splenectomy. Dominant HS is usually a mild disease characterized by splenomegaly and in some patients mild jaundice. Bilirubin cholelithiasis is a common complication. In some patients, anemia is absent due to adequate compensation by bone marrow hyperplasia. In the mildest cases, red cell fragility can be detected only by laboratory tests of osmotic fragility. More severe cases, including recessive HS, can exhibit splenic congestion, leg ulcers, transfusion-associated hemochromatosis, and additional problems. In some cases, splenectomy can remedy the anemia, hyperbilirubinemia, and reticulocytosis.

HS can arise from mutations in several proteins of the erythroid cytoskeleton: α-spectrin, β-spectrin, ankyrin-1, SLC4A1/AE1/band 3, and protein 4.2. In contrast, protein 4.1R mutations cause hereditary elliptocytosis. AE1/band 3 deficiency in red cells has been found in naturally occurring heritable anemias in an inbred mouse strain and in a family of cattle.

Table 1

Band 3	Mutation	Description	Reference(s)
(A) 4A1 (AE1) mutations associated with red cell shape change			
Southeast Asian ovalocytosis	Δ400–408	Ovalocytosis (MIM:109270,130600, 166900)	Jarolim *et al.* 1991; Mohandas *et al.* 1992; Schofield *et al.* 1992
High transport (HT)	P868L	Acanthocytosis, Increased sulfate transport	Bruce *et al.* 1993
(B) 4A1 (AE1) mutations associated with hereditary spherocytosis (MIM:109270)			
Genas	nt 89 (G→A)	From cap site in 5' untranslated region 33% decrease in mRNA	Alloisio *et al.* 1996
Neapolis	Splice donor mutation	Retains intron 2, termination after 19 neocodons Heterozygous parents with mild HS Homozygous child with severe HS 20% wt AE1 from intron 2 skipping	Miraglia del Guidice *et al.* 1997
	E40K	Recessive hemolytic anemia with protein 4.2 deficiency	Rybicki *et al.* 1993
Foggia	H55T	Frameshift 1 nt deletion	Miraglia del Guidice *et al.* 1997
Kagoshima	E56fs	1 nt deletion (A)	Kanzaki *et al.* 1997a
Hodonin	W81X	Mutant mRNA undetected	Jarolim *et al.* 1996
Capetown	E90K	Compound heterozygote with Prague III, both mRNAs present	Jarolim *et al.* 1995; Bracher *et al.* 2001
Napoli I	S100F	Frameshift, nt 447insT; mutant RNA undetected	Miraglia del Guidice *et al.* 1997
Fukuyama I	R112–113	Frameshift, 2 nt del (AC) aa	Kanzaki *et al.* 1997b
Nachod	D117–121	Intron 5 splice acceptor mutation, perhaps within ankyrin binding site	Jarolim *et al.* 1996

Table 1 Continued

Band 3	Mutation	Description	Reference(s)
Fukuoka	G130R	Asymptomatic alone, recessive, exacerbated HS when *in trans* with G714R Okinawa	Inoue *et al.* 1998
Mondego	P147S	*In cis* with E40K Montefiore and *in trans* with V488M Coimbra Enhances severity of heterozygous HS V488M Coimbra	Alloisio *et al.* 1993
Lyon, Osnabruck	R150X		Alloisio *et al.* 1996; Eber *et al.* 1996
Worcester	L170–172	Frameshift, 1 nt insertion; mutant mRNA undetected.	Jarolim *et al.* 1996
Fukuyama II	D183	Frameshift, 1 nt insertion	Inoue *et al.* 1998
Campinas		Q203 frameshift, followed by 13 aa neosequence, 1 nt insertion Incomplete dRTA by furosemide test, resiting bicarbonaturia.	Lima *et al.* 1997
Bohain	V241	Frameshift, 1 nt deletion	Dhermy *et al.* 1997
Princeton	A273–275	Frameshift, 1 nt insertion; mutant mRNA undetected	Jarolim *et al.* 1996
Boston	A285D		Jarolim *et al.* 1996
Tuscaloosa	P327R		Jarolim *et al.* 1992a
Noirterre	Q330X		Jenkins *et al.* 1996
Bruggen	P419	Frameshift, 1 nt deletion	Eber *et al.* 1996
Benesov	G455E	In TM2	Jarolim *et al.* 1996
Bicetre II	A456	Frameshift, 1 nt deletion	Dhermy *et al.* 1997

Name	Mutation	Description	Reference
Pribram	S477	Frameshift, 1 nt deletion in splice acceptor, intron 12 retention, terminates after 7 neocodons; small amount mutant mRNA detected, mutant protein not detected	Jarolim et al. 1996
Coimbra	V488M	Exacerbated by in trans double mutant E40K/P147S	Alloisio et al. 1993
Bicetre I	R490C		Dhermy et al. 1997
Evry	W492	Frameshift, 1 nt deletion	Dhermy et al. 1997
Milano	1500G	Followed by duplication of aa 478–499 (69 bp); mutant protein undetected	Bianchi et al. 1997
Dresden	R518C		Eber et al. 1996
Smichov	I616	Frameshift, 1 nt deletion; mutant mRNA undetected	Jarolim et al. 1996
Trutnov	Y628X	Mutant mRNA undetected	Jarolim et al. 1996
Hobart	R646–647	Frameshift, 1 nt deletion; mutant mRNA undetected	Jarolim et al. 1996
Osnabruck II	M663–664	Frameshift	Eber et al. 1996
Most	L707P		Jarolim et al. 1996
Okinawa	G714R	HS more severe when in trans with Band 3 Fukuoka	Kanzaki et al. 1997b
Prague II	R760Q	Mutant polypeptide not detected	Jarolim et al. 1995
Hradec Kralove	R760W		Jarolim et al. 1995
Chur	G771D	Mutant mRNA present at normal level	Maillet et al. 1995
Napoli II	I783N	Mutant mRNA present, mutant protein undetected	Miraglia del Guidice et al. 1997
Jablonec	R808C		Jarolim et al. 1995

Table 1   Continued

Band 3	Mutation	Description	Reference(s)
Nara	R808H		Kanzaki et al. 1997a
Prague I	V822	Frameshift, 10 nt duplication	Jarolim et al. 1995
Birmingham	H834P		Jarolim et al. 1996
Philadelphia	T837M		Jarolim et al. 1996
Tokyo	T837A	Mutant mRNA present at normal level	Iwase et al. 1998
Prague III	R870W	In compound heterozygote with Capetown E90K, both mRNAs present	Jarolim et al. 1995; Bracher et al. 2001
Vesuvio	894	Frameshift 1 nt deletion replaces C-terminal 18 aa with 133 aa neosequence; silent polymorphism 904 TAC-to-TAT in trans; mutant mRNA present, mutant protein undetected	Perrotta et al. 1999
(C) SLC4A1 (AE1) mutations associated with distal renal tubular acidosis (MIM:179800)			
Coimbra	V488M	Homozygote shows complete absence of erythroid AE1, with hydrops fetalis, hemolytic anemia, accompanying recessive dRTA Heterozygote exhibits dominant dRTA	Ribeiro et al. 2000
dRTA	R589H	Dominant dRTA	Bruce et al. 1997; Jarolim et al. 1998b; Karet et al. 1998
dRTA	R589C	Dominant dRTA	Bruce et al. 1997
dRTA	R589S	Dominant dRTA	Bruce et al. 1997
dRTA	R602H	Dominant dRTA detected so far only as compound heterozygote with SAO	Vasuvattakul et al. 1999
dRTA	S613F	Dominant dRTA	Bruce et al. 1997
Bangkok I	G701D	Homozygous recessive dRTA	Tanphaichitr et al. 1998

dRTA	ΔV850	Homozygous recessive dRTA, and as compound heterozygote with SAO	Bruce et al. 2000
dRTA	A858D	Homozygous recessive dRTA, and as compound heterozygote with SAO	Bruce et al. 2000
dRTA	A888L	Frameshift 889X	Cheidde et al. 2002
Walton	R901X	Dominant frameshift, 13 bp duplication from codon 896 base2 to codon 900 base 2	Karet et al. 1998
Pribram (?)	S477fs	HS with incomplete RTA by $CaCl_2$ load test; bicarbonaturia following HCO_3^- load	Rysava et al. 1997

(D) SLC4A1 (AE1) polymorphisms associated with blood group antigens

NFLD	E429D	Plus P561A	Zelinski et al. 1997
ELO	R432W		Jarolim et al. 1998a
Fr[a]	E480K		McManus et al. 2000
Rb(a)	P548L		Jarolim et al. 1997b
Tr(a)	K551N		Jarolim et al. 1997b
WARR	T552I		Jarolim et al. 1997a
Vg[a]	Y555H	Plus silent L441L	Jarolim et al. 1998a
Wd(a)	V557M		Jarolim et al. 1997b
BOW	P561S		McManus et al. 2000
Wu	G565A		Jarolim et al. 1998a; Zelinski et al. 1998
Jn[a]	P566S		Poole et al. 1997
KREP	P566A		Poole et al. 1997
Mo[a]	R656H		Jarolim et al. 1998a
Bp[a]	N569K		Jarolim et al. 1998a
Sw[a]	R646Q		Jarolim et al. 1998a
Sw!	R646W		Zelinski et al. 2000

Table 1   Continued

Band 3	Mutation	Description	Reference(s)
Hg[a]	R656C		Jarolim et al. 1998a
Wright b+	E658	Common antigen	Bruce et al. 1995
Wright a+	E658K	Rare variant (a+/b+ ~1 : 1000 in general population)	Bruce et al. 1995
Di[a], Memphis II	P854L	Diego a+ antigen	Bruce et al. 1994
Di[b]	P854	Common variant	Bruce et al. 1995
(E) SLC4A1 (AE1) asymptomatic polymorphic variants			
Bangkok II	M31T	*In cis* with Memphis I and the dRTA mutation G701D	Tanphaichitr et al. 1998
Napoli II	D38A		Miraglia del Giudice et al. 1997; NCBI Genbank
Memphis I	K56E	80% decreased phosphoenolpyruvate transport	Yannoukakos et al. 1991; NCBI Genbank
Intron 3 polymorphisms		c87t, c242t, g259a[a], a580g (numbered from 5′ end of intron 3) (kAE1 promoter region)	Bruce et al. 2000
		Found in HS association with other mutations	
Variant	E72D		NCBI Genbank
Variant	R112S		NCBI Genbank
Variant	I442F		NCBI Genbank
Variant	M586L		NCBI Genbank
Variant	I688V		NCBI Genbank
Variant	S690G		NCBI Genbank
Variant	R832H		NCBI Genbank
Variant	V862I		NCBI Genbank

(F) Animal models

B. taurus	Bovine equivalent of human R646X	Erythroid AE1 absent; also reductions in spectrin, ankyrin, actin, 4.2	Inaba *et al.* 1996
		Systemic acidosis with less than maximally acid urine	
D. rerio	ret(b245)	1.5–2 cM deletion encompassing AE1 gene	Paw *et al.* 2003
	ret(tr265)	E456G(aligns with human E472) slightly reduced mRNA in early embryo, nonfunctional protein	Paw *et al.* 2003
	ret(tr217)	Frameshift 13 nt insertion at nt 503 (creates new splice acceptor in intron 5), greatly reduced mRNA, severely truncated protein	Paw *et al.* 2003
		Zebrafish mutant retsina: all genotypes show dyserythropoietic anemia type II phenotype (binuclearity with failure of cytokinesis)	
M. musculus	Mouse AE1(−/−)	Insertion/disruption of exons 9–11	Peters *et al.* 1996
M. musculus	Mouse AE1(−/−)	Insertion/disruption of exon 3	Southgate *et al.* 1996
M. musculus	mAE1 Q85X	Missense terminator	Peters *et al.* 2001; L.L. Peters, personal communication
		Wan/Wan is severe recessive HS with absence of erythroid AE1	

[a] Linked with Memphis I polymorphism.

The missense mutants tested (L707P, R760Q, R760W, R808C, H834P, T837M, R870W) were all misfolded, as judged by loss of Cl^--displaceable binding to a stilbene disulfonate column, but at least two mutants were not destabilized (Quilty and Reithmeier 2000).

Although AE1 is also expressed in the renal type A intercalated cell, autosomal dominant hereditary spherocytosis is almost never associated with defective urinary acidification. [The only possible exception (Band 3 Pribram; Table 1) may have presented with an independent renal disorder, suggested by low threshold bicarbonaturia induced by oral HCO_3^- treatment (Rysava et al. 1997).]

At least two AE1 spherocytosis mutants are associated with protein 4.2 deficiency. The dominant mutation P327R was associated with 30% deficiency of protein 4.2 in red cells (Jarolim et al. 1992a). The severe hemolytic anemia associated with the recessive mutation E40K was accompanied by nearly 90% deficiency of protein 4.2 (Rybicki et al. 1993). In Japan, protein 4.2 mutations are the most common cause of HS, whereas spectrin deficiency is rare. The reduction in AE1 level accompanying protein 4.2 mutations reaches 50% in patients with the homozygous protein 4.2 Nippon mutation (Yawata et al. 2000). Red cells of protein 4.2 (–/–) mice show 30% decrease in AE1 polypeptide (Peters et al. 1999).

COMPLETE AE1 DEFICIENCY IN THE HUMAN AND IN ANIMALS

Complete absence of AE1 secondary to the recessive mutation AE1 V488M (Band 3 Coimbra) is manifest as hydrops fetalis with severe hemolytic anemia, respiratory failure, and distal renal tubular acidosis (see below) (Ribeiro et al. 2000). Circulating red cells were fragile and exhibited unusual membranous projections. The red cell membrane lacked not only AE1, but also protein 4.2, and exhibited major reductions in the level of glycophorin A, ankyrin, and spectrin (Ribeiro et al. 2000), as well as of the putative NH_3/NH_4^+

carrier, Rh antigen complex, CD47, LW antigen, and the major erythroid water channel aquaporin-1 (Bruce et al. 2003). These findings highlight the central importance of the AE1 polypeptide for the stabilization of the general protein composition of the red cell membrane (Peters et al. 1996).

A similar complete absence of AE1 in a bovine cohort was associated with a recessive mutation corresponding to human AE1 R646X. These animals also had severe hemolytic anemia and renal tubular acidosis. Red cells had numerous small vesicles attached, which were easily sheared by low speed centrifugation (Inaba et al. 1996).

The instability of red cells in the complete absence of AE1 was initially envisioned as resulting from detachment from the lipid bilayer of the red cell cytoskeleton in the absence of a major anchoring site. This hypothesis was based upon the abnormal erythroid cytoskeleton morphology in spherocytic red cells with heterozygous mutations in the ankyrin-1 or spectrin genes (Jarolim et al. 1996). However, creation and study of AE1 (–/–) null mice (Peters et al. 1996; Southgate et al. 1996) led to revision of this hypothesis (Peters et al. 1996). Very few of these severely anemic mice survived to adulthood. Despite extreme reticulocytosis, hematocrit was about 5%. Red cells exhibited extreme osmotic and mechanical fragility and a range of bizarre shapes. As later reported in human AE1-null red cells (Bruce et al. 2003), mouse AE1 (–/–) red cells lacked or showed reductions in several membrane proteins including, in addition to those mentioned previously, the band 3 binding proteins glycophorin A (Alper et al. 1998; Hassoun et al. 1998; Bruce et al. 2003), the GLUT1 neonatal erythroid glucose transporter, and the major DIDS-sensitive erythroid Cl^- conductance (Alper et al. 1998). Despite the absence of AE1 and many associated proteins, the ankyrin–spectrin cytoskeleton of Triton-X100-extracted red cell ghosts preserved a wild-type ultrastructure as judged by electron microscope examination (Peters et al. 1996). Thus, Peters et al. (1996) proposed a primary role for erythroid AE1 in the stabilization of the lipid bilayer itself in the face of circulatory shear and torsion, possibly via direct stabilization of boundary lipid, and possibly also through the lipid-binding of AE1-associated proteins. The anemia of the inbred mouse mutant strain *wan* has been shown to be secondary to the homozygous AE1 nonsense mutation Q85X (Peters *et al.* 2001; L.L. Peters, personal communication).

The zebrafish has proven a fruitful model system in which to study mutations which interrupt or alter hematopoiesis (Thisse and Zon 2002). Among the mutations which act between the hemangioblast and the erythroblast by blocking growth during the process of preterminal differentiation are those in genes encoding red cell membrane proteins. These include *riesling* (β-spectrin), *merlot/chablis* (protein 4.1R), and *retsina* (AE1/band 3). *Retsina*, identified as AE1 by positional cloning, has three alleles, two truncations, and a large genomic deletion (Paw *et al.* 2003). Neither truncation is expressed at the surface of *Xenopus* oocytes, and *retsina* red cells lack band 3.

Binucleated erythroblasts are frequently observed in *retsina*. The cleavage furrow initiates but fails to progress, and chromosomes do not segregate to the poles in the presence of a disordered mitotic spindle. Putative protein 4.1 binding sites in zebrafish band 3 appear to be important for transgenic rescue of *retsina*, whereas anion transport function appears not to be required (Paw *et al.* 2003). In normal zebrafish erythroblasts, band 3 is excluded from the cleavage furrow. Binucleated cells are also present in the marrow of the AE1(−/−) mouse, but much less frequently than in *retsina* (Paw *et al.* 2003). This phenotype is shared in congenital dyserythropoietic anemia type II (HEMPAS disease). However, HEMPAS has never been linked to AE1 mutations, despite occasional reports of mildly decreased band 3 content (Jarolim *et al.* 1994), anion transport activity, and polypeptide clustering in the membrane (De Francheschi *et al.* 1998).

BENIGN ERYTHROID PHENOTYPES OF VARIANT AE1 POLYPEPTIDE

Southeast Asian ovalocytosis

Southeast Asian ovalocytosis (SAO) is an apparently asymptomatic condition of altered red cell shape widely found in Malaysia, Papua New Guinea, Indonesia, and the Philippines. The oval erythrocytes often exhibit a longitudinal slit or a transverse ridge, and show several indices of increased cell rigidity, including decreased osmotic fragility, increased thermal stability, and resistance to induction of echinocytosis by various lipophilic agents and by ATP depletion (Bjork *et al.* 1997). This rigidity also correlates with resistance to in vitro invasion by *Plasmodium* species of several strains. However, cold storage of cells may produce this phenotype, as well as the cation leakiness (Bruce *et al.* 1999) and more rapid ATP depletion observed in SAO erythrocytes (Dluzewski *et al.* 1997). Nonetheless, SAO has been associated with reduced numbers of intraerythrocytic parasites in areas endemic for malaria (Cattani *et al.* 1987), and with partial protection against cerebral malaria (Allen *et al.* 1999). This protective effect has been postulated to serve as the selective force for maintenance of the mutation in the population, as has been similarly proposed for the hemoglobinopathies and erythroid enzyme disorders such as glucose-6-phosphate dehydrogenase deficiency.

The SAO abnormality is associated with heterozygosity for an in-frame deletion of 27 nucleotides encoding AE1 aa 400–408 (Jarolim *et al.* 1991; Mohandas *et al.* 1992; Schofield *et al.* 1992), located at the cytoplasmic surface of the first transmembrane span of the AE1 polypeptide (Kanki *et al.* 2003). Homozygotes have never been detected, and the carriage of two SAO alleles is likely lethal in utero. The SAO allele is usually found in *cis* with the prevalent K56E polymorphism, Band 3 Memphis (Yannoukakos *et al.* 1991; Jarolim *et al.* 1992b) (see below). AE1 SAO polypeptide is at the red cell surface in normal abundance, but it fails to bind stilbene disulfonates, and is inactive as a sulfate transporter (Schofield *et al.* 1992) and as a Cl^-/HCO_3^- exchanger (Dahl *et al.*, submitted). The AE1 wildtype and SAO polypeptides are equally abundant in the membrane and form heterodimers (Jennings and Gosselink 1995). The SAO polypeptide alters the protease susceptibility and biotinylation accessibility of most of the wild-type polypeptide in the membrane (Kanki *et al.* 2003), in concert with its tendency to promote linear aggregates of the intramembranous particles thought to include AE1 (Peters *et al.* 1996), and with the decreased lateral and rotational mobility of AE1 in the SAO membrane (Liu *et al.* 1990; Jarolim *et al.* 1991; Mohandas *et al.* 1992). Despite this evidence for altered conformation of the wildtype polypeptide in SAO erythrocytes, SAO polypeptide either has no detectable effect on wild-type AE1-mediated anion transport (Jennings and Gosselink 1995) or, measured differently, a modest effect (Kanki *et al.* 2003). SAO erythrocytes also exhibit tighter binding of AE1 to ankyrin (Liu *et al.* 1990) and increased tyrosine phosphorylation of AE1 (Jones *et al.* 1990).

Hereditary acanthocytosis

A family with hereditary acanthocytosis (crenated red cells with surface projections) was noted to have elevated anion transport activity in association with the heterozygous missense mutation P868L (Bruce *et al.* 1993). These red cells also exhibited selectively altered binding kinetics of H_2DIDS (Bruce *et al.* 1993), but not of the other stilbene disulfonate transport antagonists DIDS and DBDS (Bruce *et al.* 1993; Salhany *et al.* 1995).

Erythroid blood group antigens

Minor blood group antigens give rise to alloreactive antisera that are of little or no consequence for blood transfusion compatibility or health. One of these, the Diego blood group system, is carried by amino acid

polymorphisms in the exofacial loops of the AE1 polypeptide (Table 1). The first of these to be described was P854L, encoding the Di^a/Di^b polymorphism (Spring *et al.* 1992), which exhibits tighter binding of H_2DIDS in the context of K56E – the double mutant is known as Band 3 Memphis II (Hsu and Morrison 1985). The Wr^b antigen requires the E658K polymorphism along with presence of the AE1-binding protein, glycophorin A (Bruce *et al.* 1995). In subsequent years polymorphisms explaining additional blood group antigens were found in five exofacial loops of the AE1 polypeptide. All have provided important data for the evolving picture of AE1 topography in the lipid bilayer. Among the seven exofacial loops proposed in the original topographic model predicted from hydrophobicity plots then available (Wood 1992), only the fifth and sixth loops are not represented among the Diego blood group antigens to date (Jarolim 2003) (Figure 2).

Anti-band 3 antibodies and red cell senescence

Progression through the normal 90-day lifespan of human red cells in the circulation is associated with a gradual increase in red-cell-bound immunoglobulin (Ig). This bound Ig, thought to signal phagocytosis of aged red cells by macrophages of the reticuloendothelial system, has been shown to recognize various extracellular epitopes of AE1 (Turrini *et al.* 1993). Splenectomy is a standard treatment option for hemolytic anemias. However, splenectomy prolongs survival in the circulation to a greater degree for hereditary spherocytic anemia red cells when the disease is secondary to primary deficiency of spectrin or ankyrin, rather than to primary AE1 deficiency (Reliene *et al.* 2002). The former condition has been proposed to reflect the lower fraction of AE1 bound to a defective cytoskeleton, allowing AE1 to be lost by vesicle budding from unstable spherocytotic membrane. Conversely, the latter condition may reflect the anchoring to the cytoskeleton of a higher fraction of the

Figure 2 Mutations of the human AE1/SLC4A1 (Band 3) Cl⁻/HCO₃⁻ exchanger associated with variant blood group antigens and with distal renal tubular acidosis (dRTA). Mutations are shown as shaded amino acid residues and include missense, nonsense, and deletion mutations. Sites of variant blood group antigens (marked with asterisks) are restricted to residues accessible from the extracellular fluid. dRTA mutations lack asterisks. The dRTA nonsense mutation R901X is indicated by shading of the 11 C-terminal (deleted) residues. Met 66 (arrow) marks the start of kidney AE1.

reduced quantity of AE1 polypeptide (Reliene *et al.* 2002).

Band 3 Memphis and other asymptomatic polymorphisms

Band 3 Memphis I (K56E) (Yannoukakos *et al.* 1991) was discovered by its association with retarded electrophoretic mobility of AE1 on SDS–PAGE (Hsu and Morrison 1985). This altered conformation preserved in the presence of 1% sodium dodecylsulfate (SDS) is still not understood. The Memphis polymorphism has since been found in *cis* with several pathogenic AE1 mutations in hereditary spherocytosis and (see below) one form of renal tubular acidosis. Additional polymorphisms found in association with disease alleles include Band 3 Bangkok I M31T and Band 3 Napoli II D38A (Table 1). The growing number of amino acid polymorphisms, as well as silent nucleotide polymorphisms, will provide tools for future studies of the AE1 gene as a risk modifier in other disease states.

SLC4A1/AE1 MUTATIONS IN HEREDITARY DISTAL RENAL TUBULAR ACIDOSIS

Renal tubular acidoses (RTAs) are disorders of urinary acidification resulting in metabolic acidosis (see Box: AE1-deficient Hereditary Distal Renal Tubular Acidosis). The heritable primary renal tubular acidoses include impairment of proximal tubular bicarbonate reabsorption, impairment of collecting duct H⁺ secretion or bicarbonate reabsorption, and a mixed proximal/distal syndrome. Recessive proximal renal tubular acidosis has been associated with hypofunctional mutations in *SLC4A4/NBCe1*, the gene encoding the

AE1-DEFICIENT HEREDITARY DISTAL RENAL TUBULAR ACIDOSIS (dRTA, MIM 179800)

Metabolic acidosis is marked by low arterial pH with low plasma [HCO_3^-]. Renal tubular acidosis (RTA) is an impairment of the kidney's ability to eliminate acid in the urine. The chronic metabolic acidosis of RTA can present as growth retardation and failure to thrive in children, or in less severe cases as variably short stature in an adult. "Complete dRTA" is marked by failure maximally to acidify the urine in the presence of acidemia (operationally defined as urine pH > 5.5). "Incomplete dRTA" can be seen in the absence of acidemia, but clinical administration of an acid load can unmask inability to achieve maximum urinary acidification. dRTA is often accompanied by hypercalciuria, variably progressing to osteomalacia, nephrocalcinosis, and nephrolithiasis. Nephrocalcinosis and nephrolithiasis in turn predispose to recurrent urinary tract infections and pyelonephritis. dRTA is variably accompanied by hypocitraturia, hypokalemia, and decreased urinary excretion of total ammonium. If detected early, therapeutic correction of the metabolic acidosis by oral bicarbonate administration usually leads to improvement of biochemical abnormalities and resumption of normal growth.

Hereditary forms of RTA can arise from isolated proximal tubular dysfunction, isolated collecting duct dysfunction, or a "mixed" form involving both segments. Proximal RTA is associated with acid urine (urine pH < 5.5), and with bicarbonaturia in the treated (bicarbonate-supplemented) state. Hereditary proximal RTA has been associated with mutations in the *SLC4A4/NBCe1* gene. Heritable dRTA is transmitted in recessive and dominant forms. Many recessive forms represent mutations in subunits of the vacuolar H^+-ATPase. However, recessive dRTA in Southeast Asia and Melanesia is associated with mutations in the AE1 gene. A distinct set of AE1 mutations has been associated with dominant forms of dRTA. With few exceptions, erythroid abnormalities do not accompany heritable dRTA. AE1 mutations associated with recessive and dominant forms of dRTA are distinct from AE1 mutations associated with erythrocyte abnormalities. Hereditary mixed RTA has been associated with mutations in the carbonic anhydrase 2 (*CA2*) gene.

electrogenic Na^+/HCO_3^- cotransporter of the proximal tubular epithelial cell basolateral membrane (see Chapter 4). The recessive mixed proximal/distal syndrome has been associated with mutations in the gene encoding carbonic anhydrase II, a critical component of the mechanisms of luminal acidificaction and bicarbonate reclamation in both proximal tubule and collecting duct intercalated cells. The distal renal tubular acidoses (dRTA) exhibit dominant and recessive transmission patterns. Mutations in two genes encoding subunits of the vacuolar luminal H^+-ATPase, ATP6V1B1 and ATP6V0A4, are associated with recessive dRTA. Mutations in the gene encoding SLC4A1/AE1 are associated with both dominant and recessive forms of dRTA (Wood 1992; Shayabul and Alper 2000; Alper 2002; Karet 2002).

As noted previously, SLC4A1/AE1 is expressed in erythrocytes and collecting duct intercalated cells under control of distinct promoters. Most *AE1* mutations described to date alter the AE1 polypeptide sequence shared by both erythroid (eAE1) and renal (kAE1) polypeptides (Table 1 and Figures 2 and 3). However, the mutations associated with anemia or altered red cell shape and those associated with dRTA are for the most part mutually exclusive. Moreover, the vast majority of erythroid disorders associated with *AE1* mutations exhibit no renal component, whereas (with rare exceptions) patients with dRTA lack erythroid abnormality. The major outstanding problems in understanding the pathobiology of AE1 diseases lie in understanding the distinct tissue-specific phenotypes of identical mutations in the context of two different polypeptide isoforms (eAE1 and kAE1) expressed in two different cell types, the erythrocyte and the type A intercalated cell.

PATHOPHYSIOLOGY OF AE1/SLC4A1

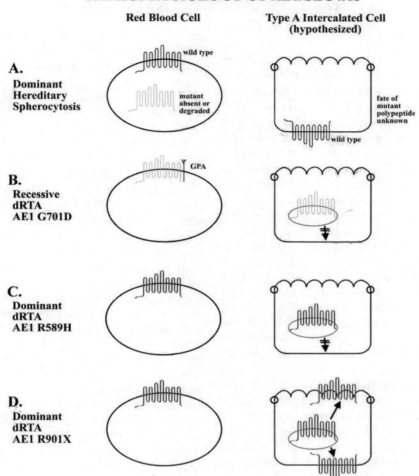

Figure 3 Pathophysiology of mutant AE1 trafficking in recessive and dominant forms of distal renal tubular acidosis (dRTA). (A) Dominant hereditary spherocytosis. In red cells, the mutant erythroid AE1 (eAE1) polypeptide is absent or degraded. The wild-type eAE1 polypeptide is at the cell surface, often at 60–80% of normal level. In type A intercalated cells, the wild-type kidney AE1 (kAE1) polypeptide is believed to be at the basolateral surface, but the surface abundance and fate of the mutant polypeptide are unknown. (B) Recessive dRTA is exemplified by AE1 G701D. In red cells, eAE1 G701D is believed to associate normally with the AE1-binding transmembrane protein, glycophorin A. This association allows normal accumulation and function in the red cell membrane. G701D and other recessive mutants fail to traffic to the surface of the *Xenopus* oocyte and (it is thought) to the surface of the intercalated cell. The coexpression of glycophorin A rescues surface expression and function of kAE1 G701D in the oocyte, a situation believed to parallel lack of surface expression in the glycophorin A⁻ intercalated cell. (C) Dominant dRTA is exemplified by AE1 R589H. In the red cell, eAE1 is of normal abundance and function, and is likely at the cell surface. In *Xenopus* oocytes, AE1 R589H exhibits only mildly diminished function. In the intercalated cell, kAE1 R589H has been postulated to be a dominant negative trafficking mutant, based on studies in HEK 293 cells. In 293 cells, eAE1 R589H exhibits wild-type behavior, in contrast with kAE1 R589H. (D) kAE1 R901X is of normal abundance and function in red cells, and exhibits near-normal function in oocytes. In transiently transfected MDCK cells, kAE1 R901X is present not only in the basolateral membrane, but also at or near the apical surface. The resultant functional short-circuiting of acid secretion could explain the dominant negative clinical phenotype.

Dominant dRTA

Dominant dRTA has been associated with a small number of heterozygous AE1 mutations. Missense substitutions at R589 have included the most common H (Bruce *et al.* 1997; Jarolim *et al.* 1998b; Karet *et al.* 1998), as well as S and C (Bruce *et al.* 1997). Additional dominant mutations have encoded the substitution S613F (Bruce *et al.* 1997), the frameshift A888L leading to immediate termination at codon 889 (Cheidde *et al.* 2002), and the 13 bp duplication leading to frameshift termination at residue R901 (Karet *et al.* 1998). Each discovery was made by testing *AE1* as a candidate gene in families with heritable distal renal tubular acidosis. In each family, erythroid indices were normal. In each family, eAE1 polypeptide abundance was normal or nearly so. eAE1 anion transport activity in red cells (measured as sulfate flux) was indistinguishable from that in unaffected family members. Cl^- influx into *Xenopus* oocytes mediated by either eAE1 or kAE1 was at least 50% of wild-type values. For the most common mutant polypeptide R589H (Jarolim *et al.* 1998b), as well as for R901X (Toye *et al.* 2002), AE1-mediated Cl^- efflux and AE1-mediated Cl^-/HCO_3^- exchange were also shown to be no lower than 50% of wild-type values. Mutant eAE1 expressed in oocytes retained partial or full responsiveness to coexpression of the eAE1-selective binding protein glycophorin A. None of the mutant AE1 polypeptides exhibited dominant negative behavior in *Xenopus* oocytes. Interestingly, the mutant S613F uniquely displayed increased affinity for sulfate (at a single pH) in red cells, despite normal red cell iodide transport and normal Cl^- uptake in oocytes (Bruce *et al.* 1997).

Thus the phenotypes of near-normal anion transport and surface expression exhibited by the mutant proteins in red cells and in *Xenopus* oocytes cannot explain the clinical renal phenotype of inadequate distal urinary acidification. The properties of mutant kAE1 in intercalated cells must differ in ways which can explain the renal phenotype. However, intercalated cells exist in a mosaic epithelium, and the terminally differentiated type A intercalated cell phenotype of apical vH^+-ATPase and basolateral kAE1 has not been stable in culture. In the meantime, established renal cell lines have been used to study the trafficking of kAE1, eAE1, and their dRTA mutant forms.

The dRTA mutant AE1 R589H has been extensively studied in HEK 293 cells (Quilty *et al.* 2002b). Although eAE1 R589H is normally expressed at the 293 cell surface, surface expression of kAE1 R589H was severely reduced, with reportedly absent Cl^-/HCO_3^- exchange. This reduction in surface expression was accompanied by reduced and retarded exit from the endoplasmic reticulum, measured in the background of an N-glycosylation mutant in which the oligosaccharide, unlike that of wild-type AE1, undergoes conversion to an Endo H-resistant form in the Golgi. kAE1 R589H stability was apparently unaltered. Most importantly, coexpression of kAE1 R589H with wild-type kAE1 resulted in hetero-oligomer formation and was associated with complete inhibition of wild-type kAE1 surface expression. This dominant negative trafficking defect could explain the pathogenesis of dRTA (Quilty *et al.* 2002b). However, when wild-type kAE1–GFP and kAE1 R589H–GFP fusion proteins were transiently expressed in polarized MDCK cells grown on filter supports, both mutant and wild-type polypeptides exhibited indistinguishable basolateral membrane localization as detected by confocal immunofluorescence microscopy (Prabakaran and Alper, unpublished data). The reason for the apparently different properties of kAE1 R589H in oocytes, 293 cells, and MDCK cells is not clear.

As was noted for kAE1 R589H, kAE1 R901X also required mammalian cell expression to demonstrate a mutant phenotype. However, in HEK 293 cells both kAE1 and eAE1 forms of the 901X mutant exhibited intracellular retention by immunofluorescence as well as by cell surface biotinylation (Quilty *et al.* 2002a). Complete sensitivity of the mutant polypeptides to Endo H showed that they did not reach the medial Golgi compartment. Both eAE1 and kAE1 forms of 901X mutant hetero-oligomerized with wild-type AE1 polypeptide. Coexpression of mutant and wild-type AE1 polypeptides exhibited a dominant negative effect of both mutants on wild-type AE1 surface expression. Although this dominant negative trafficking effect can explain dRTA (Quilty *et al.* 2002a), it cannot explain the coexisting normal erythroid phenotype of the AE1 R901X mutation.

Stable expression of wild-type kAE1 in coverslip-grown MDCK cells led to accumulation in the lateral membranes. In contrast, stably expressed kAE1 901X was entirely retained inside the cells (Toye *et al.* 2002). Filter-grown MDCK cells transiently transfected with HA-tagged wild-type kAE1 also exhibited basolateral localization of the polypeptide, whereas transiently expressed HA-tagged kAE1 901X, though largely retained intracellularly, was also apparently expressed at both apical and basolateral surfaces (Devonald *et al.* 2003). This abnormal distribution was mimicked in HA-tagged kAE1-Y904A, suggesting an important role for the 904YDEV907 YXXΦ sorting motif in basolateral accumulation of kAE1. The hypothesis was confirmed in studies with CD8 fusion proteins in which the CD8 cytoplasmic tail was replaced with the 27 C-terminal cytoplasmic tail residues of AE1. Heterologous expression of CD8 in MDCK cells leads to apical localization. Substitution of the wild-type AE1 cytoplasmic tail redirected part of the fusion protein to the lateral membranes. Truncation of the C-terminal 11 residues severely reduced this fraction of lateral membrane expression. This lateral membrane targeting effect is also evident in LLCPK1 cells stably expressing the CD8 fusion proteins, suggesting that the μ1B adaptin absent in these cells is not required for basolateral accumulation dependent on the YXXΦ sorting signal of the kAE1 C-terminal tail (Devonald *et al.* 2003).

It is not yet known whether the depolarized phenotype of kAE1 901X reflects random sorting, impaired stabilization at the basolateral membrane, or enhanced stabilization at the apical membrane (perhaps a failure of a normal apical degradation or endocytotic mechanism). The kAE1 binding proteins that might mediate these trafficking events remain unknown. The conclusions of these trafficking studies are morphological. Unfortunately, heterologous expression of kAE1 polypeptide variants in polarized epithelial cell monolayers has been at levels inadequate to allow confirmation of these trafficking studies by cell surface biotinylation or other biochemical methods. Interestingly, polarized surface targeting in MDCK cells of a chicken kidney AE1 isoform relies, at least in part, upon sorting motifs found in the early part of its N-terminal tail (Adair-Kirk *et al.* 2002).

Recessive dRTA

Although most, if not all, recessive dRTA in America and Europe is caused by mutations in subunits of the intercalated cell vH$^+$-ATPase, in Southeast Asia recessive dRTA has been associated with several AE1 mutations. The G701D mutation was described first in a single Thai family (Tanphaichitr *et al.* 1998), and has since been shown to be a common cause of recessive dRTA in Thailand (Yenchitsomanus *et al.* 2002). As true for dominant dRTA, eAE1 abundance and function were normal in the patients. Both kAE1 and eAE1 G701D were functionally inactive in *Xenopus* oocytes due to intracellular retention of protein. Thus an explanation for normal sulfate transport in patient red cells was needed. Coexpression of the erythroid AE1 binding protein glycophorin A proved able to rescue completely the transport function of both eAE1 and kAE1 G701D in oocytes. Rescue was accompanied by increased surface expression of the mutant AE1 polypeptides. Thus the presence of glycophorin A in erythrocytes and its absence in the kidney explained the pathophysiology of recessive dRTA with a normal erythroid phenotype (Tanphaichitr *et al.* 1998).

Similar sensitivity to glycophorin A rescue was subsequently shown for the recessive mutation ΔV850 discovered in Malaysia and Papua New Guinea (Bruce *et al.* 2000). In addition to a family homozygous for ΔV850, compound heterozygotes with coexistent ovalocytosis were found with genotypes ΔV850/SAO and G701D/SAO (Bruce *et al.* 2000). The dominant mutation A858D was also found in the compound heterozygote forms A858D/SAO and A858D/ΔV850. A858D/SAO erythrocytes exhibited only 6% of normal sulfate influx, consistent with failure of glycophorin A to rescue transport function in *Xenopus* oocytes coexpressing both inactive mutant polypeptides (Bruce *et al.* 2000). Coexpression of SAO with recessive mutant AE1 G701D attenuated functional rescue of the G701D mutant by glycophorin A (Shayakui *et al.* 2001), consistent with the conformational change induced by SAO on neighboring band 3 polypeptide (Kanki *et al.* 2003).

AE1 genotype–phenotype correlations in familial RTA have been difficult to establish among the small number of reported occidental families with AE1 mutations. Both direct and indirect forms of dRTA have been found within single families carrying uniform mutations. Hypo- and normokalemia are also found in single families. Inadequate information is available about the range of urinary ammonium excretion, urine–plasma pCO_2 difference, and hypocitraturia among the genotypes. Nephrocalcinosis was originally very common, but is becoming less so in developed countries with early diagnosis. Oriental dRTA patients with recessive or compound heterozygous AE1 mutations have been more severely affected than occidental patients, with higher frequency of metabolic bone disease and severe hypokalemia (Tanphaichitr *et al.* 1998; Bruce *et al.* 2000).

Renal biopsy is not performed for isolated renal tubular acidosis in most countries. In one patient with dominant dRTA secondary to the mutation AE1 R589H, a paraffin block from a resected kidney became available. AE1 was absent while vH^+-ATPase remained

present in type A intercalated cells. However, interpretation was complicated by chronic pyelonephritis and end-stage disease with scarring, atrophy, and nephrocalcinosis in the setting of congenital double collecting system and transitional cell carcinoma (Shayakul *et al.*, submitted). AE1 immunostaining was also absent in seven of eleven renal biopsies in a Korean series of primary sporadic dRTA (Han *et al.* 2002) and from the kidney of one strain of AE1(–/–) mouse (Peters *et al.* 1996). Disruption of exon 3 in the mouse AE1 gene, upstream of the kAE1 coding sequence and its nominal promoter in intron 3, was designed to leave distal renal acidification intact in the anemic knockout mouse (Southgate *et al.* 1996). However, this introduction of the neo gene into exon 3 somehow severely decreased kAE1 mRNA and protein levels (Alper, Chishti, and Shmukler, unpublished results), perhaps because the as yet unstudied kAE1 promoter extends upstream beyond intron 3. Immunocytological study of the kidney from AE1(646X/646X) cattle has not been reported (Inaba *et al.* 1996).

Newly reported animal models of renal tubular acidosis include incomplete dRTA in mice null for the ATP6V1B1 subunit of vH^+-ATPase (Finberg *et al.* 2001) and direct dRTA in mice null for the KCC4 K^+/Cl^- cotransporter (Boetiger *et al.* 2002). A more complete understanding of the pathophysiology of dRTA secondary to AE1 mutations will benefit particularly from creation of heartier mouse models of AE1-deficient dRTA, and from studies of mutant AE1 trafficking in type A intercalated cells in vivo or in vitro.

REFERENCES

Adair-Kirk, T.L., Dorsey, F.C., and Cox, J.V. (2002). Multiple Cytoplasmic Signals Direct the Intracellular Trafficking of Chicken Kidney AE1 Anion Exchangers in MDCK Cells. *J. Cell Sci.,* **116**, 655–663.

Allen, S.J., O'Donnell, A., and Alexander, N.D. (1999). Prevention of Cerebral Malaria in Children of Papua New Guinea by Southeast Asian Ovalocytosis Band 3. *Am. J. Trop. Med. Hyg.,* **60**, 1056–1060.

Alloisio, N., Texier, P., Vallier, A., Ribeiro, M.L., Morle, L., Bozon, M., et al. (1993). Modulation of Clinical Expression and Band 3 Deficiency in Hereditary Spherocytosis. Blood, 90, 414–420.

Alloisio, N., Maillet, P., Carre, G., Texier, P., Vallier, A., Baklouti, F., et al. (1996). Hereditary Spherocytosis with Band 3 Deficiency. Association with a Nonsense Mutation of the Band 3 Gene (Allele Lyon), and Aggravation by a Low-Expression Allele Occurring In Trans (Allele Genas). Blood, 88, 1062–1069.

Alper, S.L. (2002). Genetic Diseases of Acid–Base Transport. Annu. Rev. Physiol., 64, 899–923.

Alper, S.L., Vandorpe, D.H., Stuart-Tilley, A., Rotter, M., Lux, S.E., Peters, L.L., et al. (1998). Absence of DIDS-Sensitive Cl$^-$ Conductance in Red cells of AE1 (Band 3) $-/-$ Mice (Abstract). J. Am. Soc. Nephrol., 9, 151A.

Bianchi, P., Zanella, A., Alloisio, N., Barosi, G., Bredi, E., Pelissero, G., et al. (1997). Variant of the EPB3 Gene of the Anti-Lepore Type in Hereditary Spherocytosis. Br. J. Haematol., 98, 283–288.

Bjork, J., Reardon, D.M., and Backman, L. (1997). Phosphoinositide Metabolism in Hereditary Ovalocytic Red Blood Cell Membranes. Biochim. Biophys. Acta, 1326, 342–348.

Boettger, T., Hubner, C.A., Maier, H., Rust, M.B., Beck, F.X., and Jentsch, T.J. (2002). Deafness and Renal Tubular Acidosis in Mice Lacking the K–Cl Cotransporter Kcc4. Nature, 416, 874–878.

Bracher, N.A., Lyons, C.A., Wessels, G., Mansvelt, E., and Coetzer, T.L. (2001). Band 3 Cape Town (E90K) Causes Severe Hereditary Spherocytosis in Combination with Band 3 Prague III. Br. J. Haematol., 113, 689–693.

Bruce, L.J., Kay, M.M., Lawrence, C., and Tanner, M.J. (1993). Band 3 HT, a Human Red-Cell Variant Associated with Acanthocytosis and Increased Anion Transport, Carries the Mutation Pro-868→Leu in the Membrane Domain of Band 3. Biochem. J., 293, 317–320.

Bruce, L.J., Anstee, D.J., Spring, F.A., and Tanner, M.J. (1994). Band 3 Memphis Variant II. Altered Stilbene Disulfonate Binding and the Diego (Di^a) Blood Group Antigen Are Associated with the Human Erythrocyte Band 3 Mutation Pro^{854}→Leu. J. Biol. Chem., 269, 16155–16158.

Bruce, L.J., Ring, S.M., Anstee, D.J., Reid, M.E., Wilkinson, S., and Tanner, M.J. (1995). Changes in the Blood Group Wright Antigens Are Associated with a Mutation at Amino Acid 658 in Human Erythrocyte Band 3: A Site of Interaction Between Band 3 and Glycophorin A under certain conditions. Blood, 85, 541–547.

Bruce, L.J., Cope, D.L., Jones, G.K., Schofield, A.E., Burley, M., Povey, S., et al. (1997). Familial Renal Tubular Acidosis Is Associated with Mutations in the Red Cell Anion Exchanger (Band 3, AE1) Gene. J. Clin. Invest., 100, 1693–1707.

Bruce, L.J., Ring, S.M., Ridgwell, K., Reardon, D.M., Seymour, C.A., Van Dort, H.M., et al. (1999). Southeast

Asian Ovalocytic (SAO) Erythrocytes Have a Cold Sensitive Cation Leak: Implications for In Vitro Studies on Stored SAO Red Cells. Biochim. Biophys. Acta, 1416, 258–270.

Bruce, L.J., Wrong, O., Toye, A.M., Young, M.T., Ogle, G., Ismail, Z., et al. (2000). Band 3 Mutations, Renal Tubular Acidosis and South-East Asian Ovalocytosis in Malaysia and Papua New Guinea: Loss of up to 95% Band 3 Transport in Red Cells. Biochem. J., 350, 41–51.

Bruce, L.J., Beckmann, R., Ribeiro, M.L., Peters, L.L., Chasis, J.A., Delaunay, J., et al. (2003). A Band 3-Based Macrocomplex of Integral and Peripheral Proteins in the Red Cell Membrane. Blood, 101, 4180–4188.

Cattani, J.A., Gibson, F.D., Alpers, M.P., and Crane, G.G. (1987). Hereditary Ovalocytosis and Reduced Susceptibility to Malaria in Papua New Guinea. Trans. R. Soc. Trop. Med. Hyg., 81, 705–709.

Chang, S.H. and Low, P.S. (2003). Identification of a Critical Ankyrin Binding Loop on the Cytoplasmic Domain of Erythrocyte Membrane Band 3 by Crystal Structure Analysis and Site-Directed Mutagenesis. J. Biol. Chem., 278, 6879–6884.

Cheidde, L, Vieira, T.C., Lima, P.R.M., Saad, S.T.O., and Heilberg, I.P. (2002). A Novel Mutation in the Anion Exchanger 1 Gene Associated with Distal Renal Tubular Acidosis. J. Am. Soc. Nephrol., 13, 575A.

Chen, J., Vijayakumar, S., Li, X., Al-Awqati, Q. (1998). Kanadaptin Is a Protein that Interacts with the Kidney but not the Erythroid Form of Band 3. J. Biol. Chem., 273, 1038–1043.

Chernova, M.N., Jarolim, P., Palek, J., and Alper, S.L. (1995). Overexpression of AE1 Prague, But Not of AE1 SAO, Inhibits Wild-Type AE1 Trafficking in Xenopus Oocytes. J. Membr. Biol., 148, 203–210.

Chernova, M.N., Jiang, L., Crest, M., Hand, M., Vandorpe, D.H., Strange, K., et al. (1997). Electrogenic Sulfate/Chloride Exchange in Xenopus Oocytes Mediated by Murine AE1 E699Q. J. Gen. Physiol., 109, 345–360.

Cowan, C.A., Yokoyama, N., Bianchi, L.M., Henkemeyer, M., and Fritzsch, B. (2000). EphB2 Guides Axons at the Midline and Is Necessary for Normal Vestibular Function. Neuron, 26, 417–430.

Dahl, N.K., Jiang, L., Chernova, M.N., Stuart-Tilley, A.K., Shmukler, B.E., and Alper, S.L. Rescue of AE1-Mediated HCO$_3^-$ Transport by Carbonic Anhydrase II Presented on an Adjacent AE1 Protomer. Submitted for publication

De Franceschi, L., Turrini, F., Delgiudice, E.M., Perrotta, S., Olivieri, O., Corrocher, R., et al. (1998). Decreased Band 3 Anion Transport Activity and Band 3 Clusterization in Congenital Dyserythropoietic Anemia Type II. Exp. Hematol., 26, 869–873.

Devonald, M.A., Smith, A.N., Poon, J.P., Ihrke, G., and Karet, F.E. (2003). Non-polarized Targeting of AE1 Causes Distal Autosomal Dominant Distal Renal Tubular Acidosis. Nat. Genet., 33, 125–127.

Dhermy, D., Galand, C., Bournier, O., Boulanger, L., Cynober, T., Schismanoff, P.O., et al. (1997).

Heterogenous Band 3 Deficiency in Hereditary Spherocytosis Related to Different Band 3 Gene Defects. *Br. J. Haematol.*, **98**, 32–40.

Ding, Y., Casey, J.R., and Kopito, R.R. (1994). The Major Kidney AE1 Isoform Does Not Bind Ankyrin (Ank1) In Vitro. An Essential Role for the 79 NH$_2$-Terminal Amino Acid Residues of Band 3. *J. Biol. Chem.*, **269**, 32201–32208.

Dluzewski, A.R., Nash, G.B., Wilson, R.J., Reardon, D.M., and Gratzer, W.B. (1997). Invasion of Hereditary Ovalocytes by *Plasmodium falciparum* In Vitro and Its Relation to Intracellular ATP Concentration. *Mol. Biochem. Parasitol.*, **55**, 1–7.

Eber, S.W., Gonzalez, J.M., Lux, M.L., Scarpa, A.L., Tse, W.T., Dornwell, M., *et al.* (1996). Ankyrin-1 Mutations Are a Major Cause of Dominant and Recessive Hereditary Spherocytosis. *Nat. Genet.*, **13**, 214–218.

Finberg, K.E., Wang, T., Wagner, C.A., Geibel, J.P., Dou, H., and Lifton, R.P. (2001). Generation and Characterization of H$^+$-ATPase B1 Subunit Deficient Mice. *J. Am. Soc. Nephrol.*, **12**, 3A.

Fujinaga, J., Tang, X.B., and Casey, J.R. (1999). Topology of the Membrane Domain of Human Erythrocyte Anion Exchange Protein, AE1. *J. Biol. Chem.*, **274**, 6626–6633.

Han, B.G., Nunomura, W., Takakuwa, Y., Mohandas, N., and Jap, B.K. (2000). Protein 4.1R Core Domain Structure and Insights into Regulation of Cytoskeletal Organization. *Nat. Struct. Biol.*, **7**, 871–875.

Han, J.S., Kim, G.H., Kim, J., Jeon, U.S., Joo, K.W., Na, K.Y., *et al.* (2002). Secretory-Defect Distal Renal Tubular Acidosis Is Associated with Transporter Defect in H$^+$-ATPase and Anion Exchanger-1. *J. Am. Soc. Nephrol.*, **31**, 1425–1432.

Hassoun, H., Hanada, T., Lutchman, M., Sahr, K.E., Palek, J., Hanspal, M., *et al.* (1998). Complete Deficiency of Glycophorin A in Red Blood Cells from Mice with Targeted Inactivation of the Band 3 (AE1) Gene. *Blood*, **91**, 2146–2151.

Hsu, L. and Morrison, M. (1985). A New Variant of the Anion Transport Protein in Human Erythrocytes. *Biochemistry*, **24**, 3086–3090.

Hubner, S., Jans, D.A., Xiao, C.Y., John, A.P., and Drenckhahn, D. (2002). Signal- and Importin-Dependent Nuclear Targeting of the Kidney Anion Exchanger 1-Binding protein Kanadaptin. *Biochem. J.*, **361**, 287–296.

Ideguchi, H., Okubo, K., Ishikawa, A., Futata, Y., and Hamasaki, N. (1992). Band 3 Memphis is Associated with a Lower Transport Rate of Phosphoenolpyruvate. *Br. J. Haematol.*, **82**, 122–125.

Inaba, M., Yawata, A., Koshino, I., Sato, K., Takeuchi, M., Takakuwa, Y., *et al.* (1996). Defective Anion Transport and Marked Spherocytosis with Membrane Instability Caused by Hereditary Total Deficiency of Red Cell Band 3 in Cattle due to a Nonsense Mutation. *J. Clin. Invest.*, **97**, 1804–1817.

Inoue, T., Kanzaki, A., Kaku, M., Yawata, A., Takezono, M., Okamoto, N., *et al.* (1998). Homozygous Missense Mutation (Band 3 Fukuoka: G130R): A Mild Form of Hereditary Spherocytosis with Near-Normal Band 3 Content and Minimal Changes of Membrane Ultrastructure Despite Moderate Protein 4.2 Deficiency. *Br. J. Haematol.*, **102**, 932–939.

Iwase, S., Ideguchi, H., Takao, M., Horiguchi-Yamada, J., Iwasaki, M., Takahara, S., *et al.* (1998). Band 3 Tokyo: Thr837→Ala837 Substitution in Erythrocyte Band 3 Protein Associated with Spherocytic Hemolysis. *Acta Haematol.*, **100**, 200–203.

Jarolim, P. (2003). Disorders of Band 3. In *Red Cell Membrane Transport in Health and Disease*, New York: Springer, in press.

Jarolim, P., Palek, J., Amato, D., Hassan, K., Sapak, P., Nurse, G.T., *et al.* (1991). Deletion in Erythrocyte Band 3 Gene in Malaria-Resistant Southeast Asian Ovalocytosis. *Proc. Natl. Acad. Sci. USA*, **88**, 11022–11026.

Jarolim, P., Palek, J., Rubin, H.L., Prchal, J.T., Korsgren, C., and Cohen, C.M. (1992a). Band 3 Tuscaloosa: Pro327→Arg327 Substitution in the Cytoplasmic Domain of Erythrocyte Band 3 Protein Associated with Spherocytic Hemolytic Anemia and Partial Deficiency of Protein 4.2. *Blood*, **80**, 523–529.

Jarolim, P., Rubin, H.L., Zhai, S., Sahr, K.E., Liu, S.C., Mueller, T.J., *et al.* (1992b). Band 3 Memphis: A Widespread Polymorphism with Abnormal Electrophoretic Mobility of Erythrocyte Band 3 Protein Caused by Substitution AAG→GAG (Lys→Glu) in Codon 56. *Blood*, **80**, 1592–1598.

Jarolim, P., Brabec, V., Chrobak, L., Alper, S.L., Brugnara, C., Corbett, J.D., *et al.* (1994). Decreased Band 3 Content, Sulfate Flux, and Band 3 Fractional Mobility in Congential Dyserythropoietic Anemia. *Blood*, **84** (Suppl. 1), 6a.

Jarolim, P., Rubin, H.L., Brabec, V., Chrobak, L., Zolotarev, A.S., Alper, S.L., *et al.* (1995). Mutations of Conserved Arginines in the Membrane Domain of Erythroid Band 3 Lead to a Decrease in Membrane-Associated Band 3 and to the Phenotype of Hereditary Spherocytosis. *Blood*, **85**, 634–640.

Jarolim, P., Murray, J.L., Rubin, H.L., Taylor, W.M., Prchal, J.T., Ballas, S.K., *et al.* (1996). Characterization of 13 Novel Band 3 Gene Defects in Hereditary Spherocytosis with Band 3 Deficiency. *Blood*, **88**, 4366–4374.

Jarolim, P., Murray, J.L., Rubin, H.L., Coghlan, G., and Zelinski, T. (1997a). A Thr552-Kile Substitution in Erythroid Band 3 Gives Rise to the Warrior Blood Group Antigen. *Transfusion*, **37**, 398–405.

Jarolim, P., Murray, J.L., Rubin, H.L., Smart, E., and Moulds, J.M. (1997b). Blood Group Antigens Rb^a, Tr^a, and Wd^a are Located in the Third Ectoplasmic loop of Erythocyte Band 3 Protein. *Transfusion*, **37**, 607–615.

Jarolim, P., Rubin, H.L., Zakova, D., Storry, J., and Reid, M.E. (1998a). Characterization of Seven Low

Incidence Blood Group Antigens Carried by Erythrocyte Band 3 Protein. *Blood*, **92**, 4836–43.

Jarolim, P., Shayakul, C., Prabakaran, D., Jiang, L., Stuart-Tilley, A., Rubin, H.L., *et al.* (1998b). Autosomal Dominant Distal Renal Tubular Acidosis Is Associated in Three Families with Heterozygosity for the R589H Mutation in the AE1 (band 3) Cl⁻/HCO₃⁻ Exchanger. *J. Biol. Chem.*, **273**, 6380–6388.

Jenkins, P.B., Abou-Alfa, G.K., Dhermy, D., Bursaux, E., Feo, C., Scarpa, A.L., *et al.* (1996). A Nonsense Mutation in the Erythrocyte Band 3 Gene Associated with Decreased mRNA Accumulation in a Kindred with Dominant Hereditary Spherocytosis. *J. Clin. Invest.*, **97**, 373–380.

Jennings, M.L. (1995). Rapid Electrogenic Sulfate–Chloride Exchange Mediated by Chemically Modified Band 3 in Human Erythrocytes. *J. Gen. Physiol.*, **105**, 21–47.

Jennings, M.L. and Gosselink, P.G. (1995). Anion Exchange Protein in Southeast Asian Ovalocytes: Heterodimer Formation between Normal and Variant Subunits. *Biochemistry*, **34**, 3588–3595.

Jennings, M.L. and Smith, J.S. (1992). Anion–Proton Cotransport through the Human Red Blood Cell Band 3 Protein. Role of Glutamate 681. *J. Biol. Chem.*, **267**, 13964–13971.

Jones, G.L., Edmundson, H.M., Wesche, D., and Saul, A. (1990). Human Erythrocyte Band-3 Has an Altered N Terminus in Malaria-Resistant Melanesian Ovalocytosis. *Biochim. Biophys. Acta*, **1096**, 33–40.

Kanki, T., Young, M.T., Hamasaki, N., and Tanner, M.J. (2003). The N-Terminal Region of the Transmembrane Domain of Human Erythrocyte Band 3: Residues Critical for Membrane Insertion and Transport Activity. *J. Biol. Chem.*, **278**, 5564–5573.

Kanzaki, A., Takezono, M., Kaku, M., Yawata, A., Ozcan, R., Kugler, W., *et al.* (1997a). Molecular and Genetic Characteristics in Japanese Patients with Hereditary Spherocytosis: Frequent Band 3 Mutations and Rarer Ankyrin Mutations. *Blood*, **90** (Suppl. 1), 6b.

Kanzaki, A., Hayette, S., Morle, L., Inoue, F., Matsuyama, R., Inoue, T., *et al.* (1997). Total Absence of Protein 4.2 and Partial Deficiency of Band 3 in Hereditary Sphero-cytosis. *Br. J. Haematol.*, **99**, 522–530.

Karet, F.E. (2002). Inherited Distal Renal Tubular Acidosis. *J. Am. Soc. Nephrol.*, **13**, 2178–2184.

Karet, F.E., Gainza, F.J., Gyory, A.Z., Unwin, R.J., Wrong, O., Tanner, M.J., *et al.* (1998). Mutations in the Chloride–Bicarbonate Exchanger Gene *AE1* Cause Autosomal Dominant But Not Autosomal Recessive Distal Renal Tubular Acidosis. *Proc. Natl. Acad. Sci. USA*, **95**, 6337–6342.

Knauf, P.K. and Pal, P. (2003). Band 3-Mediated Transport. In *Red Cell Membrane Transport in Health and Disease*, New York: Springer

Kudrycki, K.E., Newman, P.R., and Shull, G.E. (1990). CDNA Cloning and Tissue Distribution of mRNAs for

Two Proteins That Are Related to the Band 3 Cl⁻/HCO₃⁻ Exchanger. *J. Biol. Chem.*, **265**, 462–471.

Lemieux, M.J., Reithmeier, R.A., and Wang DN. (2002). Importance of Detergent and Phospholipids in the Crystallization of the Human Erythrocyte Anion-Exchanger Membrane Domain. *J. Struct. Biol.*, **137**, 322–332.

Lima, P.R., Gontijo, J.A., Lopes de Faria, J.B., Costa, F.F., and Saad, S.T. (1997). Band 3 Campinas: A Novel Splicing Mutation in the Band 3 Gene (*AE1*) Associated with Hereditary Spherocytosis, Hyperactivity of Na⁺/Li⁺ Countertransport and an Abnormal Renal Bicarbonate Handling. *Blood*, **90**, 2810–2818.

Liu, S.C., Zhai, S., Palek, J., Golan, D.E., Amato, D., Hassan, K., *et al.* (1990). Molecular Defect of the Band 3 Protein in Southeast Asian Ovalocytosis. *N. Engl. J. Med.*, **323**, 1530–1538.

Maillet, P., Vallier, A., Reinhart, W.H., Wyss, E.J., Ott, P., Texier, P., *et al.* (1995). Band 3 Chur: A Variant Associated with Band 3-Deficient Hereditary Sphero-cytosis and Substitution in a Highly Conserved Position of Transmembrane Segment 11. *Br. J. Haematol.*, **91**, 804–810.

McManus, K., Lupe, K., Coghlan, G., and Zelinski, T. (2000). An Amino Acid Substitution in the Putative Second Extracellular Loop of RBC Band 3 Accounts for the Froese Blood Group Polymorphism. *Transfusion*, **40**, 1246–1249.

Miraglia del Guidice, E., Vallier, A., Maillet, P., Perrotta, S., Cutillo, S., Iolascon, A., *et al.* (1997). Novel Band 3 Variants (Foggia, Napoli I and Napoli II) associated with Hereditary Spherocytosis and Band 3 Deficiency: Status of the D38A Polymorphism Within the EPB3 Locus. *Br. J. Haematol.*, **96**, 70–76.

Mohandas, N., Winardi, R., Knowles, D., Leung, A., Parra, M., George, E., *et al.* (1992). Molecular Basis for Membrane Rigidity of Hereditary Ovalocytosis. A Novel Mechanism Involving the Cytoplasmic Domain of Band 3. *J. Clin. Invest.*, **89**, 686–692.

Muller-Berger, S., Karbach, D., Kang, D., Aranibar, N., Wood, P.G., Ruterjans, H., *et al.* (1995). Roles of Histidine 752 and Glutamate 699 in the pH Dependence of Mouse Band 3 Protein-Mediated Anion Transport. *Biochemistry*, **34**, 9325–9332.

NCBI Genbank P02730 (gi:114787) Citations of LocusLink SLC4A1 dbSNP (http://www.ncbi.nlm.nih.gov/SNP/snp_ref.cgi?locusId=6521).

Paw, B.H., Davidson, A.J., Zhou, Y., Li, R., Pratt, S.J., Lee, C., *et al.* (2003). Cell-Specific Mitotic Defect and Dyserythropoiesis Associated with Erythroid Band 3 Deficiency. *Nat. Genet.*, **34**, 59–64.

Perrotta, S., Polito, F., Cone, M.L., Nobili, B., Cutillo, S., Nigro, V., *et al.* (1999). Hereditary Spherocytosis due to a Novel Frameshift Mutation in AE1 Cytoplasmic COOH Terminal Tail: Band 3 Vesuvio. *Blood*, **93**, 2131–2132.

Peters, L.L., Shivdasani, R.A., Liu, S.C., Hanspal, M., John, K.M., Gonzalez, J.M., *et al.* (1996). Anion

Exchanger 1 (Band 3) Is Required to Prevent Erythrocyte Membrane Surface Loss But Not to Form the Membrane Skeleton. *Cell*, **86**, 917–927.

Peters, L.L., Jindel, H.K., Bwynn, B., Korsgren, C., John, K.M., Lux, S.E., et al. (1999). Mild Spherocytosis and Altered Red Cell Ion Transport in Protein 4.2-Null Mice. *J. Clin. Invest.*, **103**, 1527–1537.

Peters, L.L., Andersen, S.G., Gwynn, B., Li, R., Lux, S.E., and Churchill, G.A. (2001). A QTL on Mouse Chromosome 12 Modifies the Band 3 Null Phenotype: β Spectrin Is a Candidate Gene. *Blood*, **98**, abstract 1831.

Poole, J., Hallewell, H., Bruce, L., Tanner, M.J.A., Zupanska, B., and Kusnierz-Alejska, G. (1997). Identification of Two New Jn(a+) Individuals and Assignment of Jnᵃ to Erythrocyte Band 3. *Transfusion*, **37** (Suppl.), 90S.

Popov, M., Li, J., and Reithmeier, R.A. (1999). Transmembrane Folding of the Human Erythrocyte Anion Exchanger (AE1, Band 3) Determined by Scanning and Insertional N-glycosylation Mutagenesis. *Biochem. J.*, **339**, 269–279.

Quilty, J.A. and Reithmeier, R.A. (2000). Trafficking and Folding Defects in Hereditary Spherocytosis Mutants of the Human Red Cell Anion Exchanger. *Traffic.*, **1**, 987–998.

Quilty, J.A., Cordat, E., and Reithmeier, R.A. (2002a). Impaired Trafficking of Human Kidney Anion Exchanger (kAE1) Caused by Hetero-oligomer Formation with a Truncated Mutant Associated with Distal Renal Tubular Acidosis. *Biochem. J.*, **368**, 895–903.

Quilty, J.A., Li, J., and Reithmeier, R.A. (2002b). Impaired Trafficking of Distal Renal Tubular Acidosis Mutants of the Human Kidney Anion Exchanger kAE1. *Am. J. Physiol. Renal. Physiol.*, **282**, F810–F820.

Rajendran, V.M., Black, J., Ardito, T.M., Sangan, P., Alper, S.L., Schweinfest, C., et al. (2002). Regulation of Anion Exchanger RNAs in Rat Distal Colon by Dietary Na-Depletion. *Am. J. Physiol. Gastrointestinal. Physiol.*, **279**, G931–G942.

Reliene, R., Mariani, M. Zanella, A., Reinhart, W.H., Ribeiro, M.L., del Giudice, E.M., et al. (2002). Splenectomy Prolongs In Vivo Survival of Erythrocytes Differently in Spectrin/Ankyrin- and Band 3-Deficient Hereditary Spherocytosis. *Blood*, **100**, 2208–2215.

Ribeiro, M.L., Alloisio, N., Almeida, H., Gomes, C., Texier, P., Lemos, C., et al. (2000). Severe Hereditary Spherocytosis and Distal Renal Tubular Acidosis Associated with the Total Absence of Band 3. *Blood*, **96**, 1602–1604.

Rybicki, A.C., Qiu, J.J., Musto, S., Rosen, N.L., Nagel, R.L., and Schwartz, R.S. (1993). Human Erythrocyte Protein 4.2 Deficiency Associated with Hemolytic Anemia and a Homozygous 40 Glutamic acid→Lysine Substitution in the Cytoplasmic Domain of Band 3 (Band 3 Montefiore). *Blood*, **81**, 2155–2165.

Rysava, R., Tesar, V., Jirsa, M. Jr, Brabec, V., and Jarolim, P. (1997). Incomplete Distal Renal Tubular Acidosis Coinherited with a Mutation in the Band 3 (*AE1*) Gene. *Nephrol. Dial. Transplant.*, **12**, 1869–1873.

Salhany, J.M., Schopfer, L.M., Kay, M.M.B., Gamble, D.N., and Lawrence, C. (1995). Differential Sensitivity of Stilbenedisulfonates in their Reactions with Band 3 HT (Pro868–Leu) *Proc. Natl. Acad. Sci. USA*, **92**, 11844–11848.

Schofield, A.E., Reardon, D.M., and Tanner, M.J. (1992). Defective Anion Transport Activity of the Abnormal Band-3 in Hereditary Ovalocytic Red Blood Cells. *Nature*, **355**, 836–838.

Sekler, I., Lo, R.S., and Kopito, R.R. (1995). A Conserved Glutamate Is Responsible for Ion Selectivity and pH Dependence of the Mammalian Anion Exchangers AE1 and AE2. *J. Biol. Chem.*, **270**, 28751–28758.

Shayakul, C. and Alper, S.L. (2000). Inherited Renal Tubular Acidosis. *Curr. Opin. Nephrol. Hypertens.*, **9**, 541–546.

Shayakul, C., Jariyawat, S., Kaewkaukul, N., and Sophasan, S. (2001). Functional Rescue of Anion Exchanger 1 (AE1) G701D by Glycophorin A Is Attenuated by Co-expression of AE1 Δ400–408: A Basis for Transport Defect in Autosomal Recessive Distal Renal Tubular Acidosis (dRTA). *J. Am. Soc. Nephrol.*, **12**, 10A.

Shayakul, C., Jarolim, P., Ideguchi, H., Prabakaran, D., Cortez, D., Zakova, D., et al. A Highly Polymorphic Dinucleotide Repeat Adjacent to the SLC4A1 (AE1/EPB3) Cl⁻/HCO₃⁻ Exchanger Gene: Diagnostic Validation in a Family with Distal Renal Tubular Acidosis Submitted for Publication.

Southgate, C.D., Chishti, A.H., Mitchell, B., Yi, S.J., and Palek, J. (1996). Targeted Disruption of the Murine Erythroid Band 3 Gene Results in Spherocytosis and Severe Haemolytic Anaemia Despite a Normal Membrane Skeleton. *Nat. Genet.*, **14**, 227–230.

Spring, F.A., Bruce, L.J., Anstee, D.J., and Tanner, M.J. (1992). A Red Cell Band 3 Variant with Altered Stilbene Disulphonate Binding is Associated with the Diego (Dia) Blood Group Antigen. *Biochem. J.*, **288**, 713–716.

Sterling, D. and Casey, J.R. (2002). Bicarbonate Transport Proteins. *Biochem. Cell. Biol.*, **80**, 483–497.

Sterling, D., Reithmeier, R.A., and Casey, J.R. (2001). A Transport Metabolon. Functional Interaction of Carbonic Anhydrase II and Chloride/Bicarbonate Exchangers. *J. Biol. Chem.*, **276**, 47886–47894.

Takano, J., Noguchi, K., Yasumori, M., Kobayashi, M., Gajdos, Z., Miwa, K., et al. (2002). *Arabidopsis* Boron Transporter for Xylem Loading. *Nature*, **420**, 337–340.

Tang, X.B., Fujinaga, J., Kopito, R., and Casey, J.R. (1998). Topology of the Region Surrounding Glu⁶⁸¹of Human AE1 Protein, the Erythrocyte Anion Exchanger. *J. Biol. Chem.*, **273**, 22545–22553.

Tang, X.B., Kovacs, M., Sterling, D., and Casey, J.R. (1999). Identification of Residues Lining the

Translocation Pore of Human AE1, Plasma Membrane Anion Exchange Protein. *J. Biol. Chem.*, **274**, 3557–3564.

Tanner, M.J. (2002). Band 3 Anion Exchanger and its Involvement in Erythrocyte and Kidney Disorders. *Curr. Opin. Hematol.*, **9**, 133–139.

Tanphaichitr, V.S., Sumboonnanonda, A., Ideguchi, H., Shayakul, C., Brugnara, C., Takao, M., *et al.* (1998). Novel AE1 Mutations in Recessive Distal Renal Tubular Acidosis. Loss-of-Function Is Rescued by Glycophorin, A. *J. Clin. Invest.*, **102**, 2173–2179.

Thisse, C. and Zon, L.I. (2002). Organogenesis – Heart and Blood Formation from the Zebrafish Point of View. *Science*, **295**, 457–462.

Toye, A.M., Bruce, L.J., Unwin, R.J., Wrong, O., and Tanner, M.J. (2002). Band 3 Walton, a C-Terminal Deletion Associated with Distal Renal Tubular Acidosis, Is Expressed in the Red Cell Membrane But Retained Internally in Kidney Cells. *Blood*, **99**, 342–347.

Turrini, F., Mannu, F., Arese, P., Yuan, J., and Low, P.S. (1993). Characterization of the Autologous Antibodies that Opsonize Erythrocytes with Clustered Integral Membrane Proteins. *Blood*, **81**, 3146–3152.

Vasuvattakul, S., Yenchitsomanus, P., Thuwajit, P., Kaitwatcharachai, C., Vachuanichsanong, P., Laosombat, V., *et al.* (1999). Compound Heterozygosity of AE1 Genes Causes Recessive Distal Renal Tubular Acidosis in Southeast Asian Ovalocytosis. *J. Am. Soc. Nephrol.*, **10**, 444A.

Vince, J.W. and Reithmeier, R.A. (2000). Identification of the Carbonic Anhydrase II Binding Site in the Cl(−)/HCO(3)(−) Anion Exchanger AE1. *Biochemistry*, **39**, 5527–5533.

Wang, D.N., Sarabia, V.E., Reithmeier, R.A., and Kuhlbrandt, W. (1994). Three-Dimensional Map of the Dimeric Membrane Domain of the Human Erythrocyte Anion Exchanger, Band 3. *EMBO J.*, **13**, 3230–3235.

Wood, P.G. (1992). The Anion Exchange Proteins: Homology and Secondary Structure. *Prog. Cell. Res.*, **2**, 325–352.

Wrong, O., Bruce, L.J., Unwin, R.J., Toye, A.M., and Tanner, M.J. (2002). Band 3 Mutations, Distal Renal Tubular Acidosis, and Southeast Asian Ovalocytosis. *Kidney Int.*, **62**, 10–19.

Yannoukakos, D., Vasseur, C., Driancourt, C., Blouquit, Y., Delaunay, J., Wajcman, H., *et al.* (1991). Human Erythrocyte Band 3 Polymorphism (Band 3 Memphis): Characterization of the Structural Modification (Lys 56→Glu) by Protein Chemistry Methods. *Blood*, **78**, 1117–1120.

Yawata, Y., Kanzaki, A., Doerfler, W., Ozcan, R., and Eber, S.W. (2000). Characteristic Features of the Genotype and Phenotype of Hereditary Spherocytosis in the Japanese Population. *Int. J. Hematol.*, **71**, 118–135.

Yenchitsomanus, P.T., Vasuvattakul, S., Kirdpon, S., Wasanawatana, S., Susaengrat, W., Sreethiphayawan, S., *et al.* (2002). Autosomal Recessive Distal Renal Tubular Acidosis Caused by G701D Mutation of Anion Exchanger 1 Gene. *Am. J. Kidney Dis.*, **40**, 21–29.

Zelinski, T., Pongoski, J., and Coghlan, G. (1997). The Low Incidence Erythrocyte Antigen NFLD is Associated with Membrane Protein Band 3. *Transfusion*, **37** (Suppl.), 90S.

Zelinski, T., McManus, K., Punter, F., Moulds, M., and Coghlan, G. (1998). A Gly565-Ala substitution in Human Erythroid Band 3 Accounts for the Wu Blood Group Polymorphism. *Transfusion*, **38**, 745–748.

Zelinski, T., Rusnak, A., McManus, K., and Coghlan, G. (2000). Distinctive Swann Blood Group Genotypes: Molecular investigations. *Vox Sang*, **79**, 215–218.

Zhang, D., Kiyatkin, A., Bolin, J.T., and Low, P.S. (2000). Crystallographic Structure and Functional Interpretation of the Cytoplasmic Domain of Erythrocyte Membrane Band 3. *Blood*, **96**, 2925–2933.

Zhao, R. and Reithmeier, R.A. (2001). Expression and Characterization of the Anion Transporter Homologue YNL275w in *Saccharomyces cerevisiae*. *Am. J. Physiol. Cell. Physiol.*, **281**, C33–C45.

Zhu, Q., Lee, D.W., and Casey, J.R. (2003). Novel Topology in C-terminal Region of the Human Plasma Membrane Anion Exchanger, AE1. *J. Biol. Chem.*, **278**, 3112–3120.

4

MICHAEL F. ROMERO*

Electrogenic Na^+/HCO_3^- cotransporter NBC1 (SLC4A4): proximal renal tubular acidosis and ocular pathologies

INTRODUCTION

Blood and cellular pH are tightly controlled. Normal blood pH is 7.35–7.45, that is, $[H^+]$ of 36–45 nM. In fact, humans with blood pH below 7.2 or above 7.6 are quite sick. Chronic blood pH below 7.0 ($[H^+] = 100\,nM$) or above 7.8 ($[H^+] \sim 15\,nM$) is incompatible with life.

In vertebrates acid–base balance, cellular and systemic, is maintained by the CO_2/HCO_3^- buffering system. Since mammals in particular generate a great deal of acid via metabolism, these organisms must either get rid of the excess acid (H^+) or elevate systemic base (HCO_3^-) to buffer this metabolic acid. Figure 1 illustrates that this delicate balance is maintained by complementary functions of the lungs and the kidney. Specifically, the lungs control P_{CO_2} while the kidneys have the ability to control both secreted and absorbed H^+ and HCO_3^-. It should also be noted that, from the kidney's perspective, H^+ secretion is equivalent to HCO_3^- absorption or generation.

The kidney receives about 25% of cardiac output ($\sim 5\,L/min$); that is, renal blood flow (RBF) is $\sim 1.25\,L/min$. Ten to fifteen percent of RBF is filtered by the kidneys; that is, the glomerular filtration rate is $\sim 0.12–0.15\,L/min$ ($\sim 180\,L/$ day at $\sim$ pH 7.4) For 24 mM HCO_3^- at pH 7.4, this amounts to 4,320 mmol HCO_3^- per day. This enormous fluid load allows the kidney to adjust both volume and solutes to meet the body's needs. Normal urine production is only 1–2 L/day, but the pH can range from pH 4 to pH 8. HCO_3^-, like other ions and solutes in the blood, is filtered by the kidney at the glomerulus and then absorbed by transport

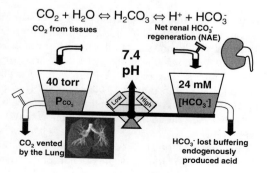

$$CO_2 + H_2O \Leftrightarrow H_2CO_3 \Leftrightarrow H^+ + HCO_3^-$$

Figure 1   Acid–base balance.

* Department of Physiology and Biophysics, Case Western Reserve University, School of Medicine, 2119 Abington Road, Cleveland, OH 44106-4970, USA

processes in the renal nephron. More than 80% of renal HCO_3^- absorption occurs isotonically in the proximal tubule: 100% filtered HCO_3^- reabsorption plus new HCO_3^- absorption. Filtered HCO_3^- reabsorption by the proximal tubule is isotonic and thus accounts for about two thirds of HCO_3^- absorption (Figure 2B). The additional 15–20% is accounted for by

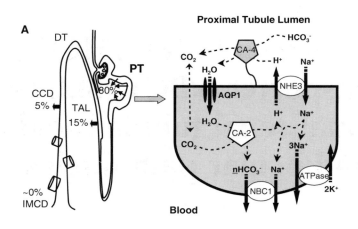

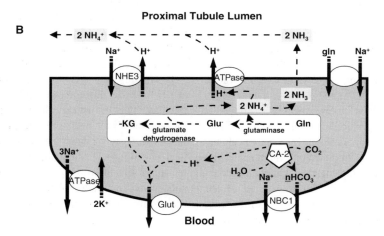

Figure 2 Cellular model of renal NBC1/SLC4a4 function. (A) Renal proximal tubule model of HCO_3^- absorption. The left panel illustrates the nephron and approximate percentages of HCO_3^- absorption. The right panel shows a cellular model of the proximal convoluted tubule where 80–90% of HCO_3^- absorption occurs. Transporters, channels, and enzymes are indicated as follows: AQP1, aquaporin 1; ATPase, Na^+/K^+ pump; CA-2, carbonic anhydrase II; CA-4, carbonic anhydrase IV; NBC1, electrogenic Na^+/HCO_3^- cotransporter; 1, NHE3, Na^+/H^+ exchanger 3. (B) Formation and absorption of "new HCO_3^-." The uptake and metabolism of glutamine (Gln/Q) by the proximal nephron allows the creation of "new HCO_3^-." Gln metabolism to glutamate (Glu$^-$/E) and then to α-ketoglutarate (α-KG) produces 2 NH_4^+. This NH_4^+ releases H^+ (transported) to form permeable NH_3. CA-2 uses cellular water to form HCO_3^- and H^+. The H^+ is consumed by the metabolism of α-KG to ultimately form half glucose, transported by a GLUT (diffusive glucose transporter). Finally, extracellular NH_3 titrates/buffers the lumenally transported H^+.

"new HCO$_3^-$" from the catabolism of glutamine and ammoniagenesis in the proximal tubule (Figure 2B) (Gebisch and Windhager 2003). Accordingly, the renal response to a metabolic acid–base disturbance is threefold: (a) alter filtered HCO$_3^-$, (b) alter H$^+$ secretion, and (c) alter NH$_4^+$ production (Gebisch and wind-hager 2003). In the case of acidosis, filtered HCO$_3^-$ would decrease, H$^+$ secretion would increase, and NH$_4^+$ production would increase.

BACKGROUND OF THE ELECTROGENIC NA$^+$/HCO$_3^-$ COTRANSPORTER NBC1

HCO$_3^-$ in ultrafiltrate luminal fluid combines with secreted H$^+$ [mostly by Na$^+$–H$^+$ exchange (Alpern and Rector 1996)] to form CO$_2$ and H$_2$O, both of which easily enter the proximal tubule cell (Figure 2B). Prior to 1980, the mechanism of HCO$_3^-$ movement from the proximal tubule cell back into the blood was unknown. A basolateral HCO$_3^-$ conductance pathway was hypothesized.

Boron and Boulpaep (1983) made the astounding discovery that this HCO$_3^-$ absorption process was coupled asymmetrically to Na$^+$ transport. This new basolateral transport activity was called the "electrogenic Na$^+$/HCO$_3^-$ cotransporter." This electrogenic Na$^+$/HCO$_3^-$ cotransporter had a "fingerprint" transport (Boron and Boulpaep 1983): multiple HCO$_3^-$ transported with each Na$^+$, and it was Cl$^-$ independent and stilbene inhibited. Later, a similar cotransporter activity was reported in mammals: bovine corneal endothelial cells (Jentsch et al. 1984), rat proximal convoluted tubules (Alpern 1985), rabbit proximal straight tubules (Sasaki et al. 1987), and several other preparations (see Boron and Boulpaep 1989).

However, the molecular nature of this cotransporter protein(s) was unknown until 1995. *Ambystoma tigrinum* (salamander) kidney was used to expression clone a renal electrogenic Na$^+$/HCO$_3^-$ cotransporter (NBC1) (Romero et al. 1996a, 1997a). As with the original characterization of the cotransporter

activity in this tissue (Boron and Boulpaep 1983), NBC1 transported Na$^+$, transported HCO$_3^-$, was electrogenic (1 Na$^+$: at least 2 HCO$_3^-$), Cl$^-$ independent, and inhibited by stilbenes (such as DIDS). Using amphibian kidney rather than mammalian tissue to clone NBC was the key to the successful NBC1 cloning (Romero et al. 1997a, 1998b). Surprisingly, at the molecular level this electrogenic NBC1 sequence was related to the electroneutral band-3-like proteins – the anion exchangers AE1, AE2, and AE3 (Romero et al. 1997a, 1998b). This molecular homology also revealed a probable bicarbonate transporter superfamily (BTS) (Romero et al. 1997a) that now has many seemingly topologically related members: electroneutral Na$^+$/HCO$_3^-$ cotransporters (NBC3), a second electrogenic Na$^+$/HCO$_3^-$ cotransporter (NBC4), and Na$^+$-driven Cl/HCO$_3$ exchangers (NDAE1 and NDCBE1). These molecular tools have helped to renew interest in HCO$_3^-$ transporters and acid–base physiology.

NBC CLONES, PROTEINS, AND GENES

The renal or "kidney" NBC1 open reading frame (kNBC1/NBCe1-A/SLC4a4-A) encodes 1035 amino acids and predicts a protein of 116 kDa (15, 16, 59, 60). The NBC1 protein is predicted to have both the N- and C-termini intracellular (Figure 3) and many potential phosphorylation sites, as well as several N-linked glycosylation sites.

A second N-terminal NBC1 isoform was cloned from pancreas (pNBC1/NBCe1-B/SLC4a4-B) (Abuladze et al. 1998) and heart (hhNBC1) (Choi et al. 1999). This clone has the first 41 amino acids replaced by 85 different amino acids (Figure 4). This pNBC1 encodes 1079 amino acids and predicts a protein of 120 kDa (Abuladze et al. 1998; Choi et al. 1999; Bevensee et al. 2000). The longer NBC1 protein also encodes similar transport (Abuladze et al. 1998; Choi et al. 1999) and is electrogenic (Choi et al. 1999).

A unique C-terminus occurs in a third NBC1 isoform (rb2NBC1/NBCe1-C/SLC4a4-C). This

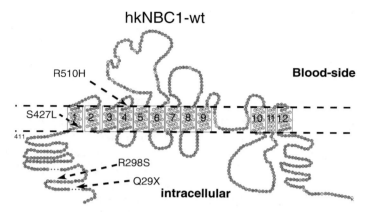

Figure 3 Molecular model of NBC1 secondary structure. Predicted secondary structure of NBC1/SLC4a4. Point mutations in NBC1 causing human pRTA are shown in white.

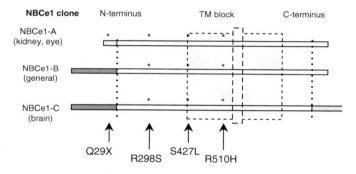

Figure 4 NBC1 isoforms.

rb2NBC1 was recently cloned and characterized from the rat brain (Bevensee *et al.* 2000). This rb2NBC1 clone is identical to pNBC1 at the N-terminus but has 61 unique COOH-terminal amino acids, the result of a 97 bp deletion and frame shift near the end of the open reading frame. The 1094 amino acid clone predicts a protein of ~130 kDa (Bevensee *et al.* 2000). This C-terminal NBC1 isoform has not yet been identified in humans. This rb2NBC1 clone mediates apparently identical transport activity to that mediated by kNBC1 and pNBC1 (Bevensee *et al.* 2000).

The human NBC1 gene (*SLC4A4*) is located at 4q21 (Abuladze *et al.* 1998; Romero *et al.* 1998a) and stretches over ~400–450 kb

(Abuladze *et al.* 2000). Both pNBC1/NBCe1-B and kNBC1/NBCe1-A are transcribed from the same gene, but kNBC1 is transcribed from an alternative internal promoter (Abuladze *et al.* 2000). Presumably, if NBCe1-C exists in humans, it is also transcribed from the common promoter.

NBC1 clones and their corresponding proteins have been identified in many tissues other than the kidney, pancreas, and brain (Table 1). Interestingly, the kidney appears to express all the identified NBC1 mRNAs and proteins. In keeping with this observation, renal disease is one of the major phenotypes of human NBC1 mutations (Igarashi *et al.* 1999, 2000; Dinour *et al.* 2000). That is, these

Table 1 Identification of the electrogenic Na$^+$/HCO$_3^-$ cotransporter (NBC1/SLC4a4) isoforms in mammalian tissues

Tissue	NBC1 isoform	Reference(s)
Brain	rb1NBC1 (pNBC1)	Bevensee *et al.* 2000; Giffard *et al.* 2000
Brain	rb2NBC1	Bevensee *et al.* 2000; Schmitt *et al.* 2000; Douglas *et al.* 2001
Colon	pNBC1	Abuladze *et al.* 1998
	kNBC1/pNBC1	Bevensee *et al.* 2000
Duodenum (small intestine)	kNBC1 & pNBC1	Praetorius *et al.* 2001
	kNBC1/pNBC1	Bevensee *et al.* 2000
Epididymis	NBC1	Jensen *et al.* 1999
Eye	kNBC1 & pNBC1	Usui *et al.* 1999
	pNBC1	Sun *et al.* 2001
Heart	hhNBC1/pNBC1	Choi *et al.* 1999
	kNBC1/pNBC1	Bevensee *et al.* 2000
Kidney	kNBC1	Burnham *et al.* 1997, 1998; Romero *et al.* 1997a, 1998b; Schmitt *et al.* 1999; Maunsbach *et al.* 2000
Kidney	pNBC1	Romero *et al.* 1998c
Kidney	rb2NBC1	Bevensee *et al.* 2000
Liver	NBC1	Abuladze *et al.* 1998; Romero *et al.* 1998b
	rb2NBC1	Bevensee *et al.* 2000
Lung	NBC1	Romero *et al.* 1998b
	kNBC1/pNBC1	Bevensee *et al.* 2000
Pancreas	pNBC1	Abuladze *et al.* 1998; Marino *et al.* 1999; Thévenod *et al.* 1999
Prostate	pNBC1	Abuladze *et al.* 1998; Romero *et al.* 1999
Salivary glands	NBC1	Roussa *et al.* 1999
Stomach	NBC1	Abuladze *et al.* 1998; Burnham *et al.* 1998; Rossmann *et al.* 1999
Testis	kNBC1/pNBC1	Bevensee *et al.* 2000
Thyroid	pNBC1	Abuladze *et al.* 1998
Vas deferens	pNBC1	Carlin *et al.* 2002

NBC1 is also known as NBCe1 or SLC4A4 in HUGO nomenclature. Note that "NBC1" as an isoform designation indicates that the exact isoform of NBC1 is currently unknown or unpublished. As noted in Figure 4, kNBC1 is designated NBCe1-A (Slc4a4-A), pNBC1 is NBCe1-B (Slc4a4-B), and rbNBC1 is NBCe1-C (Slc4a4-C).

affected patients have a permanent proximal renal tubular acidosis (type 2 RTA) manifested as blood pH ~7.05 and blood [HCO$_3^-$] ~5–8 mM, rather than the normal 7.4 and ~23–29 mM, respectively. The eye is also affected by these NBC1 point mutations, manifested as bilateral glaucoma, bilateral cataracts, and band keratopathy (Igarashi *et al.* 1999). The effects on other tissues where NBC1 isoforms are expressed (Table 1) are not obvious. Whether the mutations cause a biophysical change in cotransport activity or result in a cellular protein processing problem is not well understood and is still being investigated.

NBC1 EXPRESSION IN OOCYTES

Xenopus oocytes were used to expression clone kNBC1 (Romero *et al.* 1996a, 1997a). For these experiments, oocytes are impaled with two or more microelectrodes. The aqueous chamber is constantly perfused with the test solution. CO_2/HCO_3^- addition to the solutions causes a reduction of intracellular pH (pH_i), that is, acidification (after point a in Figure 5A), because CO_2 crosses the oocyte plasma membrane and is hydrated intracellularly to form HCO_3^- and H^+ (Figure 5B). For oocytes expressing NBC1/SLC4a4, this CO_2/HCO_3^- addition will elicit an immediate hyperpolarization (point a' in Figure 5A) due to the coupled entrance of Na^+ with multiple HCO_3^- ions ("reverse transport" as in Figure 5B). Once CO_2 has equilibrated inside and outside, pH_i achieves a steady state (before point b in Figure 5A). As illustrated in Figure 5B, Na^+ and HCO_3^- continue entering the oocyte and pH_i increases. Removal of extracellular Na^+ (replacement by an impermeant cation such as choline or *N*-methyl-D-glucamine), depolarizes the oocyte and decreases pH_i

(points b and b' in Figure 5A) ["forward transport" as for the proximal convoluted tubule (Figure 5C)]. This electrogenic HCO_3^- transport activity is unaffected by extracellular Cl^- removal (Choi *et al.* 1999; Romero 2001). The anion exchangers AE1-3 (SLC4a1-SLC4a3) do not mediate electrogenic transport and increases pH_i after extracellular Cl^- removal (Romero 2001). Similarly, these later transporters are unaffected by extracellular Na^+ replacement.

Anions transported

The NBC1 protein in the renal proximal tubule is the major, and perhaps exclusive, "HCO_3^-" exit pathway from the cell into the blood (Boron *et al.* 1997; Igarashi *et al.* 1999). Nevertheless, the chemical form (i.e., HCO_3^-, CO_3^{2-}, or the $NaCO_3^-$ ion pair) transported by the NBC1/SLC4a4 protein is still under investigation. Anions tested for NBC1 transport are indicated in Table 2.

Grichtchenko *et al.* (2000) determined the extracellular $[HCO_3^-]$ dependence of Slc4a4 clones from *Ambystoma* (aNBC1) and rat kidney (rkNBC1) expressed in *Xenopus* oocytes. By exposing injected oocytes briefly to pH 7.5 HCO_3^- solutions (0.33–99 mM, by varying $[CO_2]$ to keep pH_o constant), they measured

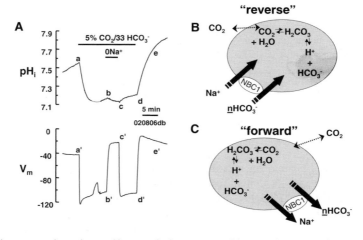

Figure 5   Physiology of human kidney NBC1 (hkNBC1/NBCe1-A/SLC4a4).

Table 2 Ion specificity of NBC1/NBCe1/SLC4a4

Cation/anion	Transported	Apparent $K_{0.5}^a$ (mM)	V_m dependence[b,c]	Reference(s)
Na$^+$	Yes	~30	Yes	Sciortino and Romero 1999
Choline$^+$	No	–	Yes, as 0Na$^+$	Romero et al. 1997a, 1998b; Sciortino and Romero 1999
NMDG$^+$	No	–	Yes, as 0Na$^+$	Romero and Boron, unpublished
Li$^+$	Minor	–	Yes, as 0Na$^+$	Sciortino and Romero 1999
K$^+$	No	–	ND	Sciortino and Romero 1999
HCO$_3^-$	Yes	~6.5	Yes	Romero et al. 1997a, 1998b; Grichtchenko et al. 1998; Choi et al. 1999; Sciortino and Romero 1999; Bevensee et al. 2000
Cl$^-$	No	–	No	Grichtchenko et al. 1998; Romero et al. 1998b; Sciortino and Romero 1999; Bevensee et al. 2000
Butyrate$^-$	No	–	No	Romero et al. 1997a, 1998b
Proprionate$^-$	No	–	No	Sciortino and Romero, unpublished
SO$_4^{2-}$	No	–	No	Grichtchenko et al. 1998
SO$_3^{2-}$	No	–	No	Grichtchenko et al. 1998
CO$_3^{2-}$	Unlikely	–	No	Grichtchenko et al. 1998
CO$_3^{2-}$	Possibly[d]	–	No	Grichtchenko and Boron 2002

ND, not determined.

[a] Ion concentration at the apparent half maximal transport rate.

[b] Membrane voltage.

[c] "Yes, as 0Na+" indicates that NBC1 exhibits voltage dependence when non-transported test ions are used in an Na$^+$ removal maneuver (i.e., reverse transport).

[d] CO$_3^{2-}$ transport by NBCe1-A required coexpression of CA-II, CA-IV, and an aquaporin.

cotransport as membrane hyperpolarization or outward current mediated by NBC1. The external HCO$_3^-$ apparent K_m, with the cotransporter running in the inward direction, was ~6–7 mM for both NBC1 clones (Grichtchenko et al. 1996, 1998, 2000). This same study revealed that SO$_4^{2-}$, SO$_3^{2-}$, and HSO$_3^-$ are not transported by NBC1 (Grichtchenko et al. 2000). In the past few years, carbonic anhydrase II has been shown to bind to the C-terminus of many SLC4 family members (Vince and Reithmeier 1998, 2000; Vince 2000; Vince et al. 2000; Sterling et al. 2001; Loiselle and Casey 2002). CA-IV binds to an extracellular loop of AE1 (Sterling et al. 2002), and a preliminary report indicates

NBC1/NBCe1 binding also (Alvarez et al. 2002). More recently, Grichtchenko and Boron (2002) found that NBCe1-A expressed in oocytes could transport CO$_3^{2-}$ if coexpressed with CA-II, CA-IV, and an aquaporin.

Even initial expression experiments with NBC1 indicated that organic anions would not substitute for the HCO$_3^-$ ion (Romero et al. 1997a, 1998b). Likewise, complete Cl$^-$ removal does not affect the activity of NBC1 (Abuladze et al. 1998; Romero et al. 1998b; Choi et al. 1999; Grichtechenko et al. 2000). NBC1 activity measured by BCECF pH measurements in transfected HEK-293 cells does not seem to require HCO$_3^-$, that is, a Na$^+$: n(OH$^-$) cotransport mode (Amlal et al. 1998).

Nevertheless, HCO_3^- is absolutely essential for electrogenic Na^+/HCO_3^- cotransport (NBC1/NBCe1-A activity) in oocyte experiments (Sciortiono and Romero 1999).

Cations transported

Experiments using ^{22}Na uptake with rabbit kidney basolateral membrane vesicles demonstrated that Li^+, K^+, and choline could partially support Na^+/HCO_3^- cotransporter activity (Soleimani et al. 1991). Later, Jensch et al. (1983) later determined electrogenic, DIDS-inhibitable Na/HCO_3 cotransporter activity in BSC-1 cells. They found an apparent K_m for Na^+ of 20–40 mM at 28 mM HCO_3^-. These investigators also found that Na^+/HCO_3^- cotransporter activity was specific for Na^+. Amlal et al. (1998) reported that transfection of HEK-293 cells with hkNBC1 reveals a low affinity for Li^+ and lesser affinity for K^+ when monitoring pH_i using BCECF. However, electrophysiology of NBC1 expressing Xenopus oocytes reveals only Na^+ transport (Sciortino and Romero 1999). Neither aNBC1 nor rkNBC1 seem to be able to transport Li^+ (Romero et al. 1996b, 1997b; Sciortino and Romero 1999).

Voltage–clamp experiments with rkNBC1 demonstrate that neither choline$^+$, Li^+, nor K^+ could substitute for Na^+ (Sciortino and Romero 1998, 1999). Table 2 summarizes cation transport by NBC1. Both influx and efflux of $NaHCO_3$ depend on extracellular Na^+ and voltage (Sciortino and Romero 1999). Regardless of extracellular $[Na^+]$, influx (outward current I increasing with depolarization) is always measured for a membrane potential V_m more positive than -40 mV, and efflux (inward current I increasing with hyperpolarization) is always measured for a membrane potential V_m more negative than -100 mV. The apparent affinity $K_{0.5}$ of extracellular Na^+ is ~30 mM between -160 and $+60$ mV (Sciortino and Romero 1999). In general, decreasing $[Na^+]_o$ in this physiologic membrane potential range enables NBC1 predominantly to efflux $NaHCO_3$ from the cell.

Stoichiometry

In their original work on the electrogenic Na/HCO_3 cotransporter of the salamander proximal tubule Boron and Boulpaep (1983) demonstrated that the cotransporter moves more HCO_3^- than Na^+. Based on measurements of pH_i, V_m and intracellular Na^+ activity, they made a thermodynamic suggestion that the stoichiometry should be at least $1Na^+ : 2HCO_3^-$. Nonetheless, they could not rule out the possibility of higher couplings (e.g., 1 : 3). Later work on Necturus proximal tubules suggested, again on thermodynamic grounds, that the $Na^+:HCO_3^-$ stoichiometry had to be at least 1 : 3 (Lopes et al. 1987).

Soleimani et al. (1987) reasoned that the net transport direction depends on both the $Na^+ : HCO_3^-$ coupling ratio and the electrochemical gradients for Na^+ and HCO_3^-. Altering these gradients allowed measurement of net transport direction. These workers concluded that the renal electrogenic Na^+/HCO_3^- cotransporter must have a 1 : 3 stoichiometry. Three models could account for this apparent 1 : 3 cotransporter stoichiometry: (a) Na^+ plus 3 HCO_3^-, (b) Na^+ plus HCO_3^- plus CO_3^{2-}, or (c) the $NaCO_3^-$ ion pair and HCO_3^-. Using isolated proximal tubules, two groups have suggested that the renal electrogenic Na^+/HCO_3^- cotransporter can alter its stoichiometry from 1 : 3 to 1 : 2, thus changing the net direction of net HCO_3^- transport (Planelles et al. 1993; Seki et al. 1983).

Even though these data on native renal cells or native-cell-derived materials at "physiological" conditions indicates a 1 : 3 stoichiometry, the $Na^+ : HCO_3^-$ coupling ratio was not directly measured. By permeabilizing the apical membrane of monolayers of proximal tubule cell-lines with nystatin, Gross and Hopfer (1996) found a linear voltage dependence on the DNDS-inhibitable short-circuit current across the epithelia basolateral membrane. Xenopus oocyte experiments with giant patches (Heyer et al. 1999) or two-electrode voltage–clamp experiments (Sciortino and Romero 1999) of rkNBC1 show a voltage dependence of both inward and outward NBC1 transport (i.e., larger outward current with depolarization, or larger inward current with hyperpolarization). Additionally, these studies revealed a 1Na$^+$: 2HCO$_3^-$ stoichiometry. This result is surprising, since human NBC1 mutations imply that NBC1 is the major HCO$_3^-$ exit pathway back to the blood for the proximal tubule and the kidney in general (Igarashi *et al.* 1999). That is, a putative accessory protein or chemical modification must modify NBC1 stoichiometry in the renal proximal tubule. Experiments by Gross *et al.* (2001a,b) have demonstrated that there is a cell context to NBC1/NBCe1-A/B expression. In fact, these investigators found that even pNBC1 expression in proximal tubule cells has a stoichiometry of 1Na$^+$: 3HCO$_3^-$ (Gross *et al.* 2001b). Mutational analysis indicates that phosphorylation of S932 via PKA is likely the "molecular switch" (Gross *et al.* 2001c). Other investigators have not found a similar dependence on PKA of NBC1 isoforms.

MUTATIONS IN HUMAN NBC1 (SLC4A4)

If NBC1 were mutated, the most obvious manifestation should be a metabolic acid–base disturbance. Igarashi *et al.* (1999) identified renal patients with mutations in NBC1/SLC4a4. The mutations (R298S and R510H; see Figure 3) were recessive. The R298S mutation (A→C transversion) resulted from consanguineous parents, while R510H (G→A transversion) were nonconsanguineous. Phenotypically, these patients are of short stature, have a permanent proximal renal tubular acidosis (type 2 RTA), mental retardation, bilateral cataracts, bilateral glaucoma, and band keratopathy. The pRTA is severe with blood pH ~7.05 and blood [HCO$_3^-$] ~5–8 mM, rather than the normal values of 7.35–7.45 and ~23–29 mM, respectively. These patients also manifest an obvious eye phenotype: bilateral glaucoma, bilateral cataracts, and band keratopathy (Igarashi *et al.* 1999). For other organ systems expressing

other NBC1 isoforms (Table 1), a disease phenotype is not obvious. In fact, amylase levels are within the normal range, indicating normal exocrine pancreas function.

Figures 3 and 4 indicate that R298S (R324S in pNBC1/NBCe1-B/C) occurs in the predicted N-terminus of NBC1. R510H (R554H in pNBC1/NBCe1-B/C) is predicted at the extracellular face of transmembrane domain TM4.

This finding by Igarashi and colleagues made several important contributions. First, as was predicted, mutations/deficiencies in NBC1 would disturb acid–base homeostasis. Secondly, the degree and chronic nature of the metabolic acidosis indicates that NBC1 is the major HCO$_3^-$ reabsorptive pathway not only of the proximal tubule but of the kidney in general. Thirdly, HCO$_3^-$ transport is crucial for normal fluid balance in the eye and clarity of the lens. Finally, there was an indication that normal NBC1 transport is important for normal CNS development and/or function.

Soon afterwards Dinour *et al.* (2000) discovered another missense mutation, S427L (1280C→T). This Israeli patient was also of short stature and had a pure pRTA, glaucoma, and cataracts. This patient was also female and the parents were nonconsanguineous.

Figures 3 and 4 indicate that S427L (or S471L in pNBC1/NBCe1-B/C) occurs at the beginning of predicted transmembrane span 1 (TM1).

A subsequent study by Igarashi *et al.* (2001) identified a novel nonsense mutation, Q29X (C→T), only found in kNBC1/NBCe1-A. This female patient had pRTA (blood pH 7.20, [HCO$_3^-$] = 9.4 mM, P CO$_2$ = 32 mmHg), short stature, mental retardation and bilateral glaucoma. Interestingly, this patient did not have cataracts or band keratopathy. Q29X removes kNBC1 while leaving pNBC1 (NBCe1-B) and the predicted brain isoforms (NBCe1-C; see Figure 4) intact. Thus, Q29X allows a separation of the NBC1-deficient phenotype.

Interestingly, the kidney seems to express all the identified NBC1 mRNAs and proteins.

In keeping with this observation, renal disease is the major phenotype of human NBC1/SLC4a4 mutations (Igarashi *et al.* 1999, 2001; Dinour *et al.* 2000). However, what is the molecular nature of the disease associated with SLC4a4? Do the mutations cause a biophysical change in cotransporter activity or do they result in a cellular protein processing problem?

The original report indicated that transfection of ECV304 cells with R298S- or R510H-NBC1 resulted in about 50% wild-type activity assayed by fluorescent monitoring of intracellular pH (Igarashi *et al.* 1999). For a more direct examination, R510H-NBC1 was expressed in *Xenopus* oocytes and activity was evaluated by electrophysiology (Sciortino 2001; Sciortino and Romero 2001). These studies indicate that R510H is a charge-dependent plasma membrane trafficking defect (Sciortino 2001). In a preliminary report, other investigators found that R554H (pNBC1/NBCe1-B) transfected into proximal tubule cells has the same DNDS-sensitive current as the wild-type pNBC1 (Gross *et al.* 2000).

As indicated, R298S apparently has reduced pH_i recovery activity (Igarashi *et al.* 1999). Again, the R342S-pNBC1 mutation in proximal tubule cells appears to have wild-type activity (Gross *et al.* 2000). More detailed information was provided in a preliminary report using *Xenopus* oocyte electrophysiology (Romero and Smith 2002). These investigators indicated that the apparent affinity for HCO_3^- is left-shifted and V_{max} is reduced. It remains unclear how this mutation in the cytoplasmic N-terminus of NBC1 modifies HCO_3^- transport. Perhaps a critical accessory protein interaction is disrupted, a protein interaction inappropriately occurs, or membrane accessibility is altered. This R→S mutation should be found in all NBC1 isoforms.

S427L occurs at the predicted start of TM1 in hkNBC1 (Figures 3 and 4). Only one preliminary study has examined transport mediated by S427L-hkNBC1 (Romero and Smith 2002). Like R298S this S427L mutation appears to change the biophysics of ion transport. S427L has very low activity, though protein synthesis is not dramatically different from wild-type hkNBC1. This mutation also appears to affect Na^+ transport, implying that at least part of the Na^+ transport conduit lies in TM1. This S→L mutation should be found in all NBC1 isoforms.

As indicated above, Q29X only affects kNBC1 (NBCe1-A/SLC4a4-A). Since this mutation results in loss of kNBC1 protein, no functional characterization is necessary.

Do any of these mutations remove NBC1 from the kidney? Since there are only four patients worldwide with these reported deficiencies, tissue is not available to answer such a question. All four mutations are rather subtle amino acid changes. Even Q29X-kNBC1 spares the other isoforms of NBC1, perhaps contributing to some minor renal function. The renal phenotype of persistent proximal renal tubular acidosis is directly related to NBC1's HCO_3^- absorptive role in the kidney. In all cases the proximal RTA is confirmed by the increased urine pH and urine bicarbonate wasting as systemic HCO_3^- is raised. Presumably the short patient stature is due to the chronic acidosis, which is generally known to promote bone loss since bone is a major buffer store (Shapiro and Kaehny 1997).

Three major eye pathologies are associated with SLC4a4 mutations (see Box: Pathophysiology of NBC1/SLC4a4): (a) bilateral glaucoma, (b) bilateral cataracts, and (c) band keratopathy. Glaucoma is increased pressure in the anterior chamber of the eye. Increased pressure damages the optic nerve and then produces characteristic defects in the peripheral vision or visual field. The increased pressure is due to inappropriate ocular fluid transport. Ocular transport studies have shown that ion-coupled fluid transport by the corneal endothelium maintains corneal hydration and transparency (Wigham and Hodson 1981). This process is dependent on the presence of HCO_3^- and is sensitive to carbonic anhydrase inhibitors (CAIs) (Fischbarg and Lim 1974; Hodson 1974; Hodson and Miller 1976; Kuang *et al.* 1990; Riley *et al.* 1995). HCO_3^-

PATHOPHYSIOLOGY OF NBC1/ SLC4A4

Persistent proximal renal tubular acidosis (pRTA):

- blood pH < 7.1
- blood [HCO$_3^-$] $= 3$–11 mM

Ocular pathologies:

- bilateral cataracts (cornea)
- bilateral glaucoma (retinal layer or endothelia)
- band keratopathy (Ca^{2+} & Mg^{2+} deposits in lens/cornea)

Short stature (acidosis related)
Enamel defects of permanent teeth (sometimes)
Calcification of basal ganglia (sometimes)
Apparently normal amylase levels = normal pancreatic function
Normal cardiac function
Neural activity (mental retardation linked to SLC4a4 in neurons?)
Hypokalemia ([K$^+$] $= 2.5$–3.5 mEq/L)

also plays a key role in regulation of endothelial cell intracellular pH$_i$ (Bonanno and Gasson 1992). That is, loss of corneal HCO$_3^-$ transport – decreased/impaired NBC1 function – is hypothesized to decrease corneal transparency (cataract) and disturb corneal pH.

Igarashi *et al.* (2002) have proposed a model accounting for the eye pathologies associated with SLC4a4 mutations. NBC1 is localized to the corneal endothelium, the conjunctiva epithelium, and the lens epithelium (Sun *et al.* 2001; Turner *et al.* 2001; Usui *et al.* 2001). These investigators hypothesize that NBC1 mediates net fluid transport in corneal and lens epithelia. Na$^+$, HCO$_3^-$, and water are transported out of the cornea into the aqueous humor (Hodson and Miller 1976; Igarashi *et al.* 2002). By inactivating or limiting the activity of NBC1, HCO$_3^-$ in the corneal stroma would increase, thereby increasing pH. The elevated pH and perhaps CO$_3^{2-}$ would facilitate Ca^{2+} and Mg^{2+} precipitation and deposition in Bowman's membrane (i.e., band keratopathy).

The Q29X nonsense mutation may prove to be quite informative. As stated, Q29X results in glaucoma without cataracts or band keratopathy. This patient is also female, has persistent pRTA, short stature, and mental retardation (Igarashi *et al.* 2001). Superficially, since Q29X only affects kNBC1/NBCe1-A, the lens opacity (cataracts) and Ca^{2+}/Mg^{2+} deposits (band keratopathy) appear to be associated with expression of pNBC1/NBCe1-B/C. NBC1 is known to form dimers (Maunsbach *et al.* 2000). In a more complicated scenario, the loss of putative heterodimeric NBC1 may be associated with the disease phenotype.

The most elusive of the associated NBC1/SLC4a4 phenotypes is mental retardation. It is tempting to speculate that there is a direct link to NBC1/SLC4a4. Transport and acid–base buffering by CNS glial cells (Chesler and Craig 1987; Deitmer and Schlue 1987; Schlue and Deitmer, 1988), and astrocytes are known to regulate nervous activity (Chesler and Kaula 1992; Deitmer 2002). This is well studied in the hippocampus where electrogenic Na$^+$/HCO$_3^-$ cotransporter activity has been measured (Bevensee *et al.* 1997a, b). Nevertheless, there was no physiologic evidence for this activity in neurons. Molecular biology and antibodies to NBC1 have revealed that this conclusion was in error (Bevensee *et al.* 2000; Schmitt *et al.* 2000). In fact, there appears to be a developmental regulation of NBC1 in the neurons and the CNS in general (Giffard *et al.* 2000; Douglas *et al.* 2001). Whether these data imply a direct link to the mental retardation observed in these patients with NBC1 mutations, is unclear. Almost certainly a better understanding of the cell and tissue specific physiology and protein interactions will be the key to answering this question.

FUTURE DIRECTIONS AND SUMMARY

As more genomes are sequenced and species variants of NBC1 are discovered, we will likely learn much more about *Homo sapiens* and the role of NBC1. Protein interactions with NBC1 will undoubtedly result in an integration of cellular transport and biochemistry. NBC1 is found throughout mammalian tissues. NBC1, like so many of our "favorite proteins," will likely be found to have several protein partners mediating specialized cellular functions. Indeed, NBC1 is postulated to have an accessory role in facilitating CFTR's role as a Cl^- and HCO_3^- channel in CaLu-3 cells (Devor *et al.* 1999). Others believe that NHERF1 might also regulate NBC1 activity (Bernardo *et al.* 1999). Most recently, carbonic anhydrases have been demonstrated to interact with discrete portions of the NBC1 sequence (Alvarez *et al.* 2002; Gross *et al.* 2002) or function (Giffard *et al.* 2000).

Additionally, molecular and immunologic reagents will allow investigators to study HCO_3^- transport processes more readily. NBC1 localization in specific tissues of a variety of species will be necessary to generate new cellular models for ion transport and acid–base homeostasis. In addition, the physiology of several tissues is being reinvestigated to integrate the role of NBC1. Importantly, we can learn from the phenotypes of those unfortunate individuals who have life-threatening and debilitating NBC1 mutations.

REFERENCES

Abuladze, N., Lee, I., Newman, D., Hwang, J., Boorer, K., Pushkin, A., *et al.* (1998). Molecular Cloning, Chromosomal Localization, Tissue Distribution, and Functional Expression of the Human Pancreatic Sodium Bicarbonate Cotransporter. *J. Biol. Chem.*, **273**, 17689–17695.

Abuladze, N., Song, M., Pushkin, A., Newman, D., Lee, I., Nicholas, S., *et al.* (2000). Structural Organization of the Human NBC1 Gene: kNBC1 Is Transcribed from an Alternative Promoter in Intron 3. *Gene*, **251**, 109–122.

Alpern, R.J. (1985). Mechanism of Basolateral Membrane $H^+/OH^-/HCO_3^-$ Transport in the Rat Proximal Convoluted Tubule. A Sodium-Coupled Electrogenic Process. *J. Gen. Physiol.*, **86**, 613–636.

Alpern, R.J., and Rector, F.C., Jr. (1996). Renal Acidification Mechanisms. In B.M. Brenner and F.C. Rector Jr (eds), *The Kidney* (5th ed.), Philadelphia, PA: W.B. Saunders.

Alvarez, B.V., Loiselle, F.B., and Casey, J.R. (2002) Direct Extracellular Interaction Between Carbonic Anhydrase IV and the Human Heart NBC1 Na^+/HCO_3^- Cotransporter. *J. Mol. Cell. Cardiol.*, **34**, A19.

Amlal, H., Wang, Z., Burnham, C., and Soleimani, M. (1998). Functional Characterization of a Cloned Human Kidney $Na^+ : HCO_3^-$ Cotransporter. *J. Biol. Chem.*, **273**, 16810–16815.

Bernardo, A.A., Kear, F.T., Santos, A.V.P., Ma, J., Steplock, D., Robey, R.B., *et al.* (1999). Basolateral Na^+/HCO_3^- Cotransport Activity is Regulated by the Dissociable Na^+/H^+ Exchanger Regulatory Factor. *J. Clin. Invest.*, **104**, 195–201.

Bevensee, M.O., Weed, R.A., and Boron, W.F. (1977a). Intracellular pH Regulation in Cultured Astrocytes from Rat Hippocampus. I. Role of HCO_3. *J. Gen. Physiol.*, **110**, 453–465.

Bevensee, M.O., Apkon, M., and Boron, W.F. (1997b). Intracellular pH Regulation in Cultured Astrocytes from Rat Hippocampus. II. Electrogenic Na/HCO_3 Cotransport. *J. Gen. Physiol.*, **110**, 467–483.

Bonanno, J.A., and Giasson, C. (1992). Intracellular pH Regulation in Fresh and Cultured Bovine Corneal Endothelium. II. $Na^+ : HCO_3^-$ cotransport and Cl^-/HCO_3^- Exchange. *Invest. Ophthalmol. Vis. Sci.*, **33**, 3068–3079.

Boron, W.F. and Boulpaep, E.L. (1983). Intracellular pH Regulation in the Renal Proximal Tubule of the Salamander. Basolateral HCO_3^- Transport. *J. Gen. Physiol.*, **81**, 53–94.

Boron W.F. and Boulpaep, E.L. (1989). The Electrogenic Na/HCO_3 Cotransporter. *Kidney Int.*, **36**, 392–402.

Boron, W.F., Fong, P., Hediger, M.A., Boulpaep, E.L., and Romero, M.F. (1997). The Electrogenic Na/HCO_3 Cotransporter. *Wien. Klin. Wochenschr.*, **109**, 445–456.

Burnham, C.E., Amlal, H., Wang, Z., Shull, G.E., and Soleimani, M. (1997). Cloning and Functional

Expression of a Human Kidney Na$^+$:HCO$_3^-$ Cotransporter. *J. Biol. Chem.*, **272**, 19111–19114.

Burnham, C.E., Flagella, M., Wang, Z., Amlal, H., Shull, G.E., and Soleimani, M. (1998). Cloning, Renal Distribution, and Regulation of the Rat Na$^+$–HCO$_3^-$ Cotransporter. *Am. J. Physiol.*, **274**, F1119–1126.

Carlin, R.W., Quesnell, R.R., Zheng, L., Mitchell, K.E., and Schultz, B.D. (2002). Functional and Molecular Evidence for Na$^+$–HCO Cotransporter in Porcine vas Deferens Epithelia. *Am. J. Physiol. Cell. Physiol.*, **283**, C1033–C1044.

Chesler, M. and Kaila, K. (1992). Modulation of pH by Neuronal Activity. *Trends Neurosci.*, **15**, 396–402.

Chesler, M. and Kraig, R.P. (1987). Intracellular pH of Astrocytes Increases Rapidly with Cortical Stimulation. *Am. J. Physiol.*, **253**, R666–R670.

Choi, I., Romero, M.F., Khandoudi, N., Bril, A., and Boron, W.F. (1999). Cloning and Characterization of an Electrogenic Na/HCO$_3$ Cotransporter from Human Heart. *Am. J. Physiol.*, **274**, C576–584.

Deitmer, J.W. (2002). A Role for CO$_2$ and Bicarbonate Transporters in Metabolic Exchanges in the Brain. *J. Neurochem.*, **80**, 721–726.

Deitmer, J.W. and Schlue, W.R. (1987). The Regulation of Intracellular pH by Identified Glial Cells and Neurones in the Central Nervous System of the Leech. *J. Physiol. (London)*, **388**, 261–283.

Devor, D.C., Singh, A.K., Lambert, L.C., DeLuca, A., Frizzell, R.A., and Bridges, R.J. (1999). Bicarbonate and Chloride Secretion in Calu-3 Human Airway Epithelial Cells. *J. Gen. Physiol.*, **113**, 743–760.

Dinour, D., Knecht, A., Serban, I., and Holtzman, E.J. (2000). A Novel Missense Mutation in the Sodium Bicarbonate Cotransporter (NBC-1) causes Congenital Proximal Renal Tubular Acidosis with Ocular Defects. *J. Am. Soc. Nephrol.*, **11**, A0012.

Douglas, R.M., Schmitt, B.M., Xia, Y., Bevensee, M.O., Biemesderfer, D., Boron, W.F., *et al.* (2001). Sodium–hydrogen Exchangers and Sodium-Bicarbonate Cotransporters: Ontogeny of Protein Expression in the Rat Brain. *Neuroscience*, **102**, 217–228.

Fischbarg, J. and Lim, J.J. (1974). Role of Cations, Anions and Carbonic Anhydrase in Fluid Transport Across Rabbit Corneal Endothelium. *J. Physiol.*, **241**, 647–675.

Giebisch, G. and Windhager, E. (2003). Transport of Acids and Bases. In W.F. Boran and E.L. Boulpaep (eds), *Medical Physiology*, Philadelphia, PA: W.B. Saunders, pp. 845–860.

Giffard, R.G., Papadopoulos, M.C., van Hooft, J.A., Xu, L., Giuffrida, R., and Monyer, H. (2000). The Electrogenic Sodium Bicarbonate Cotransporter: Developmental Expression in Rat Brain and Possible Role in Acid Vulnerability. *J. Neurosci.*, **20**, 1001–1008.

Grichtchenko, I.I. and Boron, W.F. (2002). Evidence for CO$_3^{2-}$ Transport by NBCe1, Based on Surface-pH Measurements in Voltage-Clamped *Xenopus* Oocytes Co-expressing NBCe1 and CAIV. *FASEB J.*, **16**, A795–A796.

Grichtchenko, I.I., Romero, M.F., and Boron, W.F. (1996). Extracellular Bicarbonate Dependence of aNBC, the Electrogenic Na/HCO$_3$ Cotransporter Cloned from Tiger Salamander (*Ambystoma tigrinum*). *J. Am. Soc. Nephrol.*, **7**, 1255.

Grichtchenko, I.I., Romero, M.F., and Boron, W.F. (1998). Electrogenic Na/HCO$_3$ Cotransporters (NBC) from Rat and Salamander Kidney Have Similar External HCO$_3^-$ Dependencies. *FASEB J.*, A638.

Grichtchenko, I.I., Romero, M.F., and Boron, W.F. (2000). Extracellular HCO$_3$ Dependence of Electrogenic Na/HCO$_3$ Cotransporters (NBC) Cloned from Salamander and Rat Kidney. *J. Gen. Physiol.*, **115**, 533–546.

Gross, E. and Hopfer, U. (1996). Activity and Stoichiometry of Na$^+$:HCO^{3-} Cotransport in Immortalized Renal Proximal Tubule Cells. *J. Membr. Biol.*, **152**, 245–252.

Gross, E.Z., Abuladze, N., Pushkin, A., and Kurtz, I. (2000). R342S and R554H Mutations do not Alter the Functional Activity or Stoichiometry of pNBC1. *J. Am. Soc. Nephrol.*, **11**,

Gross, E., Abuladze, N., Pushkin, A., Kurtz, I., and Cotton, C.U. (2001a). The Stoichiometry of the Electrogenic Sodium Bicarbonate Cotransporter pNBC1 in Mouse Pancreas Duct Cells is 2 HCO$_3^-$: 1 Na$^+$. *J. Physiol.*, **531**, 375–382.

Gross, E., Hawkins, K., Abuladze, N., Pushkin, A., Cotton, C.U., Hopfer, U., *et al.* (2001b). The Stoichiometry of the Electrogenic Sodium Bicarbonate Cotransporter NBC1 is Cell-Type Dependent. *J. Physiol.*, **531**, 597–603.

Gross, E., Hawkins, K., Pushkin, A., Sassani, P., Dukkipati, R., Abuladze, N., *et al.* (2001c). Phosphorylation of Ser(982) in the Sodium Bicarbonate Cotransporter kNBC1 Shifts the HCO$_3^-$: Na$^+$ Stoichiometry from 3 : 1 to 2 : 1 in Murine Proximal Tubule Cells. *J. Physiol.*, **537**, 659–665.

Gross, E., Pushkin, A., Abuladze, N., Fedotoff, O., and Kurtz, I. (2002). Regulation of the Sodium Bicarbonate Cotransporter kNBC1 Function: Role of Asp(986), Asp(988) and kNBC1-Carbonic Anhydrase II Binding. *J. Physiol.*, **544**, 679–685.

Heyer, M., Müller-Berger, S., Romero, M.F., Boron, W.F., and Frömter, E. (1999). Stoichiometry of Rat Kidney Na–HCO$_3$ Cotransporter (rkNBC) expressed in *Xenopus laevis* Oocytes. *Pflügers Arch.*, **438**, 322–329.

Hodson, S. (1974). The Regulation of Corneal Hydration by a Salt Pump Requiring the Presence of Sodium and Bicarbonate Ions. *J. Physiol.*, **236**, 271–302.

Hodson, S. and Miller, F. (1976). The Bicarbonate Ion Pump in the Endothelium Which Regulates the Hydration of Rabbit Cornea. *J. Physiol.*, **263**, 563–577.

Igarashi, T., Inatomi, J., Sekine, T., Cha, S.H., Kanai, Y., Kunimi, M., *et al.* (1999). Mutations in SLC4A4 Cause Permanent Isolated Proximal Renal Tubular

Acidosis with Ocular Abnormalities. *Nat. Genet.*, **23**, 264–266.

Igarashi, T., Inatomi, J., Sekine, T., Takeshima, Y., Yoshikawa, N., and Endou H. (2000). A Nonsense Mutation in the Na^+/HCO_3^- Cotransporter Gene (SLC4A4) in a Patient with Permanent Isolated Proximal Renal Tubular Acidosis and Bilateral Glaucoma. *J. Am. Soc. Nephrol.*, **11**, A0573.

Igarashi, T., Inatomi, J., Sekine, T., Seki, G., Shimadzu, M., Tozawa, F., *et al.* (2001). Novel Nonsense Mutation in the Na^+/HCO_3^- Cotransporter Gene (SLC4A4) in a Patient with Permanent Isolated Proximal Renal Tubular Acidosis and Bilateral Glaucoma. *J. Am. Soc. Nephrol.*, **12**, 713–718.

Igarashi, T., Sekine, T., Inatomi, J., and Seki G. (2002). Unraveling the Molecular Pathogenesis of Isolated Proximal Renal Tubular Acidosis. *J. Am. Soc. Nephrol.*, **13**, 2171–2177.

Jensen, L.J., Schmitt, B.M., Berger, U.V., Nsumu, N.N., Boron, W.F., Hediger, M.A., *et al.* (1999). Localization of Sodium Bicarbonate Cotransporter (NBC) Protein and Messenger Ribonucleic Acid in Rat Epididymis. *Biol. Reprod.*, **60**, 573–579.

Jentsch, T.J., Keller, S.K., Koch, M., and Wiederholt, M. (1984). Evidence for Coupled Transport of Bicarbonate and Sodium in Cultured Bovine Corneal Endothelial Cells. *J. Membr. Biol.*, **81**, 189–204.

Jentsch, T.J., Schill, B.S., Schwartz, P., Matthes, H., Keller, S.K., and Wiederholt, M. (1985). Kidney Epithelial Cells of Monkey Origin (BSC-1) Express a Sodium Bicarbonate Cotransport. Characterization by $^{22}Na^+$ Flux Measurements. *J. Biol. Chem.*, **260**, 15554–15560.

Kuang, K.Y., Xu, M., Koniarek, J.P., and Fischbarg, J. (1990). Effects of Ambient Bicarbonate, Phosphate and Carbonic Anhydrase Inhibitors on Fluid Transport across Rabbit Corneal Endothelium. *Exp. Eye Res.*, **50**, 487–493.

Loiselle, F.B. and Casey, J.R. (2002). Potentiation of Bicarbonate Transport Activity by Direct Interaction of NBC3 Sodium Bicarbonate Co-transporter with Carbonic Anhydrase. *Biophys. J.*, **82**, 571A.

Lopes, A.G., Siebens, A.W., Giebisch, G., and Boron, W.F. (1987). Electrogenic Na/HCO_3 Cotransport across Basolateral Membrane of Isolated Perfused *Necturus* Proximal Tubule. *Am. J. Physiol.* **253**, F340–F350.

Marino, C.R., Jeanes, V., Boron, W.F., and Schmitt, B.M. (1999). Expression and Distribution of the Na^+/HCO_3^- Cotransporter in Human Pancreas. *Am. J. Physiol.*, **277**, G487–G494.

Maunsbach, A.B., Vorum, H., Kwon, T.H., Nielsen, S., Simonsen, B., Choi, I., *et al.* (2000). Immunoelectron Microscopic Localization of the Electrogenic Na/HCO_3 Cotransporter in Rat and *Ambystoma* Kidney. *J. Am. Soc. Nephrol.*, **11**, 2179–2189.

Planelles, G., Thomas, S.R., and Anagnostopoulos, T. (1993). Change of Apparent Stoichiometry of

Proximal-Tubule $Na^+-HCO_3^-$ Cotransport Upon Experimental Reversal of its Orientation. *Proc. Natl. Acad. Sci. USA*, **90**, 7406–7410.

Praetorius, J., Hager, H., Nielsen, S., Aalkjaer, C., Friis, U.G., Ainsworth, M.A., and Johansen, T. (2001). Molecular and Functional Evidence for Electrogenic and Electroneutral $Na^+-HCO_3^-$ Cotransporters in Murine Duodenum. *Am. J. Physiol. Gastrointest. Liver Physiol.*, **280**, G332–G343.

Riley, M.V., Winkler, B.S., Czajkowski, C.A., and Peters, M.I. (1995). The Roles of Bicarbonate and CO_2 in Transendothelial Fluid Movement and Control of Corneal Thickness. *Invest. Ophthalmol. Vis. Sci.*, **36**, 103–112.

Romero, M.F. (2001). The Electrogenic Na^+/HCO_3^- Cotransporter, NBC. *JOP*, **2**, 182–191.

Romero, M.F. and Smith, B.L. (2002). Mutations in the Na^+/HCO_3^- Cotransporter Result in Biophysical Alterations Causing Renal Tubular Acidosis and Ocular Pathologies. *FASEB J.*, **16**, A52–A53.

Romero, M.F., Hediger, M.A., Boulpaep, E.L., and Boron, W.F. (1996a). Expression Cloning of the Renal Electrogenic Na/HCO_3 Cotransporter (NBC-1) from *Ambystoma tigrinum*. *FASEB J.*, **10**, A89.

Romero, M.F., Hediger, M.A., Boulpaep, E.L., and Boron, W.F. (1996b). Physiology of the Cloned *Ambystoma tigrinum* Renal Electrogenic Na/HCO_3 Cotransporter (aNBC): II. Localization and Na^+ Dependence. *J. Am. Soc. Nephrol.*, p. 1260.

Romero, M.F., Hediger, M.A., Boulpaep, E.L., and Boron, W.F. (1997a). Expression Cloning and Characterization of a Renal Electrogenic Na^+/HCO_3^- Cotransporter. *Nature*, **387**, 409–413.

Romero, M.F., Hediger, M.A., Fong, P., and Boron, W.F. (1997b). Expression of the Rat Renal Electrogenic Na/HCO_3 Cotransporter (rNBC). *Proc. 14th International Society of Nephrology*.

Romero, M.F., Davis, B.A., Sussman, C.R., Bray-Ward, P., Ward, D., and Boron, W.F. (1998a). Identification of Multiple Genes for Human Electrogenic Na/HCO^3 Cotransporters (NBC) on 4q. *J. Am. Soc. Nephrol.*, p. 11A.

Romero, M.F., Fong, P., Berger, U.V., Hediger, M.A., and Boron, W.F. (1998b). Cloning and Functional Expression of rNBC, an Electrogenic $Na^+-HCO_3^-$ Cotransporter from Rat Kidney. *Am. J. Physiol.*, **274**, F425–F432.

Romero, M.F., Sussman, C.R., Choi, I., Hediger, M.A., and Boron, W.F. (1998c). Cloning of an Electrogenic Na/HCO^3 Cotransporter (NBC) Isoform from Human Kidney and Pancreas. *J. Am. Soc. Nephrol.*, p. 11A.

Romero, M.F., Sciortino, C.M., Roussa, E., and Thévenod, F. (1999). Na^+/HCO_3^- Cotransporters in Various Species and Organs. N. DeSanto (ed.), *Proc. 3rd Annual Borelli Conf.: Acid–Base Balance*.

Rossmann, H., Bachmann, O., Vieillard-Baron, D., Gregor, M., and Seidler, U. Na^+/HCO_3^- Cotransport and

Expression of NBC1 and NBC2 in Rabbit Gastric Parietal and Mucous Cells. *Gastroenterology*, **116**, 1389–1398.

Roussa, E., Romero, M.F., Schmitt, B.M., Boron, W.F., Alper, S.L., and Thévenod, F. (1999). Immuno-localization of AE2 Anion Exchanger and Na$^+$–HCO$_3^-$ Cotransporter in Rat Parotid and Submandibular Glands. *Am. J. Physiol. – GI*, **277**, G1288.

Sasaki, S., Shiigai, T., Yoshiyama, N., and Takeuchi, J. (1987). Mechanism of Bicarbonate Exit across Basolateral Membrane of Rabbit Proximal Straight Tubule. *Am. J. Physiol.*, **252**, F11–F18.

Schlue, W.R. and Deitmer, J.W. (1988). Ionic Mechanisms of Intracellular pH Regulation in the Nervous System. *Ciba Found. Symp.*, **139**, 47–69.

Schmitt, B.M., Biemesderfer, D., Boulpaep, E.L., Romero, M.F., and Boron, W.F. (1999). Immuno-localization of the Electrogenic Na$^+$/HCO$_3^-$ Cotrans-porter in Mammalian and Amphibian Kidney. *Am. J. Physiol.*, **276**, F27–F36.

Schmitt, B.M., Berger, U.V., Douglas, R.M., Bevensee, M.O., Hediger, M.A., Haddad, G.G., *et al.* (2000). Na/HCO$_3$ Cotransporters in Rat Brain: Expression in Glia, Neurons, and Choroid Plexus. *J. Neurosci.*, **20**, 6839–6848.

Sciortino, C.M. (2001). *Characterization and Localization of the Sodium Mediated Bicarbonate Transporters NBC and NDAE1*, Unpublished Ph.D. Thesis, Case Western Reserve University, Cleveland, OH.

Sciortino, C.M., and Romero, M.F. (1998). Na$^+$ and Voltage Dependence of the Rat Kidney Electrogenic Na/HCO^3 Cotransporter (rkNBC) Expressed in *Xenopus* Oocytes. *J. Am. Soc. Nephrol.*, **9**, 12A.

Sciortino, C.M. and Romero, M.F. (1999). Cation and Voltage Dependence of Rat Kidney, Electrogenic Na$^+$/HCO$_3^-$ Cotransporter, rkNBC, Expressed in oocytes. *Am. J. Physiol.*, **277**, F611–F623.

Sciortino, C.M. and Romero, M.F. (2001). Functional Characterization of a Na$^+$/HCO$_3^-$ Cotransporter Point Mutation that Results in Type II Renal Tubular Acidosis. *FASEB J.*, A502.

Seki, G., Coppola, S., and Fromter, E. (1993). The Na$^+$–HCO$_3^-$ Cotransporter Operates with a Coupling Ratio of 2 HCO$_3^-$ to 1 Na$^+$ in Isolated Rabbit Renal Proximal Tubule. *Pflugers Arch.*, **425**, 409–416.

Shapiro, J.I. and Kaehny, W.D. (1997). Pathogenesis and Management of Metabolic Acidosis and Alkalosis. In R.W. Schrier (ed.), *Renal and Electrolyte Disorders* (5th edn), Philadelphia, PA: Lippincott–Raven, pp. 130–171.

Soleimani, M., Grassi, S.M., and Aronson, P.S. (1987). Stoichiometry of Na$^+$–HCO$_3^-$ Cotransport in Basolateral Membrane Vesicles Isolated from Rabbit Renal Cortex. *J. Clin. Invest.*, **79**, 1276–1280.

Soleimani, M., Lesoine, G.A., Bergman, J.A., and Aronson, P.S. (1991). Cation Specificity and Modes

of the Na$^+$:CO$_3^{2-}$:HCO$_3^-$ Cotransporter in Renal Basolateral Membrane Vesicles. *J. Biol. Chem.*, **266**, 8706–8710.

Sterling, D., Reithmeier, R.A., and Casey, J.R. (2001). A Transport Metabolon. Functional Interaction of Carbonic Anhydrase II and Chloride/Bicarbonate Exchangers. *J. Biol. Chem.*, **276**, 47886–47894.

Sterling, D., Alvarez, B.V., and Casey, J.R. (2002). The Extracellular Component of a Transport Metabolon. Extracellular Loop 4 of the Human AE1 Cl$^-$/HCO$_3^-$ Exchanger Binds Carbonic Anhydrase IV. *J. Biol. Chem.*, **277**, 25239–25246.

Sun, X.C., Bonanno, J.A., Jelamskii, S., and Xie, Q. (2000). Expression and Localization of Na$^+$–HCO$_3^-$ Cotransporter in Bovine Corneal Endothelium. *Am. J. Physiol. Cell. Physiol.*, **279**, C1648–C1655.

Sun, X.C., Sciortino, C.M., Sato, J-I., Romero, M.F., and Bonanno, J.A. (2001). Identification and cloning of the Na/HCO$_3^-$ cotransporter (NBC) in human corneal endothelium. *Invest. Ophthalmol. Vis. Sci.*, submitted for publication.

Thévenod, F., Roussa, E., Schmitt, B.M., and Romero, M.F. (1999). Cloning and Immunolocalization of a Rat Pancreatic Na$^+$ Bicarbonate Cotransporter. *Biochem. Biophys. Res. Commun.*, **264**, 291–298.

Turner, H.C., Alvarez, L.J., and Candia, O.A. (2001). Identification and Localization of Acid–Base Transporters in the Conjunctival Epithelium. *Exp. Eye Res.*, **72**, 519–531.

Usui, T., Seki, G., Amano, S., Oshika, T., Miyata, K., Kunimi, M., *et al.* (1999). Functional and Molecular Evidence for Na$^+$–HCO$_3^-$ Cotransporter in Human Corneal Endothelial cells. *Pflügers Arch.*, **438**, 458–462.

Usui, T., Hara, M., Satoh, H., Moriyama, N., Kagaya, H., Amano, S., *et al.* (2001). Molecular Basis of Ocular Abnormalities Associated with Proximal Renal Tubular Acidosis. *J. Clin. Invest.*, **108**, 107–115.

Vince, J.W. (2000). *Interaction of Chloride/Bicarbonate Anion Exchangers with Carbonic Anhydrase II*. Unpublished Ph.D. Thesis, University of Toronto.

Vince, J.W. and Reithmeier, R.A. (2000). Identification of the Carbonic Anhydrase II Binding Site in the Cl$^-$/HCO$_3^-$ Anion Exchanger AE1. *Biochemistry*, **39**, 5527–5533.

Vince, J.W. and Reithmeier, R.A.F. (1998). Carbonic Anhydrase II Binds to the Carboxyl Terminus of Human Band 3, the Erythrocyte Cl$^-$/HCO$_3^-$ Exchanger. *J. Biol. Chem.*, **273**, 28430–28437.

Vince, J.W., Carlsson, U., and Reithmeier, R.A. (2000). Localization of the Cl$^-$/HCO$_3^-$ Anion Exchanger Binding Site to the Amino-Terminal Region of Carbonic Anhydrase, II. *Biochemistry*, **39**, 13344–13349.

Wigham, C. and Hodson, S. (1981). Bicarbonate and the Trans-endothelial Short Circuit Current of Human Cornea. *Curr. Eye Res.*, **1**, 285–290.

TRANSPORTERS OF TRACE ELEMENTS AND BIOMINERALS

5

CARSTEN A. WAGNER*

Introduction

Small inorganic molecules like sulfate and phosphate or trace elements, mainly metals including copper, iron, zinc, selenium, chromium, and iodine, are taken up from the food in only small quantities but are essential components of enzymes, hormones, energy substrates, and nucleic acids, or for the formation of extracellular matrix in bone (phosphate) or cartilage (sulfate). The absolute requirement for these substrates and their often low concentrations in food has led to the evolution of very efficient transport systems into the body and for the distribution to their target organs. Loss of these substrates into urine is prevented by reabsorptive transport systems localized in the kidney. However, the uncontrolled accumulation and rise in intra- and extracellular concentration of many of these substrates is also associated with toxicity, making it necessary to regulate tightly the uptake and distribution in the body. Dysfunction of transport systems or receptors involved in uptake, distribution, or sensing/controlling can lead either to a deficiency or intoxication with the respective substrates, as exemplified below for the metals copper and iron. The identification of several candidate genes involved in inherited diseases of trace element homeostasis has led to the discovery of new transporter proteins such as pendrin, a Cl^-/anion exchanger involved in iodide transport, or the copper-transporting ATPases ATP7A and

ATP7B in Menke's disease and Wilson's disease, respectively. The investigation of their function and regulation gave new insights into the complex regulatory systems governing the homeostasis of trace elements.

This introductory chapter will describe three examples of transport systems, their physiological role and regulation, as well as the pathophysiology caused by inherited mutations. The examples include the systems involved in the transport of iodide, namely pendrin (SLC26A4) (Chapter 6) and the Na/I symporter NIS (SLC5A5), which are both defective in different subsets of inherited syndromes of *hypothyroidism*, transport ATPases for copper mutated in *Menke's disease* or *Wilson's disease* (ATP7A, ATP7B), and transporters involved in iron homeostasis as identified in patients or animal models (SLC11A2, SLC11A3) (Chapters 7 and 8).

PHYSIOLOGY OF IODIDE TRANSPORT

Iodine (I) is an essential component of the thyroid peptide hormones triiodothyronine (T_3) containing three iodine atoms per molecule, and thyroxin (tetraiodothyronine, T_4), containing four iodine atoms per molecule. About 120–150 μg of iodide need to be taken up from food daily. The actions of thyroid hormones include an increase in basal cellular and organ metabolic activity, that is, increased protein and cholesterol synthesis, stimulation of gluconeogenesis, and stimulation of cardiac

* University of Zürich, Institute of Physiology, Winterthurerstrasse 190, Zürich 8057, Switzerland

function, and, very importantly, T_3 and T_4 are essential for normal physical and mental development. Deficiency of both T_3 and T_4 during development leads to retardation of growth and intellectual functions as seen in cretinism (profound mental retardation, short stature, retarded motor development). Hypothyroidism due to dietary iodide deficiency is endemic in many parts of the world and is associated in mild cases with the development of goiter due to increased levels of thyroid-stimulating hormone (TSH) stimulating thyroid gland growth. In severe cases the classical symptoms of hypothyroidism may develop, such as goiter, skin symptoms, peripheral edema, constipation, fatigue, reduced cardiac function, and infertility in women.

Iodide (I^-) is taken up from the food, most likely involving a Na^+-dependent transport system in the small intestine, and reaches the thyroid gland. Uptake from the blood (defined as basolateral side) into the thyrocytes occurs via the Na^+-dependent iodide transporter NIS (SLC5A5) accumulating iodide intracellularly 10–100 fold (De La Vieja *et al.* 2000) (Figure 1).

Iodide then is released into the lumen of follicles formed by the thyrocytes (defined as apical membrane) where most steps of the synthesis of T_3 and T_4 occur. The transport into the follicular lumen is achieved by the Cl^-/anion exchanger pendrin (SLC26A4; see Chapter 6) (Scott *et al.* 1999). In the follicular lumen, iodide is oxidized to iodine (I) by the thyroid peroxidase and incorporated into tyrosines bound to thyroglobulin that has been synthesized by the thyrocytes and is secreted into the follicular lumen. Upon iodination of the tyrosine residues, monoiodotyrosine and diiodotyrosines are formed, which by coupling to each other via an ether bond form either triiodothyronine or tetraiodothyronine still bound to thyroglobulin. The thyroglobulin T_3 or T_4 complex is stable and also provides the storage form for thyroid hormones. In order to be secreted, the T_3 and T_4 thyroglobulin complex undergoes endocytosis at the apical membrane of the thyrocytes, the thyroglobulin backbone is cleaved off, and T_3 or T_4 are released into blood through processes which are not yet completely defined. The LAT-1

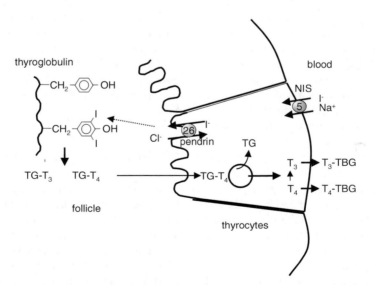

Figure 1 Iodide transport in the thyroid gland. Iodide is taken up from blood through NIS (SLC5A5), exported into the follicule via pendrin (SLC26A4), and coupled onto tyrosine residues bound to thyroglobulin (TG). After reaction of mono- and diiodotyrosines T_3 and T_4 are formed. The T_3–TG and T_4–TG complexes are endocytosed, TG is cleaved off, and T_3 and T_4 are released into blood.

amino acid transporter of the heterodimeric amino acid transporter family (see Chapter 14) has been shown to transport T_3 and T_4 with high affinity and may be involved in the transport of these hormones. Before release from the thyrocytes T_4 may be deiodinized to T_3 by a deiodinase. In blood, both T_4 and T_3 are bound to carrier proteins such as thyroxin binding globulin, prealbumin, or albumin with a ratio of $1:1000$ of free to bound T_3 and T_4. Only the free T_3 and T_4 are biologically active.

T_3 and T_4 production and systemic release are under the tight control of several hormones, the most important being thyrotropin-releasing hormone (TRH) and TSH. TRH is mainly produced in the hypothalamus and stimulates the synthesis and release of TSH from the anterior pituary gland. TSH in turn stimulates thyrocyte function and T_3 and T_4 synthesis and release in several ways, such as increasing iodide uptake, iodination of thyroglobulins, conjugation, and endocytosis, and release of T_3 and T_4 from thyrocytes into blood. The effect of TSH on iodide uptake is mediated by an increase in expression of the Na^+/iodide transporter NIS (De La Vieja et al. 2000).

Pharmacological intervention with thyroid hormone production is usually targeting either the basolateral uptake of iodide from blood into the thyrocytes through the Na^+-dependent iodide transporter NIS or the follicular oxidation of iodide to iodine. Drugs containing thiocyanate or perchlorate interact with NIS and reduce iodide transport without being transported (De La Vieja et al. 2000). Other substrates which also inhibit NIS include nitrate, rhodanite, and dinitrophenols. The use of radioactive iodine (^{131}I) that is readily taken up by thyrocytes where it accumulates and induces cell death, specifically in thyrocytes due to its short-range radiation (γ-radiation), is clinically important. This form of therapy is mainly used for cancer of the thyroid gland or in some cases of goiter.

Beside inborn defects in the genes encoding for thyroglobulin or thyroperoxidase, inherited forms of hypothyroidism due to defects in both steps of iodide transport across the thyrocytes have been identified. Mutations in either the basolateral iodide uptake system NIS (SLC5A5) or in the apical iodide release system pendrin (SLC26A4) lead to a reduction in thyroid hormone synthesis (Everett and Green 1999; Scott et al. 1999; De La Vieja et al. 2000). However, despite involvement in the same metabolic pathway, the two diseases differ in their symptoms due to differential localization and function of the transporters in other organs. Patients carrying mutations in the NIS transporter suffer mainly from *congenital hypothyroidism* with elevated levels of TSH leading to goiter. The condition of congenital hypothyroidism due to mutations in the NIS gene is rare and inherited in an autosomal recessive pattern. Treatment of patients includes supplementation with thyroid hormones; in some cases high doses of iodide may be helpful (De La Vieja et al. 2000). In contrast, patients with mutations in the pendrin gene are characterized by the combination of *hypothyroidism* and *inner ear deafness* (Everett et al. 1997; Scott et al. 1999). Studies in a mouse model deficient for the pendrin gene show an early defect in inner ear development (Everett et al. 2001). In addition, pendrin expression is also found in some specialized cells of the renal cortical collecting duct and may be involved there in bicarbonate secretion (Royaux et al. 2001). However, no defect in kidney function has been described in patients to date, which may be due to the fact that bicarbonate secretion is only activated in metabolic alkalosis. The function of pendrin is described further in Chapter 6.

THE PHYSIOLOGY OF COPPER AND ITS TRANSPORT

Copper is an essential trace element and is required for the activity of several cuproproteins (copper-containing proteins). All cuproproteins are cuproenzymes involved in transfer of electrons to oxygen, such as cytochrome *c* oxidase (respiratory chain), ascorbic acid oxidase, and superoxide dismutase (SOD), producing peroxides which are then disprotonated to form water and oxygen by the copper-containing protein catalase. Furthermore, cuproenzymes are involved in linking collagen and elastin (lysyl-6-oxidase) which contribute to the elasticity and stability of connective tissue, catecholamine synthesis (dopamine-β-hydroxylase), or pigmentation of skin (tyrosinase) (Llanos and Mercer 2002; Lutsenko and Petris 2003). The daily requirement for copper is 1–2 mg, which is reabsorbed in the upper part of the small intestine through mechanisms which have not yet been fully elucidated. After absorption, copper is predominantly bound to ceruloplasmin and albumin in blood. Ceruloplasmin is not only the main copper-binding protein in blood but is activated by copper and also has an important role in iron metabolism. A reduction in ceruloplasmin activity due to copper deficiency may cause iron deficiency and microcytic anaemia, as seen in *Wilson's disease*. The mechanisms of cellular copper uptake have not been fully elucidated. Recently, CTR1 (SLC31A1), a copper-transporting protein, was cloned by complementation in yeast, and was shown to transport copper and to be essential for copper transport in a mouse knockout model. Mice lacking CTR1 die during embryonic development (Zhou and Gitschier 1997; Lee *et al.* 2001). After cellular uptake, copper is bound to so-called copper chaperones, such as ATOX1, CCS, or COX17, which shuttle copper to mitochondria or the trans-Golgi network (TGN) (Llanos and Mercer 2002; LutsenKo and Petris 2003). Binding to chaperones reduces the toxicity of copper. Copper is then taken up into the TGN and incorporated into cuproproteins. Excess copper is secreted from the liver into bile and is excreted with feces. Both intestinal uptake and biliary excretion are regulated according to copper availability, with increased intestinal uptake with low copper and stimulation of copper excretion with high body copper concentrations.

Mutations in the two highly homologous copper-transporting ATPases ATP7A (MNKP) and ATP7B (WNDP) cause two diseases with distinct pathology: Menke's disease and Wilson's disease, characterized by copper deficiency in Menke's disease and copper overload of some organs in Wilson's disease (Bull *et al.* 1993; Tanzi *et al.* 1993; Vulpe *et al.* 1993; Kaler *et al.* 1994).

Menke's disease is a rare X-linked recessive inherited condition caused by mutations in the gene encoding for ATP7A (Vulpe *et al.* 1993; Kaler *et al.* 1994). Intestinal copper absorption as well as transport across the blood–brain barrier is impaired. As ATP7A is also expressed in the placenta, the copper deficiency affects fetal development. Patients suffer from severe copper deficiency and usually die early in childhood due to neurological deficits (head control, visual tracking), convulsions, skeletal abnormalities, skin laxity, and brain blood vessel malformations. However, there are two milder variants of Menke's disease called *occipital horn syndrome* (OHS) and *mild Menke's disease* where patients survive into adulthood but suffer from several symptoms such as skeletal abnormalities, weakness of connective tissue (skin, blood vessels), and reduced intelligence in OHS, and mild neurological symptoms in mild Menke's disease (Kaler *et al.* 1994; Llanos and Mercer 2002; Lutsenko and Petris 2003).

According to the phenotype seen in patients, ATP7A is predominantly expressed on the basolateral side of enterocytes along the small intestine, in the endothelial cells of the blood–brain barrier, in the placenta, and to a much lower extent in many other tissues (Figure 2A). On the cellular level, ATP7A is found in two distinct cellular compartments, the TGN and the plasma membrane (Figure 2B). The subcellular localization is copper dependent. At high copper levels, the ATP7A pump traffics into the plasma membrane where it mediates copper efflux; at low copper levels ATP7A resides in the TGN. The expression of ATP7A in the TGN is thought to mediate copper transport into the secretory pathway, and it has been shown that the synthesis and activity of copper-containing secreted proteins is reduced in cells derived from patients with ATP7A mutations (Lutsenko and Petris 2003). Thus it appears that ATP7A serves two distinct functions: providing copper for the synthesis of secreted copper-containing proteins and controlling intracellular copper concentrations exporting copper upon a rise of intracellular copper. These two functions would indeed

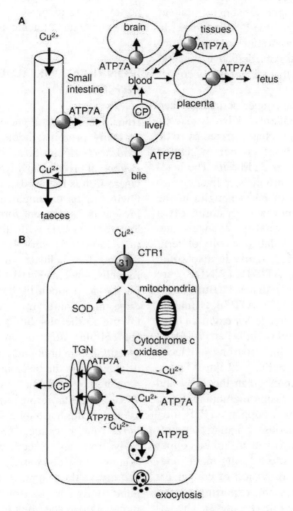

Figure 2 Role of ATP7A and ATP7B in copper transport and metabolism. (A) Role of ATP7A and ATP7B in intestinal copper absorption and redistribution between several organs. The liver plays a central role as it controls overall copper levels by adjusting biliary excretion of copper to intestinal uptake. Copper is transported in blood bound to ceruloplasmin to its target organs or incorporated into cuproproteins (CP) and secreted into blood. (B) Intracellular trafficking of copper after uptake via the CTR1 transporter. Copper is bound to copper chaperones and redistributed to several cuproenzymes, the mitochondria, or the TGN where it is taken up by ATP7A and ATP7B and incorporated into cuproproteins (CP) to be secreted. Rising intracellular copper concentrations cause trafficking of ATP7A to the plasma membrane and of ATP7B to vesicles. Copper is either pumped out of the cell via ATP7A or released from vesicles via exocytosis after intravesicular accumulation by ATP7B. Adapted from Llanos and Mercer (2002) and Lutsenko and Petris (2003).

explain many of the symptoms seen in patients suffering from Menke's disease. Systemic and brain copper deficiency are caused by a reduced reabsorption of copper in the small intestine due to a defective release of copper into blood, and this is aggravated in the brain as copper cannot cross the blood–brain barrier. In addition, copper incorporation into cuproproteins is reduced because of a defective transport step in the TGN such that the necessary copper is not provided.

In contrast with the copper deficiency seen in Menke's disease patients, *Wilson's disease* (autosomal recessive) is characterized by toxic effects of high levels of copper in several organs, particularly liver and brain. The leading symptoms include progressive liver cirrhosis and degeneration of basal ganglia in the brain, inducing tremor and dystonia often associated with personality changes and depression. Owing to reduced activity of ceruloplasmin, microcytic anemia is also often seen. Mutations in ATP7B (WNDP) are causative for Wilson's disease (Bull *et al.* 1993; Tanzi *et al.* 1993). ATP7B is mainly expressed in liver and to a lesser extent in most other organs (Lutsenko and Petris 2003). Copper excretion into bile is strongly reduced in Wilson's disease patients. Like ATP7A, ATP7B is mainly expressed in the TGN and possibly also on the plasma membrane. Upon increases of intracellular copper, ATP7B travels to a submembraneous vesicular compartment that may be involved in either copper storage or secretion into bile. Its main functions seem to be the provision of copper for cuproprotein synthesis and exporting copper from hepatocytes into bile (Figure 2). The latter function is important to control copper levels and keep it below toxic levels (Llanos and Mercer 2002; Lutsenko and Petris 2003).

Thus it appears that ATP7A and ATP7B serve complementary functions: ATP7A provides for reabsorption of copper, whereas ATP7B keeps copper levels below toxic concentrations by excreting it into bile. In addition, both proteins are involved in the synthesis of copper-containing proteins, but with differing patterns of organ expression. Menke's and Wilson's diseases may therefore be seen as the two sides of a balanced system controlling systemic and cellular copper levels, and provide a good example of two similar transporters being involved in the transport of the same substrate but achieving distinct effects.

IRON TRANSPORT AND IRON PHYSIOLOGY

Iron, like copper, is an essential component of many proteins: hemoglobin, myoglobin, oxidases, cytochromes (respiratory chain), or in general proteins involved in redox reactions where iron is involved in the electron transfer. Owing to its chemical nature, iron may be present in its divalent form Fe(II) or its trivalent form Fe(III) with distinct chemical features. Fe(II) is very reactive with oxygen, forming free radicals which are very harmful for cells, whereas Fe(III) is much less reactive. Fe(II) has a much higher solubility and can cross membranes more easily than Fe(III). Owing to the absolute requirement for iron for cellular function and its high toxicity potential in an unbound free form, iron uptake, distribution in the body, and storage are tightly regulated by several mechanisms.

Daily dietary iron intake ranges between 10 and 15 mg but only about 1 mg is absorbed in the small intestine. After having entered the body, iron can be recycled between different organs. Total body iron bound to proteins is usually in the range of 3–5 g. On the other hand, daily loss of iron is also about 1 mg, mainly due to shedding of epithelial cells containing iron into the small intestine. In menstruating women, iron loss may be in the range of 4–5 mg/day. In contrast with copper metabolism, there is no regulated process to excrete iron from the body, presumably because of the constant loss of iron. Thus iron metabolism must be regulated at the level of uptake.

Iron can be absorbed in the small intestine in several chemical forms: as free divalent Fe(II), bound to amino acids, bile acids, and organic acids, or even as part of larger protein complexes such as heme. After passing through the stomach (acidic pH), most of the iron reaching the duodenum – the major site of iron absorption – is in the oxidized trivalent form which can be only poorly absorbed. Iron bound to proteins (heme) can be absorbed through a receptor-mediated endocytotic process. Trivalent iron is then released through the action of the heme oxygenase. However, this mechanism accounts for only a small fraction of iron absorption. The majority of iron uptake occurs through a transporter-mediated process involving at least three components on the apical membrane: iron-binding proteins, iron-reducing enzymes, and iron transporters. The so-called paraferritin proteins (β-integrin, mobilferrin, and a flavin monooxygenase) present in the apical brush border membrane of duodenal enterocytes may take up iron from the digestive tract. The bound iron can either be taken up by endocytosis (small fraction) or is available for reduction to its divalent form by the membrane-bound enzyme ferrireductase (also known as duodenal cytochrome b or dcyt$b$). Ascorbic acid plays an important role in iron reduction, even though the exact molecular mechanism is not fully understood. Unlike rodents, humans have an absolute requirement for ascorbic acid from their diet (rodents are able to synthesize ascorbic acid), a fact that has to be kept in mind when comparing iron absorption in rodents and humans. Thus ascorbic acid deficiency in humans promotes iron deficiency by reducing intestinal iron absorption. After reduction, Fe(II) is available for transport by the apical divalent metal ion transporter DMT1 (SLC11A2; also known as Nramp2 or DCT1) (Figure 3A). The function and associated diseases of this transporter are described in more detail in Chapter 7. On the basolateral side the ferroportin transporter (FPN1, SLC11A3; also known as SLC40A1) mediates release of Fe^{2+} into blood (Figure 3A). Mutations in the ferroportin transporter are associated with *hemochromatosis type IV* (HFE4) as described in Chapter 8. Intracellularly, iron may be stored in enterocytes by binding to apoferritin, yielding ferritin. Intestinal iron absorption is regulated by iron availability on several levels. The abundance of the apical transport system is iron regulated, being increased in iron deficiency [increased translation due to iron responsive elements (IREs)]. In addition, a small peptide, hepcidin, has been identified that binds iron and decreases intestinal iron absorption. Patients with liver adenomas producing large quantities of hepcidin develop anemia (Kaplan 2002).

Upon release into blood, Fe(II) is rapidly oxidized by either the membrane-bound hephestin (a copper-containing oxidase) or the soluble ceruloplasmin and bound for transport to apotransferrin, forming transferrin. As evident from knockout mouse models and patients suffering from mutations in the ceruloplasmin gene, loss of hephestin or ceruloplasmin results in severe iron deficiency, underlying their importance (Kaplan 2002). Only about one third of total apotransferrin is loaded with iron; the other two thirds remains free. Once bound to transferrin, iron is distributed within the body to several target and storage organs, the most important of which are the liver and the reticulo-endothelial system (Figure 3B). Eventually about 70–90% of overall iron is incorporated into hemoglobin. Most uptake into cells occurs as iron–transferrin complex recognized by a specific transferrin receptor. The iron–transferrin–transferrin receptor complex undergoes endocytosis and, owing to the acidic pH in endosomes and an endosomal ferrireductase, iron, transferrin, and the transferrin receptor dissociate. Transferrin and its receptor are recycled, whereas iron is released from endosomes via the DMT1 (SLC11A2, Nramp2) transporter, and is subsequently bound to apoferritin and stored until further use in protein synthesis. A second transferrin-receptor-independent pathway for cellular iron uptake has recently been suggested by the discovery of novel iron-binding proteins NGAL/M24p3 which are able to deliver iron to cells. NGAL/M24p3 is also taken up via endocytosis but is trafficked to late

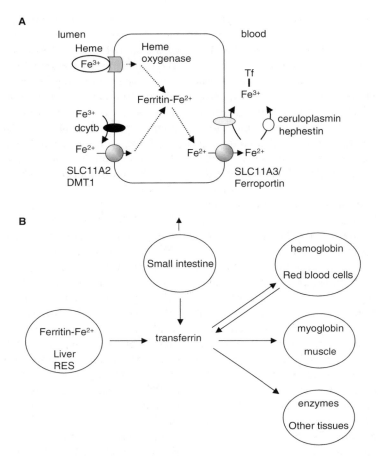

Figure 3 Iron transport and metabolism. (A) Absorption of iron across the apical membrane of duodenal enterocytes via the DMT1 transporter after reduction by the duodenal cytochrome *b* (dcyt*b*) reductase. Some iron is reabsorbed as heme complex, dissociated, and reduced by heme oxygenase. Intracellular Fe^{2+} is bound to ferritin and released into blood via the ferroportin transporter. After oxidation by hephestin (membrane-bound) and soluble ceruloplasmin, Fe^{3+} is bound to transferrin (Tf) for transport in blood. (B) Body pools of iron: iron is absorbed or lost in the small intestine, and absorbed iron is bound to transferrin in blood for distribution to target or storage organs. Most iron is stored in the liver or reticulo-endothelial system (RES), whereas functional iron is mainly incorporated into hemoglobin, myoglobin, or iron-containing proteins such as a number of enzymes. Iron released from senescent red blood cells re-enters the active pools.

endosomes in contrast with the early endosomes used by the transferrin receptor pathway (Kaplan 2000). The full physiological significance of this pathway remains to be further elucidated but could explain residual transferrin-receptor-independent iron uptake. A small amount of iron may also be stored intracellularly as hemosiderin. The amount of hemosiderin is increased in iron-overload diseases and is pathognomic for these diseases (liver biopsies are used to diagnose iron overload based on large granula containing hemosiderin). In iron overload, the largest iron deposits are usually found in liver, and to a lesser extent in pancreas and skin. Liver cirrhosis, pancreatic insufficiency (diabetes mellitus), and a hypermelanotic pigmentation of the skin resulted in the name "bronzed diabetes" for this disease.

Several forms of inherited iron overload (hemochromatosis) have been identified in the past years, affecting different components of the iron transporting and storing system. Based on clinical, biochemical, and genetic data, at least five distinct forms of inherited hemochromatosis, affecting different parts of the iron controlling system, can be distinguished at present. In addition, the very rare syndromes of aceruloplasminemia and atransferrinemia may also cause iron overload syndromes (Griffiths and Cox 2000; Pietrangelo 2002). The most common form of hemochromatosis (HFE1) is due to mutations in the *HFE* gene, an MHC class I like protein, associated with the β_2-microglobulin protein and the transferrin receptor (Griffiths and Cox 2000; Pietrangelo 2002). *HFE* mutations in the adult autosomal recessive form of hemochromatosis are the basis of the most common inherited recessive disorder in Caucasians, with an estimated heterozygosity frequency between 0.1% and 10% in different populations (Griffiths and Cox 2000). A rarer form of adult recessive hemochromatosis is due to mutations in the transferrin receptor 2 gene (*HFE*3). The more aggressive but rare form of juvenile hemochromatosis (HFE2) is caused by mutations in the hepcidin gene. Mutations in the basolateral intestinal iron exit pathway through ferroportin [SLC11A3 (SLC40A1)] underlie HFE4, an autosomal dominant form of hemochromatosis as described in Chapter 8.

The pathophysiology of iron overload due to mutations in the genes described above is not fully understood. A hallmark of iron overload syndromes is the increased intestinal absorption of iron despite full body iron stores. An increased abundance of DMT1 (SLC11A2) protein and stimulated function was noticed in HFE-induced hemochromatosis along with a reduced iron content of enteroyctes, suggesting that a regulatory signal may be missing that would limit iron absorption (Griffiths and Cox 2000). Indeed, the ratio of iron-loaded to iron-free transferrin seems to convey an important signal to enterocytes regulating iron absorption. This signal is sensed by the transferrin receptor in enterocytes. HFE interacts with the transferrin receptor, and the affinity for the iron–transferrin complex is much lower in the presence of HFE, suggesting that loss of HFE may alter the interaction between transferrin and its receptor. Loss of HFE would signal to enterocytes that iron deficiency exists and stimulate DMT1-mediated iron absorption (Pietrangelo 2002).

A second interesting topic in iron transport, which is not related to the classic function of iron, should be briefly mentioned: the potential role of an iron transporter Nramp1 (SLC1A1) in immune defense. Nramp1 was originally identified as a mouse gene conferring resistance of macrophages against infection with *Mycobacterium tuberculosis* (hence the name *n*atural *r*esistance-*a*ssociated *m*acrophage *p*rotein). Nramp1 is strongly expressed on the phagosome membrane of macrophages and may be involved in the export of iron and other metal ions from the bacteria containing phagosome. It has been proposed that Nramp1 would thereby restrict essential metal ions for bacteria survival. After phagocytosis, the macrophage produces reactive oxygen and nitrogen intermediates toxic for the bacteria. In order to neutralize these toxins, the bacteria must use metal-containing enzymes. Nramp1 would deplete bacteria from these metals (Nelson 1999). No disease due to dysfunction of Nramp1 has yet been identified. However, there is great deal of interest in utilizing Nramp1 function as an alternative means to combat the spreading resistance of bacteria to conventional treatments.

REFERENCES

Bull, P.C., Thomas, G.R., Rommens, J.M., Forbes, J.R., and Cox, D.W. (1993). The Wilson Disease Gene Is a Putative Copper Transporting P-Type ATPase Similar to the Menkes Gene. *Nat. Genet.*, **5**, 327–337.

De La Vieja, A., Dohan, O., Levy, O., and Carrasco, N. (2000). Molecular Analysis of the Sodium/Iodide Symporter: Impact on Thyroid and Extrathyroid Pathophysiology. *Physiol. Rev.*, **80**, 1083–1105.

Everett, L.A., Belyantseva, I.A., Noben-Trauth, K., Cantos, R., Chen, A., Thakkar, S.I., *et al.* (2001). Targeted Disruption of Mouse Pds Provides Insight About the Inner-Ear Defects Encountered in Pendred Syndrome. *Hum. Mol. Genet.*, **10**, 153–161.

Everett, L.A., Glaser, B., Beck, J.C., Idol, J.R., Buchs, A., Heyman, M., *et al.* (1947). Pendred Syndrome Is Caused by Mutations in a Putative Sulphate Transporter Gene (PDS). *Nat. Genet.*, **17**, 411–422.

Everett, L.A. and Green, E.D. A Family of Mammalian Anion Transporters and Their Involvement in Human Genetic Diseases. *Hum. Mol. Genet.*, **8**, 1883–1891.

Griffiths, W. and Cox, T. (2000). Haemochromatosis: Novel Gene Discovery and the Molecular Pathophysiology of Iron Metabolism. *Hum. Mol. Genet.*, **9**, 2377–2382.

Kaler, S.G., Gallo, L.K., Proud, V.K., Percy, A.K., Mark, Y., Segal, N.A., Goldstein, D.S., *et al.* (1994). Occipital Horn Syndrome and a Mild Menkes Phenotype Associated with Splice Site Mutations at the MNK Locus. *Nat. Genet.*, **8**, 195–202.

Kaplan, J. (2002). Mechanisms of Cellular Iron Acquisition: Another Iron in the Fire. *Cell*, **111**, 603–606.

Kuo, Y.M., Zhou, B., Cosco, D., and Gitschier, J. (2001). The Copper Transporter CTR1 Provides an Essential Function in Mammalian Embryonic Development. *Proc. Natl Acad. Sci. USA*, **98**, 6836–6841.

Lee, J., Prohaska, J.R., and Thiele, D.J. (2001). Essential Role for Mammalian Copper Transporter Ctr1 in Copper Homeostasis and Embryonic Development. *Proc. Natl Acad. Sci. USA*, **98**, 6842–6847.

Llanos, R.M. and Mercer, J.F. (2002). The Molecular Basis of Copper Homeostasis and Copper-Related Disorders. *DNA Cell Biol.*, **21**, 259–270.

Lutsenko, S. and Petris, M.J. (2003). Function and Regulation of the Mammalian Copper-Transporting ATPases: Insights from Biochemical and Cell Biological Approaches. *J. Membr. Biol.*, **191**, 1–12.

Nelson, N. (1999). Metal Ion Transporters and Homeostasis. *EMBO J.*, **18**, 4361–4371.

Pietrangelo, A. (2002). Physiology of Iron Transport and the Hemochromatosis Gene. *Am. J. Physiol. Gastrointest. Liver Physiol.*, **282**, G403–G414.

Royaux, I.E., Wall, S.M., Karniski, L.P., Everett, L.A., Suzuki, K., Knepper, M.A., *et al.* (1999). Pendrin, Encoded by the Pendred Syndrome Gene, Resides in the Apical Region of Renal Intercalated Cells and Mediates Bicarbonate Secretion. *Proc. Natl Acad. Sci. USA*, **98**, 4221–4226.

Scott, D.A., Wang, R., Kreman, T.M., Sheffield, V.C., and Karniski, L.P. (1999). The Pendred Syndrome Gene Encodes a Chloride–Iodide Transport Protein. *Nat. Genet.*, **21**, 440–443.

Tanzi, R.E., Petrukhin, K., Chernov, I., Pellequer, J.L., Wasco, W., Ross, B., *et al.* (1993). The Wilson Disease Gene is a Copper Transporting ATPase with Homology to the Menkes Disease Gene. *Nat. Genet.*, **5**, 344–350.

Vulpe, C., Levinson, B., Whitney, S., Packman, S., and Gitschier, J. (1993). Isolation of a Candidate Gene for Menkes Disease and Evidence that it Encodes a Copper-Transporting ATPase. *Nat. Genet.*, **3**, 7–13.

Zhou, B. and Gitschier, J. hCTR1: A Human Gene for Copper Uptake Identified by Complementation in Yeast. *Proc. Natl Acad. Sci. USA*, **94**, 7481–7486.

6

DANIEL MARKOVICH*

Anion exchangers DTDST (SLC26A2), DRA (SLC26A3), and pendrin (SLC26A4)

INTRODUCTION

This chapter is devoted to a group of structurally related membrane transporters that lead to a variety of different diseases affecting different organs in the body arising from defects in cell surface (plasma membrane) anion transport. This family of proteins, designated "sulfate transporters" based on their function, belongs to the Human Genome Nomenclature Committee (HGNC) designated Solute Linked Carrier (SLC) family 26, of which presently (at the time of writing) there are 11 siblings, designated SLC26 members A1–A11. These proteins encode anion exchangers that share significant sequence identity, they are expressed in a variety of organisms and tissues, including orthologs in higher and lower eukaryotes, and their functional activities seem to be very promiscuous in their substrate specificities, with some of them being able to transport substrates including sulfate, chloride, bicarbonate, iodide, oxalate, and formate (in no particular order), of which the list may not be complete. Furthermore, studies being performed at the

time of writing are demonstrating that some of these transporters may in fact be electrogenic, proposing a revision of the classical dogma that these anion exchangers may in fact not be electroneutral. This review will focus primarily on the structural and functional information of the genes encoding DTDST (SLC26A2), DRA (SLC26A3), and pendrin (SLC26A4), the functional properties of the proteins they encode, and their links to the several unrelated diseases with varying pathophysiology which result from mutations in these genes leading to defects in ion transport function.

DTDST (SLC26A2)

Physiological function of DTDST

By positional cloning, a gene was identified on human chromosome 5q to be linked to an autosomal recessive osteochondrodysplasia called diastrophic dysplasia (DTD; OMIM 222600), with clinical features (see below) including dwarfism, spinal deformation, and abnormalities of the joints (Hästbacka et al. 1994), vocal tract (Karlstedt et al. 1998), dentition (Karlstedt et al. 1997), and craniofacial features (Karlstedt et al. 1996). The human gene

---

* School of Biomedical Sciences, Department of Physiology and Pharmacology, University of Queensland, Brisbane, QLD 4072, Australia

linked to DTD was found to contain four exons and three introns spanning over 40 kb (Clines and Lovett 1996) and was localized to human chromosome 5q32 (Hästbacka et al. 1994). Sequence comparison of its coding region was found to share overall 44% amino acid identity (Table 1) with sat-1 (Markovich 2001), a gene localized on human chromosome 4p16 (Scott et al. 1990; Russel et al. 1996) encoding a high-affinity sulfate transporter strongly expressed in liver and kidney (Bissig et al. 1994; Markovich et al. 1994). Owing to the high structural similarity with sat-1 and the fact that sat-1 encodes a sulfate/bicarbonate anion exchanger (Bissig et al. 1994; Markovich et al. 1994), the exact function of the gene linked to DTD was sought. Primary skin fibroblasts obtained from normal, carrier, and DTD patients were cultured and assayed for sulfate transport activity (Hästbacka et al. 1994): cells from normal controls and carriers showed saturable sulfate transport, whereas cells from the DTD patients showed a greatly diminished sulfate uptake (Hästbacka et al. 1994), suggesting that the wild-type gene encodes a functional sulfate transporter whose transport activity is abolished when mutated, as in DTD patients. Thus the newly characterized gene (cDNA and protein product) linked to DTD became referred to as the diastrophic dysplasia sulfate transporter (DTDST) (Hästbacka et al. 1994), more recently designated SLC26A2 by HGNC. The DTDST gene product was shown to be a plasma membrane sulfate/chloride exchanger (Satoh et al. 1998), and inactivation of its function via DTDST mutations led to intracellular sulfate depletion and thereby synthesis of undersulfated proteoglycans in chondrocytes and fibroblasts. Translation of DTDST cDNA produced a protein of 739 amino acids, with a predicted molecular mass of 82 kDa, containing 12 transmembrane domains (TMDs) based on hydrophobicity analysis (Hästbacka et al. 1994). DTDST mRNA expression was ubiquitous, detected by Northern blot analysis in all tissues of the body (Hästbacka et al. 1994). The rat DTDST orthologous gene and cDNA (dtdst) were cloned more recently, and the encoding protein was shown to function as an Na^+-independent sulfate transporter, sensitive to chloride, thiosulfate, oxalate, and DIDS when expressed in Xenopus oocytes (Satoh et al. 1998). The mouse DTDST cDNA ortholog (originally named st-ob for sulfate transporter in osteoblasts) was induced in osteoblast precursor cells in association with osteoblast differentiation (Kobayashi et al. 1997). st-ob was isolated by subtractive hybridization, using immature mouse fibroblastic cells C3H10T1/2, that were either untreated or treated with bone morphogenetic protein-2 (BMP-2), which induced an osteoblast-like phenotype. st-ob was found to be induced by BMP-2 which is constantly expressed in osteoblastic cells (Kobayashi et al. 1997). Tissue distribution showed the st-ob mRNA to be strongly expressed in the thymus, testis, calvaria, and the osteoblastic MC3T3-E1 cells, having a reduced expression in the undifferentiated C3H10T1/2 cells. The expression of st-ob mRNA in C3H10T1/2 cells was shown to be increased by transforming growth factor-β_1 (TGF-β_1), retinoic acid, and dexamethasone, as well as BMP-2 (Kobayashi et al. 1997). BMP-2 also led to a twofold increase in sulfate uptake in C3H10T1/2 cells when compared with untreated (control) cells, suggesting that st-ob encodes a functional sulfate transporter whose activity was upregulated by BMP-2. Since osteoblasts actively take up sulfate to synthesize proteoglycans, being major components of the extracellular matrix of bone and cartilage, the functional significance of the st-ob/dtdst protein as a sulfate transporter may be of great importance for osteoblastic differentiation.

DTDST mRNA is expressed ubiquitously and its protein shares strong amino acid identity (44%) with the sat-1 sequence (Bissig et al. 1994; Markovich 2001; Lee et al. 2003; Regeer et al. 2003) (Table 1). Skin fibroblasts obtained from DTD patients demonstrated a reduced sulfate transport activity when compared with normal subjects, suggesting that the wild-type DTDST protein encodes a functional sulfate transporter (Hästbacka et al.

Table 1 The mammalian sulfate anion transporter family SLC26 members A1–A4

Protein	Species	Protein function	Protein size	Overall % amino acid identity[a]	Sulfate transporter signature % identity[b]	DDBJ/EMBL/ GenBank accession no.	Chromosome location	HGNC[c] gene name
rsat-1	*Rattus norvegicus*	Sulfate anion transporter (sulfate/carbonate antiporter)	703	100	100	P45380	ND	Slc26a1
msat-1	*Mus musculus*	Sulfate anion transporter (sulfate/carbonate antiporter)	704	94	100	AY093420	5	Slc26a1
hsat-1	*Homo sapiens*	Sulfate anion transporter (sulfate/carbonate antiporter)	701	77	100	AY124771	4p16	SLC26A1
DTDST	*Homo sapiens*	Sulfate transporter (diastrophic dysplasia protein)	739	44	77	P50443	5q32	SLC26A2
dtdst	*Rattus norvegicus*	Sulfate transporter (diastrophic dysplasia protein)	739	44	81	O70531	ND	Slc26a2
dtdst (st-ob)	*Mus musculus*	Sulfate transporter (diastrophic dysplasia protein)	739	44	81	Q62273	ND	Slc26a2
PENDRIN	*Homo sapiens*	Sodium-independent chloride/iodide transporter	780	32	60	O43511	7q22-q31.1	SLC26A4
pendrin	*Rattus norvegicus*	Sodium-independent chloride/iodide transporter	780	31	54	AF167412.1	ND	Slc26a4
pendrin	*Mus musculus*	Sodium-independent chloride/iodide transporter	780	30	52	AF167411.1	12	Slc26a4
hDRA	*Homo sapiens*	Downregulated in adenoma (sulfate/chloride antiporter)	764	28	59	P40879	7q22-q31.1	SLC26A3
mDRA	*Mus musculus*	Downregulated in adenoma (sulfate/chloride antiporter)	757	27	52	AF136751	12	Slc26a3

ND, not determined; r, rat; m, mouse; h, human.

[a] Amino acid identity to the rat sat-1 protein, as determined with the ClustalW alignment program.

[b] Amino acid identity to the sulfate transporter signature (Prosite PS01271) found in rat sat-1 protein, as determined with the ClustalW alignment program.

[c] HGNC approved name (http://www.gene.ucl.ac.uk/nomenclature).

1994). Previous genes linked to other chondrodysplasias were found to encode growth factor receptors, transcription factors, or structural proteins (Rimoin 1996); therefore DTDST represented a novel class of protein implicated in cartilage disorders. Owing to the important role of sulfated proteoglycans in cartilage extracellular matrix, a possible pathophysiological link was made for the DTDST gene. Earlier studies had in fact shown both a reduced rate of sulfate incorporation into chondroitin sulfate and reduced staining of sulfated proteoglycans in DTD patients (Scheck *et al.* 1978; Feshchenko *et al.* 1989). Furthermore, the identification of human and murine genes involved in sulfate activation as the cause of cartilage disorders provided further information about the importance of sulfated proteoglycans in cartilage formation (Orkin *et al.* 1976; Ul Haque *et al.* 1998). Murine models for DTDST and other disorders of sulfate metabolism are presently being developed in order to help in determining therapeutic treatments.

Pathophysiology of DTD

Diastrophic dysplasia (DTD; OMIM 222600) is a rare autosomal recessive chondrodysplasia present in all human populations, having the highest carrier frequency (1 in 70) in the Finnish population (Hästbacka *et al.* 1994). Clinical manifestations of the disorder include short-limbed stature, characteristic "hitchhiker" thumbs, a club foot, often a cleft palate, contractures, kyphoscoliosis, osteoarthritis, and multiple dislocations of joints which cause severely impaired mobility, requiring corrective orthopedic surgery (Everett and Green 1999). Owing to respiratory insufficiency and neurological complications of spinal deformities, DTD infants have a raised mortality rate; however subsequent life expectancy is normal.

So far there have been over 30 different mutations identified in the DTDST gene (Rossi and Superti-Furga 2001), resulting in a variety of skeletal dysplasias which are classified into two distinct groups according to

clinical severity: (i) two lethal disorders, atelosteogenesis type 2 (AO2; OMIM 256050) and achondrogenesis type 1B (ACG1B; OMIM 600972), both severe chondrodysplasias causing perinatal death from pulmonary hypoplasia (Rossi and Superti-Furga 2001); (ii) two nonlethal disorders, diastrophic dysplasia (DTD), and recessive multiple epiphyseal dysplasia (rMED), both resulting in lifelong skeletal disorders. The fact that the phenotypic severity is related to different mutations in the DTDST gene would suggest that genotype–phenotype correlations do exist: DTDST mutations producing a truncated protein or a nonconservative amino acid substitution in a transmembrane domain give rise to severe phenotypes, whereas nontransmembrane amino acid substitutions and splice-site mutations give rise to the milder phenotypes. Mutations in the DTDST gene coding region have usually been found in only one allele (true for the majority of Finnish DTD patients), with the other allele remaining intact. In a common Finnish mutation, the splice donor site of a 5′-untranslated exon is mutated (Hästbacka *et al.* 1999), leading to severely reduced mRNA levels, translating into reduced levels of the DTDST protein and thereby giving rise to the mild DTD phenotype. In most Finnish DTD patients, Northern blot analyses revealed markedly reduced expression of the DTDST mRNA. Of the currently known DTDST mutations, the most frequent is 862C > T (R279W) which gives rise to the rMED phenotype when homozygous and mostly DTD when compounded. It occurs at a CpG dinucleotide and its panethnic distribution may suggest independent recurrence. The second most common mutation is IVS1 + 2T > C which is very frequent in Finland, producing low levels of correctly spliced mRNA resulting in DTD when homozygous. Two other mutations, 1045–1047delGTT (V340del) and 558C > T (R178X), are associated with severe phenotypes and have been observed in several patients. All the other mutations are rare and heterozygotes with only one allele mutated are clinically unaffected. Screening of clinical samples for radiologic and histologic features compatible with the ACG1B/AO2/DTD/rMED spectrum disorders prior to analysis allows mutations to be detected at a success rate of over 90% per allele. Mutation analysis has largely replaced sulfate uptake or sulfate incorporation assays in cultured fibroblasts, but the latter techniques are still used in cases where mutation analysis is not informative. Although supplementation of patients' cultured cells with thiols may bypass the transporter defect and enhance sulfation of proteoglycans, no therapeutic approaches are yet available.

In ACG1B, both alleles have mutations that presumably abolish protein function, whereas in AO2 at least one allele has been suggested to harbor a missense mutation resulting in residual protein expression (Superti-Furga *et al.* 1996a, b). Biochemical studies of patients with ACG1B identified undersulfated proteoglycans, suggesting that a defect in sulfated proteoglycan biosynthesis occurs in patients with the more severe chondrodysplasias (Superti-Furga 1994; Hästbacka *et al.* 1996; Rossi *et al.* 1996a,b, 1997; Superti-Furga *et al.* 1996a). Furthermore, in patients with DTD, ACG1B, and AO2, the levels of proteoglycan sulfation closely correlate with specific mutations and clinical severity (Rossi *et al.* 1998).

In all clinical conditions, the general phenotypic defect has been demonstrated to be a reduced sulfate transport leading to undersulfation of cartilage proteoglycans (Hästbacka *et al.* 1996; Rossi *et al.* 1996a, b, 1997, 1998). In order to further characterize the nature of several common DTD mutations and their clinical phenotypes, a recent study extracted chondroitin sulfate proteoglycans from cartilage of 12 patients with known chondrodysplasias (Rossi *et al.* 1998). The amount of nonsulfated disaccharides was found to be elevated in patients when compared with normal subjects, with the highest amount present in ACG1B patients (Rossi *et al.* 1998), suggesting that undersulfation of chondroitin sulfate proteoglycans occured in cartilage in vivo and that it was correlated with the clinical severity. Futhermore, by [^{35}S]sulfate and

[^{3}H]glucosamine double-labeling of fibroblast cultures from patients with DTD, AO2, and ACG1B and from controls, the uptake of extracellular sulfate was found to be reduced in all patients' cells, with lowest values in ACG1B cells. However, disaccharide analysis of chondroitin sulfate proteoglycans derived from these patients showed that they were normally sulfated or only moderately undersulfated, suggesting that extracellular sulfate uptake was impaired but fibroblasts were still able to replenish their reduced intracellular sulfate pools by oxidizing sulfur-containing compounds (i.e., cysteine and methionine) and thus partially rescuing proteoglycan sulfation under basal conditions (Rossi *et al.* 1998). Clearly, further work is needed to determine the functional significance of the sulfate transport defects due to specific mutations in the DTDST gene and their contribution(s) to the observed clinical phenotypes.

DRA (SLC26A3)

Physiological function of DRA

By subtractive hybridization using cDNA libraries from normal colon and adenocarcinoma tissues, a human cDNA which was downregulated in adenomas (DRA) was isolated (Schweinfest *et al.* 1993). DRA mRNA was found to be exclusively expressed in normal colon tissues, with its expression significantly decreased in colonic adenomas (polyps) and andenocarcinomas. The expression pattern of DRA in intestinal tumors was dependent on the differentiation state, with strong expression observed in epithelial polyps with no or minor dysplasia, whereas adenocarcinomas were completely devoid of DRA (Haila *et al.* 2000). The human DRA (hDRA) gene (SLC26A3) was mapped to chromosome 7q22-q31.1 (Taguchi *et al.* 1994). It spans approximately 39 kb, comprising 21 exons (Haila *et al.* 1998). Comparison of the location of the predicted transmembrane domains to exon boundaries showed some correlation with the 14 TMD protein model. Analysis of the putative promoter region (570 bp) showed putative TATA and CCAAT boxes and multiple transcription factor binding sites for AP-1 and GATA-1 (Haila *et al.* 1998). hDRA encodes an 84.5 kDa protein with 14 putative TMDs, having charged clusters of amino acids at its NH_2 and COOH termini, potential nuclear targeting motifs, an acidic transcriptional activation domain, and a homeobox domain (Schweinfest *et al.* 1993). These putative domains in hDRA protein sequence could suggest possible interaction with a transcription factor, which would be consistent with DRA having a role in tissue-specific gene expression and/or as a candidate tumor suppressor gene. By northern blot analysis, a low level of DRA mRNA expression was detected in the mouse colon at birth, which increased in the first two postnatal weeks, whereas in the small intestine there was a lower level of DRA expression, which remained constant in the postnatal period (Schweinfest *et al.* 1993). In the differentiating human colon carcinoma cell line Caco-2, DRA mRNA was undetected in the preconfluent (undifferentiated) state, but became highly expressed in the postconfluent (differentiated) state (Schweinfest *et al.* 1993). By northern blot analysis, mDRA mRNA was detected at high levels in cecum and colon and at lower levels in small intestine (Melvin *et al.* 1999). hDRA mRNA was detected in eccrine sweat glands and seminal vesicles (Haila *et al.* 2000) and throughout the intestinal tract (duodenum, ileum, cecum, distal colon), at the mucosal epithelium, in particular to the Brush Border Membrane (BBM) of columnar epithelial cells, but not in the esophagus or stomach (Byeon *et al.* 1996). Consistent with its expression in the differentiated columnar epithelium of the adult human colon, DRA was first expressed in the midgut of developing mouse embryos at day 16.5, corresponding to the time of differentiation of the epithelium of the small intestine (Byeon *et al.* 1996). Colon epithelial hDRA mRNA expression in inflammatory bowel disease and ischemic colitis was found to be similar to that in normal colon epithelium, but the DRA protein was detected deeper in crypts, including the proliferative epithelial cells.

As with DTDST, the DRA protein was found to show significant amino acid identities (Table 1) with both DTDST (33%) and rsat-1 (28%), now forming a link to two functionally characterized sulfate anion transporters (Markovich 2001; Lee *et al.* 2003; Regeer *et al.* 2003). hDRA was expressed in insect Sf9 cells where it was targeted to the cell membrane and led to a greater than threefold increase in sulfate uptake when compared with control cells (Byeon *et al.* 1996). When expressed in *Xenopus* oocytes, hDRA was shown to function as a DIDS-sensitive Na^+-independent sulfate/oxalate transporter (Silberg *et al.* 1995). DRA was initially proposed to be involved in colonic oncogenesis due to its downregulation in colonic adenomas; the DRA protein was later shown to be a membrane glycoprotein located in the apical brush border membrane of the colonic mucosal epithelial cells responsible for transporting sulfate and oxalate (Silberg *et al.* 1995). DRA was shown also to function as a Cl^-/OH^- or Cl^-/HCO_3^- antiporter (Moseley *et al.* 1999). The mouse DRA (mDRA) cDNA ortholog (Melvin *et al.* 1999), when expressed in human embryonic kidney HEK293 cells, conferred Na^+-independent electroneutral chloride/bicarbonate exchange activity.

Since the expression of DRA mRNA was restricted to intestinal tissue, with strongest expression in the colon, it became an attractive candidate gene for congenital chloride diarrhea (CLD) syndrome, a recessively inherited defect of intestinal chloride/bicarbonate exchange (see below). Subsequently, hDRA was found to be defective in CLD patients (Moseley *et al.* 1999). CLD (OMIM 214700) is a potentially fatal diarrhea having high chloride content (Holmerg 1986; Kere *et al.* 1999). The locus linked to CLD was initially mapped to 7q31 (Kere *et al.* 1993), adjacent to the cystic fibrosis transmembrane regulator (CFTR) gene, which itself was a viable candidate gene for CLD owing to its involvement in chloride transport. Linkage disequilibrium and haplotype analyses localized the CLD locus to 7q22-q31.1 which corresponded with the location of the DRA gene (Hoglund *et al.* 1995, 1996a). Thus DRA became implicated as the positional and functional candidate for CLD and became known as the CLD (or SLC26A3) gene.

Pathophysiology of DRA/CLD

Congenital chloride diarrhea (CLD; OMIM 214700) is a rare autosomal recessive disease, reported around the world, but with a higher incidence in Finland, Poland, Kuwait, and Saudi Arabia (Everett and Green 1999). The main clinical feature of CLD is prenatal onset of watery diarrhea that in utero leads to polyhydramnion and often premature birth (Kere *et al.* 1999). CLD manifests as a lifelong high-volume watery diarrhea with an elevated chloride content (>90 mmol) (Holmberg 1986).

Physiologically, the defect in CLD was initially suggested to be due to aberrant Cl^-/HCO_3^- exchange in the distal segments of the ileum and colon, where Cl^- is normally absorbed from the lumen by a Cl^-/HCO_3^- exchanger coupled to Na^+ absorption via the Na^+/H^+ exchanger (Turnberg 1971; Holmberg *et al.* 1975). The secreted HCO_3^- and H^+ ions form CO_2 and H_2O which are both absorbed, resulting in net absorption of Na^+, Cl^-, and H_2O, and it is the lack of this process in CLD that leads to elevated levels of Na^+ and Cl^- in the intestinal fluid and osmotic diarrhea. Furthermore, the lack of HCO_3^- secretion leads to increased acidity in the intestinal lumen, thereby inhibiting the Na^+/H^+ exchanger. Initially, metabolic alkalosis, hyponatremia, and hypochloremia are observed systemically, leading to hyperaldosteronism and increased Na^+ reabsorption/K^+ secretion in the renal distal tubule and colon, improving the hyponatremia but producing hypokalemia.

CLD infants are usually born premature due to polyhydramnion caused by fetal diarrhea. CLD newborns usually have distended abdomens and hyperbilirubinemia, and develop metabolic alkalosis. If left untreated, CLD can result in severe dehydration, electrolyte disturbances and development of early renal complications, poor psychomotor development and retarded growth, and ultimately death (Everett and Green 1999). With appropriate therapy, involving long-term treatment by intravenous fluid and electrolyte replacement (in newborns) or by oral replacement of lost fluid and electrolytes (in older children and adults), CLD patients develop normally and have a normal life expectancy (Holmberg 1986). Current treatment for adults includes oral supplements of Na^+, Cl^-, K^+, and H_2O, which reduces electrolyte depletion but also exacerbates the diarrhea, since much of the chloride is not absorbed by CLD patients. Currently, no treatment is available which enhances intestinal chloride absorption and the common antidiarrheal drugs are ineffective; however H^+,K^+-pump inhibitors, which reduce gastric chloride secretion and thereby enhance chloride reaching the colon, have been found to be effective (Aichbichler *et al.* 1997).

To date, over 30 different mutations have been identified in CLD patients, arising in various ethnic populations (Kere *et al.* 1999). Many of these mutations are frameshifts leading to nonsense changes followed by a stop codon. Two common mutated alleles (see below) are responsible for CLD in the Finnish and Arabic populations; however, a variety of other mutations have been identified in other populations. Unlike the DTDST gene, there seem to be no obvious genotype–phenotype correlations for CLD mutations detected to date, although some clustering of mutations in the gene have been observed (Kere *et al.* 1999). The cellular defects arising from these mutations have not been fully characterized. One study looked at two mutations (ΔV317 or C307W) found in 32 CLD patients from Finland (Hoglund *et al.* 1996b) in order to determine whether these mutations may lead to a loss of protein function (Moseley *et al.* 1999). *Xenopus* oocytes were injected with either wild-type or mutagenized (ΔV317 or C307W) DRA cRNA, followed by measurement of radiotracer Cl^- and SO_4^{2-} uptakes. Both Cl^- and SO_4^{2-} transport were induced by wild-type DRA, whereas both transport activities were completely abolished in the $\Delta V317$ mutation, with no change observed for the C307W mutation (Moseley et al. 1999). These results suggest that DRA is a functional chloride/sulfate transporter which is defective in CLD patients carrying the $\Delta V317$ mutation, whereas the C307W mutation was shown to be a silent polymorphism (Moseley et al. 1999). Currently, ongoing studies in several laboratories are beginning to unravel the functional activities of the various CLD mutations in order to gain a better understanding of the mechanisms of the genetic mutations in DRA which lead to the diseased state.

PENDRIN (SLC26A4)

Physiological function of pendrin

By positional cloning, a gene named pendrin or PDS (SLC26A4) was identified, which was found to be linked to Pendred syndrome (PS; OMIM 274600), a recessively inherited disorder of syndromic deafness characterized by congenital sensorineural hearing loss and thyroid goiter (Pendred 1896; Everett et al. 1997). Previous genetic linkage studies localized the gene to human chromosome 7q, a region to which a known nonsyndromic deafness locus (DFNB4) was mapped previously (Baldwin et al. 1995; Coyle et al. 1996). In fact, the original family reported in describing DFNB4 was subsequently shown to have PS (Li et al. 1998). PDS produces a transcript of approximately 5 kb that is expressed primarily in the thyroid, with weaker signals in the kidney and brain (Everett et al. 1997). PDS encodes a putative 11 TMD protein designated pendrin (Everett et al. 1997), which shares significant amino acid identities with several sulfate transport proteins (Table 1), including DRA (45%), DTDST (32%), and rsat-1 (32%). Based on this sequence similarity, pendrin was originally proposed to function as a sulfate transporter (Everett et al. 1997). However, this initial study did not demonstrate pendrin's ability to transport sulfate, nor did it make the link between the pathogenesis of congenital deafness and the role of altered sulfate transport in human disease. Subsequent studies demonstrated that pendrin was in fact unable to induce sulfate transport in Xenopus oocytes or Sf9 cells, but rather that it could induce iodide, chloride (Scott et al. 1999) and formate (Scott and Karniski 2000) uptake, suggesting that pendrin may in fact function as a transporter of chloride, iodide, and formate, but not sulfate. However, it is quite intriguing and still unknown why pendrin shares such high structural similarity to other well-established sulfate transporters (DRA, DTDST, and sat-1), belonging to the same protein family (SLC26), yet is unable to mediate any sulfate transport.

Pathophysiology of pendrin

Pendred syndrome (PS; OMIM 274600) is a fairly common autosomal recessive disorder which is characterized by two distinct physiological features, namely adenomatous goiter and sensorineural deafness (Pendred 1896). PS was first reported over 100 years ago (Pendred 1896) and was found to occur at an incidence of 7.5 per 100,000 in many populations (Kopp 1999), hence representing one of the most common forms of syndromic deafness. PS deafness is usually severe, with its presentation being quite variable and often arising due to head trauma in late-onset cases. Typically, structural abnormalities are observed in the inner ear, with the cochlea missing its apical turn and also having an underdeveloped modiolus and reduced hair and spiral ganglion cell numbers (Everett and Green 1999), which most likely account for the hearing loss. Such malformations can often be associated with other cochlear defects, including an enlargement of vestibular aqueducts, suggestive of embryonic developmental arrest at 7 weeks (Everett and Green 1999).

The goiter observed in PS patients is quite variable in expression, typically appearing around puberty, and is generally associated with a positive perchlorate discharge test, in which, following a perchlorate challenge, an increased amount of unincorporated iodide is released from the thyroid. This suggests that the PS thyroid defect relates to the organification of iodide, either in the transport of iodide across the apical membrane of follicular cells into the follicular lumen or in its binding to thyroglobulin. A possible mechanism for the former hypothesis is outlined as follows. Upon entry from the bloodstream into the thyrocyte via a sodium/iodide cotransporter, iodide is transported into the colloid and then becomes bound to the tyrosine residues in thyroglobulin and thereafter degraded in lysosomes to yield thyroid hormones. Since PS patients release iodide inappropriately from thyrocytes following perchlorate administration, normal iodide "trapping" by thyroglobulin does not take place. The fact that pendrin transports iodide would propose a possible link to this defective role in the thyroid. Pendrin has recently been shown to function as an iodide transporter at the apical membrane in the thyrocyte (Rodriguez *et al.* 2002; Yoshida *et al.* 2002), and in several PS-causing mutations in pendrin, incorrect sorting of the mutated protein (to the apical cell surface) has been demonstrated (Rotman-Pikielny *et al.* 2002), thus resulting in iodide not being properly transported into the follicular lumen and thereafter insufficient iodide being incorporated into thyroglobulin.

To date, over 80 different mutations in the PDS gene have been identified in PS patients, ranging from small deletions and insertions to splice-site and missense mutations (Everett *et al.* 1997; Kere *et al.* 1999). Certain ethnic populations appear to have common mutations, with four mutations (including three missense mutations L236P, E384G, and T416P, and one splice-site mutation at exon 8) accounting for 72% of mutant alleles in Northern Europeans (Everett and Green 1999), suggesting a founder effect of their ancestry. PDS mutations have also been found in families suffering from deafness but in the absence of any thyroid defect (Li *et al.* 1998), which may suggest possible genotype–phenotype correlations. A likely

explanation for this pathophysiological condition could be that residual pendrin function may be sufficient to compensate for the mild biochemical defect in the thyroid, but insufficient to prevent the developmental defect from occurring in the inner ear. To further corroborated this theory, many of these PS families have one allele with a mild (nonconserved) missense mutation (Usami *et al.* 1999). Since substantial intrafamilial and interfamilial variation exists in PS with respect to disease severity and age of goiter onset (Everett and Green 1999), the precise contribution of the genetic background or environmental factors attributing to the differences in PS severity are presently unknown.

Compared with the proposed model for pendrin function in the thyroid gland, its role(s) in the inner ear is less well understood. In situ hybridization has identified the mouse pendrin (Pds) mRNA to be expressed throughout the endolymphatic duct and sac and in several other discrete regions of the inner ear (Everett *et al.* 1999). A common feature of the Pds-expressing cells in the inner ear is their putative role in fluid homeostasis, especially fluid reabsorption (Everett and Green 1999). Therefore, it could be postulated that, in the absence of pendrin, normal anion transport in the inner ear could be lost, resulting in a perturbed osmotic state which may then lead to abnormal hydrostatic effects resulting in a widened endolymphatic duct and a malformed cochlea (Everett *et al.* 1999). Furthermore, the sensory cell defect could also occur as a consequence of the altered osmotic environment (Everett and Green 1999). Obviously, more research is required to determine the precise mechanisms by which the PDS mutations lead to the pathophysiology of Pendred syndrome.

SUMMARY: DTDST, DRA, AND PENDRIN

The DTDST (SLC26A2), DRA/CLD (SLC26A3), and pendrin/PDS (SLC26A4) genes encode three structurally related anion transporters which play important roles in the etiology of several distinct genetic diseases. Several key similarities and differences exist among these genes and their respective proteins, which will be summarized below.

The physical organization of the human PDS and DRA genes is highly similar at the genomic level. They both reside within an approximate 150 kb interval on chromosome 7q22-q31.1, and both are present in a tail-to-tail configuration and contain 21 exons spanning about 40 kb, with highly similar intron–exon organization. Owing to their high sequence similarity (45% amino acid identity) and physical proximity to one another, the two paralogous genes most likely arose from a recent chromosomal duplication event. However, PDS and DRA do have markedly different tissue expression patterns, suggesting that the regulatory elements controlling their expression must be different. This could have arisen either as a consequence of sequence divergence in their promoter sequences after a gene duplication event or the juxtaposition of a different promoter next to one or both of the genes at the time of the duplication. On the other hand, DTDST is more distantly related to PDS and DRA, with respect to both nucleotide identity and genomic structure, containing four exons spanning about 15 kb on human chromosome 5q32.

The proteins encoded by DTDST, DRA, and pendrin are highly related in structure; in fact they are so similar that they have been classified by the HGNC to belong to the same protein family (SLC26). The highest sequence similarity occurs at the TMD segments and in the sulfate transporter signature (Prosite PS01271) domain (Table 1), which is highly conserved among related proteins in the whole animal kingdom, but for which the functional significance is yet unknown (Markovich 2001). Human DTDST and DRA share 33% amino acid identity, whereas human pendrin shares 45% and 32% amino acid identity with DRA and DTDST, respectively (Table 1). There may be two potential

explanations for how such structurally related proteins could be linked to such varied diseases.

1. The proteins have distinct tissue distributions, with DTDST being expressed ubiquitously. However, the pathology seems to be restricted to cartilage, possibly due to a consequence of the sensitivity to the levels of proteoglycan sulfation in cartilage or a reflection of a compensatory mechanism present in other tissues (Everett and Green 1999). On the other hand, DRA and PDS have a much more restricted tissue expression pattern that correlates well with the affected organs, primarily the colon for DRA/CLD and thyroid/inner ear for PDS.
2. These proteins have varied substrate preferences/specificities for different anions that have been physiologically associated with their respective disorders, namely sulfate for DTD, chloride for DRA/CLD, and iodide for PS.

Sequence analyses of the various protein databases reveals the presence of a large family of anion transporters that belong to the SLC26 gene family, including DTDST, DRA, and pendrin. Presently, 11 mammalian members have been isolated within this gene family, whose functional characterization is presently being determined. When one looks at a broader taxonomic span, including species from bacteria, yeast, plants, insects, nematodes, and mammals, there are over 50 proteins sharing significant structural identity with this gene family (Everett and Green 1999). For the mammalian SLC26 members, the proteins have been suggested to have a late divergence (Everett and Green 1999), possibly explaining the reason for the nonlethal diseases associated with defective family members (DTDST, CLD, and pendrin) and a nonessential function for general cellular activity, but, rather, a more likely function to serve some phylum-specific role (Everett and Green 1999). This proposal is in agreement with the fact that most of the SLC26 family of proteins have highly restricted tissue expression patterns (e.g., PDS, CLD, and sat-1), with the exception of DTDST (Markovich 2001).

Clearly, further work, presently being undertaken in numerous laboratories around the world, is essential to identify the precise roles of these important genes/proteins in the body, using various systems including animal models in order to determine precisely their physiological importance and significance during body development.

REFERENCES

Aichbichler, B.W., Zerr, C.H., Santa, A.C., Porter, J.L., and Fordtran, J.S. (1997). Proton-Pump Inhibition of Gastric Chloride Secretion in Congenital Chloridorrhea. *N. Engl. J. Med.*, **336**, 106–109.

Baldwin, C.T., Weiss, S., Farrer, L.A., De Stefano, A.L., Adair, R., Franklyn, B., *et al.* (1995). Linkage of Congenital, Recessive Deafness (DFNB4) to Chromosome 7q31 and Evidence for Genetic Heterogeneity in the Middle Eastern Druze Population. *Hum. Mol. Genet.*, **4**, 1637–1642.

Bissig, M., Hagenbuch, B., Stieger, B., Koller, T., and Meier, P.J. (1994). Functional Expression Cloning of the Canalicular Sulfate Transport System of Rat Hepatocytes. *J. Biol. Chem.*, **269**, 3017–3021.

Byeon, M.K., Westerman, M.A., Maroulakou, I.G., Henderson, K.W., Suster, S., Zhang, X.K., *et al.* (1996). The Down-Regulated in Adenoma (DRA) Gene Encodes an Intestine-Specific Membrane Glycoprotein. *Oncogene*, **12**, 387–396.

Clines, G. and Lovett, M. (1996). The Full Length Sequence, Genomic Structure and Identification of a Novel Splice Junction within the Diastrophic Dysplasia Sulfate Transporter Gene. *Am. J. Hum. Genet.*, **59**(Suppl.), A148.

Coyle, B., Coffey, R., Armour, J.A., Gausden, E., Hochberg, Z., Grossman, A., *et al.* (1996). Pendred Syndrome (Goitre and Sensorineural Hearing Loss) Maps to Chromosome 7 in the Region Containing the Nonsyndromic Deafness Gene DFNB4. *Nat. Genet.*, **12**, 421–423.

Everett, L.A. and Green, E.D. (1999). A Family of Mammalian Anion Transporters and Their Involvement in Human Genetic Diseases. *Hum. Mol. Genet.*, **8**, 1883–1891.

Everett, L.A., Glaser, B., Beck, J.C., Idol, J.R., Buchs, A., Heyman, M., *et al.* (1997). Pendred Syndrome is Caused by Mutations in a Putative Sulphate Transporter Gene (PDS). *Nat. Genet.*, **17**, 411–422.

Everett, L.A., Morsli, H., Wu, D.K., and Green, E.D. (1999). Expression Pattern of the Mouse Ortholog of

the Pendred's Syndrome Gene (Pds) Suggests a Key Role for Pendrin in the Inner Ear. *Proc. Natl Acad. Sci. USA*, **96**, 9727–9732.

Feshchenko, S.P., Krasnopol'skaia, K.D., Rebrin, I.A., and Rudakov, S.S. (1989). Molecular Heterogeneity of Proteoglycan Aggregates of Human Hyalin Cartilage in Normal Conditions and in Systemic Bone Dysplasia. *Vopr. Med. Khim.*, **35**, 24–33.

Haila, S., Saarialho-Kere, U., Karjalainen-Lindsberg, M.L., Lohi, H., Airola, K., Holmberg, C., *et al.* (2000). The Congenital Chloride Diarrhea Gene Is Expressed in Seminal Vesicle, Sweat Gland, Inflammatory Colon Epithelium, and in Some Dysplastic Colon Cells. *Histochem. Cell. Biol.*, **113**, 279–286.

Haila, S., Hoglund, P., Scherer, S.W., Lee, J.R., Kristo, P., Coyle, B., *et al.* (1998). Genomic Structure of the Human Congenital Chloride Diarrhea (CLD) Gene. *Gene*, **214**, 87–93.

HäStbacka, J., de la Chapelle, A., Mahtani, M.M., Clines, G., Reeve-Daly, M.P., Daly, M., *et al.* (1994). The Diastrophic Dysplasia Gene Encodes a Novel Sulfate Transporter: Positional Cloning by Fine-Structure Linkage Disequilibrium Mapping. *Cell*, **78**, 1073–1087.

HäStbacka, J., Superti-Furga, A., Wilcox, W.R., Rimoin, D.L., Cohn, D.H., and Lander, E.S. (1996). Atelosteogenesis Type II Is Caused by Mutations in the Diastrophic Dysplasia Sulfate-Transporter Gene (DTDST): Evidence for a Phenotypic Series Involving Three Chondrodysplasias. *Am. J. Hum. Genet.*, **58**, 255–262.

HäStbacka, J., Kerrebrock, A., Mokkala, K., Clines, G., Lovett, M., Kaitila, I., *et al.* (1999). Identification of the Finnish Founder Mutation for Diastrophic Dysplasia (DTD). *Eur. J. Hum. Genet.*, **7**, 664–670.

Hoglund, P., Sistonen, P., Norio, R., Holmberg, C., Dimberg, A., Gustavson, K.H., *et al.* (1995). Fine Mapping of the Congenital Chloride Diarrhea Gene by Linkage Disequilibrium. *Am. J. Hum. Genet.*, **57**, 95–102.

Hoglund, P., Haila, S., Scherer, S.W., Tsui, L.C., Green, E.D., Weissenbach, J., *et al.* (1996a). Positional Candidate Genes for Congenital Chloride Diarrhea Suggested by High-Resolution Physical Mapping in Chromosome Region 7q31. *Genome Res.*, **6**, 202–210.

Hoglund, P., Haila, S., Socha, J., Tomaszewski, L., Saarialho-Kere, U., Karjalainen-Lindsberg, M.L., *et al.* (1996b). Mutations of the Down-Regulated in Adenoma (DRA) Gene Cause Congenital Chloride Diarrhoea. *Nat. Genet.*, **14**, 316–319.

Holmberg, C. (1986). Congenital Chloride Diarrhoea. *Clin. Gastroenterol.*, **15**, 583–602.

Holmberg, C., Perheentupa, J., and Launiala, K. (1975). Colonic Electrolyte Transport in Health and in Congenital Chloride Diarrhea. *J. Clin. Invest.*, **56**, 302–310.

Karlstedt, E., Kaitila, I., and Pirinen, S. (1996). Phenotypic Features of Dentition in Diastrophic Dysplasia. *J. Craniofac. Genet. Dev. Biol.*, **16**, 164–173.

Karlstedt, E., Kaitila, I., and Pirinen, S. (1997). Craniofacial Structure in Diastrophic Dysplasia – A Cephalometric Study. *Am. J. Med. Genet.*, **72**, 266–274.

Karlstedt, E., Isotalo, E., Haapanen, M.L., Kalland, M., Pirinen, S., and Kaitila, I. (1998). Correlation between Speech Outcome and Cephalometric Dimensions in Patients with Diastrophic Dysplasia. *J. Craniofac. Genet. Dev. Biol.*, **18**, 38–43.

Kere, J., Sistonen, P., Holmberg, C., and de la Chapelle, A. (1993). The Gene for Congenital Chloride Diarrhea Maps Close to But Is Distinct from the Gene for Cystic Fibrosis Transmembrane Conductance Regulator. *Proc. Natl Acad. Sci. USA*, **90**, 10686–10689.

Kere, J., Lohi, H., and Hoglund, P. (1999). Genetic Disorders of Membrane Transport III. Congenital Chloride Diarrhea. *Am. J. Physiol.*, **276**, G7–G13.

Kobayashi, T., Sugimoto, T., Saijoh, K., Fukase, M., and Chihara, K. (1997). Cloning of Mouse Diastrophic Dysplasia Sulfate Transporter Gene Induced During Osteoblast Differentiation by Bone Morphogenetic Protein-2. *Gene*, **198**, 341–349.

Kopp, P. (1999). Pendred's Syndrome: Identification of the Genetic Defect a Century After Its Recognition. *Thyroid*, **9**, 65–69.

Lee, A., Beck, L., and Markovich, D. (2003). The Mouse Sulfate Anion Transporter (msat-1) cDNA and Gene Sat1 (Slc26a1): Cloning, Tissue Distribution, Gene Structure, Functional Characterization and Regulation by Thyroid Hormone. *DNA Cell Biol.*, **22**, 19–31.

Li, X.C., Everett, L.A., Lalwani, A.K., Desmukh, D., Friedman, T.B., Green, E.D., and Wilcox, E.R. (1998). A Mutation in PDS causes Non-Syndromic Recessive Deafness. *Nat. Genet.*, **18**, 215–217.

Markovich, D. (2001). The Physiological Roles and Regulation of Mammalian Sulfate Transporters. *Physiol. Rev.*, **81**, 1499–1534.

Markovich, D., Bissig, M., Sorribas, V., Hagenbuch, B., Meier, P.J., and Murer, H. (1994). Expression of Rat Renal Sulfate Transport Systems in *Xenopus laevis* Oocytes. Functional Characterization and Molecular Identification. *J. Biol. Chem.*, **269**, 3022–3026.

Melvin, J.E., Park, K., Richardson, L., Schultheis, P.J., and Shull, G.E. (1999). Mouse Down-regulated in Adenoma (DRA) Is an Intestinal Cl^-/HCO_3^- Exchanger and is Up-regulated in Colon of Mice Lacking the NHE3 Na^+/H^+ Exchanger. *J. Biol. Chem.*, **274**, 22855–22861.

Moseley, R.H., Hoglund, P., Wu, G.D., Silberg, D.G., Haila, S., de la Chapelle, A., *et al.* (1999). Down-regulated in Adenoma Gene Encodes a Chloride Transporter Defective in Congenital Chloride Diarrhea. *Am. J. Physiol.*, **276**, G185–G192.

Orkin, R.W., Pratt, R.M., and Martin, G.R. (1976). Undersulfated Chondroitin Sulfate in the Cartilage Matrix of Brachymorphic Mice. *Dev. Biol.*, **50**, 82–94.

Pendred, V. (1896). Deaf-Mutism and Goitre. *Lancet*, **ii**, 532.

Regeer, R.R., Lee, A. and Markovich, D. (2003). Characterization of the Human Renal Sulfate Anion Transporter (hsat-1) protein and Gene (SAT1; SLC26A1). *DNA Cell Biol.*, **22**, 107–117.

Rimoin, D.L. (1996). Molecular Defects in the Chondrodysplasias. *Am. J. Med. Genet.*, **63**, 106–110.

Rodriguez, A.M., Perron, B., Lacroix, L., Caillou, B., Leblanc, G., Schlumberger, M., *et al.* (2002). Identification and Characterization of a Putative Human Iodide Transporter Located at the Apical Membrane of Thyrocytes. *J. Clin. Endocrinol. Metab.*, **87**, 3500–3503.

Rossi, A., Bonaventure, J., Delezoide, A.L., Cetta, G., and Superti-Furga, A. (1996a). Undersulfation of Proteoglycans Synthesized by Chondrocytes from a Patient with Achondrogenesis Type 1B Homozygous for an L483P Substitution in the Diastrophic Dysplasia Sulfate Transporter. *J. Biol. Chem.*, **271**, 18456–18464.

Rossi, A., van der Harten, H.J., Beemer, F.A., Kleijer, W.J., Gitzelmann, R., Steinmann, B., *et al.* (1996b). Phenotypic and Genotypic Overlap between Atelosteo-genesis Type 2 and Diastrophic Dysplasia. *Hum. Genet.*, **98**, 657–661.

Rossi, A., Bonaventure, J., Delezoide, A.L., Superti-Furga, A., and Cetta, G. (1997). Undersulfation of Cartilage Proteoglycans Ex Vivo and Increased Contribution of Amino Acid Sulfur to Sulfation In Vitro in McAlister Dysplasia/Atelosteogenesis Type 2. *Eur. J. Biochem.*, **248**, 741–747.

Rossi, A., Kaitila, I., Wilcox, W.R., Rimoin, D.L., Steinmann, B., Cetta, G., *et al.* (1998). Proteoglycan Sulfation in Cartilage and Cell Cultures from Patients with Sulfate Transporter Chondrodysplasias: Relation-ship to Clinical Severity and Indications on the Role of Intracellular Sulfate Production. *Matrix Biol.*, **17**, 361–369.

Rossi, A. and Superti-Furga, A. (2001). Mutations in the Diastrophic Dysplasia Sulfate Transporter (DTDST) Gene (SLC26A2): 22 Novel Mutations, Mutation Review, Associated Skeletal Phenotypes, and Diagnostic Relevance. *Hum. Mutat.*, **17**, 159–171.

Rotman-Pikielny, P., Hirschberg, K., Maruvada, P., Suzuki, K., Royaux, I.E., Green, E.D., *et al.* (2002). Retention of Pendrin in the Endoplasmic Reticulum is a Major Mechanism for Pendred Syndrome. *Hum. Mol. Genet.*, **11**, 2625–2633.

Russel, C., Warrington, C., Gustafson, R., and Clarke, L.A. (1996). Organization of the Murine and Human Alpha-L-iduronidase locus: Evidence for Overlapping Genes. *Am. J. Hum. Genet.*, **59**(Suppl.), A159.

Satoh, H., Susaki, M., Shukunami, C., Iyama, K., Negoro, T., and Hiraki, Y. (1998). Functional Analysis of Diastrophic Dysplasia Sulfate Transporter. Its Involvement in Growth Regulation of Chondrocytes Mediated by Sulfated Proteoglycans. *J. Biol. Chem.*, **273**, 12307–12315.

Scheck, M., Parker, J., and Daentl, D. (1978). Hyaline Cartilage Changes in Diastrophic Dwarfism. *Virchows Arch. A Pathol. Anat. Histol.*, **378**, 347–359.

Schweinfest, C., Henderson, K., Suster, S., Kondoh, N., and Papas, T. (1993). Identification of a Colon Mucosa Gene that is Down-regulated in Colon Adenomas and Adenocarcinomas. *Proc. Natl Acad. Sci. USA*, **90**, 4166–4170.

Scott, D.A. and Karniski, L.P. (2000). Human Pendrin Expressed in *Xenopus laevis* Oocytes Mediates Chloride/Formate Exchange. *Am. J. Physiol. Cell Physiol.*, **278**, C207–C211.

Scott, H.S., Ashton, L.J., Eyre, H.J., Baker, E., Brooks, D.A., Callen, D.F., *et al.* (1990). Chromosomal Localization of the Human Alpha-L-iduronidase Gene (IDUA) to 4p16.3. *Am. J. Hum. Genet.*, **47**, 802–807.

Scott, D.A., Wang, R., Kreman, T.M., Sheffield, V.C., and Karniski, L.P. (1999). The Pendred Syndrome Gene Encodes a Chloride–Iodide Transport Protein. *Nat. Genet.*, **21**, 440–443.

Silberg, D.G., Wang, W., Moseley, R.H., and Traber, P.T. (1995). The Down-Regulated in Adenoma (DRA) Gene Encodes an Intestine-Specific Membrane Sulfate Transport Protein. *J. Biol. Chem.*, **270**, 11897–11902.

Superti-Furga, A. (1994). A Defect in the Metabolic Activation of Sulfate in a Patient with Achondrogenesis Type IB. *Am. J. Hum. Genet.*, **55**, 1137–1145.

Superti-Furga, A., Hastbacka, J., Wilcox, W.R., Cohn, D.H., van der Harten, H.J., Rossi, A., *et al.* (1996a). Achondrogenesis Type IB is caused by Mutations in the Diastrophic Dysplasia Sulphate Transporter Gene. *Nat. Genet.*, **12**, 100–102.

Superti-Furga, A., Rossi, A., Steinmann, B., and Gitzelmann, R. (1996b). A Chondrodysplasia Family Produced by Mutations in the Diastrophic Dysplasia Sulfate Transporter Gene: Genotype/Phenotype Correlations. *Am. J. Med. Genet.*, **63**, 144–147.

Taguchi, T., Testa, R.T., Papas, T.S., and Schweinfest, C. (1994). Localization of a Candidate Colon Tumor Suppressor Gene (DRA) to 7q22-q31.1 by Fluorescence In Situ Hybridization. *Genomics*, **20**, 146–147.

Turnberg, L.A. (1971). Abnormalities in Intestinal Electrolyte Transport in Congenital Chloridiarrhea. *Gut*, **12**, 544–551.

Ul Haque, M.F., King, L.M., Krakow, D., Cantor, R.M., Rusiniak, M.E., Swank, R.T., *et al.* (1998). Mutations in Orthologous Genes in Human Spondyloepimetaphyseal Dysplasia and the Brachymorphic Mouse. *Nat. Genet.*, **20**, 157–162.

Usami, S., Abe, S., Weston, M.D., Shinkawa, H., Van Camp, G., and Kimberling, W.J. (1999). Non-Syndromic Hearing Loss Associated with Enlarged Vestibular Aqueduct is Caused by PDS Mutations. *Hum. Genet.*, **104**, 188–192.

Yoshida, A., Taniguchi, S., Hisatome, I., Royaux, I.E., Green, E.D., Kohn, L.D., *et al.* (2002). Pendrin is an Iodide-specific Apical Porter Responsible for Iodide Efflux from Thyroid Cells. *J. Clin. Endocrinol. Metab.*, **87**, 3356–3361.

7

MICHAEL D. GARRICK* AND LAURA M. GARRICK*

Divalent metal transporter DMT1 (SLC11A2)

This chapter will discuss the properties of DMT1 and its relationship to iron uptake and iron trafficking (and the metabolism of other metals as well); we will describe the existing mutants representing specific transporter diseases in rodents. In addition, the chapter will further describe the potential contribution of DMT1 to diseases where *SLC11A2* is not the gene mutated. Lastly, we will present insights derived from or to be gained from in vitro mutagenesis including a mutant in *SLC11A1*. DMT1 is the focus of two recent reviews (Andrews 1999; M.D. Garrick *et al.* 2003).

DMT1, Divalent Metal Transporter 1, was first called Nramp2 (Vidal *et al.* 1995) and then DCT1 (Gunshin *et al.* 1997), and is also classified as SLC11A2 under OMIM 604653. Computer analysis predicts 12 transmembrane domains (Gunshin *et al.* 1997) as illustrated in Figure 1. Four isoforms are expected from the currently known mRNAs. Two of these differ at the N-terminal region; two differ in their C-termini. All forms share motifs that should allow glycosylation, expanding the potential for apparent variation in cells and tissues. They also contain a predicted transport motif, reflecting the function as a transporter of Fe^{2+}

and probably of other divalent metal ions, and other motifs that should affect protein targeting. Referring to these splice variants by the presence of N-terminal peptides encoded by specific exons (1A and 2) and C-terminal peptides derived from mRNA exons with or without a potential regulatory IRE (Iron Responsive Element – thus +IRE and −IRE, respectively), they could be called 1A+IRE, 1A−IRE, 2+IRE, and 2−IRE. One would not be surprised to find that the different isoforms have different intracellular localizations because there is the potential for each to carry distinct targeting motifs. Nevertheless, it was a surprise to find that the −IRE DMT1 isoform localizes to the nucleus of some cells (Roth *et al.* 2000; M.D. Garrick *et al.* 2003). The first report was prior to the awareness that the 1A/2 isoforms occurred, and that exon 1A contains a potential nuclear localization sequence (M.D. Garrick *et al.* 2003). The association of exon 1A mRNA isoform with polarized cells (Hubert and Hentze 2002) has also led to speculation that 1A DMT1 has a specific localization and function in polarized cells. Certainly differential localization and function is a fertile area for future research and one should keep in mind when considering the phenotype of mutants in DMT1 that the multiple functions of DMT1 may have specific associations with particular forms of the transporter. Consistent with the prediction that

* State University of New York, 140 Farber Hall, Buffalo, NY 14214-3000, USA

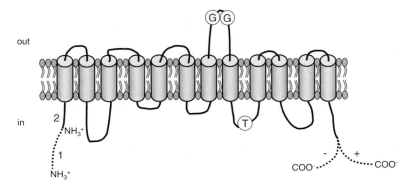

Figure 1 Predicted structural features for DMT1. The polypeptide chain starts at the amino terminal end (NH_3^+ lower left) and runs continuously to the carboxyterminal end (COO^- lower right): 1A and 2 represent alternative exons for the N-terminal end (Hubert and Hentze 2002), cylinders represent putative transmembrane domains, G represents a potential site for glycosylation, T represents a potential transport motif, and − and + represent alternative exons for the C-terminal end translated from the −IRE and +IRE forms of DMT1 mRNA (Gunshin *et al.* 1997; M.D. Fleming *et al.* 1998; Lee *et al.* 1998). Continuous lines indicate primary structure common to all four known isoforms, while dashed lines indicate alternative sequences found in only two of the four isoforms. Orientation relative to membranes is also depicted.

isoforms will have distinct subcellular localization, GFP-tagged +IRE DMT1 is present in late endosomes, while −IRE DMT1 is enriched in early endosomes (Tabuchi *et al.* 2002). Isoforms were also expressed to different degrees in different types of cells and localized specifically in polarized cells.

WHAT IS DMT1 AND WHAT DOES IT DO?

The putative structure of DMT1 (Figure 1) suggests that it is a transporter. The defects in the Belgrade (*b/b*) rat (M.D. Fleming *et al.* 1998) and microcytic (*mk/mk*) mouse (M.D. Fleming *et al.* 1997) support the notion that DMT1 is an iron transporter, but one should consider the formal possibility that it could still be a membrane protein essential for iron uptake and trafficking which strategically regulates an iron transporter similar to CFTR and chloride channels (Schwiebert *et al.* 1995).

Figure 2 illustrates DMT1's role for duodenal iron uptake in the enterocyte. Prior to uptake, Fe^{3+}, after solubilization due to gastric acidity, is reduced to Fe^{2+} by duodenal cytochrome *b* (dcyt*b*) (McKie *et al.* 2001) or a similar reductase on the apical surface. Fe^{2+} enters the apical surface via DMT1. DMT1's location on the brush border of the enterocyte (Canonne-Hergaux *et al.* 1999; Trinder *et al.* 2000) is consistent with this role as there is a divalent cation transport activity associated with intestinal brush border membrane vesicles (Knöpfel *et al.* 2000). Iron uptake by DMT1 is optimal at a mildly acidic pH (Gunshin *et al.* 1997; M.D. Garrick and Dolan 2002), like the conditions in the proximal duodenum. How iron gets across the enterocyte starting from the apical surface is unclear, but there are arguments that DMT1 mediates transcytosis (Yeh *et al.* 2000; Ma *et al.* 2002). Iron is exported into the serum by SLC11A3, a related transporter (also called ferroportin, IREG1, and MTP1) and the subject of Chapter 8. Apo-transferrin (Tf) collects the iron, probably after endocytosis into the enterocyte from the basolateral surface (Nuñez *et al.* 1997; Nuñez and Tapia 1999; Alvarez-Hernandez

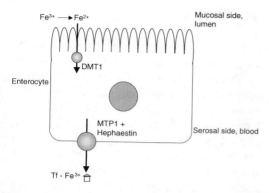

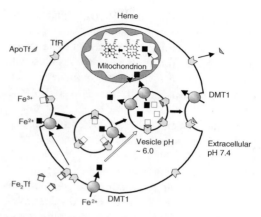

Figure 2   DMT1 in enterocytes. DMT1 is ordinarily present on the luminal side (apical/mucosal – the brush border) where it is responsible for Fe(II) uptake. After transcytosing the cell, Fe(II) exits via MTP1 (also known as ferroportin1/Ireg1/SLC11A3) to load Tf (the triangle, with the square for Fe³⁺), probably after oxidation via hephaestin or ceruloplasmin. This exchange may occur on the basolateral side as drawn or could take place perinuclearly. Adapted from M.D. Garrick *et al.* (2003) with permission of Kluwer Academic Publishers.

Figure 3   DMT1 and the transferrin (Tf) cycle. Squares represent Fe, filled for Fe^{2+} and open for Fe^{3+}. The triangles are Tf and the hexagons are their receptors. Constitutive endocytosis brings the Fe–Tf–TfR complex into vesicles. Although the pH at the cell surface is 7.4, the pH of the endosomes drops to ~6.0 as protons are pumped into the vesicles by the v-ATPase. At the lower pH, the Fe–Tf–TfR complex cooperatively releases Fe(III) so that a ferrireductase can reduce it. DMT1 is on the surface of the vesicles so Fe(II) exits, also probably driven by the pH gradient. Fe^{2+} may transit to the mitochondrion via the cytosol and could rely on carriers yet to be identified, or it may enter the mitochondrion directly. DMT1, also present on the cell surface, is responsible for a portion of Tf-independent Fe(II) uptake as well. Adapted from M.D. Garrick *et al.* (2003) with permission of Kluwer Academic Publishers.

et al. 2000). Because Fe^{2+} must be converted to Fe^{3+} to bind to Tf, hephaestin (Vulpe *et al.* 1999), a multicopper oxidase, or circulating ceruloplasmin (Yoshida *et al.* 1995) also participates in iron export.

DMT1 is as important for iron trafficking as it is for intestinal iron uptake. Its function in the Tf cycle and its possible role in non-Tf bound iron (NTBI) uptake are illustrated in Figure 3. Circulating Fe^{3+}-Tf delivers most iron to cells as reviewed by Ponka (1997). After Fe-Tf binds to the Tf receptor (TfR), endocytosis leads to vesicle formation. The endosomes acidify due primarily to proton pumping by v-ATPase; then the resulting acidification releases Fe^{3+} from the complex by cooperatively affecting the Tf and TfR (Nuñez *et al.* 1990; Bali *et al.* 1991; Sipe and Murphy 1991; Watkins *et al.* 1991). As DMT1 transports Fe^{2+} but not Fe^{3+} (Gunshin *et al.* 1997; Conrad *et al.* 2000), exit from the endosome requires a ferrireductase (Nuñez *et al.* 1990;

Sipe and Murphy 1991), an activity that may also contribute to release from Tf. Iron may be routed to ferritin storage or to mitochondria for heme and FeS cluster synthesis. Apo-Tf stays on TfR at endosomal pH, but is released after reaching the cell surface.

Figure 3 also illustrates the potential for DMT1 to participate in NTBI transport. Although there are likely to be multiple means for NTBI uptake, a high affinity pathway is defective in *b/b* rats (Farcich and Morgan 1992a; L.M. Garrick *et al.* 1999). This defect clearly indicates that DMT1 is one means for NTBI uptake. Because the *mk/mk* mouse has the identical mutation, it is likely that it will exhibit a similar impairment. As shown in the figure, DMT1 could be involved in endocytosis or directly transport Fe^{2+} to vesicles, but either way this NTBI transport pathway would intersect in vesicles with iron delivered by the Tf cycle.

While Gunshin *et al.* (1997) reported that DMT1 transported multiple metals (Zn^{2+}, Cd^{2+}, Mn^{2+}, Cu^{2+}, Fe^{2+}, Co^{2+}, Ni^{2+}, and Pb^{2+}) and was a proton symport for each, it is now uncertain as to whether the conductance used to show metal transport for divalent metals other than Fe^{2+} actually indicates transport. Similarly, the role of proton transport is less clear than originally postulated. We refer the reader to a recent review indicating that more work is required in order to substantiate that DMT1 "is a divalent metal/proton symporter, although it is safe to state that DMT1 usually transports Fe(II) and H^+ in an associated fashion" (M.D. Garrick *et al.* 2003). There is no doubt that DMT1 is critical for Fe^{2+} transport and is unable to utilize Fe^{3+} directly (M.D. Garrick *et al.* 2003). Mn^{2+}, Co^{2+}, Ni^{2+}, and Cu^{2+} are likely to be physiologically relevant substrates, although Cu^{2+} certainly also has alternative means of transport. Cd^{2+}, Pb^{2+}, and Zn^{2+} are the least certain. Evidence at the time of the review (M.D. Garrick *et al.* 2003), suggested that Cd^{2+} is transported by DMT1 and has other physiological transporters as well, and recent publications (Park *et al.* 2002, Bannen *et al.* 2003) further strengthen this evaluation. Although evidence also suggested that Pb^{2+} could be a substrate for DMT1, another transporter may be the more relevant pathophysiological Pb^{2+} transporter (Bannon *et al.* 2002, 2003), at least in Caco-2 cells. Very recently, experiments using antisense to DMT1 showed that it is a physiologically relevant copper transporter in Caco-2 cells (Arredondo *et al.* 2003). Remarkably, evidence supported Cu^{1+} as the species transported. If this specificity is confirmed, then including the term divalent in the name of the transporter may be inappropriate.

IRON UPTAKE

Remarkably, the known mutations (*mk* in the mouse and *b* in the rat) are both G185R (M.D. Fleming *et al.* 1997, 1998). This coincidence is made even more surprising by the knowledge that *mk* mutants were independently isolated three times in the mouse and that two of the isolates had the G185R mutation (M.D. Fleming *et al.* 1997); unfortunately, the third has been lost to follow-up. This mutation converts a CpG dinucleotide sequence to TpG (on the antisense strand). CpGs represent potential hot spots for mutation and thus are under-represented in mammalian genomes, occurring at only one fifth the expected frequency (International Human Genome Sequencing Consortium 2001). The mutation abrogates nearly all of the transport activity for Fe^{2+} in both the rat (M.D. Fleming *et al.* 1998) and the mouse (Su *et al.* 1998) DMT1 constructs. [A lesser proportion for human DMT1 is lost with the G185R mutation (Worthington *et al.* 2000), an issue to which we will return later in this chapter.] If most or nearly all activity for Fe^{2+} uptake is lost in the duodenum and a similar proportion is lost for Fe^{2+} export from the endosome in these mutants, one obvious question is how the mutants survive at all. Although we will consider the possibilities of residual activity for DMT1 and alternative Fe^{2+} transporters later, here we remind the reader that there are also potential transporters for Fe^{3+} (Conrad *et al.* 2000) and heme (Worthington *et al.* 2001). Clearly, the heme uptake pathway must exist in most mammals but is currently not well defined; Worthington's paper (Worthington *et al.* 2001) is an early step in defining it. Advocates for the IMP (integrin/ mobilferrin/ paraferritin) pathway for Fe^{3+} uptake (Conrad *et al.* 1999) have recently reviewed its status in relation to the pathway for Fe^{2+} uptake (Conrad and Umbreit 2001). The Conrad group also found that DMT1 (Fe^{2+} uptake) immunoprecipitated along with mobilferrin (Fe^{3+} uptake) and paraferritin (a complex with possible reductase function), demonstrating the potential for the Fe^{2+} and Fe^{3+} pathways to intersect (Umbreit *et al.* 2000). One point from one of their reviews (Conrad and Umbreit 2001) is the reminder that rodents make their own vitamin C while humans require it in their diet and readily become scorbutic. Therefore duodenal iron absorption in humans may rely relatively more on Fe^{3+} uptake than on Fe^{2+} uptake, while in rodents Fe^{3+} uptake may be a minor pathway. If so, mutations in DMT1 could have a less striking phenotype in humans. Given that DMT1 is not restricted in location to the proximal intestine and that it also participates in iron trafficking, we should probably await the discovery of a human mutant to learn the phenotype.

The coincidence that the G185R mutation occurred independently three times in two species, each leading to a very similar phenotype, makes one wonder whether something more than CpG mutability is special about this mutation. Perhaps G185R leads to sufficient residual activity for survival or possibly impacts on some functions of DMT1, but leaves others unimpaired. Modern genetic technology offers a direct approach to this set of issues. One would like to see the phenotype of mice homozygous for abrogation of DMT1 by knockout technology. Recently, a null allele has been identified in zebrafish (Donovan *et al.* 2002). The *cdy* (chardonnay) gene was isolated by a functional genomics approach and a nonsense mutation identified in the mutant allele. The truncated mutant DMT1 protein was not functional, so the extremely pale phenotype is clearly due to greatly diminished iron uptake and trafficking in a fashion resembling the Belgrade rat and microcytic mouse.

IRON TRAFFICKING AND METABOLISM OF OTHER METALS

One can get a sense of what DMT1 does by asking where it is found. Northern blotting (Gunshin *et al.* 1997) revealed DMT1 mRNA in kidney, brain, bone marrow, heart, testis, liver, spleen, lung, and skeletal muscle. Although the mRNA was enriched and particularly responsive to iron status in the duodenum, it was also present in the jejunum, ileum, colon, and stomach, and thus practically ubiquitous in the digestive system. Quantitative RT-PCR confirms the distribution of the +IRE and −IRE forms (Tchernitchko *et al.* 2002) in kidney, brain, heart, duodenum, lung, liver, spleen, thyroid, testis, and bone marrow. Both forms are present in each tissue, with the +IRE form the predominant species in the duodenum, liver, and bone marrow, and the −IRE mRNA predominant in the other tissues. RT-PCR recently illuminated the locations of the mRNAs containing exons 1A and 1B (Hubert and Hentze 2002), with exon 1B mRNA in kidney, lung, spleen, thymus, heart, duodenum (and other parts of the digestive tract), liver, and testis, and exon 1A mRNA present in the duodenum and kidney and to a lesser degree in the testis, jejunum, and ileum.

Where mRNA is found is usually, but not always, a guide to identifying the tissues in which the transporter is expressed; however, one ultimately needs to learn finer details about where the transporter is within the tissue and even its subcellular locations to know its specific role. Immunological methods should provide some of the answers, but the mutants themselves also help identify locations by affecting tissue densities of metals. Thus studies that evaluate tissue metal densities in the *b/b* rat (M.D. Garrick *et al.* 2003) or *mk/mk* mice (Conrad *et al.* 2000) will begin to let us know where DMT1 is a critical participant in metal trafficking and for which metals. The currently available data (Conrad *et al.* 2000; M.D. Garrick *et al.* 2003) are preliminary and more data need to be collected.

The kidney is a predominant location for DMT1. Immunoreactivity is present in the collecting ducts in both principal and intercalated cells (Ferguson *et al.* 2001), particularly in distal convoluted tubules and also in Henle's loop. DMT1 was also found in S3 segments. Although this distribution is consistent with DMT1's functioning in iron reabsorption

(Wareing *et al.* 2000), it is inconsistent with the observation that *b/b* rats do not exhibit a defect in iron reabsorption (Ferguson *et al.* 2002). DMT1 immunoreactivity was greatly diminished in *b/b* kidneys and the cortical/outer medullary junction was ill defined (Ferguson *et al.* 2002). Thus one must conclude that, although the lack of DMT1 activity and protein has clearly identifiable morphological consequences for kidney microanatomy, its absence must be compensated for by a functionally redundant unidentified mechanism. In mice DMT1 is found in the cortex, not in the medulla, and at the brush border of proximal tubule epithelial cells (Canonne-Hergaux and Gros 2002). However, the antibody used for the rat kidney (Ferguson *et al.* 2001) identifies immunoreactivity in the mouse kidney in locations similar to those in the rat (C. Smith and D. Riccardi, personal communication). Thus the apparent differences are likely to depend on the specificity of the antibody preparations and immunostaining techniques used rather than representing a species difference in DMT1 location. Homozygous *mk/mk* mice also have a drastic decrease in DMT1 in the kidney (Canonne-Hergaux and Gros 2002).

DMT1 also appears to be important in the transfer of iron into the central nervous system. The *b* mutation leads to diminished uptake of iron and Tf into the brain of 2-week-old rats (Farcich and Morgan 1992b). Although iron was affected to a larger extent, this result suggests that the Tf cycle could be involved in movement of iron across the blood–brain barrier and clearly indicates that DMT1 plays a role in trafficking of iron to the brain. The distribution of iron within the brain of *b/b* rats (Burdo *et al.* 1999) confirms and extends the earlier finding in that pyramidal neurons were iron positive in *b/b* cortex but fewer and less intensely stained than in +/*b* controls. There is dramatically less iron in *b/b* oligodendrocytes and myelin than in +/*b* controls, with the few strongly staining *b/b* oligodendrocytes associated closely with blood vessels. Astrocytes staining for iron were fewer in the cerebral

cortex of the mutant. Generally, regions that normally have considerable iron staining (like the substantia nigra and globus pallidus) exhibited the greatest decrease in iron staining. The data on iron staining correlate well with the locations of immunoreactivity for DMT1 in normal and *b/b* brains (Burdo *et al.* 2001). DMT1 is present in neurons in the striatum, cerebellum, and thalamus as well as in ependymal and vascular cells. Thus DMT1 is involved in iron's crossing the blood–brain barrier and entering via the cerebrospinal fluid, and is also important in iron trafficking to neurons possibly via a glial route. One wonders about behavioral consequences of this impairment in the two rodent models.

The presence of DMT1 in peripheral (Roth *et al.* 2000) and central (M.D. Garrick *et al.* 2003) neurons also suggests that it is critical in iron nutrition and trafficking for these cells. Thus it is not surprising to find both −IRE and +IRE isoforms on the cell surface and the +IRE form in vesicles within the cell. Nevertheless, the presence of −IRE DMT1 in the nucleus was an unexpected finding (Roth *et al.* 2000). The same pattern is present in neurons from the *b/b* rat (unpublished results). The occurrence of DMT1 in the nucleus might indicate a previously unidentified function for the transporter.

Well in advance of the identification of the G185R mutation in DMT1, *b/b* rats (Oates and Morgan 1996; M. Garrick *et al.* 1997) and *mk/mk* mice (Edwards and Hoke 1972) had exhibited diminished gastrointestinal iron uptake. Detection of DMT1 in the villus membrane of enterocytes (Canonne-Hergaux *et al.* 1999; Trinder *et al.* 2000) is consistent with the altered function. While the loss of transport activity after the G185R mutation (M.D. Fleming *et al.* 1998; Su *et al.* 1998) might explain the intestinal phenotype, there is also evidence for elevated expression in both mutants (Canonne-Hergaux *et al.* 2000; Oates *et al.* 2000), presumably due to severe iron deficiency. Increased DMT1 is accompanied by improper targeting in the mouse mutant (Canonne-Hergaux *et al.* 2000). One wonders if the Belgrade rat exhibits a similar dislocalization in view of the situation in erythroid cells (below).

The *mk/mk* mouse has decreased red cell uptake of iron (Edwards and Hoke 1975), as does the *b/b* rat (Edwards *et al.* 1978). The decrease is due to an intracellular defect (Bowen and Morgan 1987) that can be bypassed by delivering iron chelated with salicylisonicotinic hydrazide (L.M. Garrick *et al.* 1991). The intracellular defect prevents iron from reaching mitochondria (L.M. Garrick *et al.* 1993); hence the defect should be within the Tf cycle. The Tf cycle defect has been characterized as subsequent to the entry of Fe_2-Tf and prior to the exocytosis of Tf, resulting in the release of nearly diferric Tf from *b/b* reticulocytes (M.D. Garrick *et al.* 1993). Similarly, the *mk/mk* mouse can internalize Fe_2-Tf but is unable to move the iron from the Tf to distal destinations (Canonne-Hergaux *et al.* 2001). DMT1 is apparently unstable or improperly targeted in *mk/mk* reticulocytes (Canonne-Hergaux *et al.* 2001), but appears in the endosomes of *b/b* reticulocytes in a fashion indistinguishable from normal erythroid cells (M.D. Garrick *et al.* 1999a).

The lung is another tissue where DMT1 is expressed. After exposure to iron, BEAS-2B cells, a model for human bronchial epithelial cells, exhibit increased DMT1 (Wang *et al.* 2002). The increase is due to upregulation of mRNA for −IRE DMT1; the +IRE isoform does not respond to iron. A similar increase occurs in situ. Because the lung is a gateway tissue for the entry of iron and many other potentially toxic metals, we must consider that this increase could be part of a protective mechanism. After all, the lung is not normally a route for acquisition of metals as nutrients. Consistent with (but not proof of) this protective hypothesis, hypotransferrinemic mice also have an elevation in −IRE DMT1 in lung epithelia and in alveolar macrophage; moreover, this elevation is associated with decreased susceptibility to exposure challenges involving oxidative stress after increased presence of

catalytically active metal (Ghio *et al.* 2003). A better test of the hypothesis is to look at the effect of a knockout of DMT1; the *b/b* rat and *mk/mk* mouse are essentially this test as DMT1 activity is very low. Thus if DMT1 protects against metal toxicity (by removing the metal from the surface of pneumocytes to safe storage, perhaps in ferritin within the cell), *b/b* airways should be more at risk than those of +/*b* controls. If metal entry through DMT1 is part of the toxic process, then *b/b* airways should be protected by the diminished ability to transport metal. Our results indicate greater damage to lung tissue after metal provocation (A.J. Ghio, X. Wang, L.A. Dailey, J.D. Stonehueme, C.A. Piantadosi, R. Silbajoris, F. Yang, K.G. Dolan, L.M. Garrick, and M.D. Garrick, in preparation) and support the argument that upregulation of the −IRE isoform is a defense mechanism to reduce metal toxicity in this gateway organ. The results also argue that DMT1 participates in both arms of maintenance of iron homeostasis; it is critical in absorption at the enterocyte, and it aids in managing undesirable exposure at the pneumocyte.

The G185R mutation also impairs iron transfer to the liver of 2-week-old *b/b* rats (Farcich and Morgan 1992b). DMT1 is found primarily in the membranes of hepatocytes (Trinder *et al.* 2000) where the level increases during iron overload and decreases during iron deficiency. One can also think of the liver as a gateway organ for iron after absorption in the duodenum. This increase could indicate a similar protective response to that seen in the lung, and suggests that the isoform that increases might be −IRE DMT1.

Similar analyses need to be done in other tissues to learn more about the function of DMT1 and achieve a better understanding of the consequences of the transporter disease that results from mutation in its gene. Unfortunately, available data are limited in quantity and only modestly informative. The recent finding of DMT1 immunoreactivity associated with the phagosome of Sertoli cells (Jabado *et al.* 2002) may account for the presence of DMT1 mRNA in testes. Jabado and colleagues suggest that DMT1 participates in recycling of iron by the so-called iron shuttle (Sylvester and Griswold 1994) and that *mk/mk* mice are subfertile due to an impairment of this process (Russell *et al.* 1970a). Nevertheless, the fertility of *b/b* males is improved by iron supplementation (M. Garrick *et al.* 1997). Although it has been shown that the mutation also decreases iron uptake by the heart of 2-week-old *b/b* rats (Farcich and Morgan 1992b), the immunoreactivity of DMT1 in normal and mutant rodents still needs to be studied in that tissue. There exists even less information about other tissues, so a fertile ground exists for learning more about the physiology and pathology of DMT1.

DMT1 also transports other divalent metals. Uptake of manganese into kidney and brain of *b/b* rats is impaired (Chua and Morgan 1997). Conrad *et al.* (2000) found that *mk/mk* livers are deficient not only in iron but also in manganese. We have also examined iron (M. Garrick *et al.* 1997) and manganese (unpublished results) in multiple tissues. Eventually this approach will help us to understand the consequences of the G185R mutation, but it is necessary to bypass the initial absorption of metal into the body to eliminate this uptake step as the source of the problem before the data will be completely interpretable.

Because DMT1 occurs in the lung and can transport Mn^{2+}, it could be involved in toxicity of manganese in humans. Several occupations are at risk of chronic exposure to manganese leading to a disorder called manganism, characterized by dystonic movements reminiscent of Parkinson's disease (Gorell *et al.* 1999; Pal *et al.* 1999). Susceptibility to managanism may vary despite comparable exposures. We predict that a portion of this variation may be associated with variability in DMT1 expression. One would like to see controlled studies on the extent of polymorphism in DMT1-linked markers in patients with manganism as well as to know how much variation they exhibit in DMT1 levels due to iron status and other factors.

INTRACELLULAR LOCALIZATION OF DMT1

The phenotype of *mk/mk* mice (Edwards and Hoke 1972) and *b/b* rats (Oates and Morgan 1996; M. Garrick *et al.* 1997) implied that DMT1 would be found on the apical surface of the enterocyte. It is on the brush border membrane as shown by immunohistochemistry (Canonne-Hergaux *et al.* 1999; Trinder *et al.* 2000). By implication, the +IRE isoform is the main species in that location (Canonne-Hergaux *et al.* 1999), possibly the 1A+IRE form (Hubert and Hentze 2002). The mutant phenotype also implied that DMT1 would be found on Tf-containing endosomes, weakly for *mk/mk* mice (Edwards and Hoke 1975) and indirectly (Bowen and Morgan 1987) and then more directly for *b/b* rats (M.D. Garrick *et al.* 1993). Immunohistochemistry also verifies these locations (Su *et al.* 1998; Canonne-Hergaux *et al.* 2001). Evidence that DMT1 can transport NTBI also places it by implication at the cell surface (Farcich and Morgan 1992a; L.M. Garrick *et al.* 1999), a location verified by a more direct detection (Su *et al.* 1998). A study of the intracellular localization of the + and −IRE isoforms of DMT1 (Roth *et al.* 2000) indicates that the −IRE form may be preferentially enriched at the cell surface while the +IRE form resides preferentially in vesicles at least in several neuronal cell types. Remarkably, the −IRE form also appeared in the nucleus of the neuronal and neuronal-like cells (Roth *et al.* 2000). This unexpected location implies that we need to learn the rationale for such a location of a protein that putatively contains 12 transmembrane domains, in terms of both function and targeting mechanism. Some possible functions have been presented in an earlier review (M.D. Garrick *et al.* 2003).

TRANSPORTER DISEASE

What can one expect for the pathophysiology of DMT1-based transporter disease? The Belgrade rat and microcytic mouse are the prototypes for answering this question. The disorder in the rat was a recessively inherited anemia (Sladic-Simic *et al.* 1966), characterized by microcytosis and hypochromia. Although it appeared during a protocol for radiation-induced mutations, close inspection of the breeding scheme reveals that the mutant was very likely pre-existing and became manifest because of the breeding scheme (Sladic-Simic *et al.* 1963, 1966). The discoverers called attention to the resemblance to thalassemia (Sladic-Simic *et al.* 1969), but later studies ruled out thalassemia and pointed to iron transport as the likely metabolic step(s) affected by the mutation (Edwards *et al.* 1978, 1986). The *mk/mk* mouse was also initially detected as a recessively inherited, microcytic, hypochromic anemia (Russell *et al.* 1970b), and then identified to have a defect in iron transport (Edwards and Hoke 1972, 1975). Thus, if one were seeking the human version of this transporter disease, one should look for a phenotype similar to the rodent models.

The mutation in DMT1 has interesting consequences in reticulocytes of the Belgrade rat. Iron uptake is only about 20% of normal in Belgrade reticulocytes (Edwards *et al.* 1978; Bowen and Morgan 1987) because DMT1 fails to export iron from the endosome (Bowen and Morgan 1987; M.D. Garrick *et al.* 1993; M.D. Fleming *et al.* 1998). This iron deficiency leads to heme deficiency with *b/b* rats having only about 40% of the normal level of "free" heme (M.D. Garrick *et al.* 1999b). As predicted from a large body of work, an inhibitor of translational initiation is produced (M.D. Garrick *et al.* 1999b). Globin synthesis is reduced to about 50% of normal, although α- and β-globin chain synthesis is balanced (Edwards *et al.* 1978). Defective initiation of translation has been shown to occur with 33% of mRNA in the nonpolysomal fraction of *b/b* reticulocytes compared with less than 6% in +/+ reticulocytes (Crkvenjakov *et al.* 1976). In *b/b* reticulocytes, only 35% of the polysomes are disomes or larger, while the number is 65% in +/+ reticulocytes (Crkvenjakov *et al.* 1982). In polysomes, α- and β-globin synthesis is

balanced but in the nonpolysomal fraction excess α mRNA is found (Crkvenjakov *et al.* 1982). Chu *et al.* (1978) showed that the amount of globin mRNA was actually decreased in *b/b* reticulocytes to 39–46% of normal, perhaps due to degradation of message upon lack of translation or an effect of heme deficiency on transcription or processing of mRNA.

Shortly before the mutations in the rodent models were identified, Hartman and Barker (1996) reported two siblings with severe microcytic anemia and iron malabsorption, and called attention to the resemblance to the *mk* phenotype. Hartman had earlier discussed the resemblance with us, trying to decide whether the mouse or rat model was closer. Now of course we know that the two models have identical G185R mutations (M.D. Fleming *et al.* 1997, 1998). Hartman (personal communication) informs us that the DMT1 genes from the two patients have been screened for a mutation that could account for the phenotype without success. Three kindreds that display similar propositi are also reported in the literature (Shahidi *et al.* 1964; Stavem *et al.* 1973; Buchanan and Sheehan 1991); if materials from the patients are still available for study, the DMT1 genes should be scanned for mutations.

Two arguments could mean that a search for human mutants in DMT1 with a phenotype resembling *mk/mk* mice and *b/b* rats is likely to be fruitless: One argument is based on the identical G185R mutations (M.D. Fleming *et al.* 1997, 1998) – the coincidence makes one wonder if there is something special about this particular mutation and that many other mutations may even be lethal, not an unreasonable speculation considering the multiple functions assigned to DMT1. This argument would be strengthened if abrogation of the DMT1 locus in mice leads to lethality when made homozygous. The other argument (Conrad and Umbreit 2001) is that the necessity for primates to obtain vitamin C through their diet decreases dependence of intestinal absorption of iron on the DMT1 pathway and increases the utilization of ferric and heme iron. This rationale might mean that the iron deficiency phenotype seen in rodents after DMT1 mutation might be less severe in humans. With one argument suggesting that human mutants might have a lethal transporter disease and another implying that the disease might be mild (perhaps resembling thalassemia trait), the best way to address the possibilities is still to characterize the extent of variation at the *SLC11A2* locus in humans and to see if variants are associated with disease.

LESSONS FROM MUTAGENESIS AND A RELATED TRANSPORTER

Although it seemed unlikely that the same G185R mutation is a polymorphism that occurs in both rats (M.D. Fleming *et al.* 1998) and mice (M.D. Fleming *et al.* 1997), we used mutagenesis and transient expression in HEK293T cells to verify that this mutation abrogated iron uptake by rat DMT1 with residual activity below 5% for Fe^{2+} (M.D. Fleming *et al.* 1998). Subsequently, the same assay demonstrated that this mutation led to even less residual activity when the construct was murine DMT1 (Su *et al.* 1998) and that the main effect was on uptake activity, although localization and stability of the protein might also be affected. Later, a similar assay also showed that rat DMT1 lost detectable uptake activity for manganese (M.D. Garrick and Dolan 2002). As mentioned earlier, Worthington *et al.* (2000) showed that the G185R mutation in the context of human DMT1 left about 20% residual activity in a related expression assay. The G185R mutant in murine DMT1 is also one of a series examined by Gros's group (Lam-Yuk-Sung *et al.* 2003) by complementation of yeast smf1/smf2 mutants and by permanent expression in CHO cells. Although 185R did not complement, 185G did. Yet in transfected CHO cells, a calcein-based assay for uptake indicated that the construct with the *mk* mutation retained 50% of uptake activity for Fe^{2+} and 30% for Co^{2+}. Clearly, the variation in the residuum points to a need for an understanding of how context and assay affect the results.

Su *et al.* (1998) also examined a variety of other mutations. G185S, a less drastic change at the same residue, led to a threefold decrease in activity (where G185R had led to a 35-fold decrease). Helical wheel analysis led them to query the role of D192, but the mutation D192G only decreased the activity about threefold, while the more conservative alterations D192N and D192E had even less effect. They also investigated G184D because it corresponds to a natural variant G169D that occurs in Nramp1 (SLC11A1, discussed below). It retained about a seventh of the wild type activity for Fe^{2+} uptake.

In a survey of mutagenesis of several residues of interest (Lam-Yuk-Sung *et al.* 2003), H267 and H272 turned out to be critical for metal uptake. Lowering the pH for the assay rescued the uptake activity, indicating that the two His residues affect the role of protons in metal transport. This alteration reminds us that mutagenesis should help us to understand this transporter's function, the role of variants occurring in human and other mammalian populations, and transporter disease.

For nearly 30 years, Nramp1 (Natural Resistance Associated Macrophage Protein 1; earlier *Bcg*, *Ity*, *Lsh*) has been studied to learn what underlies the natural resistance phenomenon. This topic has recently been reviewed (Wiley *et al.* 2002), so we will focus only on how a mutation in Nramp1 can be seen as another transporter disease, how it relates to DMT1, and information appearing after that review. *Nramp1* was identified by comparing inbred strains of mice for their inherent resistance to a series of infections. Positional cloning of the gene by Gros's group revealed that resistant strains had G169 while susceptible strains had D169. After Nramp2 (DMT1) was also cloned and its function identified, the extensive sequence similarity aligned G169 in the former with G184 in the latter, placing it only one residue before the G185R mutation in DMT1. Nramp1 is found only in phagosomes, although DMT1 has a widespread cellular distribution.

Identification of DMT1 as an iron transporter and divalent metal transporter led to two schools of thought on the function of Nramp1: perhaps it pumps iron and other potential nutrient metal ions out of the phagosome, depriving intraphagosomal parasites of needed material, or it pumps Fe^{2+} into the vesicle to generate reactive oxygen species to kill the invaders. In either case, the mutation would decrease transport, weakening the effect. Obviously the answer to this controversy is tied to whether Nramp1 moves metal ions in or out. Because DMT1 is also present in phagosomes (Jabado et al. 2002), we also need to know whether the two transporters act in concert or move metals in opposite directions. If they act in concert, then one would predict that diminished resistance to disease is another consequence of the b and mk mutations; however, if they oppose each other, then the mk mutation might improve disease resistance in G164D (sensitive) strains of mice if the general effects of the severe iron deficiency do not obscure an effect on the parasite resistance phenotype. Either way, it is intriguing to think about how a transporter disease can be linked intellectually to a natural resistance phenomenon.

DMT1 IN OTHER DISEASES

In a very real sense, dysregulation of a transporter is another way that mutation can lead to a pathological state. Therefore reports that DMT1 levels are elevated in enterocytes from patients with hereditary hemochromatosis (Zoller et al. 1999; Rolfs et al. 2002), a disorder involving mutation(s) in the HFE gene (Feder et al. 1996), could indict DMT1 overexpression as responsible for the iron overload that characterizes this disorder. This possibility receives support from similar observations in a murine model in which the hfe locus has been ablated (R.E. Fleming 1999; Levy et al. 1999; Griffiths et al. 2001). It is also supported by the mk gene's being epistatic to the hfe locus (Levy et al. 2000), a genetic relation that indicates that HFE expression requires

DMT1 expression. These observations do not necessarily mean that the etiology of the HFE mutations directly involves DMT1 dysregulation, because DMT1 overexpression could be secondary to an effect on another part of iron trafficking and MTP1 is also elevated in hereditary hemochromatosis (Rolfs et al. 2002).

The HFE protein forms a heterodimer with β_2-microglobulin (Feder et al. 1996), so it is not surprising that β_2-microglobulin-deficient mice also exhibit iron overload (de Sousa et al. 1994) and that the mk mutation is also epistatic to this mutation (Levy et al. 2000). The effect on DMT1 levels in these mice may be subtle, and the response to iron deficiency and overload appears to be retained although probably displaced to favor higher levels of gastrointestinal iron uptake (Moos et al. 2002). One should not forget that the HFE protein also interacts with the TfR (Lebron et al. 1998). This interaction focused attention on how increased expression of HFE decreased the efficiency of the Tf cycle (e.g., Feder et al. 1998), an apparently contradictory response. In contrast, Waheed et al. (2002) showed recently that co-expressing HFE and β_2-microglobulin increased the efficiency of the Tf cycle. This effect was mediated by an increased rate of TfR exocytosis, leading to higher levels of TfR on the cell surface. Clearly, mutation in either β_2-microglobulin or HFE should impair this increase without causing a total collapse of the Tf cycle. In crypt cell progenitors of enterocytes, this effect might "convince" the progenitor that body iron stores were lower than they really are so that it would produce enterocytes with mildly upregulated DMT1 and MTP1. This upregulation would eventually lead to iron overload because there is no way to adjust iron excretion. Interestingly, the authors noted that a protein of size about 65 kDa was consistently associated with TfR (Waheed et al. 2002); if this protein is DMT1 (it is the right size), then DMT1 might participate in its own dysregulation.

It would also be of interest to learn if DMT1 is elevated in enterocytes from patients with other forms of hereditary iron overload, such as mutations in TfR2 (Roetto *et al.* 2001), or from mice in which the TfR2 gene is knocked out (R.E. Fleming *et al.* 2002). Nevertheless, what is critical is the balance between iron uptake (DMT1) and iron export (MTP1), illustrated by dominantly inherited iron overload in patients with MTP1 mutations, as reviewed in Chapter 8.

SUMMARY

DMT1 (OMIM 604653, also called DCT1 and Nramp2) has 12 transmembrane domains and very likely transports ferrous iron and as many as seven other divalent metal ions. It is responsible for absorption of Fe(II) from the lumen of the duodenum into the enterocyte. It also plays a major role in iron trafficking via both the transferrin cycle, where export of Fe(II) from endosomes depends on DMT1 activity, and one type of non-transferrin bound iron (NTBI) uptake that is also due to DMT1. The only known transporter disease is due to a G185R mutation found, remarkably, in both microcytic (*mk*) mice and Belgrade (*b*) rats. Both rodent mutants have decreased gastrointestinal iron uptake and diminished endosomal iron exit. The Belgrade rat also has reduced NTBI uptake, a defect likely also to occur in the *mk* mouse. Both the mutants exhibit microcytic hypochromic anemia, attributable in part to poor iron nutrition and in part to inadequate delivery of iron to erythroid precursors. There is no doubt that iron delivery is inadequate for many, probably most, tissues. While the G185R mutation clearly diminishes transport activity dependent on DMT1, it may also render the transporter unstable or cause it to be improperly targeted. A human equivalent of the two rodent mutants is yet to be identified, but one can predict the phenotype of the resulting disorder and debate how much difference the role of ascorbic acid will make (as a required vitamin in humans and a natural product in rodents). DMT1 expression likely plays a role in iron overload diseases in humans and models for these diseases in mice.

Mutagenesis provides additional insight into the role of specific amino acid residues in DMT1 function and in the function of Nramp1, a related protein. Nramp1 (Natural Resistance Associated Macrophage Protein 1; SLC11A1) is associated with natural resistance to a number of parasites that infect macrophage. A similar G169D Nramp mutation (G184D if one aligns the two proteins) impairs the resistance by altering this protein's more specialized transport activity.

REFERENCES

Alvarez-Hernandez, X., Smith, M., and Glass, J. (2000). Apo-transferrin Is Internalized and Routed Differently from Fe-transferrin by Caco-2 Cells: A Confocal Microscopy Study of Vesicular Transport in Intestinal Cells. *Blood*, **95**, 721–723.

Andrews, N.C. (1999). The Iron Transporter DMT1. *Int. J. Biochem. Cell Biol.*, **31**, 991–994.

Arredondo, M., Muñoz, P., Mura, C.T., and Núñez, M.T. (2003). DMT1, a Physiologically Relevant Apical Cu^{1+} Transporter of Intestinal Cells. *Am. J. Physiol. Cell Physiol.*, **284**, C1525–C1530.

Bali, P.K., Zak, O., and Aisen, P. (1991). A New Role for the Transferrin Receptor in the Release of Iron from Transferrin. *Biochemistry*, **30**, 324–328.

Bannon, D.I., Portnoy, M.E., Olivi, L., Lees, P., Culotta, V.C., and Bressler, J.P. (2002). Uptake of Lead and Iron by Divalent Metal Transporter 1 in Yeast and Mammalian Cells. *Biochem. Biophys. Res. Commun.*, **295**, 978–984.

Bannon, D.I., Abounader, R., Lees, P., and Bressler, J.P. (2003). Effect of DMT1 Knockdown on Iron, Cadmium and Lead Uptake in Caco-2 Cells. *Am. J. Physiol. Cell Physiol.*, **284**, C44–C50.

Bowen, B.J. and Morgan, E.H. (1987). Anemia of the Belgrade Rat: Evidence for Defective Membrane Transport of Iron. *Blood*, **70**, 38–44.

Buchanan, G.R. and Sheehan, R.G. (1991). Malabsorption and Defective Utilization of Iron in Three Siblings. *J. Pediatr.*, **98**, 723–728.

Burdo, J.R., Martin, J., Menzies, S.L., Dolan, K.G., Romano, M.A., Fletcher, R.J., *et al.* (1999). Cellular Distribution of Iron in the Brain of the Belgrade Rat. *Neuroscience*, **93**, 1189–1196.

Burdo, J.R., Menzies, S.L., Simpson, I.A., Garrick, L.M., Garrick, M.D., Dolan, K.G., *et al.* (2001). Distribution of Divalent Metal Transporter 1 and Metal Transport Protein 1 in the Normal and Belgrade Rat. *J. Neurosci. Res.*, **66**, 1198–1207.

Canonne-Hergaux, F.S. and Gros, P. (2002). Expression of the Iron Transporter DMT1 in Kidney from Normal and Anemic *mk* Mice. *Kidney Int.*, **62**, 147–156.

Canonne-Hergaux, F., Gruenheid, S., Ponka, P., and Gros, P. (1999). Cellular and Subcellular Localization of the Nramp2 Iron Transporter in the Intestinal Brush Border and Regulation by Dietary Iron. *Blood*, **93**, 4406–4417.

Canonne-Hergaux, F., Fleming, M.D., Levy, J.E., Gauthier, S., Ralph, T., Picard, V., *et al.* (2000). The Nramp2/DMT1 Iron Transporter is Induced in the Duodenum of Microcytic Anemia *mk* Mice but is not Properly Targeted to the Intestinal Brush Border. *Blood*, **96**, 3964–3970.

Canonne-Hergaux, F., Zhang, A.S., Ponka, P., and Gros, P. (2001). Characterization of the Iron Transporter DMT1 (NRAMP2/DCT1) in Red Blood Cells of Normal and Anemic *mk/mk* Mice. *Blood*, **98**, 3823–3830.

Chu, M.-L., Garrick, L.M., and Garrick, M.D. (1978). Deficiency of Globin Messenger RNA in Reticulocytes of the Belgrade Rat. *Biochemistry*, **17**, 5128–5133.

Chua, A. and Morgan, E.H. (1997). Manganese Metabolism Is Impaired in the Belgrade Laboratory Rat. *J. Comp. Physiol. B.*, **167**, 361–369.

Conrad, M.E. and Umbreit, J.N. (2001). Iron Absorption: Relative Importance of Iron Transport Pathways. *Am. J. Hematol.*, **67**, 215.

Conrad, M.E., Umbreit, J.N., and Moore, E.G. (1999). Iron Absorption and Transport. *Am. J. Med. Sci.*, **318**, 213–229.

Conrad, M.E., Umbreit, J.N., Moore, E.G., Hainsworth, L.N., Porubcin, M., Simovich, M.J., *et al.* (2000). Separate Pathways for Cellular Uptake of Ferric and Ferrous Iron. *Am. J. Physiol. Gastrointest. Liver. Physiol.*, **279**, G767–G774.

Crkvenjakov, R., Cusic, S., Ivanovic, I., and Glisin, V. (1976). Rat *b/b* Anemia: Translation of Normal and Anemic mRNA in Wheat-Germ Cell-Free System. *Eur. J. Biochem.*, **71**, 85–91.

Crkvenjakov, R., Maksimovic, R., and Glisin, V. (1982). A Pool of Nonpolysomal Globin mRNAs in Globin Deficient Reticulocytes of the Anemic Belgrade Rat. *Biochem. Biophys. Res. Commun.*, **105**, 1524–1531.

de Sousa, M., Reimao, R., Lacerda, R., Hugo, P., Kaufmann, S., and Porto, G. (1994). Iron Overload in β_2-Microglobulin-Deficient Mice. *Immunol. Lett.*, **39**, 105–111.

Donovan, A., Brownlie, A., Dorschner, M.O., Zhou, Y., Pratt, S.J., Paw, B.H., *et al.* (2002). The Zebrafish Mutant Gene *chardonnay* (*cdy*) Encodes Divalent Metal Transporter 1 (DMT1). *Blood*, **100**, 4655–4659.

Edwards, J.A. and Hoke, J.E. (1972). Defect of Intestinal Mucosal Iron Uptake in Mice with Hereditary Microcytic Anemia. *Proc. Soc. Exp. Biol. Med.*, **141**, 81–84.

Edwards, J.A. and Hoke, J.E. (1975). Red Cell Iron Uptake in Hereditary Microcytic Anemia. *Blood*, **46**, 381–388.

Edwards, J.A., Garrick, L.M., and Hoke, J.E. (1978). Defective Iron Uptake and Globin Synthesis by Erythroid Cells in the Anemia of the Belgrade Laboratory Rat. *Blood*, **51**, 347–357.

Edwards, J.A., Huebers, H., Kunzler, C., and Finch, C. (1986). Iron Metabolism in the Belgrade Rat. *Blood*, **67**, 623–628.

Farcich, E.A. and Morgan, E.H. (1992a). Uptake of Transferrin-Bound and Nontransferrin-Bound Iron by Reticulocytes from the Belgrade Laboratory Rat: Comparison with Wistar Rat Transferrin and Reticulocytes. *Am. J. Hematol.*, **39**, 9–14.

Farcich, E.A. and Morgan, E.H. (1992b). Diminished Iron Acquisition by Cells and Tissues of Belgrade Laboratory Rats. *Am. J. Physiol. Regul. Integr. Comp. Physiol.*, **262**, R220–R224.

Feder, J.N., Gnirke, A., Thomas, W., Tsuchihashi, Z., Ruddy, D.A., Basava, A., *et al.* (1996). A Novel MHC Class I-Like Gene Is Mutated in Patients with Hereditary Haemochromatosis. *Nat. Genet.*, **13**, 399–408.

Feder, J.N., Penny, D.M., Irrinki, A., Lee, V.K., Lebron, J.A., Watson, N., *et al.* (1998). The Hemochromatosis Gene Product Complexes with the Transferrin Receptor and Lowers its Affinity for Ligand Binding. *Proc. Natl. Acad. Sci. USA*, **95**, 1472–1477.

Ferguson, C.J., Wareing, M., Ward, D.T., Green, R., Smith, C.P., and Riccardi, D. (2001). Cellular Localization of Divalent Metal Transporter DMT-1 in Rat Kidney. *Am. J. Physiol. Renal. Physiol.*, **280**, F803–F814.

Ferguson, C.J., Delannoy, M., Wareing, M., Green, R., Smith, C.P., and Riccardi, D. (2002). Expression of the Metal Transporter DMT-1 in the Kidney of Belgrade Rats. *J. Clin. Gastroenterol.*, **34**, 323.

Fleming, M.D., Trenor, C.I., Su, M.A., Foernzler, D., Beier, D.R., Dietrich, W.F., *et al.* (1997). Microcytic Anaemia Mice have a Mutation in *Nramp2*, a Candidate Iron Transporter Gene. *Nat. Genet.*, **16**, 383–386.

Fleming, M.D., Romano, M.A., Su, M.A., Garrick, L.M., Garrick, M.D., and Andrews, N.C. *Nramp2* is Mutated in the Anemic Belgrade (*b*) Rat: Evidence of a Role for Nramp2 in Endosomal Iron Transport. *Proc. Natl Acad. Sci. USA*, **95**, 1148–1153.

Fleming, R.E., Migas, M.C., Zhou, X.Y., Jiang, J.X., Britton, R.S., Brunt, E.M., *et al.* (1999). Mechanism of Increased Iron Absorption in Murine Model of Hereditary Hemochromatosis: Increased Duodenal Expression of the Iron Transporter DMT1. *Proc. Natl Acad. Sci. USA*, **96**, 3143–3148.

Fleming, R.E., Ahmann, J.R., Migas, M.C., Waheed, A., Koeffler, H.P., Kawabata, H., *et al.* (2002). Targeted Mutagenesis of the Murine Transferrin Receptor-2 Gene Produces Hemochromatosis. *Proc. Natl Acad. Sci. USA*, **99**, 10653–10658.

Garrick, L.M., Gniecko, K., Hoke, J.E., Al-Nakeeb, A., Ponka, P., and Garrick, M.D. (1991). Ferric-salicylaldehyde Isonicotinoyl Hydrazone, a Synthetic Iron Chelate, Alleviates Defective Iron Utilization by Reticulocytes of the Belgrade Rat. *J. Cell. Physiol.*, **146**, 460–465.

Garrick, L.M., Gniecko, K., Liu, Y., Cohan, D.S., Grasso, J.A., and Garrick, M.D. (1993). Iron Distribution in

Belgrade Rat Reticulocytes After Inhibition of Heme Synthesis with Succinylacetone. *Blood*, **81**, 3414–3421.

Garrick, L.M., Dolan, K.G., Romano, M.A., and Garrick, M.D. (1999). Non-Transferrin-Bound Iron Uptake in Belgrade and Normal Rat Erythroid Cells. *J. Cell. Physiol.*, **178**, 349–358.

Garrick, M.D. and Dolan, K.G. (2002). Ultrastructure and Molecular Biology Protocols for Oxidant and AntiOxidants, Methods in Molecular Biology. **196**, 147–154.

Garrick, M.D., Gniecko, K., Liu, Y., Cohan, D.S., and Garrick, L.M. (1993). Transferrin and the Transferrin Cycle in Belgrade Rat Reticulocytes. *J. Biol. Chem.*, **268**, 14867–14874.

Garrick, M., Scott, D., Walpole, S., Finkelstein, E., Whitbred, J., Chopra, S., *et al.* (1997). Iron Supplementation Moderates but does not Cure the Belgrade Anemia. *BioMetals*, **10**, 65–76.

Garrick, M.D., Fletcher, R.J., Dolan, K.G., Romano, M.A., Walowitz, J., Roth, J.A., *et al.* (1999a). The Iron Transporter DMT1 (or Nramp2 or DCT1) is Located Mostly in Endosomes in Normal and Belgrade Rat Reticulocytes. *Blood*, **94**, 403a.

Garrick, M.D., Scott, D., Kulju, D., Romano, M.A., Dolan, K.G., and Garrick, L.M. (1999b). Evidence for and Consequences of Chronic Heme Deficiency in Belgrade Rat Reticulocytes. *Biochim. Biophys. Acta Mol. Cell Res.*, **1449**, 125–136.

Garrick, M.D., Dolan, K.G., Ghio, A., Horbinski, C., Higgins, D., Porubcin, M., *et al.* (2003). DMT1 (Divalent Metal Transporter 1): A Mammalian Transporter for Multiple Metals. *BioMetals*, **16**, 41–54.

Ghio, A.J., Wang, X., Silbajoris, R., Garrick, M.D., Piantadosi, C.A., and Yang, F. (2003). DMT1 Expression is Increased in the Lungs of Hypotransferrinemic Mice. *Am. J. Physiol. Lung. Cell. Mol. Physiol.*, **284**, L938–L944.

Gorell, J.M., Johnson, C.C., Rybicki, B.A., Peterson, E.L., Kortsha, G.X., Brown, G.G., *et al.* (1999). Occupational Exposure to Manganese, Copper, Lead, Iron, Mercury and Zinc and the Risk of Parkinson's Disease. *Neurotoxicology*, **20**, 239–247.

Griffiths, W., Sly, W.S., and Cox, T.M. (2001). Intestinal Iron Uptake Determined by Divalent Metal Transporter is Enhanced in HFE-Deficient Mice with Hemochromatosis. *Gastroenterology*, **120**, 1420–1429.

Gunshin, H., Mackenzie, B., Berger, U.V., Gunshin, Y., Romero, M.F., Boron, W.F., *et al.* (1997). Cloning and Characterization of a Mammalian Proton-Coupled Metal-Ion Transporter. *Nature*, **388**, 482–488.

Hartman, K.R. and Barker, J.A. (1996). Microcytic Anemia with Iron Malabsorption: An Inherited Disorder of Iron Metabolism. *Am. J. Hematol.*, **51**, 269–275.

Hubert, N.H. and Hentze, M.W. (2002). Previously Uncharacterized Isoforms of Divalent Metal Transporter (DMT)-1: Implications for Regulation and Cellular Function. *Proc. Natl. Acad. Sci. USA*, **99**, 12345–12350.

International Human Genome Sequencing Consortium. (2001). Initial Sequencing and Analysis of the Human Genome. *Nature*, **409**, 860–921.

Jabado, N., Canonne-Hergaux, F., Gruenheid, S., Picard, V., and Gros, P. (2002). Iron Transporter Nramp2/DMT-1 is Associated with the Membrane of Phagosomes in Macrophages and Sertoli Cells. *Blood*, **100**, 2617–2621.

Knöpfel, M., Schulthess, G., Funk, F., and Hauser, H. (2000). Characterization of an Integral Protein of the Brush Border Membrane Mediating the Transport of Divalent Metal Ions. *Biophys. J.*, **79**, 874–884.

Lam-Yuk-Sung, S., Govoni, G., Forbes, J., and Gros, P. (2003). Iron Transport by Nramp2/DMT1: pH Regulation of Transport by Two Histidines in Transmembrane Domain 6. *Blood*, **101**, 3699–3707.

Lebron, J.A., Bennett, M.J., Vaughn, D.E., Chirino, A.J., Snow, P.M., Mintier, G.A., *et al.* (1998). Crystal Structure of the Hemochromatosis Protein HFE and Characterization of its Interaction with Transferrin Receptor. *Cell*, **93**, 111–123.

Lee, P.L., Gelbart, T., Halloran, C., and Beutler, E. (1998). The Human *Nramp2* Gene: Characterization of the Gene Structure, Alternative Splicing, Promoter Region and Polymorphisms. *Blood Cells, Mol. Dis.*, **24**, 199–215.

Levy, J.E., Montross, L.K., Cohen, D.E., Fleming, M.D., and Andrews, N.C. (1999). The C282Y Mutation Causing Hereditary Hemochromatosis does not Produce a Null Allele. *Blood*, **94**, 9–11.

Levy, J.E., Montross, L.K., and Andrews, N.C. (2000). Genes that Modify the Hemochromatosis Phenotype in Mice. *J. Clin. Invest.*, **105**, 1209–1216.

Ma, Y.X., Specian, R.D., Yeh, K.Y., Yeh, M., Rodriguez, P.J., and Glass, J. (2002). The Transcytosis of Divalent Metal Transporter 1 and Apo-transferrin during Iron Uptake in Intestinal Epithelium. *Am. J. Physiol. Gastrointest. Liver Physiol.*, **283**, G965–G974.

McKie, A.T., Barrow, D., Latunde-Dada, G.O., Rolfs, A., Sager, G., Mudaly, E., *et al.* (2001). An Iron-Regulated Ferric Reductase Associated with the Absorption of Dietary Iron. *Science*, **291**, 1755–1759.

Moos, T., Trinder, D., and Morgan, E.H. (2002). Effect of Iron Status on DMT1 Expression in Duodenal Enterocytes from β_2-Microglobulin Knockout Mice. *Am. J. Physiol. Gastrointest. Liver. Physiol.*, **283**, G687–G694.

Nuñez, M.T. and Tapia, V. (1999). Transferrin stimulates Iron Absorption, Exocytosis, and Secretion in Cultured Intestinal Cells. *Am. J. Physiol. Cell. Physiol.*, **276**, C1085–C1090.

Nuñez, M.T., Gaete, V., Watkins, J.A., and Glass, J. (1990). Mobilization of Iron from Endocytic Vesicles: The Effects of Acidification and Reduction. *J. Biol. Chem.*, **265**, 6688–6692.

Nuñez, M.T., Nuñez-Millacura, C., Beltran, M., Tapia, V., and Alvarez, H.X. (1997). Apotransferrin and Holotransferrin Undergo Different Endocytic Cycles in Intestinal Epithelia (Caco-2) Cells. *J. Biol. Chem.*, **272**, 19425–19428.

Oates, P.S. and Morgan, E.H. (1996). Defective Iron Uptake by the Duodenum of Belgrade Rats Fed Diets of Different Iron Contents. *Am. J. Physiol. Gastrointest. Liver. Physiol.*, **270**, G826–G832.

Oates, P.S., Thomas, C., Freitas, E., Callow, M.J., and Morgan, E.H. (2000). Gene Expression of Divalent Metal Transporter 1 and Transferrin Receptor in Duodenum of Belgrade Rats. *Am. J. Physiol. Gastrointest. Liver. Physiol.*, **278**, G930–G936.

Pal, P.K., Samii, A., and Calne, D.B. (1999). Manganese Neurotoxicity: A Review of Clinical Features, Imaging and Pathology. *Neurotoxicology*, **20**, 227–238.

Park, J.D., Cherrington, N.J., and Klaassen, C.D. (2002). Intestinal Absorption of Cadmium is Associated with Divalent Metal Transporter 1 in Rats. *Toxicol. Sci.*, **68**, 288–294.

Ponka, P. (1997). Tissue-Specific Regulation of Iron Metabolism and Heme Synthesis: Distinct Control Mechanisms in Erythroid Cells. *Blood*, **89**, 1–18.

Roetto, A., Totaro, A., Piperno, A., Piga, A., Longo, F., Garozzo, G., *et al.* (2001). New Mutations Inactivating Transferrin Receptor 2 in Hemochromatosis Type 3. *Blood*, **97**, 2555–2560.

Rolfs, A., Bonkovsky, H.L., Kohlroser, J.G., McNeal, K., Sharma, A., Berger, U.V., *et al.* (2002). Intestinal Expression of Genes Involved in Iron Absorption in Humans. *Am. J. Physiol. Gastrointest. Liver Physiol.*, **282**, G598–G607.

Roth, J.A., Horbinski, C., Feng, L., Dolan, K.G., Higgins, D., and Garrick, M.D. (2000). Differential Localization of Divalent Metal Transporter 1 with and without Iron Response Element in Rat PC12 and Sympathetic Neuronal Cells. *J. Neurosci.*, **20**, 7595–7601.

Russell, E.S., McFarland, E.C., and Kent, E.L. (1970a). Low Viability, Skin Lesions and Reduced Fertility Associated with Microcytic Anemia in the Mouse. *Transplant Proc.*, **2**, 144–151.

Russell, E.S., Nash, D.J., Bernstein, S.E., Kent, E.L., McFarland, E.C., Matthews, S.M., *et al.* (1970b). Characterization and Genetic Studies of Microcytic Anemia in House Mouse. *Blood*, **35**, 838–850.

Schwiebert, E.M., Egan, M.E., Hwang, T.-H., Fulmer, S.B., Allen, S.S., Cutting, G.R., *et al.* (1995). CFTR Regulates Outwardly Rectifying Chloride Channels through an Autocrine Mechanism Involving ATP. *Cell*, **81**, 1063–1073.

Shahidi, N.T., Nathan, D.G., and Diamond, L.K. (1964). Iron Deficiency Anemia Associated with an Error of Iron Metabolism in Two Siblings. *J. Clin. Inv.*, **43**, 510–521.

Sipe, D.M. and Murphy, R.F. (1991). Binding to Cellular Receptors Results in Increased Iron Release from Transferrin at Mildly Acidic pH. *J. Biol. Chem.*, **266**, 8002–8007.

Sladic-Simic, D., Pavic, D., Zivkovic, N., Marinkovic, D., and Martinovitch, P.N. (1963). Changes in the Offspring of Female Rats Exposed to 50 R of X Rays when Eight Days Old. *Br. J. Radiol.*, **36**, 542–543.

Sladic-Simic, D., Zivkovic, N., Pavic, D., Marinkovic, D., Martinovic, J., and Martinovitch, P.N. (1966). Hereditary Hypochromic Microcytic Anemia in the Laboratory Rat. *Genetics*, **53**, 1079–1089.

Sladic-Simic, D., Martinovitch, P.N., Zivkovic, N., Pavic, D., Martinovic, J., Kahn, M., *et al.* (1969). A Thalassemic-Like Disorder in Belgrade Laboratory Rats. *Ann. NY Acad. Sci.*, **165**, 93–99.

Stavem, P., Saltveldt, E., Elgjo, K., and Rootwelt, K. (1973). Congenital Hypochromic Microcytic Anaemia with Iron Overload of the Liver and Hyperferraemia. *Scand. J. Haematol.*, **10**, 153–160.

Su, M.A., Trenor, C.I., Fleming, J.C., Fleming, M.D., and Andrews, N.C. (1998). The G185R Mutation Disrupts Functions of the Iron Transporter Nramp2. *Blood*, **92**, 2157–2163.

Sylvester, S.R. and Griswold, M.D. (1994). The Testicular Iron Shuttle: A Nurse Function of the Sertoli Cells. *J. Androl.*, **15**, 381–385.

Tabuchi, M., Tanaka, N., Nishida-Kitayama, J.O., and Kishi, F. (2002). Alternative Splicing Regulates the Subcellular Localization of Divalent Metal Transporter 1 Isoforms. *Mol. Biol. Cell*, **13**, 4371–4387.

Tchernitchko, D., Bourgeois, M., Martin, M.E., and Beaumont, C. (2002). Expression of the Two mRNA Isoforms of the Iron Transporter Nrmap2/DMTI in Mice and Function of the Iron Responsive Element. *Biochem. J.*, **363**, 449–455. (Note: "Nrmap2" should be Nramp2 but the error is in the title as published.)

Trinder, D., Oates, P.S., Thomas, C., Sadleir, J., and Morgan, E.H. (2000). Localization of Divalent Metal Transporter 1 (DMT1) to the Microvillus Membrane of Rat Duodenal Enterocytes in Iron Deficiency, but to Hepatocytes in Iron Overload. *Gut*, **46**, 270–276.

Umbreit, J.N., Conrad, M.E., Hainsworth, L.N., and Simovich, M. (2002). The Ferrireductase Paraferritin Contains Divalent Metal Transporter as Well as Mobilferrin. *Am. J. Physiol. Gastrointest. Liver. Physiol.*, **282**, G534–G539.

Vidal, S.M., Belouchi, A., Cellier, M., Bewaty, B., and Gros, P. (1995). Cloning and Characterization of a Second Human NRAMP Gene on Chromosome 12q13. *Mamm. Genome*, **6**, 224–230.

Vulpe, C.D., Kuo, Y.M., Murphy, T.L., Cowley, L., Askwith, C., Libina, N., *et al.* (1999). Hephaestin, a Ceruloplasmin Homologue Implicated in Intestinal Iron Transport, is Defective in the *sla* Mouse. *Nat. Genet.*, **21**, 195–199.

Waheed, A., Grubb, J.H., Zhou, X.Y., Tomatsu, S., Fleming, R.E., Costaldi, M.E., *et al.* (2002). Regulation of Transferrin-Mediated Iron Uptake by HFE, the Protein Defective in Hereditary Hemochromatosis. *Proc. Natl. Acad. Sci. USA*, **99**, 3117–3122.

Wang, X., Ghio, A.J., Yang, F., Dolan, K.G., Garrick, M.D., and Piantadosi, C.A. (2002). Iron Uptake and Nramp2/DMT1/DCT1 in Human Bronchial Epithelial Cells. *Am. J. Physiol. Lung. Cell. Mol. Physiol.*, **282**, L987–L995.

Wareing, M., Ferguson, C.J., Green, R., Riccardi, D., and Smith, C.P. (2000). In Vivo Characterization of Renal Iron Transport in the Anaesthetized Rat. *J. Physiol.*, **524**, 581–586.

Watkins, J.A., Nuñez, M.T., Gaete, V., Alvarez, O., and Glass, J. (1991). Kinetics of Iron Passage through Subcellular Compartments of Rabbit Reticulocytes. *J. Membr. Biol.*, **119**, 141–149.

Worthington, M.T., Browne, L., Battle, E.H., and Luo, R.Q. (2000). Functional Properties of Transfected Human DMT1 Iron Transporter. *Am. J. Physiol. Gastrointest. Liver Physiol.*, **279**, G1265–G1273.

Worthington, M.T., Cohn, S.M., Miller, S.K., Luo, R.Q., and Berg, C.L. (2001). Characterization of a Human Plasma Membrane Heme Transporter in Intestinal and Hepatocyte Cell Lines. *Am. J. Physiol. Gastrointest. Liver Physiol.*, **280**, G1172–G1177.

Wyllie, S., Seu, P., and Goss, J.A. (2002). The Natural Resistance-Associated Macrophage Protein 1 Slc11a1 (Formerly Nramp1) and Iron Metabolism in Macrophages. *Microbes Infect.*, **4**, 351–359.

Yeh, K.Y., Yeh, M., Watkins, J.A., Rodriguez, P.J., and Glass, J. (2000). Dietary Iron Induces Rapid Changes in Rat Intestinal Divalent Metal Transporter Expression. *Am. J. Physiol. Gastrointest. Liver. Physiol.*, **279**, G1070–G1079.

Yoshida, K., Furihata, K., Takeda, S., Nakamura, A., Yamamoto, K., Morita, H., *et al.* (1995). A Mutation in the Ceruloplasmin Gene is Associated with Systemic Hemosiderosis in Humans. *Nat. Genet.*, **9**, 267–272.

Zoller, H., Pietrangelo, A., Vogel, W., and Weiss, G. (1999). Duodenal Metal-Transporter (DMT-1, NRAMP-2) Expression in Patients with Hereditary Haemochromatosis. *Lancet*, **353**, 2120–2123.

8

ANTONELLO PIETRANGELO*

Iron transporter ferroportin FPN1

INTRODUCTION

Iron is a major component of the Earth's crust, but its own chemistry greatly limits its utilization and also sets the basis for its toxicity. The capacity of readily exchanging electrons in aerobic conditions makes iron essential for fundamental cell functions, such as DNA synthesis, transport of oxygen and electrons, and cell respiration. In fact, iron deprivation threatens cell survival, thus making iron deficiency in humans a public health problem throughout the world and iron supplementation the only therapeutic option. On the other hand, since humans have no means of controlling iron excretion, excess iron, regardless of the route of entry, accumulates in parenchymal organs and threatens cell viability. In fact, a number of disease states are pathogenetically linked to excess body iron stores and iron removal therapy is an effective life-saving strategy in such circumstances.

Environmental iron exists in the ferric (Fe^{3+}) form, which is almost insoluble in water at neutral pH. In addition, iron, as polar hydrophilic metal ion, is unable to cross membranes. Therefore specialized transport systems and membrane carriers have evolved in humans to maintain iron in a "soluble" state suitable for circulation in the bloodstream [i.e., bound to serum transferrin (Tf)] or to transfer it across cell membranes [through metal transporters such as divalent metal transporter 1 (DMT-1) or ferroportin1/IREG1/MTP-1, the product of the *SLC11A3* gene] for tissue utilization.

As mentioned above, iron toxicity arises from its own chemistry. In every cell during its normal life in aerobic conditions, a small amount of the consumed oxygen is reduced in a specific way, yielding a variety of highly reactive chemical entities collectively called reactive oxygen species (ROS). Transition metal ions such as iron, which frequently have unpaired electrons, are excellent catalysts and play a decisive role in the generation of very reactive species from less reactive ones, for instance by catalyzing the formation of hydroxyl radicals from reduced forms of O_2 (Cadenas 1989). Therefore the modulation of "free iron" availability and appropriate sequestration of iron is the main means by which cells keep ROS levels under strict control. In this vein, to avoid oxidative stress and cell damage, cells have developed systems to adjust intracellular iron concentration to levels that are adequate for their metabolic needs but are below the toxicity threshold.

Although in recent years we have learnt much about the way that iron enter cells, very

*Unit for the study of Disorders of Iron Metabolism, Department of Internal Medicine, University of Modena and Reggio Emilia, Via del Pozzo 71, 41100 Modena, Italy

little is known about the route(s) used by iron to exit cells, whether they are specialized for iron transport (i.e., villus intestinal cells), storage (e.g., hepatocytes), or recycling (macrophagic/reticuloendothelial cells). This chapter describes a novel hereditary iron overload disease associated with the malfunction of the main iron export protein in mammals (ferroportin1/IREG1/MTP-1), the product of the *SLC11A3* gene.

THE DISEASE DUE TO MUTATION OF THE *SLC11A3* GENE

Historical description

In 1999 an autosomal dominant form of hereditary iron overload similar to classical hemochromatosis was described (Pietrangelo *et al.* 1999). Classical hemochromatosis (HC), synonymous with genetic hereditary primary iron overload, represents one of the most common single-gene hereditary diseases and is the paradigm of a genetic disorder leading to body iron overload and multiorgan failure. HC is due to inappropriately increased iron absorption in which iron loading of parenchymal cells in the liver, pancreas, heart, and other organs impairs the function and damages the structure of these organs (EASL 2000). A cornerstone of HC genetics was laid in 1996, with the isolation of the hemochromatosis gene, now called *HFE* (Feder *et al.* 1996). The majority of HC patients (83–100% in different series) carry the same mutation, resulting in a change from cysteine at position 282 to tyrosine (*C282Y*) in the HFE protein (Merryweather-Clarke *et al.* 1997). The human HFE protein is closely related to the major histocompatibility complex (MHC) class I molecules. The *C282Y* mutation disrupts a critical disulfide bond in the α_3 domain of the *HFE* protein and abrogates binding of the mutant *HFE* protein to β_2-microglobulin (Feder *et al.* 1997). This results in reduced transport to and expression on the cell surface. HFE can influence iron homeostasis by interacting with transferrin

receptor 1 (TfR1), the main protein devoted to iron uptake in most cells (Lebron *et al.* 1998; Bennett *et al.* 2000). As demonstrated in macrophages from HC patients (Montosi *et al.* 2002), it is likely that a mutated HFE impairs Tf-iron accumulation in cells important for iron homeostasis, such as intestinal crypt cells and macrophages, secondarily leading to intracellular iron deficiency and uncontrolled intestinal iron absorption (Pietrangelo 2002).

However, the hereditary iron overload disease described by Pietrangelo *et al.* (1999) was not due to the presence of the C282Y mutation of the *HFE* gene. Direct gene sequencing ruled out the possibility of other pathogenic mutations in the coding regions as well as the intron–exon boundaries of the *HFE* gene. Finally, the "hemochromatosis" phenotype did not cosegregate with specific haplotypes of the short arm of chromosome 6, where the *HFE* gene lies, excluding the possibility that the disease was due to mutation(s) in the enhancer–promoter regions of the *HFE* gene or other genes lying on 6p. Phenotypically, the disease had very unusual features compared with classical HC (see below). In conclusion, this study indicated the existence of a distinct genetic disease(s) causing adult hereditary iron overload beyond the *HFE*-associated form, possibly due to mutations in a gene involved in iron recycling in reticuloendothelial macrophagic cells.

In order to identify the responsible gene, a genome-wide screen was performed analyzing 375 markers at approximately 10 cM intervals. Evidence of linkage for two markers (D2S364 and D2S117) from 2q32 was found and a region of approximately 5 cM was defined (Montosi *et al.* 2001). Most of the known genes within this region had no apparent connection with iron homeostasis. However, one compelling candidate gene, *SLC11A3* (encoding ferroportin, also known as IREG1, MTP1; DDBJ/EMBL/GenBank accession number NM_014585) (Donovan *et al.* 2000; Mckie *et al.* 2000; Abboud and Haile 2000), was located within the candidate interval. The *SLC11A3* gene was analyzed by amplifying

Table 1 Distinguishing clinical features of the *SLC11A3* associated hereditary iron overload disease

Symptoms and signs	*SLC11A3* associated disease	HFE-associated hemochromatosis
Serum ferritin	High since childhood	Normal at early stages; high after second to third decade
Transferrin saturation	Low in early stages; increasing after second to third decade	Increasing from early stages
Hypochromic anemia	Present in young females	Absent
Organ disease	Mild to moderate	Severe
Treatment at diagnosis	Soft phlebotomy regimen; possibly iron chelation therapy	Intense phlebotomy regimen

each exon and determining their DNA sequences. All affected patients were heterozygous for a C to A substitution in exon 3 that results in replacement of alanine 77 with aspartate. The A77D substitution was not found in 25 healthy family members or in 100 normal blood donors from the same geographic area.

To date, two additional mutations in the *SLC11A3* gene have been identified in subjects with an iron overload condition identical with that described by Pietrangelo *et al.* (1999): a N144H change in a Dutch pedigree (Njajon *et al.* 2001), and a 3 bp deletion in exon 5 leading to a Val162 deletion (Devalia *et al.* 2002; Wallace *et al.* 2002).

Clinical features

The disease due to mutations of the *SLC11A3* gene has distinctive features compared with classical hemochromatosis (Table 1) (Pietrangelo *et al.* 1999; Montosi *et al.* 2001). First, inheritance is of the autosomal dominant form as opposed to the autosomal recessive pattern of HC. In the reported pedigrees, all affected individuals carry a *SLC11A3* change at the heterozygote state in each generation consistent with a dominant transmission.

Biochemically, the earliest abnormality is an increase in serum ferritin levels in childhood with normal or low transferrin saturation rate. In young females, a hypochromic anemia

is a consistent finding and often requires oral iron supplementation. In the second to third decades the transferrin saturation rate also increases. The pattern of serum ferritin and transferrin saturation changes in classical HC during the course of the disease is usually the exact opposite. The biochemical penetrance of the genetic defect appears to be complete. Clinically, the disease is milder than HC; although the full spectrum of clinical symptoms typical of hemochromatosis has been reported in this disease, the phenotype is usually less severe. Despite severe iron burden, liver disease is limited to a precirrhotic stage with various degrees of sinusoidal and portal fibrosis. Glucose intolerance and arthropathy have been also described. More observations are needed to understand fully the clinical penetrance of the mutations, but it is likely that additional environmental (i.e., alcohol abuse, hepatitis viruses, etc.) and genetic (HFE and non-HFE gene status) factors may influence the final clinical manifestation.

Although, phlebotomy is the current treatment of the disease, as for HC, some individuals cannot tolerate a weekly phlebotomy program and slight anemia and low transferrin saturation may rapidly be reached despite the serum ferritin level still being elevated. With a less aggressive phlebotomy regimen, they can also be iron depleted, although serum ferritin levels may stay at the upper normal laboratory limit.

The *SLC11A3* gene product: ferroportin/Ireg/MTP

In 2000, three groups, using different strategies, isolated and characterized the product of the *SLC11A3* gene referred to as ferroportin1/IREG1/MTP1 (Donovan *et al.* 2000; Mckie *et al.* 2000; Abboud *et al.* 2000). Donovan and coworkers a positional cloning strategy to identify the gene responsible for the hypochromic anemia of the zebrafish mutant *weissherbst*. The identified gene encoded a multiple transmembrane domain protein, ferroportin1, expressed in the yolk sac and required for the transport of iron from maternally derived yolk stores to the circulation; it functioned as an iron exporter when expressed in *Xenopus* oocytes. In order to identify new genes that are important in the regulation of iron metabolism, Abboud and Haile constructed a library of mRNA sequences enriched for iron regulatory protein-1 (IRP-1) binding (see below) using SELEX technology. A 200 bp sequence corresponding to the 5′-UTR of metal transporter protein 1 (MTP 1) was isolated from this library and used to probe a day-15.5 mouse embryo cDNA library and subsequently to isolate and complete an MTP1 clone. Interestingly, HEK293T cells transiently transfected with a MTP-1 cDNA showed a significant decrease of intracellular ferritin and activation of the IRP-1 system, indicating that the MTP-1 overexpression leads to intracellular and cytosolic iron depletion. McKie and coworkers used a subtracted cDNA library from duodenal mucosa of hypotransferrinemic (hpx/hpx) mice to identify genes important for intestinal iron absorption. They isolated a cDNA clone particularly expressed during iron deficiency (Ireg, for iron-regulated mRNA). Ireg1 was localized basolaterally in polarized epithelial cells, was able to induce iron efflux in the *Xenopus* oocyte system, was dramatically induced in the duodenum by iron deficiency and hypoxia, and was developmentally controlled (i.e., lower in the neonatal duodenum and progressively higher during growth and development).

The localization of ferroportin1/Ireg1/MTP1 in cells and tissues is consistent with its proposed function of exporting iron from cells. It is expressed in several cell types that play critical roles in mammalian iron metabolism, including placental syncytiotrophoblasts, duodenal enterocytes, hepatocytes, and reticuloendothelial macrophages. In fact, it has been postulated to function in placental maternofetal iron transfer, intestinal iron absorption, and in release of iron from hepatocytes and reticuloendothelial macrophages (reviewed by Andrews 2000). In humans it is highly expressed in conditions of enhanced iron absorption, such as anemia and HC (Zoller *et al.* 2001). Recently, a key role of the protein in the pathogenesis of the hypoferremia of anemia of chronic disease, in which several perturbations of iron homeostasis take place including iron trapping in reticulo endothelial (RE) cells, has been proposed (Yang *et al.* 2002).

How is the *SLC11A3* gene controlled? The ferroportin1/IREG1/MTP1 mRNA possesses an iron-responsive element in the 5′ untranslated region that binds IRPs and confers iron-dependent regulation of luciferase in desferoxamine-treated COS7 cells (Abboud and Haile 2000). The IRPs represent the *sensors* of cytoplasmic iron and the putative *controllers* of the main proteins involved in iron homeostasis: ferritin (Ft), TfR1, ferroportin1/Ireg1/MTP1, and DMT1 (previously known as Nramp2 and DCT1) (Fleming *et al.* 1998). These proteins are used by cells to adjust intracellular iron concentration to levels that are adequate for their metabolic needs but below the toxicity threshold. The target proteins contain discrete non-coding sequences [i.e., the iron-responsive elements (IREs)] at the untranslated region (UTR) of the mRNA that are recognized by two IRPs (IRP-1 and IRP-2). The IRP belong to the aconitase superfamily: by means of an Fe–S cluster-dependent switch, IRP-1 can function as an mRNA binding protein or as an enzyme that converts citrate to isocitrate (Cairo and Pietrangelo 2000). Although structurally and functionally similar to IRP-1, IRP-2 does not seem to assemble a cluster or to possess aconitase activity; moreover, it has a distinct pattern of tissue expression and is modulated by means of proteasome-mediated degradation. In response to fluctuations in the level of the "free iron pool," IRPs act as key regulators of cellular iron homeostasis as a result of the translational control of the expression of a number of iron genes. When iron is scarce, Ft and TfR1 mRNAs are specifically recognized and bound by the active form of IRP that blocks Ft translation (IRE at the 5′ UTR of the mRNA) and stabilizes TfR1 mRNAs (IRE at the 3′ UTR of the mRNA), respectively. On the contrary, when iron is abundant, IRPs are devoid of mRNA binding activity and target transcripts are freely accessible to translation complexes (ferritin mRNA) or nucleases for degradation (TfR1 mRNA). Therefore IRPs control cell iron status by means of divergent but coordinated regulation of iron storage (Ft) and uptake (TfR1). Seemingly, since ferroportin1/Ireg1/MTP1 possesses an IRE at the 5′ UTR, low iron should activate IRP and the latter should block ferroportin1/Ireg1/MTP1 mRNA translation, resulting in a decrease in iron export, whereas high iron should lead to enhanced mRNA translation and protein accumulation, leading to increased iron export. However, the protein levels found in vivo under varying conditions of iron repletion are not entirely consistent with this type of regulation. In a recent study (Abboud and Haile 2000), immunohistochemical staining for the iron exporter revealed strong reactivity in the Kupffer cells of iron-replete mice with weaker staining in iron-depleted mice, consistent with regulation through the IRPs. However, the opposite result was found in the duodenal epithelium: immunohistochemical staining was strong in iron-depleted mice and weaker in iron-replete mice. However, it should be noted that the mechanisms controlling duodenal iron absorption are complex and only partially understood (Pietrangelo 2002). It is likely that other stimuli, such as hypoxia or erythron demands, may be operative in vivo and overcome an IRP-dependent translational control in the duodenum. These additional mechanisms may be particularly important in anemic situations (Mckie et al. 2000; Abboud and Haile 2000). Indeed, this is the case for Tfr1, a gene usually controlled post-trancriptionally by IRP, which is also controlled transcriptionally under hypoxic conditions (Tacchini et al. 1999).

Molecular pathogenesis

Consensus structural predictions for ferroportin1/IREG1/MTP1 protein have been performed (Devalia et al. 2000). The structure of the protein shown in Figure 1 has nine

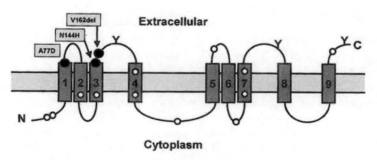

Figure 1   Predicted structure of ferroportin1/IREG1/MTP1 with the positions of the three mutations reported so far in humans (black circles), 12 Cys residues (white circles), and three putative exposed N-linked oligosaccharide sites (Y) marked as shown. The N- and C-termini are denoted by N and C, respectively. Modified from Wallace et al. (2002).

transmembrane helices. The distribution of 12 Cys residues in human, mouse, and rat ferroportin1 within the lipid bilayer and on the two surfaces is consistent with the formation of disulfide bridges between pairs of Cys residues within each of these regions. Six prediction methods for the helix topology favored the location of the N-terminus of ferroportin1 to be within the cell cytoplasm (Devalia et al. 2000). This is supported by the location of three of the four putative N-glycosylation sites on the external loops of ferroportin1, as expected for a membrane protein (the fourth is predicted to be inside transmembrane 2 and is assumed to be unglycosylated).

The reported mutations of the SLC11A3 gene product are all located in a cluster at either the end of the transmembrane helices 1 and 3 or at the extracellular loop at the C-terminal end of helix 3. The A77D mutation results in replacement of alanine 77, a small hydrophobic amino acid, by aspartate, a large negatively charged amino acid. A77 lies close to or within the first predicted transmembrane domain and is conserved across vertebrate species. This mutation modifies a predicted myristoylation site and may affect the secondary structure of the protein (Montosi et al. 2001). The N144H mutation leads to the replacement of Asp (a neutral amino acid) by His (a polar residue) and is predicted to reduce the hydrophobicity of the transmembrane domain (Njajou et al. 2001). The Val162del is likely due to mispairing by a slipped strand. It seems to be a common mutation in the SLC11A3 gene, being reported in families of different ethnic origin (Devalia et al. 2002; Wallace et al. 2002; Roetto et al. 2002), suggesting that it results from multiple deletion events. The region directly adjacent to the mutant residues described so far shows homology with several other divalent metal transporters, suggesting that it may be involved in iron binding and/or transport properties or may define a functional binding site for a protein that is important for the export of iron from the cell.

The proposed function of the SLC11A3 gene product in iron export is consistent with the phenotype of the disease described early and with the original hypothesis of a selective disturbance of iron recycling in RE cells (Pietrangelo et al. 1999). A gain-of-function mutation might augment transfer of iron across the absorptive epithelium of the intestine, resulting in increased accumulation of dietary iron. This model is most consistent with the dominant inheritance of this disorder. However, the data collected so far, and particularly the phenotype described with selective iron accumulation in RE cells, are better explained by a mutation resulting in a loss of protein function. In this context, a loss-of-function mutation could lead to a perturbation of iron distribution within the body. It might cause a mild but significant impairment of iron recycling by RE macrophages, which normally must process and release a large quantity of iron derived from the lysis of senescent erythrocytes (Figure 2). A defective iron exit from RE cells may explain the early rise in serum ferritin levels and low transferrin saturation; the latter is due to inability of RE cells to load circulating transferrin with iron. As a consequence, iron retention by macrophages would then lead to decreased availability of iron for the hematopoietic system and anemia. This will then result in activation of feedback mechanisms to increase intestinal absorption, compensating for the latent iron deficiency, but contributing to iron overload (Figure 2). This model is consistent with the finding that patients with the SLC11A3 mutations in ferroportin have much larger reticuloendothelial iron stores than do patients with other forms of hemochromatosis. Although the patients are not anemic in adulthood, indicating that adequate iron is available for normal erythropoiesis, they may be anemic earlier in life, particularly young females (Pietrangelo et al. 1999; Montosi et al. 2001). Furthermore, patients may show a reduced tolerance to phlebotomy and become anemic on therapy despite persistently elevated serum ferritin values. The presence of a mutated ferroportin1/

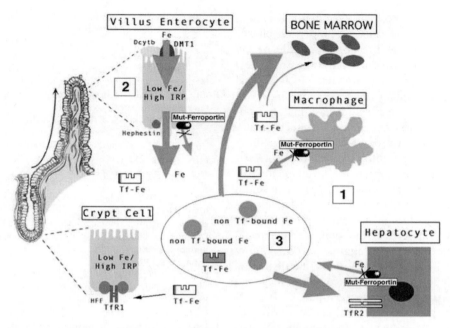

Figure 2 Pathophysiology of the *SLC11A3* disease. (1) The first anomaly is a defective release of iron, particularly from reticuloendothelial cells that normally release iron to circulating transferrin to meet the bone marrow's needs for erythropoiesis. The clinical consequences in affected individuals are hypochromic anemia with high serum ferritin and low transferrin saturation. At the tissue level this will reflect iron overload, preferentially in reticuloendothelial cells. (2) Inadequate transferrin-iron supply to the erythron and low transferrin-iron delivery to crypt cells will lead to enhanced iron absorption [see also Pietrangelo (2002) for regulatory mechanisms involved in iron absorption]. (3) Increased intestinal iron absorption will compensate for the relative erythron iron deficiency and reverses anemia, while contributing to tissue iron overload. Biochemically, this stage is characterized by increasing transferrin saturation, whereas, at the tissue level, parenchymal iron overload (e.g., hepatocytes in the liver) will be more evident.

IREG1/MTP1 at the basolateral membrane of enterocytes is not in contradiction with the putative role of the protein in intestinal iron transfer or with the proposed pathogenesis of the disease. In fact, the possibility exists that other iron export proteins may overcome the ferroportin defect or, more likely, the protein haploinsufficiency could represent a critical threshold for iron export in RE cells but still be sufficient to allow iron export from enterocytes. It is likely that in the near future the availability of transgenic animal models reproducing the human SLC11A3 disease will allow us to prove or disprove this model.

SUMMARY

The *SLC11A3* gene encodes for a main iron export protein in mammals: ferroportin1/IREG1/MTP-1. This multiple transmembrane domain protein is likely to function in placental materno-fetal iron transfer, in intestinal iron absorption, and in release of iron from hepatocytes and reticuloendothelial macrophages. Recently, a disease associated with systemic iron overload and due to mutations in the *SLC11A3* gene has been described. It is an autosomal dominant iron overload condition characterized by iron accumulation in several organs, mainly those involved in iron recycling such as liver and spleen. The disease phenotype is linked to the dominant role of the *SLC11A3* gene product, ferroportin1/IREG1/MTP-1, in iron export from reticuloendothelial cells. The clinical phenotype of this novel hereditary iron overload disease may be confused with classical hereditary hemochromatosis, but the disorder shows distinguishing features such as autosomal dominant inheritance, early and preferential accumulation of iron in tissue macrophages, early rise in serum ferritin despite normal or low transferrin saturation, hypochromic anemia despite normal or high serum ferritin in young females, and reduced tolerance to aggressive phlebotomy regimens. The pathogenesis of the disease is due to a selective disturbance of iron recycling and trafficking due to ferroportin1/IREG1/MTP-1 protein malfunction. In particular, a defective iron egress from reticuloendothelial macrophages is responsible for tissue iron accumulation, reduced transferrin-iron delivery to the erythron and possibly compensatory enhanced intestinal iron absorption. Several families presenting with the *SLC11A3* associated disease are now reported worldwide. This disease is likely to represent the most common hereditary iron overload disease apart from classical HFE-associated hemochromatosis.

REFERENCES

Abboud, S. and Haile, D.J. (2000). A Novel Mammalian Iron-Regulated Protein Involved in Intracellular Iron Metabolism. *J. Biol. Chem.*, **275**, 19906–19912.

Andrews, N.C. (2000). Iron Homeostasis: Insights from Genetics and Animal Models. *Nat. Rev. Genet.*, **1**, 208–217.

Bennett, M.J., Lebron, J.A., and Bjorkman, P.J. (2000). Crystal Structure of the Hereditary Haemochromatosis Protein HFE Complexed with Transferrin Receptor. *Nature*, **403**, 46–53.

Cadenas, E. (1989). Biochemistry of Oxygen Toxicity. *Annu. Rev. Biochem.*, **58**, 79–110.

Cairo, G. and Pietrangelo, A. (2000). Iron Regulatory Proteins in Pathobiology. *Biochem. J.*, **352**, 241–250.

Devalia, V., Carter, K., Walker, A.P., Perkins, S.J., Worwood, M., May, A., *et al.* (2002). Autosomal Dominant Reticuloendothelial Iron Overload Associated

with a 3-Base Pair Deletion in the Ferroportin 1 Gene (SLC11A3). *Blood*, **100**, 695–697.

Donovan, A., Brownlie, A., Zhou, Y., Shepard, J., Pratt, S.J., Moynihan, J., *et al.* (2000). Positional Cloning of Zebrafish Ferroportin1 Identifies a Conserved Vertebrate Iron Exporter. *Nature*, **403**, 776–781.

EASL (2000). EASL International Consensus Conference on Haemochromatosis – Part III. Jury Document. *J. Hepatol.*, **33**, 496–504. (FTXT Journals@OVID)

Feder, J.N., Gnirke, A., Thomas, W., Tsuchihashi, Z., Ruddy, D.A., Basava, A., *et al.* (1996). A Novel MHC Class I-Like Gene Is Mutated in Patients with Hereditary Haemochromatosis. *Nat. Gen.*, **13**, 399–408.

Feder, J.N., Tsuchihashi, Z., Irrinki, A., Lee, V.K., Mapa, F.A., Morikang, E., *et al.* (1997). The Hemochromatosis Founder Mutation in HLA-H Disrupts Beta2-Microglobulin Interaction and Cell Surface Expression. *J. Biol. Chem.*, **272**, 14025–14028.

Fleming, M.D., Romano, M.A., Su, M.A., Garrick, L.M., Garrick, M.D., and Andrews, N.C. (1998). Nramp2 is Mutated in the Anemic Belgrade (b) Rat: Evidence of a Role for Nramp2 in Endosomal Iron Transport. *Proc. Natl Acad. Sci. USA*, **95**, 1148–1153.

Lebron, J.A., Bennett, M.J., Vaughn, D.E., Chirino, A.J., Snow, P.M., Mintier, G.A., *et al.* (1998). Crystal Structure of the Hemochromatosis Protein HFE and Characterization of its Interaction with Transferrin Receptor. *Cell*, **93**, 111–123.

McKie, A.T., Marciani, P., Rolfs, A., Brennan, K., Wehr, K., Barrow, D., *et al.* (2000). A Novel Duodenal Iron-Regulated Transporter, IREG1, Implicated in the Basolateral Transfer of Iron to the Circulation. *Mol. Cell*, **5**, 299–309.

Merryweather-Clarke, A.T., Pointon, J.J., Shearman, J.D., and Robson, K.J.H. (1997). Global Prevalence of Putative Haemochromatosis Mutations. *J. Med. Gen.*, **34**, 275–278.

Montosi, G., Paglia, P., Garuti, C., Guzman, C.A., Bastin, J.M., Colombo, M.P., *et al.* (2000). Wild-Type HFE Protein Normalizes Transferrin Iron Accumulation in Macrophages from Subjects with Hereditary Hemochromatosis. *Blood*, **96**, 1125–1129.

Montosi, G., Donovan, A., Totaro, A., Garuti, C., Pignatti, E., Cassanelli, S., *et al.* (2001). Autosomal-Dominant Hemochromatosis is Associated with a Mutation in the Ferroportin (SLC11A3) Gene. *J. Clin. Invest.*, **108**, 619–623.

Njajou, O.T., Vaessen, N., Joosse, M., Berghuis, B., van Dongen, J.W., Breuning, M.H., *et al.* (2001). A Mutation in SLC11A3 is Associated with Autosomal Dominant Hemochromatosis. *Nat. Genet.*, **28**, 213–214.

Pietrangelo, A. (2002). Physiology of Iron Transport and the Hemochromatosis Gene. *Am. J. Physiol. Gastrointest. Liver Physiol.*, **282**, G403–G414.

Pietrangelo, A., Montosi, G., Totaro, A., Garuti, C., Conte, D., Cassanelli, S., *et al.* (1999). Hereditary

Hemochromatosis in Adults without Pathogenic Mutations in the Hemochromatosis Gene. (See Comments.) *N. Engl. J. Med.*, **341**, 725–732.

Roetto, A., Merryweather-Clarke, A.T., Daraio, F., Livesey, K., Pointon, J.J., Barbabietola, G., *et al.* (2002). A Valine Deletion of Ferroportin 1: A Common Mutation in Hemochromastosis Type 4. *Blood*, **100**, 733–734.

Tacchini, L., Bianchi, L., Bernelli-Zazzera, A., and Cairo, G. (1999). Transferrin Receptor Induction by Hypoxia. HIF-1-Mediated Transcriptional Activation and Cell-Specific Post-transcriptional Regulation. *J. Biol. Chem.*, **274**, 24142–24146.

Wallace, D.F., Pedersen, P., Dixon, J.L., Stephenson, P., Searle, J.W., Powell, L.W., *et al.* (2002). Novel Mutation in Ferroportin1 is Associated with Autosomal Dominant Hemochromatosis. *Blood*, **100**, 692–694.

Yang, F., Liu, X.B., Quinones, M., Melby, P.C., Ghio, A., and Haile, D.J. (2002). Regulation of Reticuloendothelial Iron Transporter MTP1 (Slc11a3) by Inflammation. *J. Biol. Chem.*, **277**, 39786–39791.

Zoller, H., Koch, R.O., Theurl, I., Obrist, P., Pietrangelo, A., Montosi, G., *et al.* (2001). Expression of the Duodenal Iron Transporters Divalent-Metal Transporter 1 and Ferroportin 1 in Iron Deficiency and Iron Overload. *Gastroenterology*, **120**, 1412–1419.

TRANSPORTERS OF ENERGY METABOLITES AND BUILDING BLOCKS

9

STEFAN BRÖER*

Introduction

Two major biochemical functions of mammalian cells are discussed in this chapter. One is the degradation of metabolites to generate energy for cellular processes and the second is the transport of building blocks during synthesis and breakdown of macromolecules. The two functions are discussed together because there are significant overlaps between them. For example, glucose is the main cellular energy metabolite but in addition is a building block of glycogen, and glucose derivatives are building blocks of glycoproteins; amino acids are the building blocks of proteins and are the main energy metabolites during fasting and starvation. The overlap between energy metabolism and synthesis/degradation of macromolecules also becomes apparent in the phenotype of transporter pathologies, which frequently affect both energy metabolism and synthesis of cellular components.

ENERGY METABOLISM

Energy metabolism in mammalia, including humans, occurs at two levels, the cellular and the interorgan level. At the cellular level four main classes of metabolites are used to meet cellular energy demands, namely *glucose*, *fatty acids*, *ketone bodies*, and *amino acids*. At the level of tissues and organs an interorgan

traffic of energy metabolites is observed and the plasma levels of most metabolites are well regulated. Both levels of energy metabolism are discussed in detail in biochemistry textbooks. In the following a short review with an emphasis on transport processes required for a functional energy metabolism is presented.

Cellular energy metabolism

Glucose is the main energy metabolite for most cells under normal physiological conditions, and it is the sole source of energy for the brain under all circumstances. It enters the cell by uniporters of the SLC2 family (Figure 1) (Barrett *et al.*, 1999). Subsequently, glucose is metabolized to pyruvate by the glycolytic pathway. Pyruvate enters the mitochondrion via a novel member of the SLC25 family. Inside the mitochondrion it is converted into acetyl-CoA, which is fully oxidized in the TCA cycle, generating three molecules of $NADH+H^+$, one molecule of $FADH_2$, and two molecules of CO_2. $NADH+H^+$ and $FADH_2$ are oxidized by the respiratory chain, which generates a proton-motive force across the inner mitochondrial membrane during this process. In turn, the proton-motive force is used to synthesize ATP inside the mitochondrion by F_1F_0-ATPase. However, the majority of the generated ATP is used in the cytosol for biosynthetic purposes. It is transported out of the mitochondria by the ATP/ADP antiporter (SLC25 family), which at the same time imports ADP generated from ATP hydrolysis.

* School of Biochemistry and Molecular Biology, Faculty of Science, Australian National University, Canberra, ACT 0200, Australia

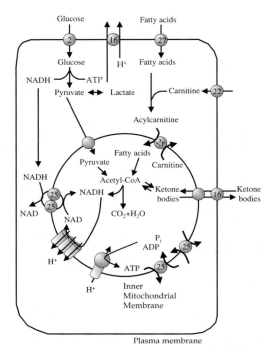

Figure 1 Transporters of energy metabolism. Overview of transport processes involved in the energy metabolism of mammalian cells. Transporters are represented by circles. The number in the circle indicates the solute carrier family.

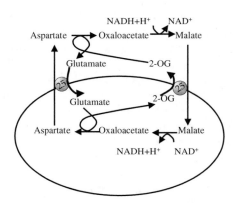

Figure 2 The mitochondrial malate–aspartate shuttle. The malate–aspartate shuttle results in a net translocation of the reducing equivalents of NADH+H$^+$ from the cytosol into the mitochondrion. It involves two transporters of the SLC25 family, the aspartate–glutamate antiporter and the oxoglutarate–malate antiporter. 2-OG, 2-oxoglutarate.

The phosphate, required for synthesis of ATP from ADP, is imported by a separate phosphate/OH$^-$ antiporter (SLC25 family). During glycolysis an additional molecule of NADH+H$^+$ is produced in the cytosol, but its oxidation takes place in the mitochondrion. There is no transporter that accepts NADH as a substrate. Instead, its redox equivalents are transferred into the mitochondrion by the malate–aspartate shuttle (Figure 2). NADH+H$^+$ is used to reduce oxaloacetate to malate, which is transported into mitochondria in exchange for 2-oxoglutarate by the oxoglutarate carrier (SLC25 family). Malate is converted back to oxaloacetate inside the mitochondria, thereby regenerating NADH+H$^+$. To generate a catalytic cycle that only transports reducing equivalents of NADH+H$^+$ into mitochondria, all substrates have to be recycled. This is achieved by converting oxaloacetate into aspartate by transamination, a reaction that at the same time converts glutamate into 2-oxoglutarate. Aspartate is exchanged against cytosolic glutamate by the aspartate–glutamate exchanger (SLC25 family) (see Chapter 10). In the cytosol, aspartate is converted back to oxaloacetate. This transamination regenerates glutamate from 2-oxoglutarate (Figure 2). The malate–aspartate and glycerolphosphate shuttles (Chapter 10) transfer reducing equivalents from the cytosol into the mitochondria for respiration. Although the malate–aspartate shuttle appears to be critical for energy metabolism, mutations in the aspartate–glutamate carrier affect metabolism in a much more complex way. The aspartate–glutamate carrier also provides cytosolic aspartate as a substrate for the urea cycle, for protein synthesis, and as a precursor for gluconeogenesis. Hence the aspartate–glutamate carrier plays an essential role in several metabolic pathways which are discussed in more detail in Chapter 10.

Instead of being oxidized in the mitochondria, pyruvate can also be reduced to lactate, thereby recycling the NADH+H$^+$ that is generated during glycolysis in the cytosol. Lactate is subsequently released from cells by a H$^+$-lactate symporter (SLC16 family)

(Halestrap and Price 1999), which at the same time disposes of the proton that is being generated during glycolysis (Figure 1). Conversion of pyruvate to lactate occurs not only under anaerobic conditions in muscle cells but also under aerobic conditions in certain cell types, such as brain astrocytes. Monocarboxylate transporters of the SLC16 family also mediate the uptake of *ketone bodies*, such as hydroxybutyrate and acetoacetate, into cells. These compounds are generated from acetyl-CoA as a result of fatty acid metabolism in the liver (see below). Lactate and ketone bodies are used as energy metabolites proportionally to their concentration in the blood by all extrahepatic tissues except the brain. They are oxidized inside mitochondria to produce $NADH+H^+$ and acetyl-CoA. As products of fatty acid metabolism they are produced mainly during fasting (between meals) and starvation.

Fatty acids contribute over 40% of the caloric intake in Western diets and become a major energy metabolite for the body when glycogen stores are used up during exercise. After being released from the adipose tissue, fatty acids are largely bound to serum protein in the plasma because of the hydrophobic nature of the molecule. As a result, the concentration of free fatty acids in the plasma is exceedingly low (about 7.5 nM) (Hajri and Abumrad 2002). Diffusion of free fatty acids across lipid bilayers is a very slow process at this concentration because it requires flipping of this amphiphilic molecule from one leaflet to the opposite one. This reaction is facilitated by fatty acid transporters (SLC27) (Figure 1) (Stahl *et al.* 2001). The molecular identity of fatty acid transporters is not fully resolved. Members of the SLC27 family have some similarity with acyl-CoA synthetases and might combine transport with further metabolism. Another molecule that has been implicated in fatty acid transport is the two-transmembrane-helix protein CD36, which might act as a receptor for fatty acids. Expression of both proteins in mammalian cells increases the uptake of free fatty acids, although the mechanism of uptake remains to be defined.

Ultimately, fatty acids are broken down in mitochondria by β-oxidation. This cyclic process oxidizes C3 of fatty acids, thereby generating $NADH+H^+$. During each cycle one acetyl-CoA unit is generated from C1 and C2 of the fatty acid, which is finally oxidized in the mitochondrial tricarboxylic acid (TCA) cycle and the respiratory chain. An unusual transport principle is used to translocate fatty acids across the inner mitochondrial membrane. Fatty acids are esterified to carnitine, resulting in acyl-carnitine residues which are the substrate of the acyl-carnitine/carnitine exchanger of mitochondria (SLC25) (see Chapter 11, Figure 1). Carnitine synthesis by the body is not sufficient and thus needs to be supplemented by nutritional sources. Carnitine uptake across the plasma membrane is mediated by a transporter of the organic cation transporter family (SLC22). Mutations in this transporter result in *carnitine deficiency*, which in turn generates problems of fatty acid metabolism, which are described in detail in Chapter 11.

Very-long-chain fatty acids and branched fatty acids ingested from plant nutrition cannot be digested in the mitochondria but require predigestion in the peroxisome. Transport of these compounds across the peroxisomal membrane is mediated by ABC transporters (see Chapter 17).

Amino acids are the building blocks for protein synthesis, but at the same time are valuable energy metabolites and the main precursors for gluconeogenesis. Amino acids are transported by a combination of uniporters, antiporters, and cotransporters (SLC1, SLC3, SLC6, SLC7, SLC36, and SLC38) (Bröer 2002). Most amino acid transporters are antiporters. In order to achieve a net transport into the cell for protein synthesis and energy metabolism, antiporters have overlapping substrate specificity with the uniporters and cotransporters. Thus uptake of an amino acid, which is lacking in the cytosol, can occur only at the expense of another intracellular amino acid that is in excess. Net accumulation of amino acids is carried out by Na^+-amino acid cotransporters (SLC38), which are activated on demand (i.e., when amino acids are lacking). The large number of amino acid transporters reflects the diverse usage of amino acids in metabolism and the active interorgan transport of these compounds.

When amino acids are used as energy metabolites or precursors for gluconeogenesis it is necessary to dispose of ammonia derived from the amino group before the carbon skeleton can be used for further metabolism. Ultimately, ammonia is detoxified by urea synthesis in the liver in the urea cycle. Urea is subsequently released by the kidney (urea transporters, SLC14) (Sands 2003). The urea cycle itself takes place in two compartments, the cytosol and the mitochondrion (see Chapter 10, Figure 4). Carbamoylphosphate is synthesized from ammonium ions and CO_2 inside the mitochondrion. Ornithine carbamoyltransferase subsequently catalyzes the transfer of the carbamoyl moiety to ornithine, thus generating citrulline. All other steps of the cycle occur in the cytosol. The import of ornithine and the export of citrulline occurs concomitantly by the ornithine/citrulline exchanger (SLC25). An additional influx of ammonium into the urea cycle occurs via aspartate. Aspartate is transferred onto citrulline by argininosuccinate synthase. Since fumarate is eliminated in the subsequent step, forming arginine, the amino group of aspartate is incorporated into arginine. Finally, arginine is hydrolyzed to form urea and ornithine, thus closing the cycle. Urea contains the amino groups of both carbamoylphosphate and aspartate. Mutations in the ornithine/citrulline antiporter (SLC25) interrupt the urea cycle and result in increased plasma concentrations of urea cycle intermediates [hyperornithinemia–hyperammonemia–homocitrullinuria (HHH) syndrome] (Camacho et al. 1999). The clinical symptoms are mainly caused by hyperammonemia and resemble those of other urea cycle disorders. The brain is particularly sensitive to elevated plasma ammonia concentrations. Patients have an aversion to high-protein diet and may develop mental retardation if the disorder is left untreated. Mutations of the mitochondrial aspartate–glutamate carrier also affect the urea cycle because aspartate is the substrate of the argininosuccinate synthase (see Chapter 10).

Interorgan transfer of energy metabolites

After a meal, macromolecules are broken down in the stomach and intestine to small metabolites, which are quickly absorbed. The resulting increase of the plasma glucose concentration is downregulated by insulin action. The plasma glucose concentration is tightly regulated and kept at around 4 mM. An increase in the plasma glucose concentration above this level results in the secretion of insulin by pancreatic β cells (Figure 3). Insulin is stored in vesicles inside β cells, which fuse with the membrane and release insulin in response to elevated glucose concentration. The glucose-sensing mechanism is tightly coupled to the energy metabolism of β cells (Maechler and Wollheim 2000). In contrast with many other cells, β cells are unable to convert pyruvate into lactate. As a result, glucose is completely converted into pyruvate, which is subsequently oxidized in the mitochondria. Lactate is not produced by β cells, nor do they possess a lactate transporter. Similar to liver cells, glucose is phosphorylated to glucose-6-phosphate in β cells by the low-affinity enzyme glucokinase and not by the high-affinity enzyme hexokinase which catalyzes this step in most cells. As a result, the metabolic flux through glycolysis and the respiratory rate of β cells is proportional to the intracellular glucose concentration over a wide range. Glucose is transported into β cells by the glucose transporter GLUT2 (Chapter 12), which rapidly equilibrates the intracellular glucose concentration with the extracellular concentration because of its uniport mechanism. Thus changes in the extracellular glucose concentration are translated into changes in the glycolytic flux and consequently into changes in the respiratory rate. Therefore ATP production in β cells reflects the extracellular glucose concentration. In turn, the intracellular

ATP/ADP ratio is sensed by ATP-binding potassium channels (Figure 3). The ATP-sensitive K^+ channels of β cells possess an ATP binding site on the channel subunit. Moreover, they are associated with a regulatory ABC-transporter-like subunit, which is the target of sulfonylurea compounds used in the treatment of type II diabetes [sulfonylurea receptor

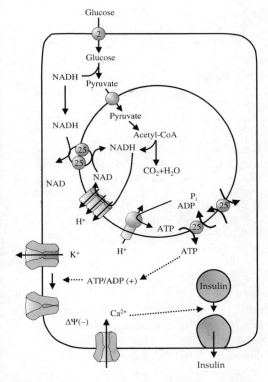

Figure 3 Glucose sensing and insulin release in pancreatic β cells. See text for explanation.

(SUR)]. Binding of ATP to these K^+ channels causes the channel to close, which results in depolarization of the β cells. The ratio of open to closed channels, and hence the extent of depolarization, is proportional to the glucose concentration. In turn, the depolarization causes voltage-dependent Ca^{2+} channels to open, thereby increasing the intracellular Ca^{2+} concentration (the role of calcium transporters in calcium signaling is discussed in more detail in Chapter 24). Calcium acts as a trigger for the fusion of insulin-containing vesicles with the plasma membrane resulting in insulin release.

The metabolic effects of insulin are exerted in the insulin-responsive tissues liver, muscle, and adipose tissue. The main actions of insulin in these tissues are listed in Table 1. The increase in glucose uptake observed in muscle cells and adipose tissue is caused by an increased surface expression of the glucose uniporter GLUT4 (SLC2). GLUT4 is held in stock by insulin-responsive cells in vesicles underneath the plasma membrane. Activation of the insulin receptor on these cells causes fusion of the vesicles with the plasma membrane, thereby increasing the number of glucose transporters at the surface (Bryant *et al.* 2002).

Plasma glucose is converted into fatty acids and triacylglycerols in adipose tissue or to glycogen in the liver. Synthesis of fatty acids, although occurring in the cytosol, involves mitochondrial transporters. Acetyl-CoA, which is the starting material for fatty acid synthesis, is generated inside the mitochondria. Similar to NADH, acetyl-CoA is not transported as such

Table 1 Effects of insulin on energy metabolism

Tissue	Insulin action	Comment
Liver	Switch from gluconeogenesis to glycolysis Glycogen synthesis increased	
Muscle	Increase of glucose transport Glycogen liguthesis increased	GLUT4 trafficking to the plasma membrane increased
Adipose tissue	Increase of glucose transport Stimulation of lipogenesis	GLUT4 trafficking to the plasma membrane increased

across the inner mitochondrial membrane but is exported by a metabolic shuttle. Using the first step of the TCA cycle, citrate is formed from acetyl-CoA and oxaloacetate by the action of citrate synthase. Citrate is released from mitochondria by the citrate carrier (SLC25) (see Chapter 10, Figure 5). Cytosolic ATP-citrate lyase then regenerates oxaloacetate and acetyl-CoA at the expense of ATP hydrolysis.

Between meals and during fasting and starvation plasma levels of glucose are held constant by mechanisms that generate glucose. Glucose is initially replenished by breakdown of glycogen in the liver. Glycogen breakdown generates glucose-6-phosphate, which is subsequently converted by the glucose-6-phosphatase system to glucose. Glucose is then released by GLUT2 and other routes into the plasma (see Chapter 12). Several transport steps are involved in this pathway because the catalytic center of glucose-6-phosphatase is located in the lumen of the endoplasmic reticulum (see Chapter 13). The homeostasis of plasma glucose levels is severely disturbed in *glycogen storage disease Type 1b*, which is caused by mutations in the glucose-6-phosphate transporter. This transporter catalyzes the translocation of glucose-6-phosphate from the cytosol into the endoplasmic reticulum (ER) as described in Chapter 13. A similar phenotype is found in patients suffering from *Fanconi–Bickel syndrome* as a result of mutations in the GLUT2 transporter (Chapter 12).

Once glycogen stores are empty, glucose levels are maintained by gluconeogenesis in liver and kidney using amino acids derived from cellular protein mainly from muscle. Therefore muscle cells not only express amino acid transporters that mediate the uptake of amino acids but can also mediate the release of amino acids.

During fasting energy metabolism switches from using glucose to alternative fuels such as fatty acids and ketone bodies. Oxidation of fatty acids in liver mitochondria fosters the switch from glycolysis to gluconeogenesis by increasing the concentration of acetyl-CoA (Figure 4). Acetyl-CoA stimulates pyruvate

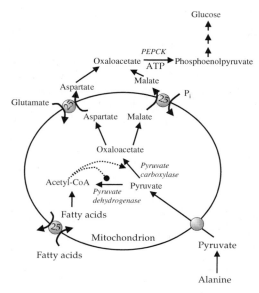

Figure 4 Transport of precursors for gluconeogenesis across the mitochondrial membrane. Gluconeogenesis uses cytosolic oxaloacetate as a precursor. Oxaloacetate can be generated from aspartate by transamination or from malate by oxidation. Both are generated from mitochondrial oxaloacetate. PEPCK, phosphoenolpyruvate carboxykinase.

carboxylase (synthesizing oxaloacetate) and inhibits pyruvate dehydrogenase which are key enzymes of gluconeogenesis and glucose breakdown, respectively. In addition, the oxidation of acetyl-CoA generates ATP for the phosphoenolpyruvate carboxykinase (PEPCK) reaction, another key enzyme of gluconeogenesis that converts oxaloacetate into phosphoenolpyruvate. Gluconeogenesis from phosphoenolpyruvate occurs in the cytosol, whereas the pyruvate carboxylase reaction occurs in the mitochondria. However, mitochondria do not possess transporters for oxaloacetate. Oxaloacetete can be converted to malate or to aspartate, both of which can be transported across the mitochondrial membrane by specific transporters (see Chapter 10, Figure 4). Once in the cytosol, aspartate or malate are converted back to oxaloacetate which is used as a precursor for gluconeogenesis (see Chapter 10). Thus hypoglycemia is observed not only when glucose release from the liver is disturbed (GLUT2 defect; see Chapter 12), but also when gluconeogenesis is not induced because fatty acids cannot be oxidized to acetyl-CoA (*carnitine deficiency*; see Chapter 11) or because precursors cannot be released from mitochondria (aspartate–glutamate transporter deficiency; see Chapter 10).

Resorption of metabolites in epithelial cells

Epithelial cells are involved in interorgan transfer of metabolites in the intestine and the kidney. In the intestine, macromolecules are broken down into building blocks to allow resorption by specific transporters. In the kidney, metabolites are reabsorbed to avoid loss of valuable nutrients from the glomerular ultrafiltrate. Nutrients are accumulated inside epithelial cells by transporters on the apical side and subsequently are released into the blood by different transporters at the basolateral membrane (Figure 5). Glucose transport in the kidney and intestine has served as a model system for developing concepts of epithelial transport. In the kidney, glucose is transported across the apical membrane by two different Na^+-glucose cotransporters (SLC5) (Wright 2001). In the early segments of the proximal tubule (S1, S2; see Chapter 2 for details about nephron organization), where glucose concentration is high, glucose is transported by $1Na^+$-glucose cotransporter (SGLT2). The transporter has a low affinity (2 mM) but a high capacity for glucose transport. In the more distal parts of the proximal tubule (S3), where most of the glucose has already been absorbed, a $2Na^+$-glucose cotransporter is expressed (SGLT1). This transporter has a higher affinity for glucose ($K_m = 0.5$ mM) and is also expressed in the intestine. The $1Na^+$-cotransport mechanism provides the driving force for an approximately 100-fold accumulation of glucose inside the cell, whereas the cotransport of $2Na^+$ generates a

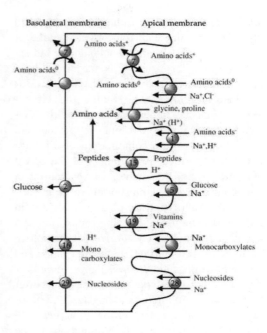

Figure 5 Resorption of metabolites by epithelial cells. Major epithelial transport systems for general metabolites. The specifc set of transporters differs between renal and intestinal cells and also varies in different parts of each epithelium.

much higher driving force, allowing an accumulation of up to 10,000-fold. On the basolateral side, glucose is released by the glucose uniporter GLUT2 (Chapter 12). Mutations in the SGLT1 transporter result in an inability to absorb glucose and galactose in the intestine [*glucose–galactose malabsorption* (Wright 1998)], but have little effect on the spillover of glucose in the urine because the main glucose load is reabsorbed by the SGLT2 isoform. The disease is associated with severe diarrhea in infants, when lactose (galactose–glucose) is a major component of the nutrition. Absorption of glucose and galactose together with Na^+ in the intestine generates an osmotic gradient which aids in the resorption of water.

The concept of a Na^+-driven uptake on the apical membrane followed by a release by uniporters does not apply to all nutrients (Figure 5). The intestine not only has a large Na^+ gradient across the apical membrane but also a significant pH gradient because of an acidic microclimate in the lumen. Uptake of peptides and certain amino acids in the kidney and intestine, for example, is carried out by a H^+ cotransport mechanism (SLC15, SLC36) (Thwaites *et al.* 1995; Daniel and Herget 1997). Transport of cationic amino acids deviates from the glucose concept even further and relies largely on the combination of different antiporters formed by members of the SLC3 and SLC7 families (Chapter 14). Mutations in epithelial amino acid transporters result in variety of disorders that are characterized by an increased secretion of certain groups of amino acids in the urine. *Cystinuria* (described in Chapter 14) is a failure to reabsorb cystine and cationic amino acids. *Lysinuric protein intolerance* results from mutations in the transporter that releases cationic amino acids on the basolateral side of the epithelium (Chapter 14). *Hartnup disease* is caused by mutations in the epithelial transporter for neutral amino acids. The transporter has not yet been identified. *Iminoglycinuria*, which is characterized by a loss of proline and glycine in the urine, may be caused by mutations of proton-coupled amino acid transporters (SLC36). Mutations of apical amino acid transporters result in relatively mild disorders which can usually be treated by dietary changes – avoiding amino acids which cannot be absorbed. Nutritional deficits usually do not occur because of the presence of active peptide transporters and other routes of amino acid uptake in the intestine. Peptide transport cannot compensate nonfunctional amino acid transporters in the basolateral membrane. As a result, mutations of basolateral transporters have a more severe phenotype (Chapter 14).

METABOLITE TRANSPORTERS OF BUILDING BLOCKS DURING BREAKDOWN AND SYNTHESIS OF MACROMOLECULES

Synthesis of membrane and secretory *proteins* creates a number of topological problems that require transport of building blocks. The synthesis of the peptide backbone occurs at the ribosomes of the rough endoplasmic reticulum using cytosolic amino acids. Essential amino acids cannot be synthesized by mammalian cells and therefore must be recruited from nutritional sources. It is very likely that mutations in plasma membrane transporters for essential amino acids are lethal, thereby explaining why amino acid transport defects are rarely observed in nonepithelial cells. Knockout mice of the cationic amino acid transporter 1, for example, die shortly after birth (Perkins *et al.* 1997).

Synthesis of membrane and secretory proteins, in addition, requires post-translational modifications. Membrane proteins and secretory proteins, for example, are heavily glycosylated. Glycosylation occurs in several stages in the lumen of the endoplasmic reticulum and the Golgi apparatus. The lipid dolichol plays an essential role in this process. Initially, an oligosaccharide is assembled on the lipid carrier dolichol while located in the cytoplasmic leaflet of the ER. To transfer the oligosaccharide to the nascent proteins the glycosylated dolichol must be flipped from one side of the membrane to the other. The mechanism underlying this process has not yet been elucidated; however, it has been proposed that the SL15/Lec35 protein, which is the homolog of a lysosomal cystine transporter (cystinosin), may be involved (Kalatzis *et al.* 2001). Inside the lumen of the Golgi apparatus, the initial glycosylation is modified by adding and removing carbohydrates. This requires the import of activated carbohydrates, such as UDP-*N*-acetylglucosamine and CMP-*N*-acetylneuraminic acid (sialic acid) into the lumen of this compartment (Figure 6). This transport is carried out by antiporters that import the nucleotide-coupled carbohydrate and export nucleoside monophosphates, which are the product of the enzymatic reaction that transfers the carbohydrate onto the protein (SLC35) (Gerardy-Schahn *et al.* 2001). Mutations in the GDP-fucosyltransporter have been associated with *congenital glycosylation disorder type IIc* (CDG IIc), also known as

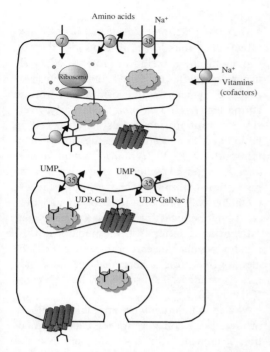

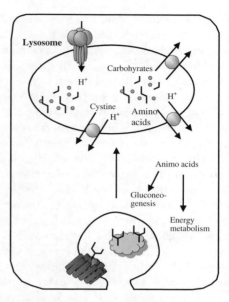

Figure 7 Breakdown of glycoproteins in the lysosomes. Glycoproteins and glycolipids are broken down inside lysosomes. The resulting metabolites and inorganic ions have to be released from the lysosome by specific transporters.

Figure 6 Transport of building blocks for protein synthesis. Synthesis of membrane proteins and secretory proteins (cloud-shaped) occurs at the rough endoplasmic reticulum. Both types of proteins are modified by glycosylation. Both the glycosylation core structure and carbohydrates for further modifcation are imported into the lumen of the endoplasmic reticulum and Golgi apparatus. Secretory and membrane proteins are delivered to their final destination by vesicular trafficking.

leukocyte adhesion deficiency II (Luhn *et al.* 2001). CDG IIc patients suffer from immunodeficiency and severe mental and psychomotor defects. It also appears that the enzymes that carry out the transfer of these carbohydrates require Mn^{2+} or Ca^{2+} as a cofactor, which is imported into the Golgi apparatus by a subclass of Ca^{2+} ATPases (see Chapter 24). Some sialic acids become further modified by O-acetylation. This function requires the import of acetyl-CoA into the lumen of the Golgi apparatus, which is thought to be mediated by specific acetyl-CoA transporters (SLC33).

Breakdown of proteins occurs in the cytosol and the lysosomes. Digestion in the cytosol is carried out by the proteasome. Specialized proteasomes (immunoproteasomes) degrade cytosolic proteins to oligopeptides, which are presented to the immune system. This pathway is discussed in more detail in Chapter 21.

The lysosome is connected to the endocytotic pathway, which delivers ingested plasma membrane lipids, plasma membrane proteins, and extracellular proteins to this compartment (Figure 7). The lysosome is the major organelle for the turnover of glycolipids, membrane proteins, and extracellular proteins. It contains a large variety of hydrolytic enzymes that are active in the acidic environment (pH 5.0) of the lysosome. The acidic pH, generated by vacuolar ATPases, is important for the release of ligands from their receptors (e.g., iron from transferring; see Chapter 7) and may also protect the cell against accidental release of lysosomal enzymes. As a result of the hydrolysis of macromolecules, the lysosome accumulates amino acids, carbohydrates (from glycolipids, glycosylated proteins, and glycogen), lipids, and inorganic ions. Apart from the membrane-embedded part of the lipids, all these compounds have to be released from the lysosome by special transport proteins (Figure 7). A congestion of the lysosome by undigested cellular material results in a number of severe lysosomal storage diseases, which are mostly associated with enzymatic defects of glycolipid degradation. The importance of metabolite release from the lysosome is illustrated by a number of *lysosomal transport disorders*. Sialic acids and glucuronic acid, generated during breakdown of glycoproteins and glycolipids, are released from the lysosome by a sialic acid-H^+ cotransporter. Mutations in this transporter cause two related diseases, namely *sialic acid storage disease* and *Salla disease* as described in Chapter 15. Both diseases are autosomal recessive neurodegenerative disorders. Sialic acid storage disease is a severe infantile form and Salla disease is a slowly progressive adult form. Symptoms include hypotonia, cerebellar ataxis, and mental retardation. Progressive cerebellar atrophy and dysmyelination have been documented. The symptoms are likely to be caused by a general failure to break down glycoproteins and glycolipids, which affects the turnover of myelin in the brain.

Cystinosis is a disorder in which cystine export from the lysosome is impaired. The function is carried out by cystinosin, which is a specific cystine transporter (Kalatzis *et al.* 2001). The disease is associated with proximal renal tubulopathy and subsequent renal failure, which illustrates the active role of the kidney in the resorption of small proteins and peptides from the urine. Members of the SLC36 family of H^+-amino acid cotransporters are thought to reside in both the lysosomal and plasma membranes and are likely to release most other amino acids from the lysosome (Sagne *et al.* 2001).

Breakdown of proteins generates amino acids as energy metabolites. The use of amino acids as energy metabolites requires a functional urea cycle. Thus defects of the urea cycle and its associated reactions (discussed above) also affect the breakdown of proteins, thereby explaining the aversion to protein-rich meals in patients suffering from mutations in urea-cycle enzymes or the ornithine/citrulline antiporter (*HHH syndrome*). Similarly, aversion to protein-rich meals is observed in patients with mutations in the amino acid transporter y$^+$LAT1 (*lysinuric protein intolerance*, Chapter 14), which is important for the resorption of arginine and ornithine across epithelial cells and thus fails to provide sufficient intermediates for the urea cycle. In contrast, patients suffering from mutations in the aspartate–glutamate carrier (or with citrin deficiency) like protein-rich meals because they lack aspartate as a substrate for the urea cycle (Chapter 10).

Water-soluble *vitamins* are building blocks of proteins as cofactors for the catalytic function of enzymes. Since vitamins are essential nutrients, they have to be absorbed in the intestine and must subsequently be delivered to individual cells. Lack of a particular vitamin affects all enzymes that require it as a cofactor, and by the same token affects the corresponding metabolic pathways. Thiamine is an essential cofactor of pyruvate dehydrogenase. Patients carrying mutations in the thiamine transporter (SLC19) gene suffer, among other symptoms described in Chapter 16, from non-type I diabetes (insulin-dependent diabetes). As described above, pancreatic β cells only use aerobic metabolism to break down glucose and generate ATP. Thus impaired function of the pyruvate dehydrogenase might result in an inability of the β cells to sense glucose because not enough ATP is produced to close ATP-binding K^+ channels.

Synthesis of *lipids (membranes)* occurs in the ER and in the Golgi apparatus. The synthesis itself creates a topological problem because it occurs in only one leaflet of the membrane. For example, sphingomyelin is synthesized from ceramide on the luminal side of the Golgi apparatus, whereas glycerol phospholipids, such as phosphatidylcholine and phosphatidylethanolamine, are synthesized in the ER membrane leaflet facing the cytosol. Some of the asymmetry that is created by the synthesis is retained. However, phosphatidylcholine is enriched in the outer leaflet of plasma membranes and therefore the bulk of this lipid must be transported to the opposite leaflet of the membrane. It appears that this process is carried out by P-type ATPases (ATP8 family) (Daleke and Lyles 2000). However, secretion of phospholipids into the bile or loading them onto plasma lipoproteins is mediated by ABC proteins (see Chapters 17, 18, and 19).

Synthesis and degradation of *polynucleotides* occurs within the cytosol and hence is less affected by disorders of biomembrane transport. Nevertheless, transport proteins for nucleosides (SLC28, SLC29) are expressed in all mammalian cells to reduce the effort of synthesizing nucleosides as precursors for nucleotides (Cabrita *et al.* 2002).

REFERENCES

Barrett, M.P., Walmsley, A.R., and Gould, G.W. (1999). Structure and Function of Facilitative Sugar Transporters. *Curr. Opin. Cell Biol.*, **11**, 496–502.

Broer, S. (2002). Adaptation of Plasma Membrane Amino Acid Transport Mechanisms to Physiological Demands. *Pflügers Arch.*, **444**, 457–466.

Bryant, N.J., Govers, R., and James, D.E. (2002). Regulated Transport of the Glucose Transporter GLUT4. *Nat. Rev. Mol. Cell Biol.*, **3**, 267–277.

Cabrita, M.A., Baldwin, S.A., Young, J.D., and Cass, C.E. (2002). Molecular Biology and Regulation of Nucleoside and Nucleobase Transporter Proteins in Eukaryotes and Prokaryotes. *Biochem. Cell Biol.*, **80**, 623–638.

Camacho, J.A., Obie, C., Biery, B., Goodman, B.K., Hu, C.A., Almashanu, S., *et al.* (1999). Hyperornithinaemia–Hyperammonaemia–Homocitrullinuria Syndrome is caused by Mutations in a Gene Encoding a Mitochondrial Ornithine Transporter. *Nat. Genet.*, **22**, 151–158.

Daleke, D.L. and Lyles, J.V. (2000). Identification and Purification of Aminophospholipid Flippases. *Biochim. Biophys. Acta*, **1486**, 108–127.

Daniel, H. and Herget, M. (1997). Cellular and Molecular Mechanisms of Renal Peptide Transport. *Am. J. Physiol.*, **273**, F1–F8.

Gerardy-Schahn, R., Oelmann, S., and Bakker, H. (2001). Nucleotide Sugar Transporters: Biological and Functional Aspects. *Biochimie*, **83**, 775–782.

Hajri, T. and Abumrad, N.A. (2002). Fatty Acid Transport Across Membranes: Relevance to Nutrition and Metabolic Pathology. *Annu. Rev. Nutr.*, **22**, 383–415.

Halestrap, A.P., and Price, N.T. (1999). The Proton-Linked Monocarboxylate Transporter (MCT) Family: Structure, Function and Regulation. *Biochem. J.*, **343**, 281–299.

Kalatzis, V., Cherqui, S., Antignac, C., and Gasnier, B. (2001). Cystinosin, the Protein Defective in Cystinosis, Is a H(+)-Driven Lysosomal Cystine Transporter. *EMBO J.*, **20**, 5940–5949.

Luhn, K., Wild, M.K., Eckhardt, M., Gerardy-Schahn, R., and Vestweber, D. (2001). The Gene Defective in Leukocyte Adhesion Deficiency II Encodes a Putative GDP-Fucose Transporter. *Nat. Genet.*, **28**, 69–72.

Maechler, P. and Wollheim, C.B. (2000). Mitochondrial Signals in Glucose-Stimulated Insulin Secretion in the Beta Cell. *J. Physiol.*, **529**, 49–56.

Perkins, C., Mar, V., Shutter, J., del Castillo, J., Danilenko, D., Medlock, E., *et al.* (1997). Anemia and Perinatal Death Result from Loss of the Murine Ecotropic Retrovirus Receptor mCAT-1. *Genes Dev.*, **11**, 914–925.

Sagne, C., Agulhon, C., Ravassard, P., Darmon, M., Hamon, M., El Mestikawy, S., *et al.* (2001). Identification and Characterization of a Lysosomal Transporter for Small Neutral Amino Acids. *Proc. Natl Acad. Sci. USA*, **98**, 7206–7211.

Sands, J.M. (2003). Mammalian Urea Transporters. *Annu. Rev. Physiol.*, 65, 543–566.

Stahl, A., Gimeno, R.E., Tartaglia, L.A., and Lodish, H.F. (2001). Fatty Acid Transport Proteins: A Current View of a Growing Family. *Trends Endocrinol. Metab.*, **12**, 266–273.

Thwaites, D.T., McEwan, G.T., and Simmons, N.L. (1995). The Role of the Proton Electrochemical Gradient in the Transepithelial Absorption of Amino Acids by Human Intestinal Caco-2 Cell Monolayers. *J. Membr. Biol.*, **145**, 245–256.

Wright, E.M. (1998). Glucose Galactose Malabsorption. *Am. J. Physiol.*, **275**, G879–G882.

Wright, E.M. (2001). Renal Na$^+$-Glucose Cotransporters. *Am. J. Physiol. Renal Physiol.*, **280**, F10–F18.

10

KEIKO KOBAYASHI* AND TAKEYORI SAHEKI*

Aspartate glutamate carrier (citrin) deficiency

INTRODUCTION: CHARACTERISTICS OF SLC25 MEMBERS

The mitochondrial inner membrane harbors a set of carrier proteins for metabolite transport that constitute a superfamily of related proteins (Krämer and Palmieri 1992; Walker 1992; Palmieri 1994; Palmieri and van Ommen 1999). Many metabolite transporters or carriers with the features of the mitochondrial carrier family are now classified as solute carrier family 25 (SLC25). The members of SLC25 are each composed of approximately 300 amino acids comprising three homologous domains repeated in tandem, each about 100 amino acids long, that contain a conserved sequence motif, each with two hydrophobic α-helical segments connected by an extensive hydrophilic sequence and two putative mitochondrial transmembrane (TM) spanners (Krämer and Palmieri 1992; Walker 1992; Palmieri 1994). A topological model predicts six TM spanners with amino and carboxyl termini on the cytosolic side and hydrophilic loops on the matrix side. Functional studies of intact mitochondria have indicated that more than 20 SLC25 proteins catalyze the net flux or the exchange of physiologically important metabolites, nucleotides, and cofactors

between the cytosol and the matrix (Krämer and Palmieri 1992; Palmieri and van Ommen, 1999). More recently, many proteins of unknown function with the same characteristic features of this family have emerged from genome sequencing of various organisms. In Table 1, the classification and characteristics of SLC25 proteins found by searching the OMIM database in the National Center for Biotechnology Information (http://www.ncbi.nlm.nih.gov/Omim/) are summarized. Most SLC25 members are small proteins with a molecular mass of about 30 kDa, and for some of them the deficiency in humans is known. Progressive external ophthalomolegia (PEO) is associated with a lack of ANT1 (SLC25A4), hyperornithinemia–hyperammonemia–homocitrullinemia (HHH) syndrome is associated with a lack of ORNT1 (SLC25A15) and carnitine-acylcarnitine translocase deficiency is observed in CACT (SLC25A20).

In contrast, two SLC25 members, A13 and A12, have a much larger molecular mass of 74 kDa (Table 1): citrin encoded by SLC25A13 on chromosome (chr.) 7q21.3 consists of 675 amino acids (Kobayashi et al. 1999; Sinasac et al. 1999) and aralar encoded by SLC25A12 on chr. 2q24 consists of 678 amino acids (del Arco and Satrústegui 1998; Crackower et al. 1999). Citrin and aralar have 77.8 percent identity in amino acid sequences and a bipartite structure: a C-terminal half with characteristic

*Department of Molecular Metabolism and Biochemical Genetics, Kagoshima University Graduate School of Medical and Dental Sciences, Kagoshima

Table 1 Solute carrier family 25

SLC25	Alternative titles (Acronym)	Protein (aa)	Locus	Diseases (Candidate diseases)
A1	Tricarboxylate transport protein, Citrate transport protein (CTP), formerly SLC20A3	311	22q11	(DiGeorge syndrome: DGCR) (Velocardiofacial syndrome: VCFS)
A3	Phosphate carrier (PHC)	361	12q23	
A4	Adenine nucleotide translocator 1 (ANT1) (ANT-muscle), ADP/ATP translocase 1	298	4q35	(Facioscapulohumeral muscle dystrophy: FSHD) Progressive external ophthalmoplegia: PEO, OMIM 157640
A5	Adenine nucleotide translocator 2 (ANT2) (ANT-fibroblast), ADP/ATP translocase 2	298	Xq24–26	
A6	Adenine nucleotide translocator 3 (ANT3) (ANT-liver), ADP/ATP translocase 3	298	Xp22.32	
A7	Uncoupling protein 1 (UCP1), Brown adipose tissue uncoupling protein, thermogenin	307	4q31	
A8	Uncoupling protein 2 (UCP2)	309	11q13	(Linked to obesity and hyperinsulinemia)
A9	Uncoupling protein 3 (UCP3)	312	11q13	(Juvenile-onset obesity, severe obesity, type 2 diabetes)
A10	Dicarboxylate ion carrier (DIC)	287	17q25.3	
A11	Oxoglutarate carrier (OGC), Oxoglutarate/ malate carrier, formerly SLC20A4	314	17p13.3	
A12	Aralar (Ca^{2+}-dependent) aspartate glutamate carrier (AGC)	678	2q24	
A13	Citrin (Ca^{2+}-dependent) aspartate glutamate carrier (AGC)	675	7q21.3	Adult-onset type II citrullinemia: CTLN2, OMIM 603471, Neonatal hepatitis, Neonatal intrahepatic cholestasis caused by citrin deficiency: NICCD, OMIM 605814
A14	Brain mitochondrial carrier protein 1 (BMCP1), Uncoupling protein 5: UCP5	325	Xq24	

Table 1   Continued

SLC25	Alternative titles (Acronym)	Protein (aa)	Locus	Diseases (Candidate diseases)
A15	Ornithine transporter 1 (ORNT1)	301	13q14	Hyperornithine–hyperammonemia–homocitrullinuria (HHH) syndrome, OMIM 238970
A16	Graves disease autoantigen (GDA), Graves disease carrier protein (GDC)	332	10q21.3 ~q22.1	(Associated with Graves disease)
A17	Peroxisomal membrane protein (PMP34)	307	22q13	
A18		315	22q11.2	(Cat eye syndrome critical region)
A19	Deoxynucleotide carrier (DNC)	320	17q25.3	
A20	Carnitine-acylcarnitine translocase (CACT), Carnitine-acylcarnitine carrier (CAC)	301	3p21.31	Carnitine-acylcarnitine translocase deficiency
A21	Oxodicarboxylate carrier	299	14q11.2	
A22	Glutamate carrier	323	11	

features including six TM spanners as found in other SLC25 proteins of about 30 kDa, and an N-terminal extension harboring four EF-hand domains to bind calcium ions, as shown in Figure 1. Overlay assay of calcium revealed that the N-terminal half of each protein is able to bind calcium and that the first EF-hand domain is required for such binding (del Arco and Satrústegui 1998; del Arco *et al*. 2000; Kobayashi *et al*., 2000). These results indicate that citrin and aralar belong to a calcium-binding mitochondrial solute carrier. Recently, we found that citrin and aralar are isoforms of an aspartate glutamate carrier (AGC) (Palmieri *et al*. 2001). Recombinant citrin and aralar reconstituted into liposome showed electrogenic exchange of aspartate for glutamate and H^+. The substrate specificity and kinetic parameters are identical to those of natural AGC (Palmieri 1994). The transport activity of citrin and aralar are entirely accounted for by their C-terminal half regions (Palmieri *et al*. 2001). The turnover number of citrin is about four times greater than that of aralar. The activity of citrin and aralar transfected into mammalian cells was stimulated by calcium ions on the external side of the inner mitochondrial membrane, where the calcium ion-binding domains of these proteins are located. We speculate that the known activation of the malate aspartate (MA)-shuttle by calcium ions is based on the calcium-binding property of citrin and aralar and that AGC is one of the most important targets of calcium-signaling hormones (Saheki and Kobayashi 2002).

In citrin and aralar, the structure and function are similar, but the chromosome locus and expression patterns are quite different (Figure 1). Citrin was expressed predominantly in the liver, kidney, and heart, whereas aralar was not observed in the liver but was more prevalent in the skeletal muscle, brain, and heart (del Arco and Satrústegui 1998; Kobayashi *et al*. 1999; del Arco *et al*. 2000; Iijima *et al*. 2001; Begum *et al*. 2002; del Arco *et al*. 2002). Citrin and aralar are differentially distributed among the

[A]

	SLC25A13	SLC25A12
Chromosome	7q21.3	2q24
Gene (kb)	>200 18 exons	(18 exons)
mRNA (kb)	3.4 liver kidney heart intestine	3.3, 2.8, 6.6 muscle brain heart kidney
Protein	Citrin 675 aa (74 kD)	Aralar 678 aa
Deficiency	CTLN2 NICCD	unknown

[B]

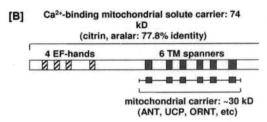

Ca^{2+}-binding mitochondrial solute carrier: 74 kD
(citrin, aralar: 77.8% identity)

4 EF-hands   6 TM spanners

mitochondrial carrier: ~30 kD
(ANT, UCP, ORNT, etc)

Figure 1 (A) Comparison of SLC25A13 and SLC25A12 and (B) schematic protein structure of Ca^{2+}-binding mitochondrial carrier and other mitochondrial carriers. The figure is slightly modified from a previous report (Kobayashi et al. 2000), because of additional recent data (Iijima et al. 2001; Begum et al. 2002; Saheki and Kobayashi 2002). CTLN2, adult-onset type II citrullinemia; NICCD, neonatal intrahepatic cholestasis caused by citrin deficiency; ANT, adenine nucleotide translocator; UCP, uncoupling protein; and ORNT, ornithine transporter (see Table 1).

organs, suggesting functional sharing between the isoforms of AGC. This is important for citrin deficiency, a liver-specific disorder, although aralar deficiency is not known in humans. Therefore, we devote this chapter to adult-onset type II citrullinemia and a type of neonatal hepatitis with intrahepatic cholestasis caused by citrin deficiency.

DISCOVERY OF SLC25A13 AND CITRIN

Citrullinemia based on the accumulation of citrulline in the body fluid is caused by a deficiency of argininosuccinate synthetase (OMIM #603470), the third enzyme of the urea cycle (see Figure 4A), which catalyzes the formation of argininosuccinate from citrulline and aspartate at the expense of ATP breakdown. We have analyzed over 200 patients with citrullinemia and have classified them into three types according to the enzyme abnormality and into two forms according to pathogenesis (Saheki et al. 1981, 1982, 1985, 1987a; Kobayashi et al. 1993, 1994, 1995a). Now, citrullinemia can be clearly divided into classical citrullinemia types I/III with generalized argininosuccinate synthetase deficiency (CTLN1: OMIM #215700, abnormality in argininosuccinate synthetase gene on chr. 9q34.1), and adult-onset type II citrullinemia with a liver-specific argininosuccinate synthetase deficiency (CTLN2: OMIM #603471, abnormality in SLC25A13 gene on chr. 7q21.3), on the basis of our discovery of the gene responsible for CTLN2 Kobayashi et al. 1999; Sinasac et al. 1999; Kobayashi et al. 2000).

CTLN2, as described in the special topic box Clinical Features and Characteristics of CTLN2, is characterized by a decrease in argininosuccinate synthetase protein in the liver, but there are no abnormalities in argininosuccinate synthetase mRNA in the liver of CTLN2 patients concerning amount, translational activity, gross structure, and nucleotide sequence (Sase et al. 1985; Kobayashi et al. 1986, 1993). Since 1977, we have analyzed over 150 patients with CTLN2 and have found that approximately 20 percent of the patients are from consanguineous parents and that siblings are sometimes affected (Tsujii et al. 1976; Saheki et al. 1987a; Kobayashi et al. 1993), suggesting that CTLN2 is an autosomal recessive disorder. RFLP analysis in argininosuccinate synthetase gene of 16 patients with CTLN2 from consanguineous marriages shows that the frequency of the heterozygous haplotype is not different from control (Kobayashi et al. 1993), suggesting that the abnormality is not

CLINICAL FEATURES AND CHARACTERISTICS OF CTLN2

The most characteristic feature of CTLN2 is the late onset of serious and recurring symptoms. The onset is sudden and usually between the ages of 20 and 50. Patients diagnosed were from 11–79 years old with a mean age of 34.4 ± 12.8 ($n = 103$), and the ratio of male to female was 77 to 26 (Kobayashi et al. 1997, 2000; Yasuda et al. 2000). CTLN2 patients show citrullinemia, hyperammonemia, and neuropsychiatric symptoms such as disorientation, delirium, aberrant behavior, delusion, and disturbance of consciousness, often leading to rapid death (Saheki et al. 1987a; Kobayashi et al. 2000). Although the prognosis is bad, liver transplantation is remarkably effective (Todo et al. 1992; Yazaki et al. 1996; Kawamoto et al. 1997; Onuki et al. 2000; Takenaka et al. 2000; Ikeda et al. 2001; Kasahara et al. 2001; Takashima et al. 2002). Pathological findings in the liver include fatty infiltration and mild fibrosis, but mild or no liver dysfunction. Symptoms are often provoked by medication, infection, and/or alcohol intake. Many patients have a peculiar fondness for beans and peanuts. They also like to eat high protein diets such as egg, fish, and meat, and dislike carbohydrates such as rice and sweets. Most of the patients are thin, more than 90 percent showing a body mass index less than 20, and about 40 percent showing an index less than 17. Hepatoma, pancreatitis, and hyperlipidemia are the major complications of CTLN2 (Kobayashi et al. 2000; Ikeda et al. 2001).

CTLN2 can be distinguished from CTLN1 metabolically. CTLN2 patients show a moderate increase of serum citrulline (521 ± 290, control: 20–40 nmol/ml), a slight increase of serum arginine (232 ± 167, control: 80–130 nmol/ml), and a significant elevation of urinary argininosuccinate excretion (Saheki et al. 1986, 1987a,b; Kobayashi et al. 1997, 2000), whereas CTLN1 patients show a marked elevation of serum citrulline (2500 ± 1040) but a decrease of serum arginine (58 ± 31). This is in accordance with the liver-specific argininosuccinate synthetase deficiency in CTLN2, which contrasts with the generalized argininosuccinate synthetase deficiency in CTLN1. Liver-specific argininosuccinate synthetase deficiency, the main characteristic of CTLN2, was found to be secondary to the defect of citrin (Kobayashi et al. 1999, 2000; Yasuda et al., 2000). However, it is difficult to imagine how a defect of citrin causes a decrease in argininosuccinate synthetase protein. The liver-specific argininosuccinate synthetase decrease may derive from the fact that hepatic AGC is composed only of citrin and the defect of citrin causes a complete loss of AGC function in the liver. Purified argininosuccinate synthetase is stabilized by the substrates, citrulline and aspartate, and the product, argininosuccinate, but needs high concentrations (Saheki et al. 1983a). A direct protein–protein interaction between argininosuccinate synthetase in the cytosol and citrin in the mitochondrial inner membrane (Iijima et al. 2001; Palmieri et al. 2001) is difficult to imagine, because they are separated by the outer membrane. Further research is needed.

The remaining mechanisms to be clarified in CTLN2 patients are the increased ratio of threonine to serine in the serum (Saheki et al. 1986; Kobayashi et al. 1997), the enhanced expression of pancreatic secretory trypsin inhibitor (PSTI) gene in the liver (Kobayashi et al. 1995b, 1997) and the unique uneven distribution of argininosuccinate synthetase protein in the liver lobulus (Saheki et al. 1983b, 1987a; Yagi et al., 1988). These phenomena are useful for diagnosis of CTLN2 (Yajima et al. 1981; Saheki et al., 1983b, 1986, 1987a; Yagi et al. 1988; Todo et al. 1992; Kobayashi et al. 1995b; Chow et al. 1996; Yazaki et al. 1996; Kawamoto et al. 1997; Kobayashi et al. 1997, 2000; Onuki et al. 2000; Takenaka et al. 2000; Hwu et al. 2001; Ikeda et al. 2001; Kasahara et al. 2001; Maruyama et al. 2001; Tsuboi et al. 2001; Oshiro et al. 2002; Takashima et al. 2002; Imamura et al., 2003) and to predict the prognosis of the patients (Yagi et al., 1988), and may be related to the finding that about 10 percent of the CTLN2 patients suffer from pancreatitis or hepatoma without liver cirrhosis at relatively young ages (Tsujii et al. 1976; Kobayashi et al. 2000; Ikeda et al. 2001; Tsuboi et al., 2001).

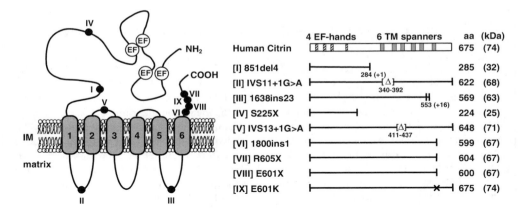

Figure 2   Predicted topological model of citrin (left) and SLC25A13 mutations (right). Sites of nine mutations, [I]–[IX], are shown in the predicted citrin structure. EF, calcium-binding EF-hand motif; TM (1–6 in left), mitochondrial transmembrane spanner; and IM, inner membrane. The figure was made from data reported previously (Kobayashi *et al.* 1999; Yasuda *et al.* 2000; Yamaguchi *et al.* 2002).

within the argininosuccinate synthetase gene locus.

We collected DNA samples of CTLN2 patients from 18 consanguineous families, which allowed us to perform homozygosity mapping to delimit the critical region for the disease on chr. 7q21.3. SLC25A13, which encodes a putative calcium-binding mitochondrial solute carrier protein, was identified to be the responsible gene for CTLN2 by positional cloning and mutation search (Kobayashi *et al.* 1999). The human *SLC25A13* gene spanned > 200 kb of genomic DNA organized into 18 exons (Figure 1). We have named the protein with a molecular weight of 74 kDa encoded by SLC25A13, "citrin." We first identified five mutations in *SLC25A13* gene of 18 consanguineous CTLN2 patients (Kobayashi *et al.* 1999), and then four others (Yasuda *et al.* 2000; Yamaguchi *et al.* 2002) as shown in Figure 2. Except for the ninth mutation, E601K, all eight mutations cause truncation of citrin protein by nonsense or frame shift mutations, or disruption of the membrane structure by splicing mutations. No cross-reactive immune material with anti-human citrin antibody was detected in the liver of CTLN2

patients with the first seven mutations (Yasuda *et al.* 2000). We have established DNA diagnosis methods for the nine mutations by using PCR and gel electrophoresis (Kobayashi *et al.* 1999; Yasuda *et al.* 2000; Yamaguchi *et al.* 2002) and multiple DNA diagnosis methods by using a genetic analyzer with GeneScan and the single nucleotide primer extension procedure (SNaPshot) (Yamaguchi *et al.* 2002).

CLINICAL FEATURES AND CHARACTERISTICS OF NICCD

Until recently, very little was known about clinical symptoms or abnormal episodes in the neonatal/infantile period of CTLN2 patients. However, establishment of DNA diagnosis for citrin deficiency revealed that SLC25A13 mutation is the cause of one type of neonatal hepatitis with intrahepatic cholestasis (Ohura *et al.* 2001; Tazawa *et al.* 2001; Tomomasa *et al.* 2001; Ben-Shalom *et al.* 2002; Naito *et al.* 2002; Saheki and Kobayashi, 2002; Tamamori *et al.* 2002; Yamaguchi *et al.* 2002; Lee *et al.* 2002). Since the symptoms of the neonates are different from the more severe CTLN2, we

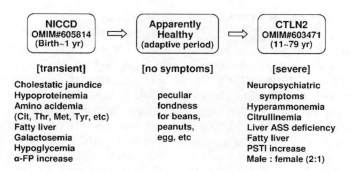

Figure 3 Adaptation and decompensation in citrin deficiency. The figure is slightly modified from a previous report (Saheki and Kobayashi 2000), since numbers of NICCD patients diagnosed in our laboratory have increased. Cit, citrulline; Thr, threonine; Met, methionine; Tyr, tyrosine; α-FP, α-fetoprotein; ASS, argininosuccinate synthetase; and PSTI, pancreatic secretory trypsin inhibitor.

designated the disorder as neonatal intrahepatic cholestasis caused by citrin deficiency (NICCD: OMIM #605814) (Ohura *et al.* 2001; Yamaguchi *et al.* 2002). So far (August, 2002), we have diagnosed over 80 NICCD cases of neonates or infants as having *SLC25A13* gene mutations (Ohura *et al.* 2001; Tazawa *et al.* 2001; Tomomasa *et al.* 2001; Ben-Shalom *et al.* 2002; Naito *et al.* 2002; Saheki and Kobayashi 2002; Tamamori *et al.* 2002; Yamaguchi *et al.* 2002; Lee *et al.* 2002). NICCD patients have been found in neonates/infants who presented with neonatal hepatitis in association with hypergalactosemia, hypermethioninemia, and/or hyperphenylalaninemia detected by neonatal mass screening, or who presented with persistent infantile cholestatic jaundice at 1–4 months of age. In addition, they show a variety of symptoms and findings such as elevation of bile acid and bilirubin, mild liver dysfunction, hypoproteinemia, hypoglycemia, and multiple aminoacidemia including citrulline, threonine, methionine, lysine, arginine, and tyrosine (Figure 3). Liver histology shows a diffuse fatty change and portal fibrosis. High levels of α-fetoprotein have been detected in the early infancy of NICCD patients (Ben-Shalom *et al.* 2002; Tamamori *et al.* 2002), but not in CTLN2 patients (Kobayashi *et al.* 1997; Hwu *et al.* 2001).

Most NICCD patients show symptoms that ameliorate by 12 months of age without special treatment other than feeding programs; formulas containing middle-chain triglyceride or lactose-free formula and supplementation of fat-soluble vitamins have been used (Ohura *et al.* 2001; Tazawa *et al.* 2001; Tomomasa *et al.* 2001; Ben-Shalom *et al.* 2002; Naito *et al.* 2002; Tamamori *et al.* 2002; Lee *et al.* 2002). It is noteworthy that the patient who had shown transient NICCD episodes in early infancy lived in apparent health and then suddenly suffered from typical CTLN2 at the age of 16 years (Kasahara *et al.* 2001; Tomomasa *et al.* 2001). As shown in Figure 3, more than 10 years or even several decades later, some patients developed CTLN2 (Saheki and Kobayashi 2002), although we do not know which NICCD patients or what percentage developed CTLN2. On the other hand, most of the 80 NICCD patients diagnosed showed relatively mild and transient symptoms. However, three cases with NICCD, who were diagnosed as having hypertyrosinemia of an unknown cause, had a severe phenotype such as liver dysfunction, and they had been treated with liver transplantation at the age of 10–12 months (Tamamori *et al.* 2002). We need to follow up NICCD patients carefully, even during infancy, to detect the mechanisms and factors for onset of CTLN2 and also severe NICCD.

FREQUENCY OF CITRIN DEFICIENCY

We have analyzed the frequency of the nine SLC25A13 mutations among 120 Japanese patients with CTLN2, 70 with NICCD, and 1372 Japanese controls (Saheki and Kobayashi 2002). Mutations [I] and [II] are most frequent at 82 percent and 80 percent in CTLN2 and NICCD, respectively. There are twice as many male CTLN2 patients as female (Kobayashi et al. 1997, 2000; Yasuda et al. 2000), but there was no such gender difference in NICCD (Yamaguchi et al. 2002), suggesting that sex hormones may affect the onset of CTLN2. Twenty out of 1372 controls had a mutated SLC25A13 gene in one allele, which implies that the frequency of heterozygotes is approximately 1 in 70 of the Japanese population (Saheki and Kobayashi 2002). Thus, the frequency of homozygotes with SLC25A13 mutations is calculated to be 1 in 20,000 as a minimal estimate from the frequency of carriers (Saheki and Kobayashi 2002; Yamaguchi et al. 2002; Yasuda et al. 2000). The frequency of homozygotes with mutated SLC25A13 is higher than the incidence of CTLN2, estimated to be 1/230,000 from a survey of patients (Nagata et al. 1991) or calculated to be 1/100,000 from consanguinity (Kobayashi et al. 1993). This discrepancy is difficult to explain. Homozygotes with mutated SLC25A13 may be classified at a certain time as one of the following: (1) diagnosed as suffering from CTLN2, (2) diagnosed as having other diseases such as hepatitis, hepatoma, pancreatitis, or hyperlipidemia, (3) misdiagnosed as having a psychosis such as epilepsy, schizophrenia or depression, or (4) apparently healthy. We have already detected several non-CTLN2 individuals with SLC25A13 mutations in both alleles (Onuki et al. 2000; Imamura, 2003; Kawata et al. unpublished data). Others are individuals who recovered from NICCD (Ohura et al. 2001; Tazawa et al. 2001; Tomomasa et al. 2001; Ben-Shalom et al. 2002; Naito et al. 2002; Saheki and Kobayashi 2002; Tamamori et al. 2002; Yamaguchi et al. 2002; Lee et al. 2002).

It is important to predict whether those who have recovered from NICCD will suffer from CTLN2 or not. We postulate that the probability is not so high because the gender ratio is different between CTLN2 and NICCD patients (Yamaguchi et al. 2002).

Until quite recently, citrin deficiency had been found mostly among Japanese, and there were no reports of CTLN2 and NICCD cases from other countries except for one Chinese from Singapore who was clinically diagnosed as suffering from CTLN2 (Chow et al. 1996). However, we recently found three Chinese CTLN2 patients from Taiwan (Hwu et al. 2001) and from Beijing (Yang et al. unpublished data), a Vietnamese infant suffering from NICCD in Australia (Lee et al. 2002), and a Chinese NICCD case from Taiwan (Hwu et al. unpublished data). It is interesting that they had the same SLC25A13 mutations as those identified in Japanese patients. These results suggest that the common mutations are old enough to go back to a common ancestor and prevail at least in East Asia. Furthermore, we found a Palestinian NICCD patient with a novel mutation, duplication of exon 15 in SLC25A13 gene (Ben-Shalom et al. 2002) and an Ashkenazi Jewish NICCD patient with another novel mutation (Elpeleg et al. unpublished data), suggesting a wide distribution of citrin deficiency among races.

CITRIN DEFICIENCY AND AGC FUNCTION

Citrin as AGC plays a role in the supply of aspartate from mitochondria to the cytosol and is a member of the MA-shuttle, as shown in Figures 4 and 5. Supply of aspartate formed in the mitochondria to the cytosol via AGC is essential for urea synthesis from ammonia (Figure 4A) and also important for gluconeogenesis from lactate (Figure 4B), because oxaloacetate can be formed in mitochondria without requiring NADH formation in the cytosol. Gluconeogenesis from alanine also

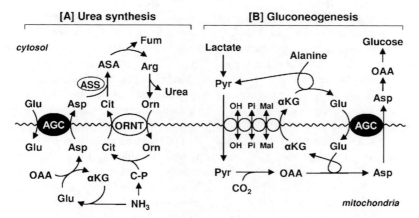

Figure 4   Role of citrin as aspartate glutamate carrier (AGC) in (A) urea synthesis and (B) gluconeogenesis. Glu, glutamate; Asp, aspartate; Cit, citrulline; ASS, argininosuccinate synthetase; ASA, argininosuccinate; Fum, fumarate; Arg, arginine; Orn, ornithine; ORNT, ornithine transporter (SLC25A15); C-P, carbamoyl phosphate; αKG, α-ketoglutarate; OAA, oxaloacetate; Pyr, pyruvate; and Mal, malate.

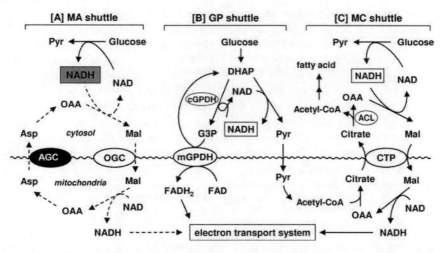

Figure 5   Malate aspartate (MA)-shuttle and predicted metabolic pathway for compensation in citrin deficiency. (A) MA-shuttle, (B) glycerophosphate (GP)-shuttle, and (C) malate citrate (MC)-shuttle. OGC, oxoglutarate malate carrier (SLC25A11); DHAP, dihydroxyacetone phosphate; G3P, glycerol-3-phosphate; cGPDH and mGPDH, cytosolic and mitochondrial glycerophosphate dehydrogenase; ACL, ATP citrate lyase; CTP, citrate transport protein (SLC25A1); and others, see Figure 4.

requires the AGC for nitrogen disposal via the urea cycle. Lack of aspartate for argininosuccinate synthetase reaction probably causes accumulation of citrulline, resulting in transient citrullinemia in NICCD without hepatic argininosuccinate synthetase deficiency (Tazawa *et al.* 2001). Aspartate is low in the plasma and the uptake by the liver is very low, as judged from the rate of gluconeogenesis (Ross *et al.* 1967), suggesting that the requirement of aspartate in cytosol largely depends upon the supply from mitochondria. Therefore, citrin deficiency probably causes deficiency of aspartate in the cytosol of liver, followed by inhibition of protein and nucleotide synthesis, resulting in hypoproteinemia in NICCD. Aspartate and/or asparagine in beans and peanuts are the highest among dietary proteins. This may be the reason why patients with citrin deficiency show an extraordinary liking for beans and peanuts.

Citrin as AGC is a member of the MA shuttle, which plays a role in the transport of cytosolic reducing-equivalent into mitochondria (see Introduction to Chapter 9). Citrin deficiency blocks the MA-shuttle, which may increase the $NADH/NAD^+$ ratio in the cytosol (Figure 5A). The increased $NADH/NAD^+$ ratio inhibits glycolysis and makes alcohol metabolism difficult. This may be the reason why the patients dislike carbohydrates and alcohol, and why alcohol intake often causes psychiatric symptoms in CTLN2. Increased $NADH/NAD^+$ ratio also inhibits gluconeogenesis from reduced substrates such as lactate, glycerol, and sorbitol (Saheki and Kobayashi 2002). This may cause hypoglycemia in NICCD. Although NICCD patients suffer from galactosemia sometimes even leading to cataract, no abnormality in the enzymes for galactose metabolism has been found (Ohura et al. 2001; Naito et al. 2002; Tamamori et al. 2002). Since UDP-glucose epimerase requires NAD as a cofactor and is strongly inhibited by NADH, galactosemia in NICCD may result from high NADH levels in the cytosol of the liver.

Another system for the transport of cytosolic reducing-equivalent into mitochondria is the glycerophosphate (GP)-shuttle (Figure 5B), which consists of mitochondrial and cytosolic glycerophosphate dehydrogenase (mGPDH and cGPDH). As mentioned above, the symptoms caused by the elevated cytosolic NADH level suggest low activity of the GP-shuttle in human liver. The amelioration of NICCD symptoms within 1 year of age suggests some adaptation, compensation, or metabolic change during development (see Figure 3). One such system may be the GP-shuttle. The other possibility is the malate citrate (MC)-shuttle (Figure 5C), which produces

oxaloacetate from citrate in the cytosol by ATP citrate lyase (ACL) reaction, reduces oxaloacetate concomitantly with consumption of NADH and transports malate into the mitochondria. Mitochondrial malate is converted to oxaloacetate to produce NADH and is condensed with acetyl-CoA to form citrate, which is transported to cytosol. As a result, cytosolic NADH or the reducing equivalent is transported into the mitochondria. However, since this system plays a role in fatty acid synthesis, it also produces acetyl-CoA in the cytosol and may stimulate fatty acid synthesis. If the GP-shuttle is less active, triglyceride synthesis may be activated by supplying glycerol-3-phosphate (G3P), which is produced by cytosolic glycerol-3-phosphate dehydrogenase. All these promote fatty liver and hyperlipidemia. In fact, Imamura et al. (Imamura et al. 2003) found a CTLN2 patient who showed a dramatic increase in triglyceride after a high carbohydrate diet, which was normalized after changing to a low carbohydrate diet. We presume that the increased cytosolic NADH may cause so-called reductive stress and subsequent oxidative stress, leading to liver damage (Saheki and Kobayashi 2002). However, this hypothesis needs more thorough examination.

POSSIBLE THERAPY BASED ON A NEW CONCEPT

CTLN2 patients have been treated with two kinds of therapeutic procedures: symptomatic therapy against hyperammonemia and liver transplantation (Takenaka et al. 2000). Low protein diets are the first choice for hyperammonemia. Various medicines that decrease blood ammonia, such as lactulose, sodium benzoate, citrate, and arginine, have been used but have failed to improve the long-term prognosis (Takenaka et al. 2000). On the other hand, liver transplantation has been shown to be very effective. Following two cases of orthotopic liver transplantation performed at Pittsburgh in 1988 (Todo et al. 1992) and at Brisbane in 1993 (Kawamoto *et al.* 1997), 20 cases of CTLN2 in Japan have been treated with living related partial liver transplantation (Yazaki *et al.* 1996; Onuki *et al.* 2000; Takenaka *et al.* 2000; Ikeda *et al.* 2001; Kasahara *et al.* 2001; Takashima *et al.* 2002). Liver transplantation normalizes most of the metabolic disturbances and symptoms, except those from brain atrophy, which have already occurred. It is curious that liver transplantation is so effective even though citrin is distributed not specifically in the liver, but also in the kidney, the heart, and so on (Kobayashi *et al.* 1999; del Arco *et al.* 2000; Iijima *et al.* 2001; Begum *et al.* 2002; del Arco *et al.* 2002). This is probably because the liver plays a major role in metabolism and is the only major organ that expresses citrin, but not aralar which might deputize as an AGC isoform. This treatment, however, raises the problems of financial support and liver donors in Japan.

Now, since we know the functions of citrin and characteristics of NICCD and CTLN2, we are in a position to establish new therapeutic procedures based on the functions of citrin (Saheki and Kobayashi 2002). There are five possible treatment procedures: (1) aspartate could be supplied in the cytosol, (2) cytosolic NADH/NAD$^+$ ratio could be decreased, (3) dehydroepiandrosterone and peroxisome proliferators may be effective to activate alternative NADH shuttle, (4) sex hormone therapy may have some effect on preventing or delaying the onset of CTLN2, and (5) gene therapy for hepatocytes may be effective.

Supply of aspartate or oxaloacetate in the cytosol may be effective, but as described above, aspartate may be taken up into the hepatocytes only slowly, and oxaloacetate is very unstable. Asparagine may be substituted for them, although it should be noted that asparagine loads additional nitrogen onto the urea cycle. Citrate has been used effectively in some cases for the therapy of CTLN2 (Yajima *et al.* 1981). Citrate supplies oxaloacetate by acetyl-CoA citrate lyase, but it may also supply acetyl-CoA for fatty acid synthesis, resulting in fatty liver and hyperlipidemia. Again, it

is important to note that patients with citrin deficiency like beans and peanuts and dislike carbohydrates. Beans and peanuts probably provide aspartate and asparagine, which may support the above stated processes, but it is also important to note that they load the liver with excessive nitrogen. Arginine may be effective (Imamura *et al.* 2003). Beans and peanuts contain much arginine. Care should be taken not to supply high-carbohydrate diets, which may lead to high NADH/NAD$^+$ ratio in the liver. Alcohol too, should be avoided. It is known that mitochondrial glycerol-3-phosphate dehydrogenase, a member of the GP-shuttle is induced by thyroid hormone, dehydroepiandrosterone and peroxisome proliferators. These may be effective to activate the GP-shuttle, and to decrease the cytosolic NADH/NAD$^+$ ratio. To establish proper therapeutic procedures for CTLN2, animal models are essential.

SUMMARY

Mitochondrial transporters or carriers belong to the solute carrier family 25 (SLC25). Among them, citrin and aralar having aspartate glutamate carrier activity are unique, because they are Ca^{2+}-activated and larger (74 kDa) than the other SLC25 members (30 kDa). In this chapter, we focus on the deficiency of citrin (SLC25A13), since a deficiency of aralar (SLC25A12) is not known in humans. *Adult-onset type II citrullinemia* (CTLN2) is characterized by a liver-specific deficiency of argininosuccinate synthetase, which is a rate-limiting enzyme of the urea cycle. We have discovered that argininosuccinate synthetase deficiency in CTLN2 is caused by mutations in SLC25A13. This has led to the further discovery that another neonatal disease, *neonatal intrahepatic cholestasis caused by citrin deficiency* (NICCD), is caused by the same gene mutations. Usually NICCD is not a severe disease except in some cases and the patients generally recover to apparently healthy states. But more than 10 years or even several decades later, some of the individuals carrying *SLC25A13* gene mutations in both alleles may develop CTLN2 with neuropsychiatric symptoms. Since citrin functions as an aspartate glutamate carrier, the various symptoms of NICCD and CTLN2 may be understood through the defects of aspartate export from mitochondria to cytosol and of the malate aspartate shuttle. It is, however, still difficult to understand how the hepatic lack of argininosuccinate synthetase develops in CTLN2. It is now important and urgent to find genetic modifiers and/or environmental factors which lead to the deterioration to CTLN2, because we have diagnosed over 80 cases to be suffering or have suffered from NICCD, and have found siblings of CTLN2 patients to be citrin deficient but without CTLN2 symptoms. Now, it is much easier than before to study the pathophysiology of citrin deficiency and develop therapies, because we now know the functions of citrin as an aspartate glutamate carrier, can use two kinds of citrin-deficient mice, and are examining them to see whether they are suitable as animal models for NICCD and/or CTLN2. Finally, we are also developing aralar-deficient mice to find out human disease caused by aralar deficiency.

REFERENCES

Begum, L., Jalil, M.A., Kobayashi, K., Iijima, M., Li, M.X., Yasuda, T., Horiuchi, M., del Arco, A., Satrústegui, J., and Saheki, T. (2002). Expression of Three Mitochondrial Solute Carriers, Citrin, Aralar1, and Ornithine Transporter, in Relation to Urea Cycle in Mice. *Biochim. Biophys. Acta.*, **1574**, 283–292.

Ben-Shalom, E., Kobayashi, K., Shaag, A., Yasuda, T., Gao, H.Z., Saheki, T., Bachmann, C., and Elpeleg, O. (2002). Infantile Citrullinemia caused by Citrin Deficiency with Increased Dibasic Amino Acids. *Mol. Genet. Metab.*, **77**, 202–208.

Chow, W.C., Ng, H.S., Tan, I.K., and Thum, T.Y. (1996). Case Report: Recurrent Hyperammonaemic Encephalopathy due to Citrullinaemia in a 52 Year Old Man. *J. Gastroenterol. Hepatol.*, **11**, 621–625.

Crackower, M.A., Sinasac, D.S., Lee, J.R., Herbrick, J.A., Tsui, L.C., and Scherer, S.W. (1999). Assignment of the *SLC25A12* Gene Coding for the Human Calcium-Binding Mitochondrial Solute Carrier Protein Aralar to Human Chromosome 2q24. *Cytogenet. Cell. Genet.*, **87**, 197–198.

del Arco, A., Agudo, M., and Satrústegui, J. (2000). Characterization of a Second Member of the Subfamily of Calcium-Binding Mitochondrial Carriers Expressed in Human Non-Excitable Tissues. *Biochem. J.*, **345**, 725–732.

del Arco, A., Morcillo, J., Martinez-Morales, J.R., Galián, C., Martos, V., Bovolenta, P., and Satrústegui, J. (2002). Expression of the Aspartate-Glutamate Mitochondrial Carriers Aralar1 and Citrin during Development and Adult Rat Tissues. *Eur. J. Biochem.*, **269**, 3313–3320.

del Arco, A. and Satrústegui, J. (1998). Molecular Cloning of Aralar, a New Member of the Mitochondrial Carrier Superfamily that Binds Calcium and is Present in Human Muscle and Brain. *J. Biol. Chem.*, **273**, 23327–23334.

Hwu, W.L., Kobayashi, K., Hu, Y.H., Yamaguchi, N., Saheki, T., Chou, S.P., and Wang, J.H. (2001). A Chinese Adult-Onset Type II Citrullinaemia Patient with 851del4/1638ins23 Mutations in the *SLC25A13* Gene. *J. Med. Genet.*, **38**, E23.

Iijima, M., Jalil, M.A., Begum, L., Yasuda, T., Yamaguchi, N., Li, M.X., Kawada, N., Endou, H., Kobayashi, K., and Saheki, T. (2001). Pathogenesis of Adult-Onset Type II Citrullinemia caused by deficiency of Citrin, a Mitochondrial Solute Carrier Protein: Tissue and Subcellular Localization of Citrin. *Adv. Enzyme. Regul.*, **41**, 325–342.

Ikeda, S., Yazaki, M., Takei, Y., Ikegami, T., Hashikura, Y., Kawasaki, S., Iwai, M., Kobayashi, K., and Saheki, T. (2002). Type II (Adult-Onset) Citrullinaemia: Clinical Pictures and the Therapeutic Effect of Liver Transplantation. *J. Neurol. Neurosurg. Psychiatry*, **71**, 663–670.

Imamura, Y., Kobayashi, K., Shibatou, T., Aburada, S., Tahara, K., Kubozono, O., and Saheki, T. (2003). Effectiveness of Carbohydrate-Restricted Diet and Arginine Granules Therapy for Adult-Onset Type II Citrullinemia: A Case Report of Siblings Showing Homozygous SLC25A13 Mutation with and without the Disease. *Hepatology Res.* **26**, 68–72.

Kasahara, M., Ohwada, S., Takeichi, T., Kaneko, H., Tomomasa, T., Morikawa, A., Yonemura, K., Asonuma, K., Tanaka, K., Kobayashi, K., Saheki, T., Takeyoshi I., and Morishita, Y. (2001). Living-Related Liver Transplantation for Type II Citrullinemia using a Graft from Heterozygote Donor. *Transplantation*, **70**, 157–159.

Kawamoto, S., Strong, R.W., Kerlin, P., Lynch, S.V., Steadman, C., Kobayashi, K., Nakagawa, S., Matsunami, H., Akatsu, T., and Saheki, T. (1997). Orthotopic Liver Transplantation for Adult-Onset Type II Citrullinaemia. *Clin. Transplantation*, **11**, 453–458.

Kobayashi, K., Horiuchi, M., and Saheki, T. (1997). Pancreatic Secretory Trypsin Inhibitor as a Diagnostic Marker for Adult-Onset Type II Citrullinemia. *Hepatology*, **25**, 1160–1165.

Kobayashi, K., Iijima, M., Yasuda, T., Sinasac, D.S., Yamaguchi, N., Tsui, L.C., Scherer, S.W., and Saheki, T. (2000). Type II Citrullinemia (Citrin Deficiency): A Mysterious Disease Caused by a Defect of Calcium-Binding Mitochondrial Carrier Protein. Calcium: The Molecular Basis of Calcium Action in Biology and Medicine. Dordrecht: Kluwer. pp. 565–587.

Kobayashi, K., Kakinoki, H., Fukushige, T., Shaheen, N., Terazono, H., and Saheki, T. (1995a). Nature and Frequency of Mutations in the Argininosuccinate Synthetase Gene that cause Classical Citrullinemia. *Hum. Genet.*, **96**, 454–463.

Kobayashi, K., Nakata, M., Terazono, H., Shinsato, T., and Saheki, T. (1995b). Pancreatic Secretory Trypsin Inhibitor Gene is Highly Expressed in the Liver of Adult-Onset Type II Citrullinemia. *FEBS Lett.*, **372**, 69–73.

Kobayashi, K., Saheki, T., Imamura, Y., Noda, T., Inoue, I., Matuo, S., Hagihara, S., Nomiyama, H., Jinno, Y., and Shimada, K. (1986). Messenger RNA Coding for Argininosuccinate Synthetase in Citrullinemia. *Am. J. Hum. Genet.*, **38**, 667–680.

Kobayashi, K., Shaheen, N., Kumashiro, R., Tanikawa, K., O'Brien, W.E., Beaudet, A.L., and Saheki, T. (1993). A Search for the Primary Abnormality in Adult-Onset Type II Citrullinemia. *Am. J. Hum. Genet.*, **53**, 1024–1030.

Kobayashi, K., Shaheen, N., Terazono, H., and Saheki, T. (1994). Mutations in argininosuccinate synthetase mRNA of Japanese patients, causing classical citrullinemia. *Am. J. Hum. Genet.*, **55**, 1103–1112.

Kobayashi, K., Sinasac, D.S., Iijima, M., Boright, A.P., Begum, L., Lee, J.R., Yasuda, T., Ikeda, S., Hirano, R., Terazono, H., Crackower, M.A., Kondo, I., Tsui, L.C., Scherer, S.W., and Saheki, T. (1999). The Gene Mutated in Adult-Onset Type II Citrullinaemia Encodes a Putative Mitochondrial Carrier Protein. *Nat. Genet.*, **22**, 159–163.

Krämer, R. and Palmieri, F. (1992). *Metabolite Carriers in Mitochondria. Molecular Mechanisms in Bioenergetics.* Amsterdam: Elsevier Science Publishers, pp. 359–384.

Lee, J., Ellaway, C., Kobayashi, K., and Wilcken, B. (2002). Citrullinaemia type II: A Rare Cause of Neonatal Hepatitis detected by Newborn Screening. *J. Inherit. Metab. Dis.* 25(suppl. 1), 29.

Maruyama, H., Ogawa, M., Nishio, T., Kobayashi, K., Saheki, T., and Sunohara, N. (2001). Citrullinemia Type II in a 64-Year-Old Man with Fluctuating Serum Citrulline Levels. *J. Neurol. Sci.*, **182**, 167–170.

Nagata, N., Matsuda, I., and Oyanagi, K. (1991). Estimated Frequency of Urea Cycle Enzymopathies in Japan. *Am. J. Med. Genet.*, **39**, 228–229.

Naito, E., Ito, M., Matsuura, S., Yokota, I., Saijo, T., Ogawa, A., Kitamura, S., Kobayashi, K., Saheki, T., Nishimura, Y., Sakura, N., and Kuroda, Y. (2002). Type II Citrullinemia (Citrin Deficiency) in a Neonate with Hypergalactosaemia Detected by Mass Screening. *J. Inherit. Metab. Dis.*, **25**, 71–76.

Ohura, T., Kobayashi, K., Tazawa, Y., Nishi, I., Abukawa, D., Sakamoto, O., Iinuma, K., and Saheki, T. (2001). Neonatal Presentation of Adult-Onset Type II Citrullinemia. *Hum. Genet.*, **108**, 87–90.

Onuki, J., Nishimura, S., Yoshino, K., Takahashi, H., Suzuki, T., Abe, K., Sakurabayashi, S., Iwase, T., Hirano, M., Kobayashi, K., and Saheki, T. (2000). Genetic Abnormality in 2 Brothers of a Case With Adult-Onset Type II Citrullinemia: Trial of Pre-Onset Genetic Diagnosis. *Acta. Hepatologica. Jpn. (Japanese)*, **41**, 555–560.

Oshiro, S., Kochinda, T., Tana, T., Yamazato, M., Kobayashi, K., Komine, Y., Muratani, H., Saheki, T., Iseki, K., and Takishita, S. (2002). A Patient with Adult-Onset Type I Citrullinemia on Long-Term Hemodialysis: Reversal of Clinical Symptoms and Brain MRI Findings. *Am. J. Kidney Dis.*, **39**, 189–192.

Palmieri, F. (1994). Mitochondrial Carrier Proteins. *FEBS. Lett.*, **346**, 48–54.

Palmieri, F. and van Ommen, B. (1999). *The Mitochondrial Carrier Protein Family. Frontiers of Cellular Bioenergetics.* New York: Kluwer Academic/Plenum Publishers. pp. 489–519.

Palmieri, L., Pardo, B., Lasorsa, F.M., del Arco, A., Kobayashi, K., Iijima, M., Runswick, M.J., Walker, J.E., Saheki, T., Satrústegui, J., and Palmieri, F. (2001). Citrin and Aralar1 are Ca^{2+}-Stimulated Aspartate/Glutamate Transporters in Mitochondria. *EMBO. J.*, **20**, 5060–5069.

Ross, B.D., Hems, R., and Krebs, H.A. (1967). The Rate of Gluconeogenesis from Various Precursors in the Perfused Rat Liver. *Biochem. J.*, **102**, 942–951.

Saheki, T. and Kobayashi, K. (2002) Mitochondrial Aspartate Glutamate Carrier (Citrin) Deficiency as the Cause of Adult-Onset Type II Citrullinemia (CTLN2) and Idiopathic Neonatal Hepatitis (NICCD). *J. Hum. Genet.*, **47**, 333–341.

Saheki, T., Kobayashi, K., and Inoue, I. (1987a). Hereditary Disorders of the Urea Cycle in Man: Biochemical and Molecular Approaches. *Rev. Physiol. Biochem. Pharmacol.*, **108**, 21–68.

Saheki, T., Kobayashi, K., Inoue, I., Matuo, S., Hagihara, S., and Noda, T. (1987b). Increased Urinary Excretion of Argininosuccinate in Type II Citrullinemia. *Clin. Chim. Acta.*, **170**, 297–304.

Saheki, T., Kobayashi, K., Miura, T., Hashimoto, S., Ueno, Y., Yamasaki, T., Araki, H., Nara, H., Shiozaki, Y., Sameshima, Y., Suzuki, M., Yamauchi, Y., Sakazume, Y., Akiyama, K., and Yamamura, Y. (1986). Serum Amino Acid Pattern of Type II Citrullinemic Patients and Effect of Oral Administration of Citrulline. *J. Clin. Biochem. Nutr.*, **1**, 129–142.

Saheki, T., Nakano, K., Kobayashi, K., Imamura, Y., Itakura, Y., Sase, M., Hagihara, S., and Matuo, S. (1985). Analysis of the Enzyme Abnormality in Eight

Cases of Neonatal and Infantile Citrullinemia in Japan. *J. Inherit. Metab. Dis.*, **8**, 155–156.

Saheki, T., Sase, M., Nakano, K., Azuma, F., and Katsunuma, T. (1983a). Some Properties of Argininosuccinate Synthetase Purified from Human Liver and a Comparison with the Rat Liver Enzyme. *J. Biochem.*, **93**, 1531–1537.

Saheki, T., Ueda, A., Hosoya, M., Kusumi, K., Takada, S., Tsuda, M., and Katsunuma, T. (1981). Qualitative and Quantitative Abnormalities of Argininosuccinate Synthetase in Citrullinemia. *Clin. Chim. Acta.*, **109**, 325–335.

Saheki, T., Ueda, A., Iizima, K., Yamada, N., Kobayashi, K., Takahashi, K., and Katsunuma, T. (1982). Argininosuccinate Synthetase Activity in Cultured Skin Fibroblasts of Citrullinemic Patients. *Clin. Chim. Acta.*, **118**, 93–97.

Saheki, T., Yagi, Y., Sase, M., Nakano, K., and Sato, E. (1983b). Immunohistochemical Localization of Argininosuccinate Synthetase in the Liver of Control and Citrullinemic Patients. *Biomed. Res.*, **4**, 235–238.

Sase, M., Kobayashi, K., Imamura, Y., Saheki, T., Nakano, K., Miura, S., and Mori, M. (1985). Level of Translatable Messenger RNA Coding for Argininosuccinate Synthetase in the Liver of the Patients with Quantitative-Type Citrullinemia. *Hum. Genet.*, **69**, 130–134.

Sinasac, D.S., Crackower, M.A., Lee, J.R., Kobayashi, K., Saheki, T., Scherer, S.W., and Tsui, L.C. (1999). Genomic Structure of the Adult-Onset Type II Citrullinemia Gene, *SLC25A13*, and Cloning and Expression of its Mouse Homologue. *Genomics*, **62**, 289–292.

Takashima, Y., Koide, M., Fukunaga, H., Iwai, M., Miura, M., Yoneda, R., Fukuda, T., Kobayashi, K., and Saheki, T. (2002). Recovery from Marked Altered Consciousness in a Patient with Adult-Onset Type II Citrullinemia Diagnosed by DNA Analysis and Treated with a Living Related Partial Liver Transplantation. *Internal Medicine*, **41**, 555–560.

Takenaka, K., Yasuda, I., Araki, H., Naito, T., Fukutomi, Y., Ohnishi, H., Yamakita, N., Hasegawa, T., Sato, H., Shimizu, Y., Matsunami, H., and Moriwaki, H. (2000). Type II Citrullinemia in an Elderly Patient Treated with Living Related Partial Liver Transplantation. *Internal Medicine*, **39**, 553–558.

Tamamori, A., Okano, Y., Ozaki, H., Fujimoto, A., Kajiwara, M., Fukuda, K., Kobayashi, K., Saheki, T., Tagami Y., and Yamano, T. (2002). Neonatal Intrahepatic Cholestasis Caused by Citrin Deficiency: Severe Hepatic Dysfunction in an Infant Requiring Liver Transplantation. *Eur. J. Pediatr.*, **161**, 609–613.

Tazawa, Y., Kobayashi, K., Ohura, T., Abukawa, D., Nishinomiya, F., Hosoda, Y., Yamashita, M., Nagata, I., Kono, Y., Yasuda, T., Yamaguchi, N., and Saheki, T.

(2001). Infantile Cholestatic Jaundice Associated With Adult-Onset Type II Citrullinemia. *J. Pediatr.*, **138**, 735–740.

Todo, S., Starzl, T.E., Tzakis, A., Benkov, K.J., Kalousek, F., Saheki, T., Tanikawa, K., and Fenton, W.A. (1992). Orthotopic Liver Transplantation for Urea Cycle Enzyme Deficiency. *Hepatology*, **15**, 419–422.

Tomomasa, T., Kobayashi, K., Kaneko, H., Shimura, H., Fukusato, T., Tabata, M., Inoue, Y., Ohwada, S., Kasahara, M., Morishita, Y., Kimura, M., Saheki, T., and Morikawa, A. (2001). Possible Clinical and Histologic Manifestations of Adult-Onset Type II Citrullinemia in Early Infancy. *J. Pediatr.*, **138**, 741–743.

Tsuboi, Y., Fujino, Y., Kobayashi, K., Saheki, T., and Yamada, T. (2001). High Serum Pancreatic Secretory Trypsin Inhibitor Before Onset of Type II Citrullinemia. *Neurology*, **57**, 933.

Tsujii, T., Morita, T., Matsuyama, Y., Matsui, T., Tamura, M., and Matsuoka, Y. (1976). Sibling Cases of Chronic Recurrent Hepatocerebral Disease with Hypercitrullinemia. *Gastroenterologia. Jpn.*, **11**, 328–340.

Walker, J.E. (1992). The Mitochondrial Transporter Family. *Curr. Opin. Struct. Biol.*, **2**, 519–526.

Yagi, Y., Saheki, T., Imamura, Y., Kobayashi, K., Sase, M., Nakano, K., Matuo, S., Inoue, I., Hagihara, S., and Noda, T. (1988). The Heterogeneous Distribution of Argininosuccinate Synthetase in the Liver of Type II Citrullinemic Patients: Its Specificity and Possible Clinical Implications. *Am. J. Clin. Pathol.*, **89**, 735–741.

Yajima, Y., Hirasawa, T., and Saheki, T.(1981). Treatment of Adult-Type Citrullinemia with Administration of Citrate. *Tohoku. J. Med.*, **134**, 321–330.

Yamaguchi, N., Kobayashi, K., Yasuda, T., Nishi, I., Iijima, M., Nakagawa, M., Osame, M., Kondo, I., and Saheki, T. (2002). Screening of SLC25A13 Mutations in Early and Late Onset Patients with Citrin Deficiency and in the Japanese Population: Identification of Two Novel Mutations and Establishment of Multiple DNA Diagnosis Methods for Nine Mutations. *Hum. Mutat.*, **19**, 122–130.

Yasuda, T., Yamaguchi, N., Kobayashi, K., Nishi, I., Horinouchi, H., Jalil, M.A., Li, M.X., Ushikai, M., Iijima, M., Kondo, I., and Saheki, T. (2000). Identification of Two Novel Mutations in the *SLC25A13* Gene and Detection of Seven Mutations in 102 Patients with Adult-Onset Type II Citrullinemia. *Hum. Genet.*, **107**, 537–545.

Yazaki, M., Ikeda, S., Takei, Y., Yanagisawa, N., Matsunami, H., Hashikura, Y., Kawasaki, S., Makuuchi, M., Kobayashi, K., and Saheki, T. (1996). Complete Neurological Recovery of an Adult Patient With Type II Citrullinaemia after Living Related Partial Liver Transplantation. *Transplantation*, **62**, 1679–1681.

11

NICOLA LONGO*, CRISTINA AMAT DI SAN FILIPPO*,
AND MARZIA PASQUALI**

The OCTN2 carnitine transporter and fatty acid oxidation

INTRODUCTION

Carnitine (3-hydroxy-4-trimethylammonium butyrate) is a hydrophilic molecule that plays an essential role in the transfer of long-chain fatty acids into mitochondria for β-oxidation (Scaglia and Longo 1999). Carnitine also binds acyl residues and helps in their elimination. This decreases the number of acyl residues conjugated with Coenzyme A (CoA) and increases the ratio between free and acylated CoA (Bieber 1988). Less defined functions of carnitine include the shuttling of fatty acids between different intracellular organelles (peroxisomes, microsomes, and mitochondria) involved in fatty acid metabolism (Bieber 1988). Carnitine deficiency has been known for several years in humans, but the difference between primary and secondary carnitine deficiency has only been fully defined in recent years. This chapter will review the structure and function of the OCTN2 carnitine transporter defective in primary carnitine deficiency.

Carnitine can be synthesized by the human body or assumed by diet. In mammals, carnitine is synthesized by liver and kidney, and in smaller amounts by the brain from methionine and protein bound lysine (Bieber 1988; Scaglia and Longo 1999). Several cofactors, such as pyridoxal phosphate, niacin, vitamin C, and iron, are required for carnitine biosynthesis (Scaglia and Longo 1999). However, the majority of carnitine is supplied to the organism by exogenous sources. In utero, carnitine is transferred across the placenta to provide significant carnitine stores to the growing human (Schmidt-Sommerfeld et al. 1985). Milk contains sufficient carnitine for the growing child, while meat products are the major source of carnitine for adults. The average adult diet provides about 75% of daily carnitine requirements (Borum 1995). Strict vegetarians maintain normal carnitine levels (Lombard et al. 1989), indicating that humans not only synthesize carnitine but also effectively conserve it. Normal infants of vegetarian mothers do not need supplemental carnitine. However, a case of sudden death in the neonatal period has been described in a child with primary carnitine deficiency born to a vegetarian mother (Rinaldo et al. 1997). Carnitine is not metabolized by humans and is excreted in free or conjugated form in urine and bile. The high concentration of carnitine and its conjugates in

* Division of Medical Genetics, Department of Pediatrics, University of Utah, 2C412 SOM, 50 North Medical Drive, Salt Lake City, UT 84132, USA
** Department of Pathology, University of Utah, 500 Chipeta Way, Salt Lake City, UT 84108, USA

the bile can be used clinically for the post-mortem diagnosis of infants with sudden infant death caused by inherited disorders of fatty acid oxidation (Rashed *et al.* 1995). The quantitative contribution of biliary excretion to total carnitine losses in humans remains unclear.

CARNITINE AND FATTY ACID OXIDATION

Carnitine is required for the entry of long-chain fatty acids into the mitochondrial matrix for subsequent β-oxidation (Scaglia and Longo 1999). During periods of fasting, fatty acids turn into the predominant substrate for energy production via oxidation in the liver, cardiac muscle, and skeletal muscle. The brain does not directly utilize fatty acids for oxidative metabolism, but oxidizes ketone bodies derived from acetyl CoA and acetoacetyl CoA produced by fatty acid oxidation in the liver. Fatty acids are mobilized from the adipose tissue and transported in the circulation primarily bound to albumin. After their entry into the cells by a specific membrane transporter, fatty

acids are conjugated to CoA by acyl CoA synthase (Figure 1). Fatty acids must then be conjugated to carnitine to enter mitochondria. Carnitine is accumulated inside the cell by OCTN2, the high-affinity carnitine transporter in the heart, muscle, kidney, and other tissues. Carnitine forms a high-energy ester bond with long-chain carboxylic acids at its β-hydroxyl position by the action of carnitine palmitoyl transferase 1 (CPT-1), located in the inner aspect of the outer mitochondrial membrane. Acylcarnitine is then translocated across the inner mitochondrial membrane by the carnitine–acylcarnitine translocator and cleaved by CPT-2 in the inner aspect of the inner mitochondrial membrane. Carnitine is released in the mitochondrial matrix and can then return to the cytoplasm for another cycle using the carnitine–acylcarnitine translocator, while the fatty acid is conjugated back to CoA in the mitochondrial matrix and can undergo (in aerobic conditions and in the presence of low levels of ATP) β-oxidation with production of acetyl-CoA for oxidative phosphorylation or production of ketone bodies in the liver. Inherited defects of all known steps in fatty acid oxidation have been reported in humans

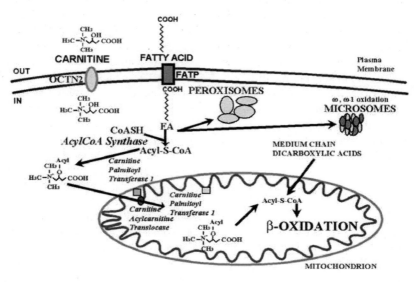

Figure 1   The carnitine cycle in fatty acid oxidation. Modified from Scaglia and Longo (1999).

(Scaglia and Longo 1999; Roe and Ding 2001). Ketone bodies can be used as an alternate energy source by the heart, skeletal muscle, and brain, sparing glucose. In the liver, acetyl-CoA activates pyruvate carboxylase to favor gluconeogenesis. The net result of both actions is glucose sparing and production (although the latter cannot occur from fat itself). If carnitine is missing, fat cannot be utilized, glucose is consumed without regeneration via gluconeogenesis, and there is a drop in glucose levels (hypoglycemia).

FUNCTIONAL CHARACTERIZATION OF THE CARNITINE TRANSPORTER

The OCTN2 carnitine transporter defective in primary carnitine deficiency has been characterized in human fibroblasts (Scaglia et al. 1999) and in cell lines after overexpression of the OCTN2 cDNA (Ohashi et al. 1999, 2001, 2002; Wu et al. 1999). OCTN2 is energized by the sodium electrochemical gradient. OCTN2 operates as an Na^+/carnitine cotransporter with a $1:1$ stoichiometry in both fibroblasts and CHO cells overexpressing OCTN2 (Scaglia et al. 1999; Wang et al. 2000b). Direct measurement of carnitine-induced currents in Xenopus oocytes expressing OCTN2 is also consistent with this stoichiometry (Wagner et al. 2000). The transporter is sensitive to the membrane potential (Scaglia et al. 1999). Ouabain inhibits OCTN2 activity as a result of its effects on the membrane potential and the transmembrane Na^+ gradient (Scaglia et al. 1999). Carnitine transport does not require chloride (Scaglia et al. 1999) or H^+, although the activity of the transporter is pH sensitive (Wagner et al. 2000).

OCTN2 functions as both a sodium-dependent carnitine transporter and a sodium-independent cation transporter (Wu et al. 1999). OCTN2 has a high affinity toward carnitine ($K_m = 4\ \mu M$) and is physiologically saturated by Na^+ in the physiologic range ($K_{[Na]} = 12$ mM) (Scaglia et al. 1999; Wang et al. 2000b). Recent data suggest that carnitine

and organic cations bind to different sites on the transporter molecule (Seth et al. 1999), with an anionic binding site needed only for carnitine transport (Ohashi et al. 2002). For Na^+-independent organic cation transport, OCTN2 functions as other organic cation transporters (Wu et al. 1999; Ohashi et al. 2001). Accumulated organic cations can exchange with extracellular carnitine producing trans-stimulation (Ohashi et al. 2002). The influx of carnitine stimulates the efflux of organic cations, suggesting that this exchange mechanism might help in the excretion of organic compounds in the kidney (Ohashi et al. 2002).

Carnitine transport is not affected by amino acids (including trimethyllysine), but is competitively inhibited by carnitine conjugates (with palmitoylcarnitine being the most powerful), butyrobetaine (a metabolic precursor), betaine (with low affinity), and high concentrations of choline (Scaglia et al. 1999). Quinidine and verapamil cause a noncompetitive inhibition of carnitine transport in human fibroblasts (Scaglia et al. 1999). In cells overexpressing OCTN2, an increased transport of the organic cations quinidine and verapamil is observed, although to a minimal extent (Ohashi et al. 2002). Therefore some degree of competitive inhibition by these two compounds cannot be excluded (Ohashi et al. 2002). Other organic cations that inhibit carnitine transport are cimetidine, desipramine, and emetine (Ohashi et al. 1999; Wu et al. 1999). This wide range of substrates suggests that OCTN2, together with other organic cation and anion transporters, contributes to the absorption and elimination of common pharmaceutical compounds.

MOLECULAR IDENTIFICATION OF THE OCTN2 CARNITINE TRANSPORTER

The OCTN2 carnitine transporter was initially identified for its homology with OCTN1 (Tamai et al. 1997), a novel organic cation transporter with an unknown physiological

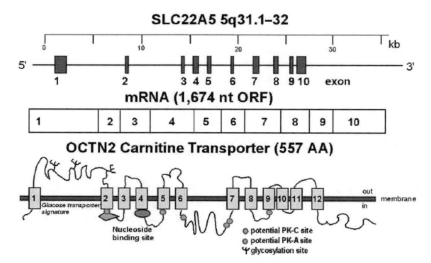

Figure 2   The carnitine transporter gene *SLC22A5* and putative structure of the OCTN2 membrane transporter.

substrate that does not transport carnitine (Wu *et al.* 1998, 2000). Subsequently, OCTN2 was shown to transport carnitine with high affinity and functional characteristics similar to those reported in human fibroblasts (Tamai *et al.* 1998). The gene mapped to 5q31.1-32, the locus for primary carnitine deficiency (Shoji *et al.* 1998). The identification of mutations in this gene confirmed its identity as the carnitine transporter responsible for primary carnitine deficiency (Nezu *et al.* 1999; Tang *et al.* 1999; Wang *et al.* 1999). The gene for primary carnitine deficiency, named *SLC22A5*, spans about 30 kb on the long arm of chromosome 5 (Figure 2). It is composed of 10 exons and generates transcripts of about 3,500 nucleotides, of which 1,674 constitute the open reading frame. The resulting protein is composed of 557 amino acids. The protein has 12 predicted transmembrane spanning domains with both the amino and carboxy terminus facing the cytoplasm (Tamai *et al.* 1998; Wu *et al.* 1998).

OCTN2 is a novel organic cation transporter and, unique among the cation transporters, is Na⁺ dependent. Novel organic cation transporters share similarity with other organic cation transporters (Figure 3). They have 12

predicted transmembrane domains, with both the N- and C-terminus located inside the cell. OCTN1 and OCTN2 possess a nucleotide-binding site, which is not present in OCT-1, OCT-2, or OCT-3. There are potential phosphorylation sites for protein kinases C and A, located on predicted intracellular loops, and potential glycosylation sites, located in the extracellular loop between the first and second transmembrane domain (Tamai *et al.* 1998; Wu *et al.* 1998). Finally, there is a conserved glucose transporter (GLUT) signature motif in the intracellular loop between transmembrane domains 2 and 3. The function of all these putative sites has not yet been defined or confirmed by direct studies, except that mutations in the GLUT signature motif (R169W, R169Q), identified in patients with primary carnitine deficiency, severely impair carnitine transport activity (Wang *et al.* 2000c).

SUBCELLULAR AND TISSUE DISTRIBUTION OF OCTN2

OCTN2 is expressed almost ubiquitously, with the highest levels detected in human heart, placenta, skeletal muscle, kidney, and pancreas

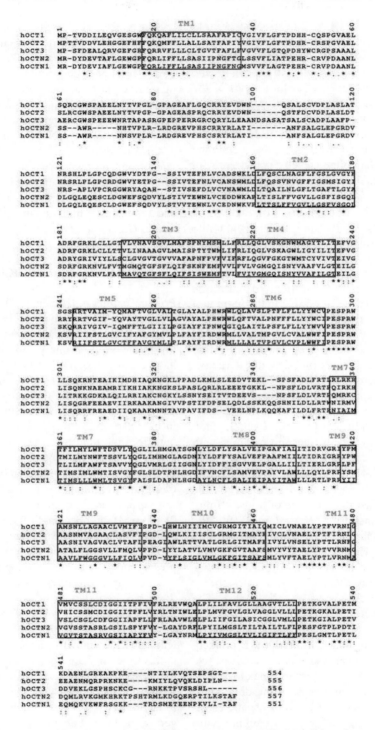

Figure 3 Alignment of human organic (OCT1, OCT2, and OCT3) and novel organic (OCTN1, OCTN2) cation transporters. Predicted transmembrane domains (TM1–TM12) are indicated as boxed regions: *, identical residues; .:, conserved residues.

(Wu *et al.* 1992). OCTN2 is also expressed in a number of cell lines (Wu *et al.* 1992) and in human fibroblasts. By in situ hybridization to rat tissues, OCTN2 is expressed in the glomeruli, proximal and distal tubules of the kidney, the heart, the placenta, and the brain, although its function in this last organ remains to be clarified (Wu *et al.* 1999). In mice and rats, OCTN2 localizes in the apical membrane of renal tubular epithelial cells, consistent with its role in renal carnitine reabsorption (Tamai *et al.* 2001).

Despite the ubiquitous distribution of OCTN2, carnitine transport differs among tissues. In vesicles from chicken and rat intestine the K_m for carnitine transport is about 10 times higher than that of OCTN2 (Duran *et al.* 2002). Intestinal absorption of carnitine remains about 50% of normal in the *jvs* mouse lacking functional carnitine transporters (Yokogawa *et al.* 1999a). Patients with primary carnitine deficiency absorb carnitine well from the intestine despite defective OCTN2. Therefore, at least in the gut, there must be multiple carnitine transporters including some with low affinity towards carnitine. The same might apply to the liver. In fact, mammalian liver cells transport carnitine with a K_m of 2–10 mM, 1,000 times greater than that of OCTN2 (Christiansen and Bremer 1976). Studies in human HepG2 liver cells have confirmed the presence of a low-affinity carnitine transporter ($K_m = 2$ mM) and the absence of a high-affinity component (Scaglia *et al.* 1999). However, hepatocytes from normal mice have both a high-affinity (indistinguishable from OCTN2) and a low-affinity component (Yokogawa *et al.* 1999b). The high-affinity component is lost in hepatocytes from the *jvs* mouse, confirming its identity as OCTN2 (Yokogawa *et al.* 1999b). It is likely that the human liver also expresses at least two different carnitine transporters, one of which is not affected in primary carnitine deficiency, explaining the normalization of hepatic carnitine content by therapy in patients with primary carnitine deficiency (Roe and Ding 2001).

The amino acid transporter $ATB^{0,+}$ can transport carnitine with low affinity ($K_m = 0.8$ mM) (Nakanishi *et al.* 2001). This transporter is expressed primarily in the lungs, mammary gland, and intestine. The last characteristic might explain residual intestinal carnitine transport in patients with primary carnitine deficiency (Nakanishi *et al.* 2001). A third carnitine transporter, CT1, was isolated from the testis (Enomoto *et al.* 2002). This transporter is homologous with cation and anion transporters, but its exclusive presence in the testis limits its importance in other tissues (Enomoto *et al.* 2002).

PRIMARY CARNITINE DEFICIENCY

Primary carnitine deficiency (OMIM 212140) is an autosomal recessive disorder of fatty acid oxidation due to the lack of functional carnitine transporters. Primary carnitine deficiency has a frequency of about 1 in 40,000 newborns in Japan (Koizumi *et al.* 1999) and 1 in 37,000 newborns in Australia (Wilcken *et al.* 2001). The frequency of primary carnitine deficiency in the USA and Europe has not been defined but, from the reported cases, it seems similar to that in Japan and Australia.

The lack of the plasma membrane carnitine transporter results in urinary carnitine wasting, low serum carnitine levels (0–3 μM; normal 25–50 μM), and decreased intracellular carnitine accumulation. Affected patients can have a predominant metabolic or cardiomyopathic presentation. The metabolic presentation is more frequent before 2 years of age. Typically, these children start to refuse food during an upper respiratory tract infection or acute gastroenteritis. Subsequently, they become lethargic and minimally responsive. If evaluated at this time, they are obtunogd and show minimal response to the examination (encephalopathy). In most cases, they have hepatomegaly in addition to signs and symptoms of the triggering condition. Laboratory evaluation usually reveals hypoglycemia with minimal or no ketones in urine and hyperammonemia with variably elevated liver function tests. If children are not treated promptly with intravenous glucose, they progress to coma and death. Cardiomyopathy associated with skeletal myopathy is more frequent in older patients. These children present with progressive respiratory compromise that does not respond to antibiotics. Chest radiograms may show an enlarged heart, and decreased ventricular ejection fraction can be seen by echocardiography. Cardiomyopathy can also be seen in older patients with a metabolic presentation, even if asymptomatic from a cardiac standpoint.

Fatty acid oxidation defects and carnitine deficiency become evident when the body must use fat to provide sufficient energy to meet physiological requirements. Breastfed infants may experience a catabolic state shortly after birth (1–4 days old), when milk production is not yet adequate to meet their nutritional requirements (Scaglia and Longo 1999). With the exception of one infant of a vegetarian mother (Rinaldo et al. 1997), children with primary carnitine deficiency do not present in the neonatal period. In our series of patients with primary carnitine deficiency with demonstrated mutations, the earliest presentation has been around 5–6 months of age (Scaglia et al. 1998; Wang et al. 1999, 2000a,c). In these patients, metabolic symptoms were predominant. In many cases, cardiac involvement was not observed when the child presented before 2 years of age, even if analysed with echocardiography. Later in life, fatty acid oxidation defects become apparent during times of increased energy demand, such as fever, sepsis, or prolonged fasting. If the system is not stressed, fatty acid oxidation defects may remain completely silent. Many children have presented with respiratory distress due to cardiomyopathy without any evident stressor (Wang et al. 2000a). A few children diagnosed because of an affected sibling had only mild developmental delays (Wang et al. 2001).

Key to the diagnosis is the measurement of plasma carnitine levels. Free and acylated carnitine are extremely reduced (free carnitine <3 μM; normal 25–50 μM). Urine organic acids do not show any consistent anomaly, although a nonspecific dicarboxylic aciduria has been reported (Scaglia et al. 1998). Diagnosis is confirmed by demonstrating reduced carnitine transport in skin fibroblasts from the patient. Heterozygous parents of affected children tend to have borderline low levels of plasma carnitine and half-normal carnitine transport in their fibroblasts (Scaglia et al. 1998). Patients with primary carnitine deficiency lose most (90–95%) of the filtered carnitine in the urine and their heterozygous parents lose two to three times the normal amount, explaining their mildly reduced plasma carnitine levels (Scaglia et al. 1998).

Primary carnitine deficiency should be differentiated from other causes of carnitine deficiency. These include a number of organic acidemias (propionic, methylmalonic, isovaleric, and glutaric acidemia, and other rarer disorders) where carnitine is conjugated with the abnormal organic acid and lost in urine. Defects of fatty acid oxidation (short-, medium-, long-, and very-long-chain acyl-CoA-dehydrogenase deficiency, and short- and long-chain 3-OH-acyl-CoA-dehydrogenase deficiency) and of the carnitine cycle (carnitine-acylcarnitine translocase and carnitine palmitoyl transferase-2 deficiency) can also cause significant carnitine deficiency (Roe and Ding 2001). In all these disorders, analysis of urine organic acids, plasma amino acids, and serum acylcarnitine profile, in conjunction with the clinical presentation, allows a definitive diagnosis. Low carnitine levels can also be seen in patients with generalized renal tubular dysfunction, such as renal Fanconi syndrome. In this case, the urinary wasting of other compounds, such as bicarbonate, phosphorus and amino acids, allows a net differentiation, since patients with primary carnitine deficiency have selective carnitine losses.

Patients with primary carnitine deficiency respond to dietary carnitine supplementation (100–400 mg/kg/day), if started before irreversible damage occurs. The dose of carnitine should be adapted to each individual patient by serial measurements of plasma carnitine levels. Carnitine has few side effects. It can cause diarrhea and intestinal discomfort at high doses. This is usually self-limiting and can be resolved by reducing carnitine dosage. Sometimes, bacterial metabolism in the intestine can result in carnitine degradation, with production of trimethylamine, a nontoxic chemical with a very unpleasant odor. This responds to oral therapy with metronidazole, an antibiotic active against anaerobic bacteria. The long-term prognosis is favorable as long as children remain on carnitine supplements. Repeated attacks of hypoglycemia or sudden death have been reported in patients discontinuing carnitine against medical advice.

Primary carnitine deficiency can be diagnosed prenatally by measuring carnitine transport in amniotic fluid cells (Christodoulou et al. 1996) and by newborn screening programs utilizing tandem mass spectrometry for the measurement of carnitine and acylcarnitine species (Wilcken et al. 2001). However, about 50% of children can be missed by newborn screening (Wilcken et al. 2001). This is because children with primary carnitine deficiency have normal levels of plasma carnitine at birth due to transplacental transfer which become severely depressed only after 2 days of life (Christodoulou et al. 1996). If newborn screening is performed before the decline in carnitine levels, a normal result can be obtained (Wilcken et al. 2001).

Heterozygotes for primary carnitine deficiency have mildly reduced plasma carnitine levels (Scaglia et al. 1998). This, compounded with the reduced capacity of organs and tissues to transport carnitine, could result in decreased carnitine content in tissues. In animal models, reduced carnitine content has been demonstrated in liver, heart, and muscle of heterozygous mice (Lahjouji et al. 2002). These animals are at increased risk of developing cardiomyopathy with advancing age (Xiaofei et al. 2002). Heterozygous humans develop benign cardiac hypertrophy (Koizumi et al. 1999), and heterozygosity for primary carnitine deficiency has been cited as a possible risk factor for cardiomyopathy (Scaglia et al. 1998).

This is not a trivial issue considering that about 1% of the population is heterozygous for this potentially treatable condition (Koizumi et al. 1999; see Box: Pathophysiology of Carnitine Deficiency).

THE *jvs* MOUSE, A MODEL OF CARNITINE DEFICIENCY

The *jvs* mouse is a natural model of primary carnitine deficiency. This mouse model was initially noted to develop a disease similar to Reye's syndrome in children, with accumulation of fat in viscera (juvenile visceral steatosis or *jvs*) (Koizumi et al. 1988). Subsequent studies indicated that these mice lack carnitine (Kuwajima et al. 1991) and that their fibroblasts have impaired carnitine transport (Kuwajima et al. 1996). Recently, the causative missense mutation (L352R) in the gene encoding for OCTN2 was identified (Lu et al. 1998; Nezu et al. 1999).

These mice are unable to burn fat and present with hypoglycemia and cardiomyopathy. In addition, they have hyperammonemia, which is caused by downregulation of the expression of urea cycle enzymes in the liver (Horiuchi et al. 1992). Hyperammonemia and urea cycle gene expression can be corrected by carnitine administration, indicating that carnitine deficiency itself or the accumulation of long-chain fatty acids can cause hyperammonemia (Horiuchi et al. 1992; Saheki et al. 2000).

In these mice, carnitine is lost in urine as in patients with primary carnitine deficiency (Yokogawa et al. 1999a). The bioavailability of carnitine from the gut is about half that in normal mice, indicating that transporters other than OCTN2 contribute to intestinal absorption (Yokogawa et al. 1999a). Mice heterozygous for the mutant phenotype have normal life expectancy. However, they develop cardiac hypertrophy and have an increased incidence of cardiomyopathy later in life (Xiaofei et al. 2002). The study of this model might help in defining whether heterozygosity for primary

PATHOPHYSIOLOGY OF CARNITINE DEFICIENCY

Primary carnitine deficiency is a fatty acid oxidation defect caused by defective carnitine transporters. There is loss of carnitine in urine and tissues are unable to accumulate carnitine. The lack of carnitine impairs the transfer of long-chain fatty acids across the inner mitochondrial membrane. The inability to utilize fat decreases the production of ketone bodies and increases the utilization of glucose, causing hypoglycemia. The skeletal muscle and heart derive a considerable amount of energy from fatty acid oxidation. Affected patients can have a predominant metabolic or cardiomyopathic presentation. The metabolic presentation is more frequent between 6 months and 2 years of age. It is triggered by fasting or infections, conditions that require an increased utilization of fat to meet energy requirements. Children refuse food and initially become irritable and then lethargic and minimally responsive. If evaluated at this time, they are obtunded and show minimal response to the examination (encephalopathy). In most cases, they have hepatomegaly in addition to signs and symptoms of the triggering condition. Laboratory evaluation usually reveals hypoglycemia with minimal or no ketones in urine and hyperammonemia with variably elevated liver function tests. If children are not treated promptly with intravenous glucose, they progress to coma and death. Cardiomyopathy associated with skeletal myopathy is more frequent in older patients. It becomes evident with progressive respiratory compromise that does not respond to antibiotics. Chest radiograms may show an enlarged heart. Echocardiography shows decreased ventricular ejection fraction. Diagnosis requires demonstration of low levels of carnitine in plasma and is confirmed by measurement of reduced carnitine transport in the patient's fibroblasts. Patients respond well to carnitine supplement that should be continued for life.

carnitine deficiency is a risk factor for adult-onset heart disease in humans.

MUTATIONS IN THE OCTN2 TRANSPORTER IN PATIENTS WITH PRIMARY CARNITINE DEFICIENCY

Mutations in the OCTN2 carnitine transporter identified in patients with primary carnitine deficiency are summarized in Table 1 and Figure 4. Although most families have distinct mutations, some mutations have been reported more than once. A few mutations have been reported in different populations (Burwinkel et al. 1999; Vaz et al. 1999; Wang et al. 1999, 2000c; Mayatepek et al. 2000). These mutations likely occur at mutation-prone DNA sequences (Wang et al. 2000c) since they have occurred on different haplotypes (Burwinkel et al. 1999). In other cases, the same mutation recurs in patients with similar genetic backgrounds (Wang et al, 1999, 2000c) and there is an unusually high frequency of homozygosity for mutations in the SLC22A5 gene in families

with no known consanguinity (Christiansen and Bremer 1976; Vaz et al. 1999; Wang et al. 1999, 2000c, 2001; Christensen et al. 2000). This phenomenon has been attributed to a founder effect in socially and geographically isolated islands or to limited mobility of families in small neighbouring geographical areas (Koizumi et al. 1999; Wang et al. 2001; Tang et al. 2002).

Mutations producing the premature insertion of a STOP codon either as a result of a direct mutational event or because of a frameshift (nonsense mutations) usually result in reduced levels of RNA for the carnitine transporter (Wang et al. 1999, 2000c, 2001). This is due to nonsense-mediated RNA decay, a process common to all eukaryotes aimed at removing messages that could result in truncated proteins capable of dominant-negative effects.

Missense mutations impair function of the OCTN2 transporter. If the topology of the transporter is correctly predicted by the current model (Tamai et al. 1998; Wu et al. 1998), some mutations (N32S, Y211C, G242V,

Table 1   Mutations in the carnitine transporter OCTN2 in patients with primary carnitine deficiency

Codon	Exon	Nucleotide change (cDNA)	Reference(s)	Transport (%)
Y4X	1	12C>G	Wang *et al.* 2001	0
R19P	1	56G>C	Wang *et al.* 2001	4
ΔF23	1	67–69delTTC	Lamhonwah *et al.* 2002	?
N32S	1	95A>G	Christensen *et al.* 2000; Lamhonwah *et al.* 2002	?
W132X	2	396G>A	Koizumi *et al.* 1999; Nezu *et al.* 1999	0
I89fsX133	1	253–264dup GGCTCGCTCGCCACC	Wang *et al.* 2001	0
R2fsX137	1	4_5insC	Nezu *et al.* 2002	0
R169Q	3	506G>A	Burwinkel *et al.* 1999	?
R169W	3	505C>T	Wang *et al.* 2000c	0
M179L	3	535A>T	Koizumi *et al.* 1999	74
Y211C	3	632A>G	Vaz *et al.* 1999	?
G242V	4	725G>T	Wang *et al.* 2000c	0
R254X	4	981C>T	Tang *et al.* 2002	0
262X	4	1027delT	Cederbaum *et al.* 2000	0
R282X	5	844C>T	Burwinkel *et al.* 1999; Vaz *et al.* 1999; Wang *et al.* 1999	0
R282fs295X	5	844delC	Lamhonwah *et al.* 2002	0
W283R	5	847T>C, 847T>A	Mayatepek *et al.* 2000	<2
W283C	5	849G>T	Koizumi *et al.* 1999	2
A301D	5	902C>A	Wang *et al.* 2000c	2
T337fs348X	6	1009delA	Lamhonwah *et al.* 2002	0
W351R	6	1051T>C	Wang *et al.* 2000c	0
Y387X	7	1161T>G	Tang *et al.* 2002	0
R399Q	7	1196G>A	Wang *et al.* 2001	4
Y401X	7	1202_1203insA	Wang *et al.* 1999	0
T440M	8	1319C>T	Lamhonwah *et al.* 2002	?
V446F	8	1336G>T	Mayatepek *et al.* 2000	0.5
E452K	8	1354G>A	Wang *et al.* 2000a	4
458X	8	1303delG	Wang *et al.* 1999	0
S467C	8	1400C>G	Koizumi *et al.* 1999	11
T468R	8	1403C>G	Lamhonwah *et al.* 2002	?
S470F	8	1409C>T	Lamhonwah *et al.* 2002	?
P478L	8	1433C>T	Tang *et al.* 1999	0

? No studies reported on the transfected DNA.

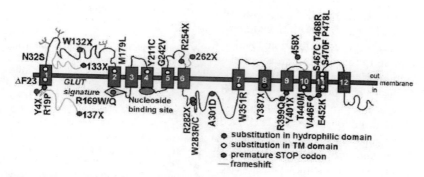

Figure 4 Mutations in the OCTN2 carnitine transporter in patients with primary carnitine deficiency. Residual carnitine transport was measured in mammalian cells after transfection with the mutant cDNA.

W351R, T440M, S467C, T468R, S470F, and P478L) are located in transmembrane domains. The majority of the other missense mutations affect intracellular, rather than extracellular, loops. This phenomenon has also been observed in the case of the catalytic subunit of the cystine transporter responsible for cystinuria (Palacin *et al.* 2001) (see Chapter 14). It is unclear whether this is due to the limited number of patients examined at a molecular level or to the fact that variations in the extracellular domain have limited functional significance or perhaps because the phenotype produced by extracellular mutations differs from the classic phenotype, with patients escaping ascertainment. However, it is possible that the intracellular loops are less tolerant to substitutions and more easily impair transport activity when substituted.

When mutations were heterologously expressed in cultured cells, some retained residual carnitine transport activity (Table 1). The presence of significant residual carnitine transport activity does not result in a different clinical presentation (Wang *et al.* 2000c, 2001; Lamhonwah *et al.* 2002). By contrast, variable expression can be seen within individual families and among patients with identical mutations, indicating the lack of genotype– phenotype correlation (Stanley *et al.* 1991; Wang *et al.* 2000c, 2001; Lamhonwah *et al.* 2002). As in other defects of fatty acid oxidation, environmental triggers

such as fasting, fever, and infections are the primary determinants of the time and type of phenotypic expression.

Some of the natural mutations identified in patients with primary carnitine deficiency have revealed the function of domains of the carnitine transporter. The P478L substitution located in transmembrane domain 11 markedly impaired carnitine transport without affecting organic cation transport (Seth *et al.* 1999). This provided genetic evidence that the recognition sites for organic cations and carnitine by OCTN2 are separated. More recently, from the analysis of the S467C substitution, it was proposed that transmembrane domain 11 contains the anion-binding site of the transporter that is obviously not needed for transporting organic cations (Ohashi *et al.* 2002).

As noted above, most of the natural mutations identified in patients with primary carnitine deficiency affect putative intracellular loops connecting different transmembrane domains (Figure 4). How modifications of such domains affect functioning of the OCTN2 transporter is still unclear. The E452K substitution located intracellularly between transmembrane domains 10 and 11 was associated with residual carnitine transport activity (Wang *et al.* 2000a, b). E452K mutant transporters normally reached the plasma membrane and had a normal K_m for carnitine (Wang *et al.* 2000b). The E452K mutation increased the K_m for Na$^+$ from the normal value of 12 mM to 187 mM. Na$^+$ helps OCTN2 to recognize carnitine (at low concentrations) and to provide the force to move carnitine across the plasma membrane (at higher sodium concentrations) (Wang et al. 2000b). The increased K_m for Na$^+$ was interpreted as an indication that the E452K mutation in OCTN2 reduced the flux of Na$^+$ into the cells that accompanies carnitine entry and indirectly retarded the change from an outward-facing to an inward-facing configuration of the OCTN2 transporter (Wang et al. 2000b). Thus the study of additional natural mutations can clarify the function of other domains of the OCTN2 transporter.

SUMMARY

Primary carnitine deficiency is an autosomal recessive disorder of fatty acid oxidation caused by mutations in the OCTN2 carnitine transporter. The Na$^+$-dependent OCTN2 carnitine transporter (encoded by the *SLC22A5* gene) is responsible for tissue carnitine uptake and renal reabsorption of filtered carnitine. Affected patients waste carnitine in urine and cannot utilize fat during periods of fasting or fever. They present with hypoketotic hypoglycemia, encephalopathy, acute liver failure, and skeletal/cardiac myopathy. Carnitine administration resolves clinical symptoms. These patients have heterogeneous mutations in the *SLC22A5* gene that result in decreased OCTN2 mRNA levels (nonsense mutations) or nonfunctional transporters (missense mutations). Some missense mutations have identified domains of the transporter essential for carnitine recognition and transmembrane transfer.

REFERENCES

Bieber, L.L. (1988). Carnitine. *Annu. Rev. Biochem.*, **57**, 261–283.

Borum, P.R. (1995). Carnitine in Neonatal Nutrition. *J. Child Neurol.* **10**(Suppl. 2), 2S25–2S31.

Burwinkel, B., Kreuder, J., Schweitzer, S., Vorgerd, M., Gempel, K., Gerbitz, K.D., *et al.* (1999). Carnitine

Transporter OCTN2 Mutations in Systemic Primary Carnitine Deficiency: A Novel Arg169Gln Mutation and a Recurrent Arg282ter Mutation Associated with an Unconventional Splicing Abnormality. *Biochem. Biophys. Res. Commun.*, **161**, 484–487.

Cederbaum, S., Dipple, K., Vilain, E., Miller, M., Koo-McCoy, S., Hsu, B.Y.L., *et al.* (2000). Clinical Follow-up and Molecular Etiology of the Original Case of Carnitine Transporter Deficiency. *J. Inherit. Metab. Dis.*, **23**(Suppl. 1), 119 (Abstract 237-O).

Christensen, E., Holm, J., Hansen, S.H., Sorensen, N., Nezu, J., Tsuji, A., *et al.* (2000). Sudden Infant Death Following Pivampicillin Treatment in a Patient with Carnitine Transporter Deficiency. *J. Inherit. Metab. Dis.*, **23**(Suppl. 1), 117 (Abstract 234-P).

Christiansen, R.Z. and Bremer, J. (1976). Active Transport of Butyrobetaine and Carnitine into Isolated Liver Cells. *Biochim. Biophys. Acta*, **448**, 562–77.

Christodoulou, J., Teo, S.H., Hammond, J., Sim, K.G., Hsu, B.Y., Stanley, C.A., *et al.* (1996). First Prenatal Diagnosis of the Carnitine Transporter Defect. *Am. J. Med. Genet.*, **66**, 21–24.

Duran, J.M., Peral, M.J., Calonge, M.L., and Ilundiin, A.A. (2002). Functional Characterization of Intestinal L-Carnitine Transport. *J. Membr. Biol.*, **185**, 65–74.

Enomoto, A., Wempe, M.F., Tsuchida, H., Shin, H.J., Cha, S.H., Anzai, N., *et al.* (2002). Molecular Identification of a Novel Carnitine Transporter Specific to Human Testis: Insights into the Mechanism of Carnitine Recognition. *J. Biol. Chem.* **277**, 36262–36271.

Horiuchi, M., Kobayashi, K., Tomomura, M., Kuwajima, M., Imamura, Y., Koizumi, T., *et al.* (1992). Carnitine Administration to Juvenile Visceral Steatosis Mice Corrects the Suppressed Expression of Urea Cycle Enzymes by Normalizing their Transcription. *J. Biol. Chem.*, **267**, 5032–5035.

Koizumi, T., Nikaido, H., Hayakawa, J., Nonomura, A., and Yoneda, T. (1988). Infantile Disease with Microvesicular Fatty Infiltration of Viscera Spontaneously Occurring in the C3H-H-2(0) Strain of Mouse with Similarities to Reye's Syndrome. *Lab. Anim.*, **122**, 83–87.

Koizumi, A., Nozaki, J., Ohura, T., Kayo, T., Wada, Y., Nezu, J., *et al.* (1999). Genetic Epidemiology of the Carnitine Transporter OCTN2 Gene in a Japanese Population and Phenotypic Characterization in Japanese Pedigrees with Primary Systemic Carnitine Deficiency. *Hum. Mol. Genet.*, **8**, 2247–2254.

Kuwajima, M., Kono, N., Horiuchi, M., Imamura, Y., Ono, A., Inui, Y., *et al.* (1991). Animal Model of Systemic Carnitine Deficiency: Analysis in C3H-H-2 Degrees Strain of Mouse Associated with Juvenile Visceral Steatosis. *Biochem. Biophys. Res. Commun.*, **174**, 1090–1094.

Kuwajima, M., Lu, K., Harashima, H., Ono, A., Sato, I., Mizuno, A., *et al.* (1996). Carnitine Transport Defect in Fibroblasts of Juvenile Visceral Steatosis (JVS) Mouse. *Biochem. Biophys. Res. Commun.*, **223**, 283–287.

Lahjouji, K., Elimrani, I., Wu, J., Mitchell, G.A., and Qureshi, I.A. (2002). A Heterozygote Phenotype Is Present in the *jvs* +/− Mutant Mouse Livers. *Mol. Genet. Metab.*, **76**, 76–80.

Lamhonwah, A.M., Olpin, S.E., Pollitt, R.J., Vianey-Saban, C., Divry, P., Guffon, N., et al. (2002). Novel OCTN2 Mutations: No Genotype–Phenotype Correlations: Early Carnitine Therapy Prevents Cardiomyopathy. *Am. J. Med. Genet.*, **111**, 271–284.

Lombard, K.A., Olson, A.L., Nelson, S.E., and Rebouche, C.J. (1989). Carnitine Status of Lacto-ovovegetarians and Strict Vegetarian Adults and Children. *Am. J. Clin. Nutr.*, **50**, 301–306.

Lu, K., Nishimori, H., Nakamura, Y., Shima, K., and Kuwajima, M. (1998). A Missense Mutation of Mouse OCTN2, a Sodium-Dependent Carnitine Cotransporter, in the Juvenile Visceral Steatosis Mouse. *Biochem. Biophys. Res. Commun.*, **252**, 590–594.

Mayatepek, E., Nezu, J., Tamai, I., Oku, A., Katsura, M., Shimane, M., et al. (2000). Two Novel Missense Mutations of the OCTN2 Gene (W283R and V446F) in a Patient with Primary Systemic Carnitine Deficiency. *Hum. Mutat.*, **15**, 118 (Online).

Nakanishi, T., Hatanaka, T., Huang, W., Prasad, P.D., Leibach, F.H., Ganapathy, M.E., et al. (2001). Na+- and Cl−-Coupled Active Transport of Carnitine by the Amino Acid Transporter ATB$^{(0,+)}$ from Mouse Colon Expressed in HRPE Cells and *Xenopus* Oocytes. *J. Physiol.*, **532**, 297–304.

Nezu, J., Tamai, I., Oku, A., Ohashi, R., Yabuuchi, H., Hashimoto, N., et al. (1999). Primary Systemic Carnitine Deficiency Is Caused by Mutations in a Gene Encoding Sodium Ion-Dependent Carnitine Transporter. *Nat. Genet.*, **21**, 91–94.

Ohashi, R., Tamai, I., Yabuuchi, H., Nezu, J.I., Oku, A., Sai, Y., et al. (1999). Na+-Dependent Carnitine Transport by Organic Cation Transporter (OCTN2): Its Pharmacological and Toxicological Relevance. *J. Pharmacol. Exp. Ther.*, **291**, 778–784.

Ohashi, R., Tamai, I., Nezu, J.I., Nikaido, H., Hashimoto, N., Oku, A., et al. (2001). Molecular and Physiological Evidence for Multifunctionality of Carnitine/Organic cation Transporter OCTN2. *Mol. Pharmacol.*, **59**, 358–366.

Ohashi, R., Tamai, I., Inano, A., Katsura, M., Sai, Y., Nezu, J., et al. (2002). Studies on Functional Sites of Organic Cation/Carnitine Transporter OCTN2 (SLC22A5) Using a Ser467Cys Mutant Protein. *J. Pharmacol. Exp. Ther.*, **302**, 1286–1294.

Palacin, M., Borsani, G., and Sebastio, G. (2001). The Molecular Bases of Cystinuria and Lysinuric Protein Intolerance. *Curr. Opin. Genet. Dev.*, **11**, 328–335.

Rashed, M.S., Ozand, P.T., Bennet, M.J., Barnard, J.J., Govindaraju, D.R., and Rinaldo, P. (1995). Diagnosis of Inborn Errors of Metabolism in Sudden Death Cases by Acylcarnitine Analysis of Postmortem Bile. *Clin. Chem.*, **41**, 1109–1114.

Rinaldo, P., Stanley, C.A., Hsu, B.Y.L., Sanchez, L.A., and Stern, H.J. (1997). Sudden Neonatal Death in Carnitine Transporter Deficiency. *J. Pediatr.*, **131**, 304–305.

Roe, C.R. and Ding, J. (2001). Mitochondrial Fatty acid Oxidation Disorders. In: C.R. Scriver, A.L. Beaudet, W.S. Sly, and D. Valle (eds), *The Metabolic and Molecular Bases of Inherited Disorders* (8th edn), New York: McGraw-Hill, pp. 2327–2356.

Saheki, T., Li, M.X., and Kobayashi, K. (2000). Antagonizing Effect of AP-1 on Glucocorticoid Induction of Urea Cycle Enzymes: A Study of Hyperammonemia in Carnitine-Deficient, Juvenile Visceral Steatosis Mice. *Mol. Genet. Metab.*, **71**, 545–551.

Scaglia, F. and Longo, N. (1999). Primary and Secondary Alterations of Neonatal Carnitine Metabolism. *Semin. Perinatol.*, **23**, 152–161.

Scaglia, F., Wang, Y., Singh, R.H., Dembure, P.P., Pasquali, M., Fernhoff, P.M., et al. (1998). Defective Urinary Carnitine Transport in Heterozygotes for Primary Carnitine Deficiency. *Genet. Med.*, **1**, 34–39.

Scaglia, F., Wang, Y., and Longo, N. (1999). Functional Characterization of the Carnitine Transporter Defective in Primary Carnitine Deficiency. *Arch. Biochem. Biophys.*, **364**, 99–106.

Seth, P., Wu, X., Huang, W., Leibach, F.H., and Ganapathy, V. (1999). Mutations in Novel Organic Cation Transporter (OCTN2), an Organic Cation/Carnitine Transporter, with Differential Effects on the Organic Cation Transport Function and the Carnitine Transport Function. *J. Biol. Chem.*, **274**, 33388–33392.

Schmidt-Sommerfeld, E., Penn, D., Sodha, R.J., Progler, M., Novak, M., and Schneider, H. (1985). Transfer and Metabolism of Carnitine and Carnitine Esters in the In Vitro Perfused Human Placenta. *Pediatr. Res.*, **19**, 700–706.

Shoji, Y., Koizumi, A., Kayo, T., Ohata, T., Takahashi, T., Harada, K., et al. (1998). Evidence for Linkage of Human Primary Carnitine Deficiency with D5S436: A Novel Gene Locus on Chromosome 5q. *Am. J. Hum. Genet.*, **63**, 101–108.

Stanley, C.A., DeLeeuw, S., Coates, P.M., Vianey-Liaud, C., Divry, P., Bonnefont, J.P., et al. (1991). Chronic Cardiomyopathy and Weakness or Acute Coma in Children with a Defect in Carnitine Uptake. *Ann. Neurol.*, **30**, 709–716.

Tamai, I., Yabuuchi, H., Nezu, J., Sai, Y., Oku, A., Shimane, M., et al. (1997). Cloning and Characterization of a Novel Human pH-Dependent Organic Cation Transporter, OCTN1. *FEBS Lett.*, **419**, 107–111.

Tamai, I., Ohashi, R., Nezu, J., Yabuuchi, H., Oku, A., Shimane, M., et al. (1998). Molecular and Functional Identification of Sodium Ion-Dependent, High Affinity Human Carnitine Transporter OCTN2. *J. Biol. Chem.*, **273**, 20378–20382.

Tamai, I., China, K., Sai, Y., Kobayashi, D., Nezu, J., Kawahara, E., et al. (2001). Na+-Coupled Transport of

L-Carnitine via High-Affinity Carnitine Transporter OCTN2 and its Subcellular Localization in Kidney. *Biochim. Biophys. Acta*, **1512**, 273–284.

Tang, N.L., Ganapathy, V., Wu, X., Hui, J., Seth, P., Yuen, P.M., *et al.* (1999). Mutations of OCTN2, an Organic Cation/Carnitine Transporter, Lead to Deficient Cellular Carnitine Uptake in Primary Carnitine Deficiency. *Hum. Mol. Genet.*, **8**, 655–660.

Tang, N.L., Hwu, W.L., Chan, R.T., Law, L.K., Fung, L.M., and Zhang, W.M. (2002). A founder mutation (R254X) of SLC22A5 (OCTN2) in Chinese primary carnitine deficiency patients. *Hum. Mutat.*, **20**, 232 (Online).

Vaz, F.M., Scholte, H.R., Ruiter, J., Hussaarts-Odijk, L.M., Pereira, R.R., Schweitzer, S., *et al.* (1999). Identification of Two Novel Mutations in OCTN2 of Three Patients with Systemic Carnitine Deficiency. *Hum. Genet.*, **105**, 157–161.

Wagner, C.A., Lukewille, U., Kaltenbach, S., Moschen, I., Broer, A., Risler, T., *et al.* (2000). Functional and Pharmacological Characterization of Human Na(+)-Carnitine Cotransporter hOCTN2. *Am. J. Physiol. Renal. Physiol.*, **279**, F584–F591.

Wang, Y., Ye, J., Ganapathy, V., and Longo, N. (1999). Mutations in the Organic Cation/Carnitine Transporter OCTN2 in Primary Carnitine Deficiency. *Proc. Natl. Acad. Sci. USA*, **96**, 2356–2360.

Wang, Y., Kelly, M.A., Cowan, T.M., and Longo, N. (2000a). A Missense Mutation in the *OCTN2* Gene Associated with Residual Carnitine Transport Activity. *Hum. Mutat.*, **15**, 238–245.

Wang, Y., Meadows, T.A., and Longo, N. (2000b). Abnormal Sodium Stimulation of Carnitine Transport in Primary Carnitine Deficiency. *J. Biol. Chem.* **275**, 20782–20786.

Wang, Y., Taroni, F., Garavaglia, B., and Longo, N. (2000c). Functional Analysis of Mutations in the OCTN2 Transporter Causing Primary Carnitine

Deficiency: Lack of Genotype–Phenotype Correlation. *Hum. Mutat.*, **16**, 401–407.

Wang, Y., Korman, S.H., Ye, J., Gargus, J.J., Gutman, A., Taroni, F., *et al.* (2001). Phenotype and Genotype Variation in Primary Carnitine Deficiency. *Genet. Med.*, **3**, 387–392.

Wilcken, B., Wiley, V., Sim, K.G., and Carpenter, K. (2001). Carnitine Transporter Defect Diagnosed by Newborn Screening with Electrospray Tandem Mass Spectrometry. *J. Pediatr.*, **138**, 581–584.

Wu, X., Prasad, P.D., Leibach, F.H., and Ganapathy, V. (1998). cDNA Sequence, Transport Function, and Genomic Organization of Human OCTN2, a New Member of the Organic Cation Transporter Family. *Biochem. Biophys. Res. Commun.*, **246**, 589–595.

Wu, X., Huang, W., Prasad, P.D., Seth, P., Rajan, D.P., Leibach, F.H., *et al.* (1999). Functional Characteristics and Tissue Distribution Pattern of Organic Cation Transporter 2 (OCTN2), an Organic Cation/Carnitine Transporter. *J. Pharmacol. Exp. Ther.*, **290**, 1482–1492.

Wu, X., George, R.L., Huang, W., Wang, H., Conway, S.J., Leibach, F.H., *et al.* (2000). Structural and Functional Characteristics and Tissue Distribution Pattern of Rat OCTN1, an Organic Cation Transporter, Cloned from Placenta. *Biochim. Biophys. Acta*, **1466**, 315–327.

Xiaofei, E., Wada, Y., Dakeishi, M., Hirasawa, F., Murata, K., Masuda, H., *et al.* (2002). Age-Associated Cardiomyopathy in Heterozygous Carrier Mice of a Pathological Mutation of Carnitine Transporter Gene, *OCTN2*. *J. Gerontol. A Biol. Sci. Med. Sci.*, **57**, B270–B278.

Yokogawa, K., Higashi, Y., Tamai, I., Nomura, M., Hashimoto, N., Nikaido, H., *et al.* (1999a). Decreased Tissue Distribution of L-Carnitine in Juvenile Visceral Steatosis Mice. *J. Pharmacol. Exp. Ther.*, **289**, 224–230.

Yokogawa, K., Miya, K., Tamai, I., Higashi, Y., Nomura, M., Miyamoto, K., *et al.* (1999b). Characteristics of L-Carnitine Transport in Cultured Human Hepatoma HLF Cells. *J. Pharm. Pharmacol.*, **51**, 935–940.

12

MARC FORETZ* AND BERNARD THORENS**

The facilitative glucose transporter 2: pathophysiological role in mouse and human

FACILITATIVE GLUCOSE TRANSPORT

Glucose is an essential substrate for the metabolism of most cells. Because glucose is a polar molecule, its transport through biological membranes requires specific transporter proteins. Transport of glucose through the apical membrane of intestinal and kidney epithelial cells depends on the presence of active Na^+/glucose symporters, SGLT-1 and SGLT-2, which concentrate glucose inside the cells using the energy provided by cotransport of Na^+ ions down their electrochemical gradient (Hediger and Rhoads 1994). Facilitated diffusion of glucose through the cellular membrane is otherwise catalyzed by glucose carriers (protein symbol GLUT, gene symbol *SLC2* for Solute Carrier Family 2) that belong to a superfamily of transport facilitators which comprises organic anion and cation transporters, yeast hexose transporters, plant hexose/proton symporters, and bacterial sugar/ proton symporters (Henderson 1993).

GLUTs are integral membrane proteins which contain 12 membrane-spanning helices with both the NH_2- and COOH-termini exposed on the cytoplasmic side of the plasma membrane. GLUT proteins transport glucose and related hexoses according to a model of alternate conformation (Oka *et al.* 1990; Hebert and Carruthers 1992; Cloherty *et al.* 1995; see also Chapter 2), which predicts that the transporter exposes a single substrate binding site toward either the outside or the inside of the cell. Binding of glucose to one site provokes a conformational change associated with transport and releases glucose to the other side of the membrane. The inner and outer glucose-binding sites are probably located in transmembrane segments 9, 10, and 11 (Hruz and Mueckler 2001); also, the QLS motif located in the seventh transmembrane segment could be involved in the selection and affinity of transported substrate (Seatter *et al.* 1998; Hruz and Mueckler 1999).

Each glucose transporter isoform plays a specific role in glucose metabolism which is determined by its pattern of tissue expression, substrate specificity, transport kinetics, and regulated expression in different physiological conditions (Thorens 1996).

To date, 13 members of the GLUT/SLC2 have been identified (Joost and Thorens 2001). On the basis of sequence similarities, the GLUT family has been divided into three subclasses. Class I comprises the well-characterized

* Institut Cochin-INSERM U567 24, Rue du Faubourg Saint-Jacques 75014 Paris – France
** Institute of Pharmacology and Toxicology, 27 rue du Bugnon, CH 1005 Lausanne, Switzerland

glucose transporters GLUT1–GLUT4 (Bell et al. 1990). GLUT1 was the first glucose transporters to be characterized (Mueckler et al. 1985). It is widely distributed in fetal tissues. In the adult it is expressed at highest levels in erythrocytes and also in the endothelial cells of barrier tissues such as the blood–brain barrier. Mutations in the *GLUT1* gene are responsible for *GLUT1 deficiency* or *De Vivo syndrome* (OMIM 606777), which is a rare autosomal dominant disorder (Seidner et al. 1998). This disease is characterized by a low cerebrospinal fluid glucose concentration (hypoglycorrhachia) which results from impaired glucose transport across the blood–brain barrier. GLUT3 is an isoform expressed mostly in neurons where it is believed to be the main glucose transporter isoform. GLUT4 is the insulin-regulated glucose transporter found in adipose tissues and muscles that is responsible for insulin-regulated glucose disposal. Class II comprises the fructose transporter GLUT5 and GLUT7, 9, and 11. Class III comprises GLUT6, 8, 10, and 12 and the H^+/myo-inositol transporter HMIT (Uldry et al. 2000). Most members of classes II and III have been identified recently in homology searches of EST databases and the sequence information provided by the various genome projects. Their characteristics are summarized in Table 1.

Most of cells are unable to produce free glucose because they lack expression of glucose-6-phosphatase and thus are only involved in glucose uptake and catabolism. Only hepatocytes and, in more severe fasting conditions, intestine and kidney are able to produce glucose following activation of gluconeogenesis and glycogenolysis.

THE FACILITATIVE GLUCOSE TRANSPORTER 2 (GLUT2)

In 1988, cDNA clones encoding human and rat GLUT2 were isolated (Fukumoto et al. 1988; Thorens et al. 1988). The predicted protein sequence was 524 amino acids long in human and 522 in rat and mouse (Figure 1). GLUT2

exhibits a high degree of homology to GLUT1 (80% homology, 55% identity between the human GLUT1 and GLUT2 isoforms). The *GLUT2* gene also referred to as the *SLC2A2* (OMIM 138160) is localized on human chromosome 3 (3q26.1-q26.3) (Fukumoto et al. 1988), and its structure consists of 11 exons and 10 introns spanning approximately 30 kb (Takeda et al. 1993). GLUT2 has the highest K_m for glucose of the facilitative glucose transporters (~17 mM). It can also transport D-galactose, D-mannose, and D-fructose. Surprisingly, it is also a high-affinity glucosamine transporter ($K_m \approx 0.8$ mM) (Uldry et al. 2002). GLUT2 is expressed predominantly in hepatocytes, pancreatic β cells, and the basolateral membranes of intestinal and renal epithelial cells (Fukumoto et al. 1988; Orci et al. 1989; Permutt et al. 1989; Thorens et al. 1990). Unlike the other glucose transporters, which deliver glucose to cells for utilization, GLUT2 is primarily involved in glucose homeostasis through its role in glucose uptake from the intestine, reabsorption by the kidney, sensing in the pancreatic β cells, and uptake and release by liver. Glucose taken up in the intestine and reabsorbed in the proximal tubule by the apically located sodium-dependent glucose transporters SGLT1 and SGLT2 (Hediger and Rhoads 1994) is released into the circulation via GLUT2 in the basal membrane of the epithelial cells. The high K_m for glucose of GLUT2 ensures a high transport capacity and that glucose transport rate by pancreatic β cells and hepatocytes increases in proportion to elevation in glycemia. GLUT2 is also expressed in different brain regions, including the hypothalamus and nucleus of the tractus solitarius (Leloup et al. 1994; Ngarmukos et al. 2001), which contain glucose-modulated neurons. These neurons may be involved in glucose homeostasis and may require GLUT2 and glucokinase to sense variations in extracellular glucose levels. The involvement of GLUT2 in glucose-sensing mechanisms has been demonstrated in the mouse pancreatic β cells (Guillam et al. 1997, 2000) and the hepatoportal sensor, which

Table 1 The extended GLUT family (SLC2)

Unigene SLC2	Protein	Class	Substrates	Expression site	Chromosomal localization	Accession no. cDNA	Reference(s)
A1	GLUT1	I	Glucose	Endothelial cells of vessels (blood-brain barrier), erythrocytes, fetal tissue	1p35-31.3	K03195	Mueckler et al. 1985
A2	GLUT2	I	Glucose, galactose, fructose, mannose	Liver, pancreatic cells, kidney, intestine	3q26.2-27	J03810	Fukumoto et al. 1988; Thorens et al. 1988; Permutt et al. 1989
A3	GLUT3	I	Glucose	Neuronal cells	12p13.3	M20681	Kayano et al. 1988
A4	GLUT4	I	Glucose	Muscle, adipose tissue, heart	17p13	M20747	Fukumoto et al. 1989
A5	GLUT5	II	Fructose	Intestine, testis, kidney	1p36.2	J05461	Kayano et al. 1990
A6	GLUT6	III	Glucose	Spleen, leukocytes, brain	9q34	Y17803	Doege et al. 2000a; Joost et al. 2001
A7	GLUT7	II	ND	ND	1p36.2	–	–
A8	GLUT8	III	Glucose, galactose, fructose	Testis, blastocyst, brain	9	Y17801	Carayannopolous et al. 2000; Doege et al. 2000b; Ibberson et al. 2000
A9	GLUT9	II	Glucose	Liver, kidney	4p15.3-16	AF210317	Tartaglia and Weng 1999; Phay et al. 2000
A10	GLUT10	III	Glucose	Liver, pancreas	20q12-13.1	AF321240	McVie-Wylie et al. 2001
A11	GLUT11	II	Glucose, fructose	Heart, muscle	22q11.2	AJ271290	Doege et al. 2001a, b
A12	GLUT12	III	ND	Heart, prostate	6q23.2	AY046419	Rogers et al. 1998
A13	HMIT	III	Myo-inositol/H+	Brain (neuronal and glial cells)	12	AJ315644	Uldry et al. 2001

ND, not determined.
Adapted from Joost and Thorens (2001).

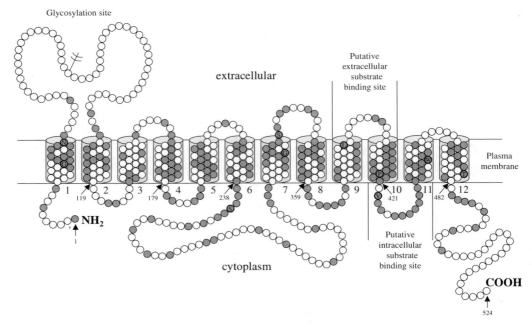

Figure 1 Conformational model of GLUT2 in the plasma membrane. The side view of the GLUT2 protein shows the 12 transmembrane helices (according to Swissprot accession P11168) with a large intracellular loop connecting helices 6 and 7, the N-glycosylation site in the first extracellular loop, and the amino and carboxy-termini located in the cytosol. Putative substrate import and export sites are also indicated. Conserved amino acids within the Class I of the GLUT family (GLUT1–4) are shown in gray. Positions of missense mutations identified in patients with FBS (Burwinkel *et al.* 1999; Sakamoto *et al.* 2000; Tsuda *et al.* 2000; Matsuura *et al.* 2002; Santer *et al.* 2002a) are indicated by filled circles and the respective amino acid code.

controls glucose utilization in brown adipose tissue and muscles (Burcelin *et al.* 2000b).

MOLECULAR GENETICS

Fanconi–Bickel Syndrome (FBS) OMIM227810 is an autosomal recessive disease, in contrast with the autosomal dominant "glucose transport defect syndrome" in which a haploinsufficiency of GLUT1 results in a severe neurological disease due to insufficient glucose transport across the blood–brain barrier (Brockmann *et al.* 2001; Klepper *et al.* 2001). However, a heterozygous GLUT2 mutation has been reported in a woman with gestational diabetes (Tanizawa *et al.* 1994). In vitro expression studies demonstrated that this mutation in GLUT2 (V197I) resulted

in an impairment of glucose transport (Mueckler *et al.* 1994), but it was not established that diabetes was caused by this mutation. A case of heterozygous missense mutation of GLUT2 has been reported in a dominant form of familial renal glucosuria (Sakamoto *et al.* 2000). Because glucose transporter proteins may function as oligomers (Pessino *et al.* 1991; Hebert and Carruthers 1992), it was proposed that this mutation generated a dominant negative form of GLUT2. Besides these case reports, there is no evidence that the GLUT2 locus is linked to pathologies such as type 2 diabetes in the general population (Baroni *et al.* 1992; Matsutani *et al.* 1990; Mutsubara *et al.* 1995; Moller *et al.* 2001) and in the great majority of FBS cases diabetes does not develop (Santer *et al.* 1998).

Manz *et al.* (1987) were the first to speculate that a defect of the facilitative glucose and galactose transport in the liver and at the basolateral membrane of the proximal kidney tubule could underlie the defects observed in FBS. Santer and coworkers noted that a functional loss of GLUT2 was compatible with the clinical symptoms observed in FBS. In 1997, they identified homozygous mutations in the *GLUT2* gene in three families with FBS (Santer *et al.* 1997), including the patient originally described by Fanconi and Bickel (1949). To date, 35 different mutations of the *GLUT2* gene have been reported in a total of 65 patients with FBS (Santer *et al.* 1997, 2002a; Matsuura *et al.* 2002). These include ten missense, eight nonsense, ten frameshift, and seven splice site mutations (Table 2). The mutations

Table 2   Thirty-five different mutations in the *GLUT2* gene identified in 65 patients with Fanconi-Bickel syndrome

Mutation type	Mutation effect	Localization	Reference(s)
Missense (10)	G20E	Exon 2	Santer *et al.* 2002a
	N32K	Exon 2	Matsuura *et al.* 2002
	S242R	Exon 5	Santer *et al.* 2002a
	G318R	Exon 6	Santer *et al.* 2002a
	S326K	Exon 7	Santer *et al.* 2002a
	L389P	Exon 8	Sakamoto *et al.* 2000
	P417L	Exon 9	Burwinkel *et al.* 1999
	V423E	Exon 9	Sakamoto *et al.* 2000
	W444R	Exon 9	Tsuda *et al.* 2000; Santer *et al.* 2002a
	T480R	Exon 10	Santer *et al.* 2002a
Nonsense (8)	K5X	Exon 1	Yoo *et al.* 2002
	R53X	Exon 3	Santer *et al.* 2002a
	S169X	Exon 4b	Santer *et al.* 2002a
	Q193X	Exon 4b	Santer *et al.* 2002a
	Q287X	Exon 6	Sakamoto *et al.* 2000
	R301X	Exon 6	Santer *et al.* 1997
	R365X	Exon 8	Santer *et al.* 1997; Tsuda *et al.* 2000
	W420X	Exon 9	Akagi *et al.* 2000
Frameshift (10)	V45 fs 74X	Exon 3	Santer *et al.* 1997
	V60 fs 83X	Exon 3	Santer *et al.* 2002a
	A105 fs 106X	Exon 3	Santer *et al.* 2002a
	I133 fs 177X	Exon 4a	Santer *et al.* 2002a
	S145 fs 177X	Exon 4a	Santer *et al.* 2002a
	S161 fs 178X	Exon 4a	Santer *et al.* 2002a
	Q232 fs 252X	Exon 5	Santer *et al.* 2002a
	M350 fs 370X	Exon 7	Santer *et al.* 2002a
	L368 fs 391X	Exon 8	Santer *et al.* 2002a
	W420 fs 426X	Exon 10	Santer *et al.* 2002a
Splice site (7)	IVS 2 − 2 a > g	5′ Exon 3	Sakamoto *et al.* 2000
	IVS 2 +2 t>c	3′ Exon 3	Santer *et al.* 2002a
	IVS 5 + 1 g>t	3′ Exon 5	Santer *et al.* 2002a
	IVS 5 + 5 g>c	3′ Exon 5	Santer *et al.* 2002a
	IVS 6 + 1 g>a	3′ Exon 6	Santer *et al.* 2002a
	IVS 6 + 1 g>c	3′ Exon 6	Santer *et al.* 2002a
	IVS 8 + 1 g>a	3′ Exon 8	Santer *et al.* 2002a

fs, frameshift; X, stop codon.

are scattered over the whole coding region, mutations have been found in all exons and also within several splicing sites (Santer *et al.* 2002a).

The functional significance of the GLUT2 mutations found in FBS have not been demonstrated by expression studies. However, the consequences of these mutations can be extrapolated from the results of in vitro mutagenesis experiments (Saravolac 1997). Most GLUT2 mutations associated with FBS result in carboxy-terminal truncated proteins. Those could be expected to generate nonfunctional transporters by analogy with GLUT1, where deletion of more than 24 amino acids from the

carboxy-terminal tail locks the transporter in an inward-facing conformation and suppresses transport activity (Oka *et al.* 1990)

However the fact that homozygous missense mutations can cause FBS can however be deduced by their presence in the affected family members but not in controls. Moreover, most of these missense mutations affect highly conserved amino acid residues characteristic of GLUT proteins which might be involved in the transport mechanism, such as the GLUT2, P417L and W444R mutations (Burwinkel *et al.* 1999; Santer *et al.* 2002a; Tsuda *et al.* 2002; see Box: *GLUT2* Gene Inactivation in Human). Proline 417, present in helix 10, is

---

GLUT2 GENE INACTIVATION IN HUMAN: THE FANCONI-BICKEL SYNDROME

The disease was first described in 1949 by Fanconi and Bickel in a Swiss boy born to consanguineous parents (Fanconi and Bickel 1949). To date, over 100 patients with FBS have been reported (Santer *et al.* 1998, 2002b). Usually patients present a combination of clinical signs including enlargement of the liver and kidney due to glycogen accumulation, severe Fanconi type renal tubulopathy with severe glucosuria, glucose, and galactose intolerance, and fasting hypoglycemia. The first symptoms are fever, vomiting, failure to thrive, and hypophosphatemic rickets at the age of 3–10 months. Later sequelae include short stature, doll face, and abdominal distension.

Kidney size is increased in most patients. Analysis of biopsy material shows glycogen accumulation in most proximal tubular cells. There is a general renal proximal tubular defect with hyperaminoaciduria, hyperuricosuria, hyperphosphaturia, hypercalciuria, renal tubular acidosis, and mild tubular proteinuria. Severe glucosuria independent of the glucose levels is the most prominent finding. Daily glucose loss reaches 40–200 g per 1.73 m^2 of body surface (Manz *et al.* 1987); a maximum of 325 g per 1.73 m^2 has been reported (Brivet *et al.* 1983). Polyuria, probably due to osmotic diuresis, is a constant finding. Glomerular filtration rate is usually normal.

The size of the liver is normal at birth and in most cases increases during infancy. Accumulation of glycogen is demonstrated in liver biopsies by electron microscopy. The liver size and glycogen content are reduced after a ketogenic diet or administration of uncooked cornstarch. However, hepatomegaly recedes after puberty.

Glucose homeostasis is also impaired in FBS patients as observed by postprandial hyperglycemia and hypergalactosemia, and fasting hypoglycemia.

In most cases of FBS, vital prognosis is favorable, with the main problem being the short stature associated with rickets and osteoporosis. Therapy consists of supplementation of water, electrolytes, vitamin D, and phosphate, adequate caloric intake to control the renal glucose losses, and restriction of galactose intake. The administration of uncooked cornstarch has been reported to be beneficial for promoting growth in young patients with FBS (Lee *et al.* 1995). Because fructose is transported by a specific transport protein, GLUT5, in epithelial cells of the intestine and kidney (Burant and Saxena 1994; Blakemore *et al.* 1995) and fructose metabolism is not affected in FBS patients (Odièvre 1966; Manz *et al.* 1987), it has been suggested that this monosaccharide might be an alternate carbohydrate source in the therapy of FBS.

conserved in all mammalian facilitative glucose transporters and even in bacterial sugar transporters and is part of a highly conserved GPXPIP motif thought to confer high flexibility to this helix. Replacement of this proline (P385) in GLUT1 with isoleucine results in a loss of transport activity (Tamori *et al.* 1994; Wellner *et al.* 1995). This proline is also close to tryptophan 420 whose counterpart in GLUT1 (W388) appears to participate in glucose binding or translocation (Garcia *et al.* 1992; Katagiri *et al.* 1993; Inukai *et al.* 1994; Kasahara *et al.* 1998). The GLUT2 W444R mutation in helix 11 (Tsuda *et al.* 2000; Santer *et al.* 2002a) is localized within a region highly conserved in all GLUTs. In vitro mutagenesis of Trp 412 of GLUT1 and Trp 410 of GLUT3, which correspond to Trp 444 of GLUT2, severely impairs glucose transport by modifying the internal glucose binding site (Katagiri *et al.* 1991; Burant and Bell 1992; Garcia *et al.* 1992; Schurmann *et al.* 1993; Inukai *et al.* 1994). Other examples of missense mutations are G20E, S326K, V423E, and T480R (Sakamoto *et al.* 2000; Santer *et al.* 2002a). These four mutations are localized within regions conserved in Class I GLUTs that comprise GLUT1 to GLUT4 (Joost and Thorens 2001). In particular, the serine residue 326 is part of an STS motif in the seventh extracellular loop, which could be involved in the glucose specificity of the transporter (Doege *et al.* 1998). The GLUT2 N32K, S242R, G318R, and L389P mutations (Sakamoto *et al.* 2000; Matsuura *et al.* 2002; Santer *et al.* 2002a) also affect highly conserved amino acid residues among the GLUT family members (Joost and Thorens 2001).

MOUSE MODEL OF GLUT2 DEFICIENCY

Mice with homozygous inactivating mutations in the *GLUT2* gene display abnormal glucose homeostasis characterized by hyperglycemia, glucosuria, relative hypoinsulinemia, and elevated plasma levels of glucagon, free fatty acids, and β-hydroxybutyrate. They usually die within the first 3 weeks of life but can be maintained alive by insulin treatment. This indicates that the primary cause of death is a lack of normal insulin secretion (Guillam *et al.* 1997).

β-cell defects, hypoinsulinemia, and hyperglycemia

Abnormal insulin secretory response to glucose can indeed be demonstrated in islet perifusion experiments which show absence of first-phase insulin secretion but a preserved second phase which, however, is of reduced amplitude compared with control islets. This secretory defect is responsible for the observed hypoinsulinemia, hyperglycemia, and elevated free fatty acid levels. The defect in secretory response is specific for glucose since amino acids normally stimulate insulin release. Furthermore, re-expression of the transporter in isolated GLUT2$^{-/-}$ islets using recombinant lentiviruses restores normal secretory response to glucose (Guillam *et al.* 2000).

In FBS patients, insulin secretion has not been systematically evaluated and therefore it is not clear whether absence of GLUT2 leads to a β-cell defect. Nevertheless reduced secretory response has been reported in a few cases (Garty *et al.* 1974; Chesney *et al.* 1980; Manz *et al.* 1987; Ogier de Baulny *et al.* 1999) and hypoinsulinemia is generally observed. The absence of a secretory defect could be explained by the reported presence in human β-cells of glucose transporter isoforms GLUT1 and GLUT3 in addition to GLUT2 (De Vos *et al.* 1995). In mice, we have demonstrated that replacement of β-cell GLUT2 by GLUT1 restores the normal secretory response with the kinetics of first and second phases indistinguishable from control islets (Thorens *et al.* 2000). This is in agreement with the proposed model for glucose-metabolism-dependent signaling in β-cells (Matschinsky 1996). In this pathway, phosphorylation of glucose by glucokinase, rather than glucose uptake, is the rate-controlling step in the generation of coupling factors triggering insulin secretion. Provided that the glucose transporter present in the β-cell plasma membrane allows a sufficiently rapid access of glucose to glucokinase, glucose signaling is preserved. Thus human islets are probably protected from the absence of GLUT2 by the expression of several transporter isoforms.

In mice with a transgenic re-expression of GLUT1 or GLUT2 in their pancreatic β-cells, the kinetics of insulin secretion is restored to normal but the plasma insulin levels are still much reduced compared with control mice and the hyperglycemia observed in GLUT2$^{-/-}$ mice is corrected. The cause of this hypoinsulinemia, which is also observed in human patients, is probably related to the severe glucosuria, which can be considered as a mechanism for glucose clearance independent of insulin action. Therefore this may reduce the insulinemic levels required to maintain normoglycemia.

Hepatic glucose metabolism

The liver of GLUT2-null mice is hyperplasic with a 40% increase in weight and a 30 % increase in DNA content, urcelin, 2000a. In addition, the glycogen content is increased as compared to control livers. These are characteristics also found in human Fanconi-Bickel liver.

In mice, we could evaluate defects in glucose metabolism due to the absence of GLUT2 more precisely. We demonstrated that GLUT2$^{-/-}$ hepatocytes have an almost suppressed glucose uptake capacity (Guillam et al. 1998). This could partially explain glucose intolerance observed in the postprandial state in FBS patients (Figure 2). However, hepatic glucose production, as measured by whole-body glucose turnover analysis or studies on isolated hepatocytes, was kinetically indistinguishable between control and mutant mice. This led us to suggest that glucose output from hepatocytes was through a membrane-traffic-based pathway. Indeed, glucose-6-phosphate (G6P) is hydrolyzed into glucose and phosphate inside the lumen of the endoplasmic reticulum, where the active site of G6Pase

is located (see Chapter 13). Thus release of glucose into the extracellular space could be through vesicles that pinch the endoplasmic reticulum (ER) and which are transported to and fuse with the plasma membrane. Alternatively, as ER tubules are often found in close apposition to the plasma membrane, transient opening of channels linking these tubules and the plasma membrane or fusion of the respective membranes to form a pore could be possible mechanisms for glucose exit. That a membrane-based pathway was involved was supported by the sensitivity of glucose release from GLUT2$^{-/-}$ hepatocytes to low temperature, progesterone, colchicine and nocodazole, but not to the glucose transporter inhibitor cytochalasin B or phloretin (Guillam et al. 1998; Hosokawa and Thorens 2002).

However, a consequence of the absence of GLUT2 was that cytosolic glucose could not diffuse out of the cells and that G6P levels were not reduced during the fed to fast transition (Burcelin et al. 2000a). This abnormal accumulation of glucose and G6P in hepatocytes was proposed to cause the observed abnormalities in glycogen breakdown and regulation of glucose-sensitive gene expression. Indeed, glucose is an activator of glycogen synthase and G6P is an inhibitor of the phosphatase, which activates glycogen phosphorylase. Thus the combined allosteric regulation of these enzymes by glucose and G6P increases glycogen accumulation. Abnormal regulation of glucose-sensitive genes such as L-type pyruvate kinase or fatty acid synthase, which are paradoxically upregulated in the fasted state in the absence of GLUT2, could also be explained by the dysregulation of G6P, a key intermediate in the stimulation of expression of these genes. Thus glycogen accumulation may be due to the provision of glucose synthesized de novo in hepatocytes and its impaired degradation by the abnormal intracellular concentrations of glucose and G6P. In fasting, when the ratio of plasma glucagon to insulin increases, there is a normal activation of the key gluconeogenic enzymes, phosphoenol pyruvate carboxykinase and G6Pase. These

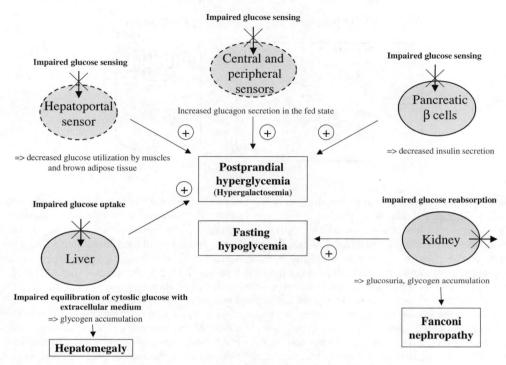

Figure 2 Pathophysiological model of Fanconi-Bickel syndrome. The circles indicate the organs affected by the loss of GLUT2 and the boxes indicate the clinical signs caused by GLUT2 loss of expression. The dotted circle indicates phenotypes found in GLUT2 knockout mice but not studied in patients with FBS. Combined physiological defects cause hyperglycemia observed in the fed state. These include the following (from left to right): defect in glucose storage by hepatocytes; impaired function of the hepatoportal sensor-stimulated glucose clearance; abnormal activation of the autonomic tone stimulating glucagon secretion as a result of impaired function of central and peripheral GLUT2-dependent sensors; impaired glucose-stimulated insulin secretion (at least in rodents). In the fasted state, hypoglycemia is probably mostly caused by the impaired glucose reabsorption in the kidney proximal convoluted tubule. This is associated with glycogen storage and Fanconi nephropathy. Hepatomegaly is caused by the abnormal equilibration of glucose between the cytosol and the extracellular milieu, despite the normal rate of hepatic glucose production through the membrane-traffic-based pathway. This leads to abnormal regulation of glycogen accumulation and hyperplasia of the liver. Galactose intolerance and galactosuria may be explained by the fact that GLUT2 is a monosaccharide transporter known to transport both glucose and galactose.

genes are regulated by hormones and not by glucose and thus are insensitive to the abnormal glucose metabolism of GLUT2-null hepatocytes.

Increase in glycemia in FBS patients following injection of glucagon or epinephrine has been observed, suggesting that hepatic glucose output may also be stimulated acutely despite abnormally elevated liver glycogen

storage. Therefore the situation in mice and humans is very comparable.

Normal intestinal glucose absorption but impaired kidney reabsorption and glucosuria

In GLUT2$^{-/-}$ mice, intestinal glucose absorption proceeds at a normal rate as observed

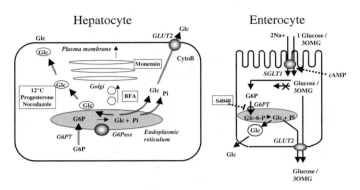

Figure 3 Glucose release from hepatocytes and transepithelial transport in intestine in the absence of GLUT2. Hepatocytes (left): glucose production requires the production of glucose-6-phosphate (G6P) originating from the gluconeogenic pathway or from glycogen degradation. G6P enters into the endoplasmic reticulum through the G6P translocase (G6PT) and is hydrolyzed into glucose and phosphate by the glucose-6-phosphatase (G6Pase). Glucose can then be released into the cytosol by an as yet unknown transporter and released out the cells through GLUT2. In the absence of GLUT2, glucose is released at a normal rate and utilizes a pathway that is inhibited by low temperature (12°C), progesterone, and nocodazole. This pathway segregates glucose from the cytosol and is probably based on a membrane traffic mechanism. This glucose release pathway is insensitive to the inhibitor of endoplasmic reticulum to Golgi trafficking Brefeldin A (BFA) and monensin. The membrane-traffic-based and GLUT2-dependent pathways coexist in normal hepatocytes as demonstrated by the additive effect of cytochalasin B (CytoB) and nocodazole on reducing the rate of glucose release from control hepatocytes. In the absence of GLUT2, in the fasted state, glucose and G6P accumulate abnormally in the cytosol and lead to increase glycogen storage and paradoxical activation of glucose-activated genes such as L-pyruvate kinase or fatty acid synthase. Intestine (right): glucose uptake through the apical membrane is catalyzed by the Na+/glucose symporter SGLT1. Glucose can then be released into the extracellular space through GLUT2. In the absence of this transporter, the transepithelial transport of glucose proceeds at a normal rate and is normally increased by cAMP stimulation of SGLT-1. Hoewever, transport is reduced in the absence, but not the presence, of GLUT2 by the G6PT inhibitor S4048. Transport of 3-O-methyl-glucose (3OMG) is not observed in the absence of GLUT2. These observations indicate that there is an alternative pathway for glucose release from enterocytes, which requires glucose phosphorylation and entry into the endoplasmic reticulum. Glucose can then be released from the cells by the same membrane-traffic-based pathway described in hepatocytes.

in oral glucose tolerance tests or perfused intestine. Using this latter technique, we could further demonstrate that glucose transepithelial transport could be suppressed by the chlorogenic acid derivative S4048, an inhibitor of the G6P translocase. In addition, 3-O-methyl-glucose, a glucose transporter substrate which cannot be phosphorylated by hexokinases, was not transported across the epithelial cells in the absence of GLUT2 (Stumpel *et al.* 2001). Together these data suggest that, for transcellular transport, glucose needs to be phosphorylated and to enter into the ER. This points to the existence in the intestine of a transport pathway similar to that described in hepatocytes for glucose release, where glucose is first phosphorylated in the cytosol, transported into the ER, and hydrolyzed to glucose before being released to the cell exterior by a membrane-traffic-based pathway (Figure 3).

Similarly to the GLUT2-null mouse situation, the absorption of glucose seems to be unimpaired in FBS patients. In contrast with the intestinal situation, reabsorption of glucose in the kidney is impaired in both human patients and mutant mice, leading to marked glucosuria. Therefore this suggests that the mechanisms for glucose reabsorption are different in intestine and kidney, and that the membrane-traffic-based pathway may not be present in the kidney nephron. This renal loss of glucose observed in mutant mice and FBS patients could be responsible for the hypoglycemia observed in the fasting periods (Figure 2). Galactosuria and also galactose intolerance in FBS patients may be explained by the fact that GLUT2 is a monosaccharide transporter known to transport both glucose and galactose.

GLUT2 in other glucose sensing units

Re-expression by transgenesis of GLUT1 or GLUT2 in the β-cells of GLUT2-null mice rescued mice from early lethality and restored glucose-stimulated insulin secretion (Thorens *et al.* 2000). These rescued mice were then used to evaluate whether other glucose sensors were GLUT2 dependent and whether they could be identified based on their requirement for GLUT2 expression.

We demonstrated that the hepatoportal vein area contains a glucose sensor which is activated by the presence of a glucose gradient established between the portal vein and the peripheral artery (Burcelin *et al.* 2000c). This sensor controls multiple physiological functions, including suppression of food intake, glucose storage in liver, and glucose utilization by a subset of muscles and by brown adipose tissue. This sensor is thought to be important in the postprandial state to increase glucose storage and terminate food intake. Investigation with GLUT2-null mice indicated that this sensor was GLUT2 dependent (Burcelin *et al.* 2000b).

No data exist yet concerning the function of this sensor in humans. Its inactivation in human patients could lead to a decrease in the efficiency of glucose clearance in the postprandial state and contribute to postprandial hyperglycemia (Figure 2).

Impaired glucagon secretion

In GLUT2$^{-/-}$ mice or mutant mice with transgenic re-expression of GLUT1 or GLUT2 in their β-cells, glucagon levels in the fed states are markedly elevated. This is observed in the presence of normoglycemia and a normal rate of hepatic glucose production and may be the cause of the elevated plasma β-hydroxybutyrate levels. We provided evidence that the fed hyperglucagonemia was due to a constant tone of the autonomic nervous system to the α cells and not to an intrinsic defect of these cells, which do not express GLUT2 (Burcelin and Thorens 2001). The explanation for this secretory defect is that extrapancreatic GLUT2-dependent glucose sensors control the tone of the autonomic nervous system to the α cells and that, in the absence of GLUT2, these sensors constantly activate the secretory activity of the α cells.

Therefore it would be of great interest to determine glucagon levels in the fed and fasted state in FBS patients to evaluate whether similar GLUT2 sensors do exist in humans.

CONCLUSIONS

There are many similarities between the physiological dysfunctions observed in GLUT2$^{-/-}$ mice and FBS patients. The major difference is that *GLUT2* gene inactivation in mice is lethal because of the loss of glucose-stimulated insulin secretion. This lethality is not observed in human patients, probably because of the expression of other transporter isoforms in their β-cells. Whether the other glucose sensors functionally identified in mice are also active in humans and whether they strictly depend on GLUT2 expression or also express other transporter isoforms remains to be established.

The efficient intestinal absorption of glucose or glucose release from hepatocytes observed in both mouse and human in the absence of GLUT2 suggests that there are common pathways, which differ from the classical hexose facilitators, to move sugar across the cells. One major area of interest is to evaluate whether this alternate pathway, based putatively on a membrane traffic mechanism, is regulated by hormone and indeed plays a role in the normal physiology of intestinal glucose absorption or in hepatic glucose production.

Despite the similarities between mice and humans with *GLUT2* gene inactivation, one possible difficulty in comparing their physiology is that the FBS patients are not only rare and difficult to study, but also have lived with their disease for many years. This may have induced adaptive changes that may obscure the defect solely due to the absence of GLUT2.

Nevertheless, it has been fascinating to be able to compare the phenotypic consequence of the genetic mutation in mouse and human. The availability of a mouse model clearly allowed us to understand in much greater detail the molecular basis of the observed physiological changes and to identify the existence of unknown transport and sensing mechanisms. The challenge is now to determine whether these are also present in humans and to evaluate how important they are in normal physiology and in the pathogenesis of diseases of glucose homeostasis.

SUMMARY

Facilitated diffusion of glucose across biological membranes is catalyzed by a family of carrier proteins (GLUTs). GLUT2 is the predominant glucose transporter isoform in hepatocytes, pancreatic β cells, and renal and intestinal epithelial cells. Because of its low affinity and high capacity, GLUT2 allows fast equilibration of glucose between the blood and the cell's cytoplasm. This is a prerequisite for efficient glucose storage in liver, for glucose absorption and reabsorption in intestine and kidney, and for normal glucose sensing by pancreatic β-cells. Inactivating mutations of the *GLUT2* gene in human causes the Fanconi-Bickel syndrome (FBS). This is a rare autosomal recessive disorder of carbohydrate metabolism characterized by hepatorenal glycogen accumulation, tubular nephropathy, glucose and galactose intolerance, and fasting hypoglycemia. To date, 35 different GLUT2 mutations have been reported in 65 patients with FBS. Inactivating mutations of GLUT2 in mouse embryonic stem cells to generate GLUT2-null mice allowed a careful analysis of the pathophysiological consequences of gene suppression of this gene on whole body glucose homeostasis. This led to the discovery of novel glucose transport and sensing mechanisms and permitted a comparison of the role of GLUT2 in mouse and human.

REFERENCES

Akagi, M., Inui, K., Nakajima, S., Shima, M., Nishigaki, T., Muramatsu, T., *et al.* (2000). Mutation Analysis of Two Japanese Patients with Fanconi-Bickel Syndrome. *J. Hum. Genet.*, **45**, 60–62.

Baroni, M.G., Alcolado, J.C., Pozzilli, P., Cavallo, M.G., Li, S.R., and Galton, D.J. (1992). Polymorphisms at the GLUT2 (Beta-Cell/Liver) Glucose Transporter Gene and Non-Insulin-Dependent Diabetes Mellitus (NIDDM): Analysis in Affected Pedigree Members. *Clin. Genet.*, **41**, 229–234.

Bell, G.I., Kayano, T., Buse, J.B., Burant, C.F., Takeda, J., Lin, D., *et al.* (1990) Molecular Biology of Mammalian Glucose Transporters. Diabetes Care, **13**, 198–208.

Blakemore, S.J., Aledo, J.C., James, J., Campbell, F.C., Lucocq, J.M., and Hundal, H.S. (1995). The GLUT5 Hexose Transporter is also Localized to the Basolateral Membrane of the Human Jejunum. *Biochem. J.*, **309**, 7–12.

Brivet, M., Moatti, N., Corriat, A. Lemonnier, A., and Odièvre, M. (1983). Defective Galactose Oxidation in a Patient with Glycogen Storage Disease and Fanconi Syndrome. *Pediatr. Res.*, **17**, 157–161.

Brockmann, K., Wang, D., Korenke, C.G., von Moers, A., Ho, Y.Y., Pascual, J.M., *et al.* (2001). Autosomal Dominant Glut-1 Deficiency Syndrome and Familial Epilepsy. *Ann. Neurol.*, **50**, 476–485.

Burant, C.F. and Bell, G.I. (1992). Mammalian Facilitative Glucose Transporters: Evidence for Similar Substrate Recognition Sites in Functionally Monomeric Proteins. *Biochemistry*, **31**, 10414–10420.

Burant, C.F. and Saxena, M. (1994). Rapid Reversible Substrate Regulation of Fructose Transporter Expression in Rat Small intestine and Kidney. *Am. J. Physiol.*, **267**, G71–G79.

Burcelin, R. and Thorens, B. (2001). Evidence that Extrapancreatic GLUT2-Dependent Glucose Sensors Control Glucagon Secretion. *Diabetes*, **50**, 1282–1289.

Burcelin, R., del Carmen Munoz, M., Guillam, M.T., and Thorens, B. (2000a). Liver Hyperplasia and Paradoxical Regulation of Glycogen Metabolism and Glucose-sensitive Gene Expression in GLUT2-Null Hepatocytes. Further Evidence for the Existence of a Membrane-Based Glucose Release Pathway. *J. Biol. Chem.*, **275**, 10930–10936.

Burcelin, R., Dolci, W., and Thorens, B. (2000b). Glucose Sensing by the Hepatoportal Sensor is GLUT2-Dependent: In Vivo Analysis in GLUT2-Null Mice. *Diabetes*, **49**, 1643–1648.

Burcelin, R., Dolci, W., and Thorens, B. (2000c). Portal Glucose Infusion in the Mouse Induces Hypoglycemia: Evidence that the Hepatoportal Glucose Sensor Stimulates Glucose Utilization. *Diabetes*, **49**, 1635–1642.

Burwinkel, B., Sanjad, S.A., Al-Sabban, E., Al-Abbad, A., and Kilimann, M.W. (1999). A Mutation in GLUT2, not in Phosphorylase Kinase Subunits, in Hepato-Renal Glycogenosis with Fanconi Syndrome and Low Phosphorylase Kinase Activity. *Hum. Genet.*, **105**, 240–243.

Carayannopoulos, M.O., Chi, M.M., Cui, Y., Pingsterhaus, J.M., McKnight, R.A., Mueckler, M., *et al.* (2000). GLUT8 is a Glucose Transporter Responsible for Insulin-Stimulated Glucose Uptake in the Blastocyst. *Proc. Natl Acad. Sci. USA*, **97**, 7313–7318.

Chesney, R.W., Kaplan, B.S., Colle, E., Scriver, C.R., McInnes, R.R., Dupont, C.H., *et al.* (1980). Abnormalities of Carbohydrate Metabolism in Idiopathic Fanconi Syndrome. *Pediatr. Res.*, **14**, 209–215.

Cloherty, E.K., Sultzman, L.A., Zottola, R.J., and Carruthers, A. (1995). Net Sugar Transport is a Multistep Process. Evidence for Cytosolic Sugar Binding Sites in Erythrocytes. *Biochemistry*, **34**, 15395–15406.

De Vos, A., Heimberg, H., Quartier, E., Huypens, P., Bouwens, L., Pipeleers, D., *et al.* (1995). Human and Rat Beta Cells Differ in Glucose Transporter but not in Glucokinase Gene Expression. *J. Clin. Invest.*, **96**, 2489–2495.

Doege, H., Schurmann, A., Ohnimus, H., Monser, V., Holman, G.D., and Joost, H.G. (1998). Serine-294 and Threonine-295 in the Exofacial Loop Domain between Helices 7 and 8 of Glucose Transporters (GLUT) Are Involved in the Conformational Alterations during the Transport Process. *Biochem. J.*, **329**, 289–293.

Doege, H., Bocianski, A., Joost, H.G., and Schurmann, A. (2000a). Activity and Genomic Organization of Human Glucose Transporter 9 (GLUT9), a Novel Member of the Family of Sugar-Transport Facilitators Predominantly Expressed in Brain and Leucocytes. *Biochem. J.*, **3**, 771–776.

Doege, H., Schurmann, A., Bahrenberg, G., Brauers, A., and Joost, H.G. (2000b). GLUT8, a novel member of the sugar transport facilitator family with glucose transport activity. *J. Biol. Chem.*, **275**, 16275–16280.

Doege, H., Bocianski, A., Scheepers, A., Axer, H., Eckel, J., Joost, H.G., *et al.* (2001a). Characterization of Human Glucose Transporter (GLUT) 11 (Encoded by *SLC2A11*), a Novel Sugar-Transport Facilitator Specifically Expressed in Heart and Skeletal Muscle. *Biochem. J.*, **359**, 443–449.

Doege, H., Bocianski, A., Scheepers, A., Axer, H., Eckel, J., Joost, H.G., (2001b). Characterization of Glucose Transporter 10 (GLUT10), a Novel Heart and Skeletal Muscle-Specific Sugar Transport Facilitator. *Diabetes*, **50**(Suppl. 2), A277–A278.

Fanconi, G. and Bickel, H. (1949). Die Chronische Aminoacidurie (Aminosäurediabetes oder Nephrotisch-Glukosurischer Zwergwuchs) bei der Glykogenose und der Cystinkrankheit. *Helv. Paediat. Acta.*, **4**, 359–396.

Fukumoto, H., Seino, S., Imura, H., Seino, Y., Eddy, R.L., Fukushima, Y., *et al.* (1988). Sequence, Tissue Distribution, and Chromosomal Localization of mRNA Encoding a Human Glucose Transporter-Like Protein. *Proc. Natl Acad. Sci. USA*, **85**, 5434–5438.

Fukumoto, H., Kayano, T., Buse, J.B., Edwards, Y., Pilch, P.F., Bell, G.I., *et al.* (1989). Cloning and Characterization of the Major Insulin-Responsive Glucose Transporter Expressed in Human Skeletal Muscle and Other Insulin-Responsive Tissues. *J. Biol. Chem.*, **264**, 7776–7779.

Garcia, J.C., Strube, M., Leingang, K., Keller, K., and Mueckler, M.M. (1992). Amino acid Substitutions at Tryptophan 388 and Tryptophan 412 of the HepG2 (Glut1) Glucose Transporter Inhibit Transport Activity and Targeting to the Plasma Membrane in *Xenopus* Oocytes. *J. Biol. Chem.*, **267**, 7770–7776.

Garty, R., Cooper, M., and Tabachnik, E. (1974). The Fanconi Syndrome Associated with Hepatic Glycogenosis and Abnormal Metabolism of Galactose. *J. Pediatr.*, **85**, 821–823.

Guillam, M.T., Hummler, E., Schaerer, E., Yeh, J.I., Birnbaum, M.J., *et al.* (1997). Early Diabetes and Abnormal Postnatal Pancreatic Islet Development in Mice Lacking Glut2. *Nat. Genet.*, **17**, 327–330.

Guillam, M.T., Burcelin, R., and Thorens, B. (1998). Normal Hepatic Glucose Production in the Absence of GLUT2 Reveals an Alternative Pathway for Glucose Release from Hepatocytes. *Proc. Natl Acad. Sci. USA*, **95**, 12317–12321.

Guillam, M.T., Dupraz, P., and Thorens, B. (2000). Glucose Uptake, Utilization, and Signaling in GLUT2-Null Islets. *Diabetes*, **49**, 1485–1491.

Hebert, D.N. and Carruthers, A. (1992). Glucose Transporter Oligomeric Structure Determines Transporter Function. Reversible Redox-Dependent Interconversions of Tetrameric and Dimeric GLUT1. *J. Biol. Chem.*, **267**, 23829–23838.

Hediger, M.A. and Rhoads, D.B. (1994). Molecular Physiology of Sodium–Glucose Cotransporters. *Physiol. Rev.*, **74**, 993–1026.

Henderson, P.J. (1993). The 12-Transmembrane Helix Transporters. Curr. *Opin. Cell Biol.*, **5**, 708–721.

Hosokawa, M. and Thorens, B. (2002). Glucose Release from GLUT2-null Hepatocytes: Characterization of a Major and a Minor Pathway. *Am. J. Physiol. Endocrinol. Metab.*, **282**, E794–E801.

Hruz, P.W. and Mueckler, M.M. (1999). Cysteine-Scanning Mutagenesis of Transmembrane Segment 7 of the GLUT1 Glucose Transporter. *J. Biol. Chem.*, **274**, 36176–36180.

Hruz, P.W. and Mueckler, M.M. (2001). Structural Analysis of the GLUT1 Facilitative Glucose Transporter (Review). *Mol. Membr. Biol.*, **18**, 183–193.

Ibberson, M., Uldry, M., and Thorens, B. (2000). GLUTX1, a Novel Mammalian Glucose Transporter Expressed in the Central Nervous System and Insulin-Sensitive Tissues. *J. Biol. Chem.*, **275**, 4607–4612.

Inukai, K., Asano, T., Katagiri, H., Anai, M., Funaki, M., Ishihara, H., *et al.* (1994). Replacement of Both Tryptophan Residues at 388 and 412 Completely Abolished Cytochalasin B Photolabelling of the GLUT1 Glucose Transporter. *Biochem. J.*, **302**, 355–361.

Joost, H.G. and Thorens, B. (2001). The Extended GLUT-Family of Sugar/Polyol Transport Facilitators: Nomenclature, Sequence Characteristics, and Potential Function of its Novel Members. *Molec. Membr. Biol.*, **18**, 247–256.

Joost, H.G., Doege, H., Bocianski, A., and Schurmann, A. (2001). Erratum: Nomenclature of the GLUT Family of Sugar Transport Facilitators. GLUT6, GLUT9, GLUT10, and GLUT11. *Biochem. J.*, **358**, 791–792.

Kasahara, T. and Kasahara, M. (1998). Tryptophan 388 in Putative Transmembrane Segment 10 of the Rat Glucose Transporter Glut1 is Essential for Glucose Transport. *J. Biol. Chem.*, **273**, 29113–29117.

Katagiri, H., Asano, T., Shibasaki, Y., Lin, J.L., Tsukuda, K., Ishihara, H., *et al.* (1991). Substitution of Leucine for Tryptophan 412 does not Abolish Cytochalasin B Labeling but Markedly Decreases the Intrinsic Activity of GLUT1 Glucose Transporter. *J. Biol. Chem.*, **266**, 7769–7773.

Katagiri, H., Asano, T., Ishihara, H., Lin, J.L., Inukai, K., Shanahan, M.F., *et al.* (1991). Role of Tryptophan-388 of GLUT1 Glucose Transporter in Glucose-Transport Activity and Photoaffinity-Labelling with Forskolin. *Biochem. J.*, **291**, 861–867.

Kayano, T., Fukumoto, H., Eddy, R.L., Fan, Y.S., Byers, M.G., Shows, T.B., *et al.* (1988). Evidence for a Family of Human Glucose Transporter-Like Proteins. Sequence and Gene Localization of a Protein Expressed in Fetal Skeletal Muscle and other Tissues. *J. Biol. Chem.*, **263**, 15245–15248.

Kayano, T., Burant, C.F., Fukumoto, H., Gould, G.W., Fan, Y.S., Eddy, R.L., *et al.* (1990). Human Facilitative Glucose Transporters. Isolation, Functional Characterization, and Gene Localization of cDNAs Encoding an Isoform (GLUT5) Expressed in Small Intestine, Kidney, Muscle, and Adipose Tissue and an Unusual Glucose Transporter Pseudogene-Like Sequence (GLUT6). *J. Biol. Chem.*, **265**, 13276–13282.

Klepper, J., Willemsen, M., Verrips, A., Guertsen, E., Herrmann, R., Kutzick, C., *et al.* (2001). Autosomal Dominant Transmission of GLUT1 Deficiency. *Hum. Mol. Genet.*, **10**, 63–68.

Lee, P.J., van't Hoff, W.G., and Leonard, J.V. (1995). Catch-up Growth in Fanconi–Bickel Syndrome with Uncooked Cornstarch. *J. Inherit. Metab. Dis.*, **18**, 153–156.

Leloup, C., Arluison, M., Lepetit, N., Cartier, N., Marfaing-Jallat, P., Ferré, P., *et al.* (1994). Glucose Transporter 2 (GLUT 2): Expression in Specific Brain Nuclei. *Brain Res.*, **638**, 221–226.

Manz, F., Bickel, H., Brodehl, J., Feist, D., Gellissen, K., Geschöll-Bauer, B., *et al.* (1987). Fanconi-Bickel Syndrome. *Pediatr. Nephrol.*, **1**, 509–518.

Matschinsky, F.M. (1996). Banting Lecture 1995. A Lesson in Metabolic Regulation Inspired by the Glucokinase Glucose Sensor Paradigm. *Diabetes*, **45**, 223–241.

Matsubara, A., Tanizawa, Y., Matsutani, A., Kaneko, T., and Kaku, K. (1995). Sequence Variations of the Pancreatic Islet/Liver Glucose Transporter (GLUT2) Gene in Japanese Subjects with Noninsulin Dependent Diabetes Mellitus. *J. Clin. Endocrinol. Metab.*, **80**, 3131–3135.

Matsutani, A., Koranyi, L., Cox, N., and Permutt, M.A. (1990). Polymorphisms of GLUT2 and GLUT4 Genes. Use in Evaluation of Genetic Susceptibility to NIDDM in Blacks. *Diabetes*, **39**, 1534–1542.

Matsuura, T., Tamura, T., Chinen, Y., and Ohta, T. (2002). A Novel Mutation (N32K) of GLUT2 Gene in a Japanese Patient with Fanconi-Bickel Syndrone. *Clin. Genet.*, **62**, 255–256.

McVie-Wylie, A.J., Lamson, D.R., and Chen, Y.T. (2001). Molecular Cloning Of a Novel Member of the GLUT Family of Transporters, SLC2a10 (GLUT10), Localized on Chromosome 20q13.1: A Candidate Gene for NIDDM Susceptibility. *Genomics*, **72**, 113–117.

Moller, A.M., Jensen, N.M., Pildal, J., Drivsholm, T., Borch-Johnsen, K., Urhammer, S.A., *et al.* (2001). Studies of Genetic Variability of the Glucose Transporter 2 Promoter in Patients with Type 2 Diabetes Mellitus. *J. Clin. Endocrinol. Metab.*, **86**, 2181–2186.

Mueckler, M., Caruso, C., Baldwin, S.A., Panico, M., Blench, I., Morris, H.R., *et al.* (1985). Sequence and Structure of a Human Glucose Transporter. *Science*, **229**, 941–945.

Mueckler, M., Kruse, M., Strube, M., Riggs, A.C., Chiu, K.C., and Permutt, M.A. (1994). A Mutation in the Glut2 Glucose Transporter Gene of a Diabetic Patient Abolishes Transport Activity. *J. Biol. Chem.*, **269**, 17765–17767.

Ngarmukos, C., Baur, E.L., and Kumagai, A.K. (2001). Co-localization of GLUT1 and GLUT4 in the Blood–Brain Barrier of the Rat Ventromedial Hypothalamus. *Brain Res.*, **900**, 1–8.

Odièvre, M. (1966). Glycogénose Hépato-Rénale avec Tubulopathie Complexe: Deux Observations d'une EntitÉ Nouvelle. *Rev. Inst. Hepat.*, **19**, 1–70.

Ogier de Baulny, H., Touati, G., Rigal, O., Loirat, C., Brivet, M., and Odièvre, M. (1999). Fanconi-Bickel Syndrome: Functional Tests and Dietetic Treatment in a Patient. *J. Inherit. Metab. Dis.*, **22**(Suppl. 1), 141.

Oka, Y., Asano, T., Shibasaki, Y., Lin, J.L., Tsukuda, K., Katagiri, H., *et al.* (1990). C-terminal Truncated Glucose Transporter is Locked into an Inward-Facing Form without Transport Activity. *Nature*, **345**, 550–553.

Orci, L., Thorens, B., Ravazzola, M., and Lodish, H.F. (1989). Localization of the Pancreatic Beta Cell Glucose Transporter to Specific Plasma Membrane Domains. *Science*, **245**, 295–297.

Permutt, M.A., Koranyi, L., Keller, K., Lacy, P.E., Scharp, D.W., and Mueckler, M. (1989). Cloning and Functional Expression of a Human Pancreatic Islet Glucose-Transporter cDNA. *Proc. Natl Acad. Sci. USA*, **86**, 8688–8692.

Pessino, A., Hebert, D.N., Woon, C.W., Harrison, S.A., Clancy, B.M., Buxton, J.M., *et al.* (1991). Evidence that Functional Erythrocyte-Type Glucose Transporters Are Oligomers. *J. Biol. Chem.*, **266**, 20213–20217.

Phay, J.E., Hussain, H.B., and Moley, J.F. (2000). Cloning and Expression Analysis of a Novel Member of the Facilitative Glucose Transporter Family, SLC2A9 (GLUT9). *Genomics*, **66**, 217–220.

Rogers, S., James, D.E., and Best, J.D. (1998). Identification of Novel Facilitative Transporter Like Protein-GLUT8. *Diabetes*, **47**(Suppl. 1), A45.

Sakamoto, O., Ogawa, E., Ohura, T., Igarashi, Y., Matsubara, Y., Narisawa, K., *et al.* (2000). Mutation Analysis of the GLUT2 Gene in Patients with Fanconi-Bickel Syndrome. *Pediatr. Res.*, **48**, 586–589.

Santer, R., Schneppenheim, R., Dombrowski, A., Götze, H., Steinmann, B., and Schaub, J. (1997). Mutations in GLUT2, the Gene for the Liver-Type Glucose Transporter, in Patients with Fanconi-Bickel Syndrome. *Nat. Genet.*, **17**, 324–326.

Santer, R., Schneppenheim, R., Suter, D., Schaub, J., and Steinmann, B. (1998). Fanconi-Bickel Syndrome – the Original Patient and his Natural History, Historical Steps Leading to the Primary Defect, and a Review of the Literature. *Eur. J. Pediatr.*, **157**, 783–797.

Santer, R., Groth, S., Kinner, M., Dombrowski, A., Berry, G.T., Brodehl, J., *et al.* (2002a). The Mutation Spectrum of the Facilitative Glucose Transporter Gene SLC2A2 (GLUT2) in Patients with Fanconi-Bickel Syndrome. *Hum. Genet.*, **110**, 21–29.

Santer, R., Steinmann, B., and Schaub, J. (2002b). Fanconi-Bickel Syndrome – a Congenital Defect of Facilitative Glucose Transport. *Curr. Mol. Med.*, **2**, 213–227.

Saravolac, E.G. and Holman, G.D. (1987). Glucose Transport: Probing the Structure/Function Relationship. In G.W. Gould (ed.), *Facilitative Glucose Transporters*, Georgetown, TX: R.G. Landes, pp. 39–66.

Schurmann, A., Keller, K., Monden, I., Brown, F.M., Wandel, S., Shanahan, M.F., *et al.* (1993). Glucose Transport Activity and Photolabelling with 3-[^{125}I]iodo-4-azidophenethylamido-7-*O*-succinyldeacetyl (IAPS)-Forskolin of Two Mutants at Tryptophan-388 and -412 of the Glucose Transporter GLUT1: Dissociation of the Binding Domains of Forskolin and Glucose. *Biochem. J.*, **290**, 497–501.

Seatter, M.J., De la Rue, S.A., Porter, L.M., and Gould, G.W. (1998). QLS Motif in Transmembrane Helix VII of the Glucose Transporter Family Interacts with the C-1 Position of D-Glucose and is Involved in Substrate Selection at the Exofacial Binding Site. *Biochemistry*, **37**, 1322–1326.

Seidner, G., Alvarez, M.G., Yeh, J.I., O'Driscoll, K.R., Klepper, J., Stump, T.S., *et al.* (1998). GLUT1 Deficiency Syndrome caused by Haploinsufficiency of the Blood–Brain Barrier Hexose Carrier. *Nat. Genet.*, **18**, 188–191.

Stumpel, F., Burcelin, R., Jungermann, K., and Thorens, B. (2001). Normal Kinetics of Intestinal Glucose Absorption in the Absence of GLUT2: Evidence for a Transport Pathway Requiring Glucose Phosphorylation and Transfer into the Endoplasmic Reticulum. *Proc. Natl Acad. Sci. USA*, **98**, 11330–11335.

Takeda, J., Kayano, T., Fukomoto, H., and Bell, G.I. (1993). Organization of the Human GLUT2 (Pancreatic Beta-Cell and Hepatocyte) Glucose Transporter Gene. *Diabetes*, **42**, 773–777.

Tamori, Y., Hashiramoto, M., Clark, A.E., Mori, H., Muraoka, A., Kadowaki, T., *et al.* (1994). Substitution at Pro385 of GLUT1 Perturbs the Glucose Transport Function by Reducing Conformational Flexibility. *J. Biol. Chem.*, **269**, 2982–2986.

Tanizawa, Y., Riggs, A.C., Chiu, K.C., Janssen, R.C., Bell, D.S., Go, R.P., *et al.* (1994). Variability of the Pancreatic Islet Beta Cell/Liver (GLUT 2) Glucose Transporter Gene in NIDDM Patients. *Diabetologia*, **37**, 420–427.

Tartaglia, L.A. and Weng, X. (1999). *Nucleic Acid Molecules Encoding GLUTX and Uses Thereof*. US Patent 5, 942, 398.

Thorens, B. (1996). Glucose Transporters in the Regulation of Intestinal, Renal, and Liver Glucose Fluxes. *Am. J. Physiol.*, **270**, G541–G553.

Thorens, B., Sarkar, H.K., Kaback, H.R., and Lodish, H.F. (1988). Cloning and Functional Expression in Bacteria of a Novel Glucose Transporter Present in Liver, Intestine, Kidney, and Beta-Pancreatic Islet Cells. *Cell*, **55**, 281–290.

Thorens, B., Cheng, Z.Q., Brown, D., and Lodish, H.F. (1990). Liver Glucose Transporter: A Basolateral Protein in Hepatocytes and Intestine and Kidney Cells. *Am. J. Physiol.*, **259**, C279–C285.

Thorens, B., Guillam, M.T., Beermann, F., Burcelin, R., and Jaquet, M. (2000). Transgenic Reexpression of GLUT1 or GLUT2 in Pancreatic Beta Cells Rescues

GLUT2-Null Mice from Early Death and Restores Normal Glucose-Stimulated Insulin Secretion. *J. Biol. Chem.*, **275**, 23751–23758.

Tsuda, M., Kitasawa, E., Ida, H., Eto, Y. and Owada, M. (2000). A Newly Recognized Missense Mutation in the GLUT2 Gene in a Patient with Fanconi-Bickel Syndrome. *Eur. J. Pediatr.*, **159**, 867.

Uldry, M., Ibberson, M., Horisberger, J.D., Chatton, J.Y., Riederer, B.M., and Thorens, B. (2001). Identification of a Mammalian H⁺-myo-inositol Symporter Expressed Predominantly in the Brain. *EMBO J.*, **20**, 4467–4477.

Uldry, M., Ibberson, M., Hosokawa, M., and Thorens, B. (2002). GLUT2 is a High Affinity Glucosamine Transporter. *FEBS Lett.*, **524**, 199–203.

Wellner, M., Monden, I., Mueckler, M.M., and Keller, K. (1995). Functional Consequences Of Proline Mutations in the Putative Transmembrane Segments 6 and 10 of the Glucose Transporter GLUT1. *Eur. J. Biochem.*, **227**, 454–458.

Yoo, H.W., Shin, Y.L., Seo, E.J., and Kim, G.H. (2002). Identification of a Novel Mutation in the GLUT2 Gene in a Patient with Fanconi-Bickel Syndrome Presenting with Neonatal Diabetes Mellitus and Galactosaemia. *Eur. J. Pediatr.*, **161**, 351–353.

13

JANICE YANG CHOU* AND BRIAN C. MANSFIELD*

Glucose-6-phosphate transporter: the key to glycogen storage disease type Ib

HISTORICAL BACKGROUND

The mammalian glucose-6-phosphate transporter (G6PT) first came to attention as a result of studies of the type I glycogen storage diseases (GSD-I). GSD-I was described first by von Gierke (1929) as hepatonephromegalia glycogenia. Later, Cori and Cori (1952) showed that the underlying condition of the disease was an absence of glucose-6-phosphatase (G6Pase) activity in hepatic tissues. As increasing numbers of patients with the clinical symptoms of GSD-I were identified, two subgroups emerged. Enzyme assays of frozen liver samples showed that one group of patients appeared to have no G6Pase activity, while the other group appeared to have retained G6Pase activity. To account for this biochemical heterogeneity, Senior and Loridan (1968) proposed that patients lacking G6Pase activity should be reclassified as GSD-Ia, while patients retaining G6Pase activity should be reclassified as GSD-Ib.

GSD-Ia was shown to be due to a loss of activity in the G6Pase catalytic unit but identification of the defect in GSD-Ib was more elusive. From 1972 to 1980, Arion and coworkers (reviewed by Arion et al. 1980; Nordlie and Sukalski 1985) performed a series of experiments measuring G6Pase activity in both intact and disrupted liver microsomes. From these studies, they proposed that the hydrolysis of glucose-6-phosphate (G6P) in vivo involves the integration of several membrane proteins, including the G6Pase catalytic unit and its associated transporters (Figure 1). This multicomponent enzyme complex was named the "G6Pase system." Refining these studies further, Narisawa et al. (1978) and Lange et al. (1980) showed that only detergent-treated microsomes from GSD-Ib patients exhibited normal G6Pase activity. They hypothesized that GSD-Ib is a membrane transport defect that can be circumvented in vitro, when the endoplasmic reticulum (ER) membrane integrity is breached. The active site of the G6Pase catalytic unit lies on the luminal side of the ER (Ghosh et al. 2002). Therefore a detect in the transport of G6P across the ER membrane would prevent the G6P substrate from reaching the enzyme, resulting in a phenotype of G6Pase deficiency in the G6Pt defective (GSD-1b) patients.

* Section of Cellular Differentiation, HDB, NICHD, NIH, Building 10, Room 9S241, National Institutes of Health, Bethesda, MD 20892-1830, USA

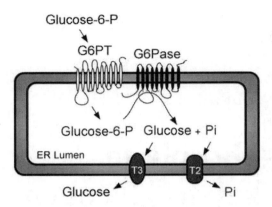

Figure 1 The G6Pase system, which catalyzes the hydrolysis of G6P to glucose and phosphate, is essential for glucose homeostasis. The system is composed of four components: G6Pase – the catalytic unit; G6PT – the G6P transporter; T2 – a putative phosphate (Pi) transporter; T3 – a putative glucose transporter. These are shown anchored in the ER membrane in contact with both the cytoplasm and ER lumen. The spatial representation is illustrative only; the proteins probably exist as a multiprotein cluster.

The G6Pase system explains the clinical heterogeneities seen in GSD-I patients (reviewed by Nordlie and Sukalski 1985; Chou and Mansfield 1999; Chou *et al.* 2002). Type I GSD is now divided into four subtypes corresponding to defects in the G6Pase catalytic unit (GSD-Ia), G6PT (GSD-Ib), a putative phosphate transporter T2 (GSD-Ic), and a putative glucose transporter T3 (GSD-Id). Of these, only GSD-Ia and GSD-Ib are well recognized and understood. Molecular genetic studies have confirmed that inactivating mutations in the *G6Pase* gene cause GSD-Ia (Lei *et al.* 1993) and inactivating mutations in the *G6PT* gene cause GSD-Ib (Hiraiwa *et al.* 1999).

There are no confirmed cases of the other putative transporter diseases GSD-Ic (Nordlie *et al.* 1983) and GSD-Id. Many initial reports of cases of GSD-Ic and GSD-Id, based on kinetic analysis of the G6Pase system in biopsy specimens, have proven incorrect. More recent genetic diagnoses have shown these to be cases of GSD-Ib (Veiga-da-Cunha *et al.* 1998, 1999).

Indeed, the existence of the phosphate and glucose transporters remains ambiguous. The patient that defined the T2 phosphate transporter disorder (Nordlie *et al.* 1983) was originally classified GSD-I by default rather than by positive scoring. The patient presented with hepatomegaly and increased hepatic glycogen deposition in the absence of disorders associated with other forms of GSD – normal hepatic branching enzyme, debranching enzyme, lysosomal α-glucosidase, and fructose–diphosphatase activities. The subsequent classification as GSD-Ic was based on kinetic analyses of the patient's hepatic G6Pase system (Nordlie *et al.* 1983) and genetic exclusion of GSD-Ia (Lei *et al.* 1995) or GSD-Ib (Lin *et al.* 1999). Confirmation of the GSD-Ic diagnosis awaits the cloning of the putative ER-associated phosphate transporter.

There is no current evidence for the existence of the putative glucose transporter or the putative defect in GSD-Id. Indeed, there is no evidence to indicate that glucose needs a specific transporter to move out of the ER into the cytoplasm for secretion. An alternative explanation is that glucose is released from the cell directly via a vesicular pathway (see Chapter 12).

PATHOPHYSIOLOGY

Mutations in the *G6PT* gene that result in a near or complete loss of G6P transport across the ER membrane prevent cytoplasmic G6P from reaching the G6Pase catalytic unit, compartmentalized within the lumen of the ER. The resulting disruption of glucose production in the terminal steps of gluconeogenesis and glycogenolysis, in gluconeogenic tissues (Figure 2), leads to a loss of blood glucose homeostasis. This results in the excessive accumulation of glycogen in these tissues, leading to the observed hepatomegaly and nephromegaly. The excess G6P in the cytoplasm also results in the hyperlipidemia, hyperuricemia, and lactic acidemia observed in GSD-I patients.

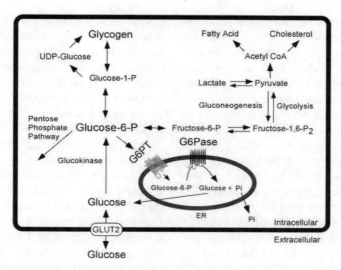

Figure 2 The primary anabolic and catabolic pathways of G6P in gluconeogenic organs. The outer solid line depicts the cell membrane. The G6Pase catalytic unit and G6PT components of the G6Pase system are shown embedded within the membrane of the ER. The GLUT2 transporter, responsible for the transport of glucose in and out of the cell, is shown embedded in the plasma membrane.

The striking difference between GSD-Ib and the other GSDs is the immune deficiency. The mechanism of the immune deficiency has not been elucidated and it is not clear whether it is due to defects in G6P/glycogen metabolism or to a completely different function of the "G6P" transporter. Neutrophils express G6PT but not G6Pase (Lin *et al*. 1998). Therefore they are not considered gluconeogenic cells. However, G6P is critical for myeloid cell function. In neutrophils and monocytes, cytoplasmic G6P stimulates both glycolysis and the hexose monophosphate shunt, which are the major sources of energy for chemotaxis and phagocytosis. Cytoplasmic G6P also enhances ATP-dependent microsomal sequestration of Ca^{2+} (Chen *et al*. 1998), which is an important second messenger for the respiratory burst and chemotaxis activities of phagocytic cells (Synderman and Uhing 1988). Moreover, any G6P deficiency in these cells is self-reinforcing because G6P accumulation into the ER is enhanced in the presence of ATP and Ca^{2+} (Chen *et al*. 1998). In this regard, it is interesting to note that neutrophils of GSD-Ib patients have impaired ability to mobilize Ca^{2+} (Kilpatrick *et al*. 1990).

The G6PT protein is most similar structurally to uhpC (Gerin *et al*. 1997), a bacterial G6P-specific receptor, which was also shown recently to mediate G6P transport (Schwoppe *et al*. 2002). This suggests that in nongluconeogenic cells, such as neutrophils and monocytes, the G6PT protein may function as an intracellular G6P sensor and/or transporter that regulates Ca^{2+} sequestration, glycolysis, and hexose monophosphate shunt activity. Alternatively, the G6PT may have a very different G6P-independent role, such as an ion or Ca^{2+} channel, consistent with the microsomal localization.

What is clear is that neutrophils lacking a functional G6PT are different from normal neutrophils. When neutrophils from normal subjects are incubated in glucose, their intracellular concentration of G6P increases 4.8-fold. In contrast, neutrophils from GSD-Ib patients have a lower endogenous level of G6P, only 73% that of wild type cells, and accumulate only 1.6-fold more when incubated in glucose (Verhoeven *et al.* 1999). While it is known that neutrophils from GSD-Ib patients have a decreased rate of glucose transport (Bashan *et al.* 1988), which could explain the reduction in G6P accumulation in response to glucose, it is unclear why a deficiency in an ER-bound G6PT reduces intracellular G6P accumulation in GSD-Ib neutrophils. One possibility is that if G6PT is a G6P sensor, it may be missensing the intracellular concentrations and misdirecting further accumulation. Although current data have only detected ER expression of the recombinant G6PT (Pan *et al.* 1999), another possibility is that G6PT is also associated with the plasma membrane and can mediate glucose transport. Alternatively, G6PT may have a very different role in which the alterations of G6P content are consequential rather than causative. While any of these possibilities could account for why GSD-1b patients manifest myeloid dysfunctions in addition to the expected phenotypic G6Pase deficiency, the role of G6PT in myeloid cells remains speculative (see Box: Clinical Features of Glycogen Storage Disease Ib).

THE GLUCOSE-6-PHOSPHATE TRANSPORTER

The identity of G6PT was debated for many years (reviewed by Nordlie and Sukalski 1985). One view held that G6Pase was a multifunction protein with both a G6Pase enzyme activity and a G6P transport function. Another view held that G6PT was a separate protein entity. Biochemical and kinetic studies were unable to resolve the issue and the resolution of the identity lay in the cloning of the gene. Two complementary methods were used to identify the GSD-Ib gene. In one approach, computer-based sequence comparisons of a liver EST database, with sequences of bacterial phosphate ester transporters, identified a putative human G6PT cDNA which was shown to be mutated in GSD-Ib patients (Gerin *et al.* 1997). In a second approach,

linkage analysis mapped the GSD-Ib locus to chromosome 11q23 (Annabi *et al.* 1998). The approaches converged when the putative G6PT cDNA was also mapped to chromosome 11q23 (Hiraiwa *et al.* 1999) and the identity of the G6PT cDNA was confirmed by mutation and expression-activity studies using the cDNA (Hiraiwa *et al.* 1999).

Human *G6PT* is a single-copy gene (Hiraiwa *et al.* 1999) consisting of nine exons (Gerin *et al.* 1999; Hiraiwa *et al.* 1999) spanning approximately 5.3 kb of DNA on chromosome 11q23 (Annabi *et al.* 1998). Transcription gives rise to two alternately spliced transcripts, G6PT (Gerin *et al.* 1997) and variant G6PT (vG6PT) (Middleditch *et al.* 1998), which differ in the absence or presence, respectively, of exon 7 (Gerin *et al.* 1999; Hiraiwa *et al.* 1999).

The G6PT cDNA from human (Gerin *et al.* 1997), mouse, and rat (Lin *et al.* 1998) are now characterized. All three proteins are predicted to be very similar, with mouse and rat proteins sharing 95% and 93% amino acid sequence identity, respectively, with the human protein. Mammalian G6PT proteins are 429 amino acid residues in length, containing a potential Asn-linked glycosylation site at amino acids 354–356 (Figure 3). However, in vitro transcription–translation in the presence of microsomal membranes shows that the 37 kDa membrane proteins are not glycosylated (Pan *et al.* 1999). The G6PT contain a C-terminal ER protein retention motif (Gerin *et al.* 1997; Lin *et al.* 1998), consistent with their expected localization. Hydropathy plots predict they are hydrophobic proteins anchored to the ER by either 10 or 12 transmembrane helices. Experimental data from protease sensitivity and glycosylation scanning analyses (Pan *et al.* 1999) using tagged proteins, favor the 10-transmembrane model with both the N- and C-termini lying in the cytoplasm (Figure 3).

Based upon amino acid sequence alignments, G6PT belongs to a family of transporters of phosphorylated metabolites that includes glycerol-3-phosphate transporter, phosphoglycerate transporter, hexose-6-phosphate

CLINICAL FEATURES OF GLYCOGEN STORAGE DISEASE TYPE IB

Until 1997, GSD-Ib was diagnosed using both clinical symptoms and biochemical profiles, then confirmed by the measurement of G6Pase activity in liver biopsy samples. With the cloning of the *G6PT* gene, noninvasive DNA-based diagnostic tests have been developed and are now used in place of kinetic studies for disease confirmation, carrier testing, and prenatal diagnosis.

While some GSD-Ib patients present in the neonatal period with hypoglycemia, most are diagnosed after birth when several months old. The initial symptoms are usually hepatomegaly and a symptomatic hypoglycemia following a short fast. Physically, affected children are of short stature with a significant accumulation of subcutaneous fat and a protuberant abdomen. They bruise easily and are prone to bleed for a prolonged time as a result of platelet dysfunction. Long-term complications include osteoporosis, gout, renal disease, pulmonary hypertension, and hepatic adenomas that may undergo malignant transformation.

GSD-Ib patients manifest the symptoms of G6Pase enzyme deficiency identical to those of GSD-Ia patients. The phenotype is characterized by growth retardation, hypoglycemia, hepatomegaly, nephromegaly, hyperlipidemia, hyperuricemia, and lactic acidemia. However, GSD-Ib patients also present with unique symptoms not so obviously associated with G6Pase activity, including neutropenia and myeloid dysfunctions which result in recurrent bacterial infections (Beaudet *et al.* 1980; Gitzelmann and Bosshard 1993; Garty *et al.* 1996). Oral and intestinal mucosal ulcerations are common and some patients also suffer chronic inflammatory bowel disease (Roe *et al.* 1986; Visser *et al.* 2000). A recent review documented the onset of neutropenia in 39 GSD-Ib patients (Visser *et al.* 2000). During their first year of life 64% of the GSD-Ib patients manifested neutropenia, with 15% of patients developing it within the first month. In 18% of patients, neutropenia was not noted until 6–9 years of age. In most cases, neutropenia was intermittent and correlated with the occurrence of infections or the onset of inflammatory bowel disease, when the neutrophil dysfunction also became apparent.

The extent of myeloid dysfunction is significant and appears to reach beyond the obvious pathways of glycogen metabolism. The polymorphonuclear leukocytes from GSD-Ib patients exhibit impairment of mobility, chemotaxis, and Ca^{2+} sequestration, and the respiratory burst, hexose monophosphate shunt, and glycolytic and phagocytotic activities are all diminished (reviewed by Gitzelmann and Bosshard 1993; Garty *et al.* 1996). So, while G6PT is a G6P transporter in the gluconeogenic tissues and is essential for glucose homeostasis, this transporter appears to have other important roles in nongluconeogenic cells, although their nature and mechanism of action remain to be elucidated.

There is no cure for GSD-Ib, but many of the disease symptoms can be managed using a dietary therapy (Greene *et al.* 1976; Chen *et al.* 1984) augmented with granulocyte colony-stimulation factor (G-CSF) therapy (Schroten *et al.* 1991; Roe *et al.* 1992). This strategy enables patients to attain near-normal growth and pubertal development, with fewer complications as they age. The dietary therapy for afflicted infants typically consists of nocturnal nasogastric infusion of glucose to avoid hypoglycemia (Greene *et al.* 1976). Older patients eat uncooked cornstarch, which acts as a slow-release glucose that prolongs euglycemia between meals (Chen *et al.* 1984). While this approach improves hypoglycemia, hyperuricemia, hyperlipidemia, and renal function, and slows the development of renal failure, it fails to normalize blood uric acid and lipids levels, especially after puberty.

The treatment of myeloid dysfunctions in GSD-Ib patients with G-CSF increases the absolute neutrophil count and reduces the frequency and severity of bacterial infection, which in turn improves the quality of daily life (Schroten *et al.* 1991; Roe *et al.* 1992; Wendel *et al.* 1993; McCawley *et al.* 1994). However it is expensive, the neutrophil chemotaxis does not improve with G-CSF therapy (Schroten *et al.* 1991; McCawley *et al.* 1994), and most patients develop splenomegaly with some requiring splenectomy (Calderwood *et al.* 2001).

Orthotopic liver transplantation (OLT) has been advocated as a potential cure of GSD-I. While OLT improves the metabolic abnormalities, it does not consistently improve neutropenia or immune function in GSD-Ib patients (Lachaux *et al.* 1993; Matern *et al.* 1999; Martinez-Olmos *et al.* 2001). Life expectancy has improved significantly with the combination therapy and with OLT therapy, but there has been insufficient time to judge their effects on long-term survival.

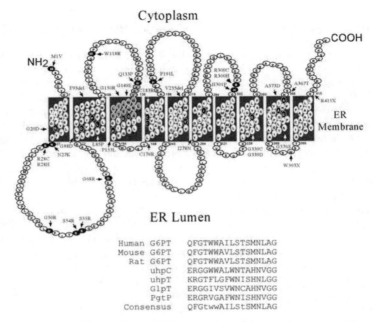

Figure 3   The topology of human G6PT. The 429 amino acid residues are denoted by circles. The 32 functionally characterized G6PT mutations identified in GSD-Ib patients are shown in black and indicated. The phosphorylated metabolite transporter signature motif, composed of amino acids 133–149, is denoted by shaded circles. Alignment of the signature motifs of mammalian G6PT, hexose-6-phosphate transporter (uhpT), G6P receptor (uhpC), glycerol-3-phosphate transporter (GlpT), and phosphoglycerate transporter (PgtP) are shown below the diagram. Modified from Pan *et al.* (1999).

transporter (uhpT), and G6P receptor (uhpC) (reviewed by Chou *et al.* 2002). These transporters share a signature motif (Figure 3), which lies between amino acids 133 and 149 in human G6PT. Two mutations, Q133P and G149E, that alter the first and last amino acids in this motif have been identified in the *G6PT* gene of GSD-Ib patients. Both are null mutations that completely abolish microsomal G6P uptake activity (Hiraiwa *et al.* 1999; Chen *et al.* 2002), suggesting that the motif is a functional component of the G6PT. However, this motif has not been functionally characterized in the other transporters. It would be of interest to conduct more mutational studies to identify the essential core residues in the motif and to determine whether the motif mediates G6P binding and/or transport.

G6PT EXPRESSION

Unlike the G6Pase catalytic unit, which is unique to the gluconeogenic tissues, namely liver, kidney, and intestine, the G6PT is expressed ubiquitously (Lin *et al.* 1998). In adult mice, G6PT transcripts are expressed abundantly in liver, kidney, large intestine, small intestine, and skeletal muscle (Figure 4A). Less abundant expression is detected in brain and heart, while a lower expression is also observed in mouse placenta, spleen, stomach, testis, and uterus. In human tissues, G6PT expression has been measured in the liver and neutrophil/monocyte cells, as well as in hepatoma (HepG2) and promonocyte (U937 and THP-1) cell lines (Figure 4A). The alternatively spliced transcript, vG6PT, is expressed exclusively in the brain, heart, and skeletal

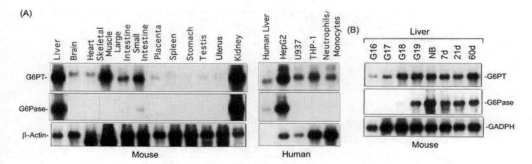

Figure 4 Expression of G6PT and G6Pase mRNA. (A) Northern blot analysis of G6PT and G6Pase expression in adult mouse and human tissues, cells, and cell lines. (B) Northern blot analysis of G6PT and G6Pase expression in mouse liver from 16 to 19 days gestation (G16–G19) and from newborn (NB), 7 day (7d), 21 day (21d) and 60 day (60d) old mice. Modified from Lin *et al.* (1998). GADPH, glyceraldehyde-3-phosphate dehydrogenase.

muscle (Lin *et al.* 2000). The role of vG6PT in these tissues remains to be determined. The patterns of G6PT and vG6PT expression support the hypothesis that the G6PT has roles beyond working with G6Pase to maintain glucose homeostasis.

The divergence in the developmental profiles of G6Pase and G6PT (Lin *et al.* 1998) also support the above hypothesis. G6PT mRNA is clearly detected in mouse liver at 16 days gestation and increases to adult levels by 18 days gestation (Figure 4B). In contrast, G6Pase mRNA is not detected until 19 days gestation, and undergoes a marked increase at parturition before leveling off to adult levels (Figure 4B). The development profile of the *G6PT* gene in neutrophils and monocytes remains to be determined and may shed more light on its role in myeloid cells.

G6P TRANSPORT ASSAYS

The transport and hydrolysis of G6P are tightly coupled processes. Hepatic microsomes from wild-type (WT) mice transport G6P efficiently (Lei *et al.* 1996), and the K_m for G6P is 1.4 mM (C.-J. Pan and J.Y. Chou, unpublished data). In contrast, hepatic microsomes from G6Pase-deficient mice transport G6P much less efficiently (Li *et al.* 1996). This coupling has been further demonstrated with functional assays of

the recombinant G6PT (Hiraiwa *et al.* 1999) based upon microsomal [U-^{14}C]G6P uptake in transfected COS-1, a monkey kidney cell line. Transfection of COS-1 with a G6Pase cDNA results in little or no microsomal G6P uptake (Figure 5A), consistent with the absence of an endogenous functional G6Pase system (Hiraiwa *et al.* 1999). In contrast, transfection with a G6PT cDNA leads to a detectable microsomal G6P transport activity (Figure 5A). Cotransfection of COS-1 cells with G6PT and G6Pase cDNAs increases the transport activity markedly (Figure 5A), consistent with the coupling observed in the G6Pase system. Using this G6PT/G6Pase cotransfection assay, mutations identified in the *G6PT* gene of GSD-Ib patients have been proven to abolish or greatly reduce microsomal G6P uptake activity (Figure 5B) (Hiraiwa *et al.* 1999; Chen *et al.* 2000), thus establishing the molecular basis of the GSD-Ib disorder.

A more sensitive transport assay for the G6PT was developed recently using an adenoviral vector-mediated expression system (Chen *et al.* 2002). The markedly enhanced expression of G6PT and G6Pase proteins in this system has allowed the identification of G6P transporter mutants with a very low residual activity and further refined our understanding of the relationships between transport activity, metabolic disorders, and

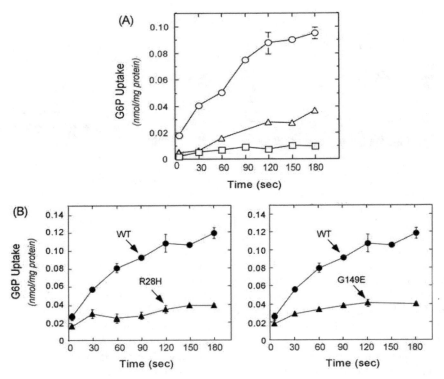

Figure 5 G6P uptake activity of wild-type (WT) and mutant G6PT. (A) Uptake of [U-^{14}C]G6P into intact microsomes prepared from COS-1 cells transfected with cDNA encoding: G6PT alone (△), G6Pase alone (□), or G6PT and G6Pase (○). (B) Microsomal G6P uptake activity in COS-1 cells cotransfected with cDNA encoding G6Pase and either WT G6PT (●), or mutant (▲) G6PT. R28H and G149E are mutant G6PT containing substitutions at R28 and G149 of the WT G6PT respectively.

myeloid dysfunctions (Chen *et al.* 2002). This more sensitive assay can now be employed to study the kinetics of microsomal G6P transport activity and to examine the binding of G6P or its analogs to the G6PT protein.

Several inhibitors are useful in dissecting the G6Pase system in intact microsomes. Chlorogenic acid and its derivatives, such as S3483, inhibit G6PT activity directly (Hemmerle *et al.* 1997; Arion *et al.* 1998). Vanadate inhibits the G6Pase catalytic unit directly and consequently inhibits G6P transport indirectly (Singh *et al.* 1981). The G6Pase catalytic activity is not the only microsomal enzyme using G6P as a substrate. In the presence of vanadate and an electron acceptor, such as NADP or metyrapone, the ER-associated

hexose-6-phosphate dehydrogenase (Hori and Takahashi 1977) will metabolize G6P in the lumen to 6-phosphogluconate (Gerin and Van Schaftingen 2002). This activity is also inhibited in the presence of S3483, reflecting its dependence on G6PT. As with G6Pase, hexose-6-phosphate dehydrogenase stimulates the G6P uptake activity of G6PT protein.

It is important to bear in mind that while G6P transport is the only known function of G6PT currently, there may be other associated activities, as discussed earlier. Such activities may have an active site that partly overlaps that of the G6P transport function, and it is important to consider this when making correlations between G6P transport activity and myeloid dysfunctions.

MOLECULAR GENETICS OF GSD-IB

G6P transporter defects (GSD-Ib) represent 15% of all GSD-I cases, and approximately 139 GSD-1b patients are available for study worldwide (reviewed by Chou and Mansfield 1999; Chou *et al.* 2002). From these patients, 69 separate mutations have been identified in the *G6PT* gene. These include 28 missense, 10 nonsense, 17 insertion/deletion (including two codon-deletion), and 14 splicing mutations, which appear scattered throughout the coding and exon–intron junction regions (Figure 6). These mutations show some ethnic variability. In Caucasian patients (216 alleles), 1042delCT (67 alleles) and G339C (33 alleles) are the prevalent mutations, accounting for over 40% of all cases. In Japanese patients (26 alleles), W118R is the prevalent mutation, accounting for 41% of the alleles. The number of known alleles in Bedouin (18 alleles), Pakistani (10 alleles), Chinese (two alleles), and Black (two alleles) patients is too low to allow clear statistical conclusions to be drawn, but some mutations are found only in these populations, suggesting that more cultural biases do exist. For instance, R28H is only present in Bedouin patients, the 936insA and IVS8+2del4 mutations are present only in a Pakistani patient, the G88D mutation is present only in an African patient, and the H191L mutation is present only in a Chinese patient.

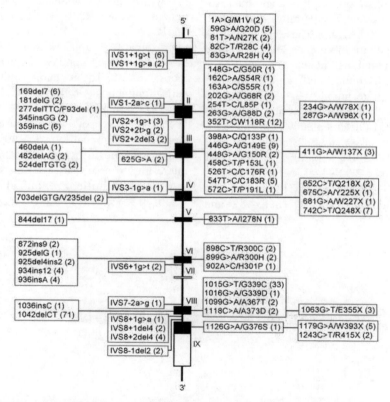

Figure 6 Mutations identified in the *G6PT* gene of GSD-Ib patients. The G6PT gene is shown as a line diagram with the nine exons marked as boxes I to IX. Black boxes represent coding regions, and white boxes the 5′ and 3′ untranslated regions of the G6PT transcript. The positions of all known mutations are listed from left to right as insertion/deletion, splicing, missense, and nonsense mutations. Numbers in parentheses represent alleles identified.

STRUCTURE–FUNCTION ANALYSIS OF G6PT

Human G6PT contains 10 helical transmembrane domains (Pan et al. 1999) separating the five loops facing into the lumen of the ER from the four loops facing into the cytoplasm (Figure 3). Two of the five luminal loops, loops 1 and 3, are of significant size, being 51 and 25 amino acids long, respectively, but the remaining three loops are small, ranging from 5 to 8 amino acids in length. The cytoplasmic loops are larger, ranging from 17 to 32 amino acids long.

To date, 32 of the mutations identified in GSD-Ib patients have been characterized functionally in transient expression assays and shown to abolish, or greatly reduce, microsomal G6P uptake activity (Hiraiwa et al. 1999; Chen et al. 2000, 2002). These 32 mutations can be grouped into four categories: signature motif (two mutations), helical (15 mutations), nonhelical (13 mutations), and N-terminal (MIV) or C-terminal (R415X) domain (Figure 7). The signature motif mutations Q133P and G149E completely abolish microsomal G6P uptake activity. Similarly, nine of the 15 helical G6PT mutations, G20D (helix 1), L85P (helix 2), F93del (helix 2), G150R (helix 3), C176R (helix 4), C183R (helix 4), V235del (helix 5), G339D (helix 8), and A373D (helix 9), lack microsomal G6P uptake activity. The remaining G88D (helix 1, 2%), P153L (helix 3, 9%), I278N (helix 5, 10%), G339C (helix 8, 5%), A367T (helix 9, 23%), and G376S (helix 9, 6%) mutations retain the indicated percentage of wild-type (WT) activity (Figure 7). Moreover, G20D (helix 1), F93del (helix 2), and I278N (helix 6) destablize the G6PT, and mutations W393X, E401X and T408X, which delete or disrupt the structure of helix 10, also abolish microsomal G6P uptake activity and result in little or no protein accumulation in COS-1 cells (Chen et al. 2000). Together, these data suggest that the structural integrity of each of the 10 transmembrane helices is critical to the protein.

Eight of the 12 missense nonhelical G6PT mutants (N27K, R28C, R28H, G50R, S54R, S55R, W118R, and P191L) are devoid of microsomal G6P uptake activity. Six of these lie in luminal loop 1 and two lie in cytoplasmic loops 1 and 2. The remaining missense nonhelical mutations (G68R, R300C, R300H, and H301P) retain about 8%, 5%, 7%, and 24% of WT activity, respectively (Figure 7). While G68R lies in luminal loop 1, R300C, R300H, and H301P lie in cytoplasmic loop 3. It is interesting to note that seven missense mutations were identified in luminal loop 1 and six of these (N27K, R28C, R28H, G50R,

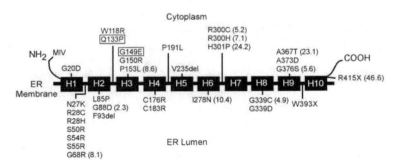

Figure 7   A summary of mutations of G6PT that affect microsomal G6P uptake activity. The G6PT protein is shown embedded in the ER membrane. The 10 helical transmembrane domains of G6PT are marked as boxes H1 to H10. Protein mutations that have been shown to obliterate G6PT activity are listed. Mutants that retain some residual activity are listed with the percentage of WT G6PT transport activity retained noted in parentheses. The two signature motif mutations Q133P and G149E are boxed.

S54R, and S55R) destroy activity. Therefore luminal loop 1 is critical for the microsomal G6P uptake activity.

The N-terminal mutation MIV (Chen *et al.* 2002), which encodes a G6PT lacking the entire N-terminal domain (amino acids 1–7) and part of helix 1 (amino acids 8–16), is devoid of microsomal G6P transport activity. The C-terminal mutation R415X (Chen *et al.* 2000) lacks the entire 15 amino acid C-terminal tail (amino acids 415–429) but retains 47% of WT activity (Figure 7). A deletion analysis of the C-terminal residues showed that the last 12 amino acids of the 15 residue C-terminal domain are not required for the G6P transport activity (Chen *et al.* 2000). Interestingly, deletion of the ER membrane protein retention motif, which lies in the C-terminal domain, does not abolish the G6PT activity either.

The G6PT and vG6PT proteins differ in the size of luminal loop 4, with 8 and 30 amino acids, respectively. Since both vG6PT and G6PT are equally active in G6P uptake (Lin *et al.* 2000), luminal loop 4 does not appear to be critical to microsomal G6P transport activity.

GENOTYPE–PHENOTYPE RELATIONSHIP IN GSD-IB

GSD-Ib is an autosomal recessive disorder. GSD-Ib carriers have 50% of normal G6PT activity and do not manifest any of the symptoms associated with G6PT deficiency. However, a correlation has emerged between the nature of the G6PT mutation and the severity of the disease. There is a report of two unrelated GSD-Ib patients who have metabolic deficiencies but apparently normal polymorphonuclear leukocyte function (Kure *et al.* 2000). This suggests that the role of G6PT in gluconeogenic tissues is much more sensitive to the level of G6PT expression than in nongluconeogenic tissues. In one of these patients a homozygous mutation (625G > A) introduces a new splice acceptor site into the G6PT transcript. If spliced, this mutation

would delete 244 bp of exon 3 and encode a truncated protein of 129 amino acids, which is predicted to be inactive. However, RT-PCR analysis of leukocyte RNA from this patient detects expression of both the mature and truncated G6PT transcripts (Kure *et al.* 2000). The mild phenotype in this patient suggests that the level of WT protein is sufficient to meet the functional needs of the polymorphonuclear leukocytes, but not the gluconeogenic tissues. The second patient is a compound heterozygote carrying the G339D and R415X mutations. While the G339D mutation is null (Chen *et al.* 2002), the R415X mutation retains 46.6% of the WT microsomal G6P uptake activity (Chen *et al.* 2000). If both alleles contribute equally, this patient would retain 23.3% of normal G6PT activity. Together these cases suggest that gluconeogenic tissues need somewhere between 23% and 50% of WT transporter activity to maintain glucose homeostasis. Assuming that the G6P transport assay reflects the activity required in myeloid cells, 23% of WT G6PT activity is sufficient to maintain the functional needs of the polymorphonuclear leukocytes. The minimal G6PT activity required to maintain normal myeloid functions remains to be elucidated. The growth of the database of residual G6P uptake activity retained by the 32 G6PT codon mutants (Chen *et al.* 2002) should facilitate future genotype–phenotype delineations and identify the threshold of G6PT activity required to prevent myeloid dysfunctions.

THE REGULATION OF *G6PT* GENE EXPRESSION

Appropriate expression of the *G6PT* gene is required for glucose homeostasis and normal myeloid functions. A clear knowledge of the regulation of G6PT expression is essential to the understanding of the metabolic implications of GSD-Ib. DNA elements and transcription factors essential for the expression of the human *G6PT* gene have been identified by transient transfection studies (Hiraiwa and

Chou 2001; Hiraiwa *et al.* 2001). The basal *G6PT* promoter is located within nucleotides -369 to -1, upstream of the translation start site at $+1$. The basal promoter contains a TATA box at nucleotides -141 to -136, and three activation elements at nucleotides -200 to -153 (AE-1), -250 to -201 (AE-2), and -369 to -251 (AE-3). HNF1α, a homeodomain-containing transcriptional activator, binds to its cognate site at nucleotides -165 to -153 within AE-1 and activates *G6PT* transcription (Hiraiwa *et al.* 2001). A glucocorticoid response element (GRE) at nucleotides -224 to -210 within AE-2 (Hiraiwa and Chou 2001) and also activates *G6PT* transcription in response to glucocorticoid hormone.

HNF1α-deficient mice (Pontoglio *et al.* 1996; Lee *et al.* 1998) share clinical phenotypes in common with both GSD-I mice and GSD-I patients, including growth retardation, hepatomegaly, hyperlipidemia, and renal dysfunctions. HNF1α positively regulates G6PT transcription, and hepatic G6PT mRNA levels and microsomal G6P transport activity in HNF1α-deficient mice are approximately 20–30% of the levels in WT or heterozygous littermates (Hiraiwa *et al.* 2001). This suggests that metabolic abnormalities in HNF1α-null mice are caused, in part, by G6PT deficiency. However, the 20–30% of normal G6PT activity remaining is able to maintain normal myeloid functions in the HNF1α-null mice.

ANIMAL MODELS FOR GSD-IB

The GSD-Ib disorder has only been described in humans and there are no known animal models for the disease. A G6PT-deficient mouse model of GSD-Ib was recently generated by gene targeting (Chen *et al.* 2003). The G6PT-deficient mice manifest all known metabolic and myeloid dysfunctions in human GSD-Ib, including impaired glucose homeostasis, neutropenia, and impaired respiratory burst of neutrophils. Using the mouse model, it was shown that G6PT deficiency perturbs growth of the hematopoietic organs and hematopoiesis, decreases local chemokine production, and impairs neutrophil recruitment during inflammation. Moreover, the defective neutrophil chemotaxis *in vitro* is exaggerated *in vivo* by the defective chemokine production. Accordingly, myeloid dysfunction in GSD-Ib is caused by transient neutropenia, resistance of neutrophils to chemotactic factors, and reduced production of neutrophil-specific chemokines at sites of inflammation. The G6PT-deficient mouse model should offer the opportunity of examining and clarifying many of the unresolved questions of GSD-Ib outlined above.

SUMMARY

The glucose-6-phosphate transporter (G6PT) is a 10-helical-domain transmembrane protein. Its main role is to translocate glucose-6-phosphate (G6P) from the cytoplasm of a cell to the lumen of the endoplasmic reticulum (ER), where it is hydrolyzed to glucose by glucose-6-phosphatase (G6Pase). Together, G6PT and G6Pase maintain glucose homeostasis. A loss of transport activity deprives G6Pase of its substrate and causes the autosomal recessive disorder glycogen storage disease type Ib (GSD-Ib, MIM232220). GSD-Ib patients manifest the metabolic characteristics of G6Pase deficiency but also exhibit neutropenia and myeloid dysfunctions. Current treatment of GSD-Ib consists of a dietary therapy to correct the G6Pase deficiency, augmented with granulocyte colony-stimulating factor therapy to restore myeloid function. While the combined therapies improve the patients' metabolic and myeloid functions, the underlying pathological processes remain untreated. As a result, long-term complications develop in adult patients. A single 5.3 kb *G6PT* gene on chromosome 11q23 encodes the transporter. Sixty-nine distinct G6PT mutations have been identified in GSD-Ib patients. Some mutations completely abolish microsomal G6P transport function, while others leave a residual transport activity which may explain the varying severity and phenotype of the GSD-Ib disorder. The other roles of G6PT are less clearly defined. Myeloid cells require a functional G6PT for efficient glucose transport, Ca^{2+} mobilization, respiratory burst, chemotaxis, and phagocytosis. Whether these are linked to G6P metabolism or reflect a completely different set of activities of G6PT remain to be elucidated. A recently developed animal model of GSD-Ib now offers opportunities to address such issues and to develop novel therapies for the disorder.

REFERENCES

Annabi, B., Hiraiwa, H., Mansfield, B.C., Lei, K.-J., Ubagai, T., Polymeropoulos, M.H., et al. (1998). The Gene for Glycogen Storage Disease Type 1b Maps to Chromosome 11q23. Am. J. Hum. Genet., 62, 400–405.

Arion, W.J., Lange, A.J., Walls, H.E., and Ballas, L.M. (1980). Evidence of the Participation of Independent Translocases for Phosphate and Glucose-6-Phosphate in the Microsomal Glucose-6-Phosphatase System. J. Biol. Chem., 255, 10396–10406.

Arion, W.J., Canfield, W.K., Ramos, F.C., Su, M.L., Burger, H.J., Hemmerle, H., et al. (1998). Chlorogenic Acid Analogue S 3483: A Potent Competitive Inhibitor of the Hepatic and Renal Glucose-6-Phosphatase Systems. Arch. Biochem. Biophys., 351, 279–285.

Bashan, N., Hagai, Y., Potashnik, R., and Moses, S.W. (1986). Impaired Carbohydrate Metabolism of Polymorphonuclear Leukocytes in Glycogen Storage Disease Type Ib. J. Clin. Invest., 81, 1317–1322.

Beaudet, A.L., Anderson, D.C., Michels, V.V., Arion, W.J., and Lange, A.J. (1980). Neutropenia and Impaired Neutrophil Migration in Type 1B Glycogen Storage Disease. J. Pediatr., 97, 906–910.

Calderwood, S., Kilpatrick, L., Douglas, S.D., Freedman, M., Smith-Whitley, K., Rolland, M., et al. (2001). Recombinant Human Granulocyte Colony-Stimulating Factor Therapy for Patients with Neutropenia and/or Neutrophil Dysfunction Secondary to Glycogen Storage Disease Type 1b. Blood, 15, 376–382.

Chen, L.-Y., Lin, B., Pan, C.-J., Hiraiwa, H., and Chou, J.Y. (2000). Structural Requirements for the Stability and Microsomal Transport Activity of the Human Glucose-6-Phosphate Transporter. J. Biol. Chem., 275, 34280–34286.

Chen, L.-Y., Pan, C.-J., Shieh, J.-J., and Chou, J.Y. (2002). Structure–Function Analysis of the Glucose 6-Phosphate Transporter Deficient in Glycogen Storage Disease Type Ib. Hum. Mol. Genet., 11, 3199–3207.

Chen, L.-Y., Shieh, J.-J., Lin, B., Pan, C.-J., Gao, J.-L., Murphy, P.M., et al. (2003) Impaired glucose homeostasis, neutrophil rafficking and function in mice lacking the glucose-6-phosphate transporter. Hum. Mol. Genet., 12, 2547–2558.

Chen, P.Y., Csutora, P., Veyna-Burke, N.A., and Marchase, R.B. (1998). Glucose-6-Phosphate and Ca^{2+} Sequestration are Mutually Enhanced in Microsomes from Liver, Brain, and Heart. Diabetes, 47, 874–881.

Chen, Y.T., Cornblath, M., and Sidbury, J.B. (1984). Cornstarch Therapy in Type I Glycogen Storage Disease. N. Engl. J. Med., 310, 171–175.

Chou, J.Y. and Mansfield, B.C. (1999). Molecular Genetics of Type 1 Glycogen Storage Diseases. Trends Endocrinol. Metab., 10, 104–113.

Chou, J.Y., Matern, D., Mansfield, B.C., and Chen, Y.-T. (2002). Type I Glycogen Storage Diseases: Disorders of the Glucose-6-Phosphatase Complex. Curr. Mol. Med., 2, 121–143.

Cori, G.T. and Cori, C.F. (1952). Glucose-6-Phosphatase of the Liver in Glycogen Storage Disease. J. Biol. Chem., 199, 661–667.

Garty, B., Douglas, S., and Danon, Y.L. (1996). Immune Deficiency in Glycogen Storage Disease Type 1b. Isr. J. Med. Sci., 32, 1276–1281.

Gerin, I., Veiga-da-Cunha, M., Achouri, Y., Collet, J.-F., and Van Schaftingen, E. (1997). Sequence of a Putative Glucose-6-Phosphate Translocase, Mutated in Glycogen Storage Disease Type 1b. FEBS Lett., 419, 235–238.

Gerin, I., Veiga-da-Cunha, M., Noel, G., and Van Schaftingen, E. (1999). Structure of the Gene Mutated in Glycogen Storage Disease Type Ib. Gene, 227, 189–195.

Gerin, I. and Van Schaftingen, E. (2002). Evidence for Glucose-6-Phosphate Transport in Rat Liver Microsomes. FEBS Lett., 517, 257–260.

Ghosh, A., Shieh, J.-J., Pan, C.-J., Sun, M.-S., and Chou, J.Y. (2002). The catalytic center of glucose-6-phosphatase: His^{176} is the nucleophile forming the phosphohistidine-enzyme intermediate during catalysis. J. Biol. Chem., 277, 32837–32842.

Gitzelmann, R. and Bosshard, N.U. (1993). Defective Neutrophil and Monocyte Functions in Glycogen Storage Disease Type Ib: A Literature Review. Eur. J. Pediatr., 152 (Suppl. 1), S33–S38.

Greene, H.L., Slonim, A.E., O'Neill, J.A., Jr, and Burr, I.M. (1976). Continuous Nocturnal Intragastric Feeding for Management of Type 1 Glycogen-Storage Disease. N. Engl. J. Med., 294, 423–425.

Hemmerle, H., Burger, H.J., Below, P., Schubert, G., Rippel, R., Schindler, P.W., et al. (1997). Chlorogenic Acid and Synthetic Chlorogenic Acid Derivatives: Novel Inhibitors of Hepatic Glucose-6-Phosphate Translocase. J. Med. Chem., 40, 137–145.

Hiraiwa, H. and Chou, J.Y. (2001). Glucocorticoids Activate Transcription of the Gene for Glucose-6-Phosphate Transporter, Deficient in Glycogen Storage Disease Type 1b. DNA Cell Biol., 20, 447–445.

Hiraiwa, H., Pan, C.-J., Lin, B., Moses, S.W., and Chou, J.Y. (1999). Inactivation of the Glucose-6-Phosphate Transporter Causes Glycogen Storage Disease Type 1b. J. Biol. Chem., 274, 5532–5536.

Hiraiwa, H., Pan, C.-J., Lin, B., Akiyama, T.E., Gonzalez, F.J., and Chou, J.Y. (2001). A Molecular Link Between the Common Phenotypes of Type 1 Glycogen Storage Disease and HNF1α-Null Mice. *J. Biol. Chem.*, **276**, 7963–7967.

Hori, S.H. and Takahashi, T. (1977). Latency of Microsomal Hexose-6-Phosphate Dehydrogenase Activity. *Biochim. Biophys. Acta*, **496**, 1–11.

Kilpatrick, L., Garty, B.-Z., Lundquist, K.F., Hunter, K., Stanley, C.A., Baker, L., *et al.* (1990). Impaired Metabolic Function and Signaling Defects in Phagocytic Cells in Glycogen Storage Disease Type 1b. *J. Clin. Invest.*, **86**, 196–202.

Kure, S., Hou, D.-C., Suzuki, Y., Yamagishi, A., Hiratsuka, M., Fukuda, T., *et al.* (2000). Glycogen Storage Disease Type 1b without Neutropenia. *J. Pediatr.*, **137**, 253–256.

Lachaux, A., Boillot, O., Stamm, D., Canterino, I., Dumontet, C., Regnier, F., *et al.* (1993). Treatment with Lenograstim (Glycosylated Recombinant Human Granulocyte Colony-Stimulating Factor) and Orthotopic Liver Transplantation for Glycogen Storage Disease Type Ib. *J. Pediatr.*, **123**, 1005–1008.

Lange, A.J., Arion, W.J., and Beaudet, A.L. (1980). Type 1b Glycogen Storage Disease Is Caused by a Defect in the Glucose-6-Phosphate Translocase of the Microsomal Glucose-6-Phosphatase System. *J. Biol. Chem.*, **255**, 8381–8384.

Lee, Y.-H., Sauer, B., and Gonzalez, F.J. (1998). Laron Dwarfism and Non-Insulin-Dependent Diabetes Mellitus in the *Hnf-1α* Knockout Mouse. *Mol. Cell. Biol.*, **18**, 3059–3068.

Lei, K.-J., Shelly, L.L., Pan, C.-J., Sidbury, J.B., and Chou, J.Y. (1993). Mutations in the Glucose-6-Phosphatase Gene that Cause Glycogen Storage Disease Type 1a. *Science*, **262**, 580–583.

Lei, K.-J., Shelly, L.L., Lin, B., Sidbury, J.B., Chen, Y.-T., Nordlie, R.C., *et al.* (1995). Mutations in the Glucose-6-Phosphatase Gene Are Associated with Glycogen Storage Disease Type 1a and 1aSP but not 1b and 1c. *J. Clin. Invest.*, **95**, 234–240.

Lei, K.-J., Chen, H., Pan, C.-J., Ward, J.M., Mosinger, B., Lee, E.J., *et al.* (1996). Glucose-6-Phosphatase Dependent Substrate Transport in the Glycogen Storage Disease Type 1a Mouse. *Nat. Genet.*, **13**, 203–209.

Lin, B., Hiraiwa, H., Annabi, B., Pan, C.-J., and Chou, J.Y. (1998). Cloning and Characterization of cDNAs Encoding a Candidate Glycogen Storage Disease Type 1b Protein in Rodents. *J. Biol. Chem.*, **273**, 31656–31670.

Lin, B., Hiraiwa, H., Pan, C.-J., Nordlie, R.C., and Chou, J.Y. (1999). Type 1c Glycogen Storage Disease is not Caused by Mutations in the Glucose-6-Phosphate Transporter Gene. *Hum. Genet.*, **105**, 515–517.

Lin, B., Pan, C.-J., and Chou, J.Y. (2000). Human Variant Glucose-6-Phosphate Transporter Is Active in Microsomal Transport. *Hum. Genet.*, **107**, 526–529.

Martinez-Olmos, M.A., Lopez-Sanroman, A., Martin-Vaquero, P., Molina-Perez, E., Barcena, R., Vicente, E., *et al.* (2001). Liver Transplantation for Type Ib

Glycogenosis with Reversal of Cyclic Neutropenia. *Clin. Nutr.*, **20**, 375–377.

Matern, D., Starzl, T.E., Arnaout, W., Barnard, J., Bynon, J.S., Dhawan, A., *et al.* (1999). Liver Transplantation for Glycogen Storage Disease Types I, III, and IV. *Eur. J. Pediatr.*, **158**(Suppl. 2), S43–S48.

McCawley, L.J., Korchak, H.M., Douglas, S.D., Campbell, D.E., Thornton, P.S., Stanley, C.A., *et al.* (1994). In Vitro and In Vivo Effects of Granulocyte Colony-Stimulating Factor on Neutrophils in Glycogen Storage Disease Type 1B: Granulocyte Colony-Stimulating Factor Therapy Corrects the Neutropenia and the Defects in Respiratory Burst Activity and Ca^{2+} Mobilization. *Pediatr. Res.*, **35**, 84–90.

Middleditch, C., Clottes, E., and Burchell, A. (1998). A Different Isoform of the Transport Protein Mutated in the Glycogen Storage Disease 1b is Expressed in Brain. *FEBS Lett.*, **433**, 31–36.

Narisawa, K., Igarashi, Y., Otomo, H., and Tada, K. (1978). A New Variant of Glycogen Storage Disease Type I Probably due to a Defect in the Glucose-6-Phosphate Transport System. *Biochem. Biophys. Res. Commun.*, **83**, 1360–1364.

Nordlie, R.C. and Sukalski, K.A. (1985). Multifunctional Glucose-6-Phosphatase: A Critical Review. In A.N. Martonosi (ed.), *The Enzymes of Biological Membranes*, New York: Plenum Press, pp. 349–398.

Nordlie, R.C., Sukalski, K.A., Munoz, J.M., and Baldwin, J.J. (1983). Type 1c, a Novel Glycogenosis. *J. Biol. Chem.*, **258**, 9739–9744.

Pan, C.-J., Lin, B., and Chou, J.Y. (1999). Transmembrane Topology of Human Glucose-6-Phosphate Transporter. *J. Biol. Chem.*, **274**, 13865–13869.

Pontoglio, M., Barra, J., Hadchouel, M., Doyen, A., Kress, C., Bach, J.P., Babinet, C., *et al.* (1996). Hepatocyte Nuclear Factor 1 Inactivation Results in Hepatic Dysfunction, Phenylketonuria, and Renal Fanconi Syndrome. *Cell*, **84**, 575–585.

Roe, T.F., Thomas, D.W., Gilsanz, V., Isaacs, H., and Atkinson, J.B. (1986). Inflammatory Bowel Disease in Glycogen Storage Disease Type Ib. *J. Pediatr.*, **109**, 55–59.

Roe, T.F., Coates, T.D., Thomas, D.W., Miller, J.H., and Gilsanz, V. (1992). Brief Report: Treatment of Chronic Inflammatory Bowel Disease in Glycogen Storage Disease Type Ib with Colony-Stimulating Factors. *N. Engl. J. Med.*, **326**, 1666–1669.

Schroten, H., Roesler, J., Breidenbach, T., Wendel, U., Elsner, J., Schweitzer, S., *et al.* (1991). Granulocyte and Granulocyte-Macrophage Colony-Stimulating Factors for Treatment of Neutropenia in Glycogen Storage Disease Type Ib. *J. Pediatr.*, **119**, 748–754.

Schwoppe, C., Winkler, H.H., and Neuhaus, H.E. (2002). Properties of the Glucose-6-Phosphate Transporter from *Chlamydia pneumoniae* (HPTcp) and the Glucose-6-Phosphate Sensor from *Escherichia coli* (UhpC). *J. Bacteriol.*, **184**, 2108–2115.

Senior, B. and Loridan, L. (1968). Studies of Liver Glycogenoses, with Particular Reference to the Metabolism of Intravenously Administered Glycerol. *N. Engl. J. Med.*, **279**, 958–965.

Singh, J., Nordlie, R.C., and Jorgenson, R.A. (1981). Vanadate: A Potent Inhibitor of Multifunctional Glucose-6-Phosphatase. *Biochim. Biophys. Acta*, **678**, 477–482.

Synderman, R. and Uhing, R.J. (1988). Phagocytic Cells: Stimulus-Response Coupling Mechanisms. In G.I. Gallin, I.M. Goldstain, and R. Snyderman (eds), *Inflammation: Basic Principles and Clinical Correlates*, New York: Raven Press, pp. 309–323.

Veiga-da-Cunha, M., Gerin, I., Chen, Y.-T., de Barsy, T., de Lonlay, P., Dionisi-Vici, C., *et al.* (1998). A Gene on Chromosome 11q23 Coding for a Putative Glucose-6-Phosphate Translocase Is Mutated in Glycogen Storage Disease Type Ib and Ic. *Am. J. Hum. Genet.*, **63**, 976–983.

Veiga-da-Cunha, M., Gerin, I., Chen, Y.-T., Lee, P.J., Leonard, J.V., Maire., I., *et al.* (1999). The Putative Glucose-6-Phosphate Translocase Is Mutated in Essentially All Cases of Glycogen Storage Disease Type I Non-a. *Eur. J. Hum. Genet.*, **7**, 717–723.

Verhoeven, A.J., Visser, G., van Zwieten, R., Gruszczynska, B., Tien Poll-The, D.W., and Smit, G.P. (1999). A Convenient Diagnostic Function Test of Peripheral Blood Neutrophils in Glycogen Storage Disease Type Ib. *Pediatr. Res.*, **45**, 881–885.

Visser, G., Rake, J.P., Fernandes, J., Labrune, P., Leonard, J.V., Moses, S., *et al.* (2002). Neutropenia, Neutrophil Dysfunction, and Inflammatory Bowel Disease in Glycogen Storage Disease Type Ib: Results of the European Study on Glycogen Storage Disease Type I. *J. Pediatr.*, **137**, 187–191.

von Gierke, E. (1929). Hepato-nephro-megalia Glycogenica (Glykogenspeicher-krankheit der Leber und Nieren). *Beitr. Pathol. Anat.*, **82**, 497–513.

Wendel, U., Schroten, H., Burdach, S., and Wahn, V. (1993). Glycogen Storage Disease Type Ib: Infectious Complications and Measures for Prevention. *Eur. J. Pediatr.*, **152**(Suppl. 1), S49–S51.

14

JOSEP CHILLARÓN*, JOAN BERTRAN**,
AND MANUEL PALACÍN***

Heteromeric amino acid transporters: cystinuria and lysinuric protein intolerance

INTRODUCTION

Six families of plasma membrane amino acid transporters have been described in mammals, one of which has a heteromeric structure (Palacín *et al.* 1998; Chillarón *et al.* 2001). These heteromeric amino acid transporters (HATs) are composed of a heavy subunit and a light subunit, linked by a disulfide bridge (Table 1, Figure 1). Two homologous heavy subunits (HSHATs) are known, rBAT (*r*elated to system $b^{0,+}$ *a*mino acid *t*ransport) and 4F2hc (heavy chain of the surface antigen 4F2, also referred to as CD98). Nine light subunits (LSHATs) have been identified. Six of them are partners of 4F2hc (LAT-1, LAT-2, y^+LAT-1, y^+LAT-2, asc-1, and xCT), one assembles with rBAT ($b^{0,+}$AT), and two (asc-2 and AGT-1) seem to interact with as yet unknown heavy subunits (Kanai *et al.* 1998; Mastroberardino *et al.* 1998; Torrents *et al.* 1998; Feliubadalo

*

**

*** Dept. Bioquimica i Biologia Molecular, Universidad Barcelona, Avenida Diagonal 645, 08028 Barcelona, Spain

et al. 1999; Pineda *et al.* 1999; Rossier *et al.* 1999; Sato *et al.* 1999; Bröer *et al.* 2000; Fukasawa *et al.* 2000; Chairoungdua *et al.* 2000; Matsuo *et al.* 2002).

The general features of HATs are as follows.

1. HSHATs are type II membrane N-glycoproteins with a single transmembrane domain, an intracellular N-terminus, and an extracellular C-terminus significantly homologous to bacterial α-glucosidases (Figure 1).
2. LSHATs (molecular weight around 50 kDa) are not glycosylated, are highly hydrophobic, and have 12 putative transmembrane (TM) domains (Figure 1). This highly hydrophobic character results in an anomalous high mobility in SDS–PAGE (35–40 kDa).
3. LSHATs are linked to the corresponding HSHAT by a disulfide bridge (Figure 1). For this reason, HATs are also named glycoprotein-associated amino acid transporters (gpaATs) (Verrey *et al.* 1999). The intervening cysteine residues are located in the putative extracellular loop II of LSHATs and a few residues apart from the TM of HSHATs (Figure 1). Evidence for this disulfide bridge has been obtained in both heterologous expression systems and tissues (Mannion *et al.* 1998; Fernández *et al.* 2002).

Table 1 LSHAT with the corresponding HSHAT form heteromeric functional amino acid transporters at the plasma membrane

Heavy chain (HSHAT)	Light chain (LSHAT)	HUGO nomenclature	Amino acid transport system	Chromosome	Inherited disease
4F2hc		SLC3A2		11q13	
	y^+LAT-1	SLC7A7	y^+L	14q11.2	LPI
	y^+LAT-2	SLC7A6	y^+L	16q22.1	
	LAT-1	SLC7A5	L	16q24.3	
	LAT-2	SLC7A8	L	14q11.2	
	asc-1	SLC7A10	asc	19q12-13	
	xCT	SLC7A11	x_c^-	4q28-q32	
rBAT		SLC3A1		2p16.3-21	Cystinuria type A
	$b^{0,+}$AT	SLC7A9	$b^{0,+}$	19q12-13	Cystinuria type B

Description of the tissue distribution and the transport function of LSHAT in mammalian species has been reviewed elsewhere (Verrey et al. 1999, 2000; Deve's and Boyd 2000); HUGO, Human Genome Organization.

4. The LSHATs need coexpression with the corresponding HSHAT to reach the plasma membrane (Mastroberardino et al. 1998; Pfeiffer et al. 1998; Torrents et al. 1998; Feliubadalo et al. 1999; Pineda et al. 1999; Sato et al. 1999; Bröer et al. 2000).

5. LSHATs confer specific amino acid transport activity to the heteromeric complex (Table 1). All these transport activities are, in general, tightly coupled amino acid antiporters (Chillarón et al. 1996, 2001; Meier et al. 2002; Reig et al. 2002).

Excellent reviews on HATs have recently been published (Palacín et al. 1998, 2000; Verrey et al. 1999, 2000; Devés and Boyd 2000; Chillarón et al. 2001). Here we will focus on the HAT subunits, which are involved in primary inherited human aminoacidurias. These are rBAT and $b^{0,+}$AT (encoded by *SLC3A1* and *SLC7A9*), the mutations of which cause type I and non-type I cystinuria, respectively (Calonge et al. 1994; Feliubadalo et al. 1999), and y^+LAT1 (encoded by *SLC7A7*), the mutations of which lead to lysinuric protein intolerance (Torrents et al. 1999) (Table 1). rBAT and $b^{0,+}$AT induce system $b^{0,+}$. This system acts through a tertiary active mechanism for renal reabsorption and intestinal

absorption of dibasic amino acids and cystine by exchange with neutral amino acids (Busch et al. 1994; Chillarón et al. 1996). y^+LAT1 (with 4F2hc) specifies system y^+L that mediates the efflux of dibasic amino acids by exchange with neutral amino acids plus sodium (Bertran et al. 1992a; Chillarón et al. 1996; Torrents et al. 1998).

THE HIGH-AFFINITY REABSORPTION SYSTEM OF CYSTINE

In the last decade, the physiology of kidney reabsorption of amino acids and other solutes entered the molecular field with the cloning of most of the transporter proteins involved (Palacín et al. 1998; Saier et al. 1999). In particular, cystine reabsorption at the level of the brush border membrane of epithelial cells of the proximal tubule has been the object of detailed physiological studies using a variety of methodologies (reviewed by Silbernagl 1988). In brush border membranes, cystine transport occurs via two different transport systems: one of low affinity unshared with dibasic amino acids and at least partially sodium dependent, and the other with high affinity shared with dibasic amino acids

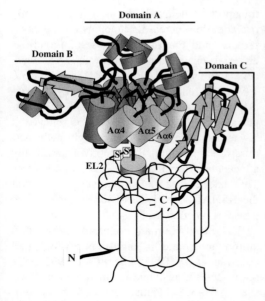

Figure 1 Schematic representation of the heteromeric amino acid transporters (HATs). The heavy subunit (gray) is linked by a disulfide bridge to the corresponding light subunit (white). The 3D-structure model for the glucosidase-like extracellular domain of rBAT is shown, based on the 3D structure of O1,6G (Watanabe *et al.* 1997; Chillarón *et al.* 2001) (see text). Sequence homology with O1,6G is poor for rBAT in the region between Aα4 and Aα7 depicted in light gray (Aα4, Aβ5, Aα5, Aβ6, Aα6, and Aβ7), and therefore this region has been tentatively assigned combining secondary structure predictions and threading techniques (Chillarón *et al.* 2001). The size of the α-helix (cylinders) and the β-strands (arrows) is proportional to the number of amino acid residues. The size of the glucosidase-like domain is comparable to the putative 12-transmembrane domain structure of the light subunit. The position and packing of the light subunit helix and the position of the glucosidase-like domain relative to the polytopic light subunit are arbitrary (with the exception of the steric restrictions that the disulfide bridge may impose on the overall structure). Loops between transmembrane domains and N- and C-terminal segments are not drawn to scale.

(reviewed by Segal and Thier 1995). The identity of the low-affinity system remains elusive and its physiological importance has been strongly questioned (Chairoungdua *et al.* 1999;

Pfeiffer *et al.* 1999a). The high-affinity system has been identified as the heterodimeric complex rBAT-b$^{0,+}$AT (Feliubadalo *et al.* 1999; Fernández *et al.* 2002). Several lines of evidence have converged in support of this.

1. Mutations in the human genes rBAT (*SLC3A1*) or b$^{0,+}$AT (*SLC7A9*) cause cystinuria (Calonge *et al.* 1994; Feliubadalo *et al.* 1999).
2. rBAT knockdown, generated with antisense technology in polarized opossum kidney (OK) cells, impairs apical transport of cystine and dibasic amino acids (Mora *et al.* 1996).
3. b$^{0,+}$AT deficient (*slc7a9* knockout) mice recapitulate human cystinuria.
4. rBAT and b$^{0,+}$AT are localized at the brush border membrane of proximal tubule epithelial cells (Furriols *et al.* 1993; Chairoungdua *et al.* 1999; Pfeiffer *et al.* 1999a), OK cells, and MDCK cells (upon stable transfection) (Mora *et al.* 1996; Bauch *et al.* 2003).
5. rBAT and b$^{0,+}$AT form a disulfide-linked heterodimer when expressed in heterologous expression systems. Both proteins must be in the same cell in order to reach the plasma membrane [the apical membrane in polarized cells (Bauch *et al.* 2003)] and elicit transport activity (Feliubadalo 1999; Chairoungdua *et al.* 1999; Pfeiffer *et al.* 1999a; Font *et al.* 2001).
6. All renal b$^{0,+}$AT protein is found as a disulfide-linked heterodimer with the rBAT protein. The expression of the heterodimer decreases from the S1 to the S3 segment of the proximal tubule, where the bulk of cystine reabsorption occurs (Furriols *et al.* 1993; Fernández *et al.* 2003).
7. Coexpression of rBAT and b$^{0,+}$AT elicits system b$^{0,+}$ (Pfeiffer *et al.* 1999a; Font *et al.* 2001; Reig *et al.* 2002; Bauch *et al.* 2003). This system, originally described in mouse blastocysts (Van Winkle 1998), is a tightly coupled exchanger with a strong preference for cystine and dibasic amino acids influx and neutral amino acid efflux

(Chillarón *et al.* 1996; Mora *et al.* 1996; Reig *et al.* 2002; Bauch *et al.* 2003). This activity fully explains the cystinuric phenotype (see section below on pathophysiology of cystinuria).

8. All naturally occurring rBAT and $b^{0,+}$AT mutations analyzed so far show a reduced transport activity due to either trafficking defects (Chillarón *et al.* 1997) or impaired catalytic transport activity (Reig *et al.* 2002).

rBAT

rBAT was cloned by functional expression of kidney amino acid transport activity in oocytes (Bertran *et al.* 1992b; Tate *et al.* 1992; Wells and Hediger 1992), where it elicits sodium-independent high affinity transport of dibasic neutral amino acids and cystine. Its mRNA is mainly expressed in kidney and intestine, and the protein has been localized to the microvilli of renal proximal tubule and small intestinal epithelial cells (Furriols *et al.* 1993; Pickel *et al.* 1993). The human and rat mRNA is barely expressed at birth and increases to adulthood (Furriols *et al.* 1993; Palacín *et al.* 2001b).

The only well-documented function assigned to rBAT is as an escort protein for $b^{0,+}$AT (Feliubadalo *et al.* 1999; Pfeiffer *et al.* 1999a). When expressed alone, rBAT has a short half-life and is localized in a pre-Golgi compartment, as shown by metabolic labeling experiments and immunolocalization and Endo-H sensitivity assays (Reig *et al.* 2002; Bauch *et al.* 2003). Coexpression of $b^{0,+}$AT stabilizes rBAT and promotes trafficking across the secretory pathway (acquisition of Endo-H resistance and plasma membrane localization) (Reig *et al.* 2002; Bauch *et al.* 2003). All cystinuria-associated mutations in *SLC3A1* are located in the extracellular domain (only one in the transmembrane segment) of rBAT (reviewed by Chillarón *et al.* 2001). The few of them that have been functionally studied in heterologous cell systems seem to display a folding defect that at least partially impairs trafficking to the plasma membrane; in oocytes, the time course of

transport is delayed, the defect is partially overcome by increasing amounts of expressed protein, and the Endo-H-resistant protein is either absent or present in lower levels when compared with wild-type rBAT (Chillarón *et al.* 1997). In mammalian cells, the cystinuria-associated mutation studied (R365W) also displays a trafficking defect, which is relieved at low temperature (33°C). Interestingly, at the permissive temperature rBAT(R365W) specifically affects the efflux of arginine, but not of leucine. This is the first indication suggesting that the HSHATs contribute to some extent to the transport function itself.

The rBAT human counterpart has 685 amino acids and is highly N-glycosylated (Bertran *et al.* 1993; Lee *et al.* 1993). Six sequences are available for mammalian rBAT, sharing 69–89% identity. Two different topologies have been proposed for rBAT. On the basis of antibody and protease accessibility studies a 4-TM helix model with the N- and C-termini located intracellularly was proposed (Mosckovitz *et al.* 1994). However, the model of a type II membrane glycoprotein with a single TM segment, originally proposed (Bertran *et al.* 1992b; Wells and Hediger 1992), has recently gained favor from the following data: (a) no similar 4-TM topology can be predicted for the homologous HSHAT subunit 4F2hc; (b) Fenczik *et al.* (2001) have shown unambiguously that the 4F2hc N- and C-termini are intracellular and extracellular, respectively; (c) from residue 117 to residue 651, the human rBAT protein (the residue number varies slightly for the mammalian homologs) displays an unexpectedly high homology with insect maltase and maltase-like precursors (35–40% identity), and with bacterial α-glucosidases (30% identity) (Bertran *et al.* 1992b; Wells and Hediger 1992; Chillarón *et al.* 2001). The homology covers more than 90% of the sequence from the TM segment. In fact, a structural model for the putative extracellular rBAT domain has been constructed based on this homology (Chillarón *et al.* 2001; Bröer and Wagner 2002).

The α-amylase family comprises a large group of enzymes with different specificities and is known as the glycosyl hydrolase family 13 (Janecek *et al.* 1997). Its members have a similar architecture, with a catalytic $(\beta/\alpha)8$-barrel or TIM-barrel (domain A), interrupted by a small calcium-binding subdomain (domain B) protruding between the third β-strand (Aβ3) and the third α-helix (Aα3), and a C-terminal domain (domain C) with an antiparallel β-Greek motif structure. Major differences in amino acid sequence among the α-amylase family members occur within domain B. Janecek *et al.* (1997) clustered the α-amylase members in five groups with more than 50% sequence identity for domain B and suggested that it varies with enzyme specificity. The group defined by *Bacillus cereus* oligo-1,6-glucosidase (O1,6G) also include the rBAT proteins.

The three-dimensional (3D) structure of O1,6G has been refined at 2.0 Å resolution (1uok in PDB) (Watanabe *et al.* 1997) and has been used as a tool for the construction of a putative 3D structure model of the rBAT extracellular domain (Figure 1) (Chillarón *et al.* 2001). Sequence homology between rBAT (and also 4F2hc) and O1,6G starts with two contiguous tryptophan residues a few amino acid residues away from the cysteine residue involved in the formation of the disulfide bridge with the corresponding LSHAT. Most of the secondary structure elements of domain A are relatively easily identified in rBAT, strongly suggesting that the structure of the bulky C-terminal domain of rBAT corresponds to a TIM-barrel. The structural features of domain B, including the C-terminal motif QPDLN, are conserved in the rBAT protein. However, only four out of the six residues involved in either catalytic activity or substrate binding in O1,6G can be identified in rBAT, suggesting that rBAT is not catalytically active as an α-glucosidase (Janecek *et al.* 1997; Chillarón *et al.* 2001). Finally, the C-terminal domain C of α-amylases, which corresponds to eight antiparallel β-strands folded in double Greek motifs, can be also found in rBAT, with the exception of Cβ6 which is not clearly predicted. It is as yet a mystery whether the high homology with glucosidases has any role in amino acid transport itself or the cell biology of the HAT transporters, or has a completely unrelated function.

$b^{0,+}AT$

The LSHAT $b^{0,+}AT$ ($b^{0,+}$ amino acid transporter) was identified by homology screening and quickly found to be the non-type I cystinuria gene (Chairoungdua *et al.* 1999; Feliubadalo *et al.* 1999; Pfeiffer *et al.* 1999a) (see section on cystinuria below). It is expressed mainly in the kidney and the small intestine, and the protein has been localized to the brush border of proximal tubule epithelial cells and at the apical pole of rBAT and $b^{0,+}AT$-transfected polarized MDCK cells (Chairoungdua *et al.* 1999; Pfeiffer *et al.* 1999a; Mizoguchi *et al.* 2001; Bauch *et al.* 2003). The $b^{0,+}AT$ protein is not N-glycosylated and is highly hydrophobic (Chairoungdua *et al.* 1999; Pfeiffer *et al.* 1999a; Mizoguch *et al.* 2001; Bauch *et al.* 2003). On the basis of homology and hydropathy plots, $b^{0,+}AT$ has 12 putative transmembrane segments with intracellular N- and C-termini. It elicits system $b^{0,+}$ activity (see above) when coexpressed with its HSHAT rBAT in mammalian cells (Font *et al.* 2001; Bauch *et al.* 2003). When expressed alone it remains intracellular but, in contrast to rBAT, it is stable (Bauch *et al.* 2003). Little information on functionally or structurally relevant residues is available. Six missense mutations (A70V, V170M, A182T, A354T, G105R, and R333W) have been tested for function after rBAT coexpression in HeLa cells (Font *et al.* 2001). A70V and A182T still have more than 50% activity, whereas the others do not elicit transport activity.

Although $b^{0,+}AT$ (and, in general, the LSHATs) was suspected to be the "catalytic subunit" for amino acid transport of the heterodimeric complex rBAT-$b^{0,+}AT$, no direct proof for this was available. Recently, Reig *et al.* (2002) convincingly demonstrated this by reconstituting a functional $b^{0,+}$ amino acid exchanger from $b^{0,+}AT$ transfected cells in the absence of rBAT. The reconstituted exchanger is asymmetric in its interaction with leucine; the apparent affinity on the extracellular side was in the micromolar range, while it was in the millimolar range on the inside. This asymmetric substrate interaction has also been shown for other LSHATs within the corresponding heterodimeric complex (Meier *et al.* 2002; Reig *et al.* 2002). With the use of the reconstituted system the A354T mutation (and not the A182T mutation), was shown to inactivate transport (Reig *et al.* 2002). Finally, the fact that the intracellularly located $b^{0,+}$AT is functional in the absence of rBAT suggests the following: (a) $b^{0,+}$AT is fully folded; (b) rBAT seems not to act as a chaperone on $b^{0,+}$AT [perhaps $b^{0,+}$AT acts as a chaperone itself for rBAT folding (Reig *et al.* 2002)]; (c) $b^{0,+}$AT may be actively retained intracellularly; (d) rBAT may relieve this retention and/or promote the export of $b^{0,+}$AT to the plasma membrane as a heterodimeric complex with rBAT.

The system $b^{0,+}$ mechanism of transport

The hyperexcretion of dibasic amino acids and cystine in the urine defines the cystinuric phenotype (Palacín *et al.* 2001b). As mutations in rBAT were shown to cause cystinuria, it was unclear how a *facilitated diffusion* carrier for neutral dibasic amino acids and cystine, as originally identified (Bertran *et al.* 1992b; Tate *et al.* 1992; Wells and Hediger 1992) would drive the accumulation of dibasic amino acids and cystine in the epithelial cells of the proximal tubule where reabsorption occurs. The solution came from the seminal observations of two groups (Busch *et al.* 1994; Coady *et al.* 1996), who reported outward positive currents associated with the hetero-exchange of neutral (efflux) and dibasic amino acids (influx) via system $b^{0,+}$ in whole or cut-open oocytes expressing rBAT. Further studies demonstrated that system $b^{0,+}$ acts as a tertiary active transporter mediating the electrogenic exchange of dibasic amino acids and cystine (influx) for neutral amino acids (efflux) with a stoichiometry of $1:1$

(Chillarón *et al.* 1996). This exchange has also been demonstrated in the apical plasma membrane of the proximal tubular cell model of OK cells by rBAT-antisense experiments (Mora *et al.* 1966), in chicken brush border jejunum vesicles (Torras-Llort *et al.* 2001), and in the reconstituted system from HeLa or MDCK cells expressing $b^{0,+}$AT (Reig *et al.* 2002). The exchange mode of transport has been shown for most of the other members of the HAT family and, when tested, the stoichiometry is $1:1$ (Sato *et al.* 1999; Meier *et al.* 2002).

Full accessibility of substrates to both sides of the membrane is required to investigate the exchange mechanism in more detail. This has been accomplished in brush border vesicles from chicken jejunum (Torras-Llort *et al.* 2001). Here, the transport mechanism of the endogenous $b^{0,+}$ system has been analyzed in detail. The results were compatible with a sequential mechanism, implying the formation of a ternary complex (the transporter bound to substrate simultaneously at both sides of the membrane). In contrast, a *ping-pong* mechanism was ruled out. The estimated dissociation constants for extracellular or intracellular substrates suggest that the binding affinity for the external amino acid is higher than for the intracellular substrate, a result consistent with the reconstitution studies (Reig *et al.* 2002) (see above). An ordered or preferential mechanism, in which the free transporter binds (or preferentially binds) first to the external amino acid and then to the internal one, may account for these results. However, a random mechanism cannot be discarded at this point.

The results in chicken brush border jejunum are compatible with a double-exchange pathway with alternating access (Figure 2), similar to the model proposed for the some mitochondrial antiporters (Dierks *et al.* 1988; Bisaccia *et al.* 1996; Schroers *et al.* 1998). These carriers are homodimeric at the membrane (Bisaccia *et al.* 1996; Schroers *et al.* 1998), and each monomer may provide one translocation pathway. It has to be kept in mind, however, that these monomers have six predicted

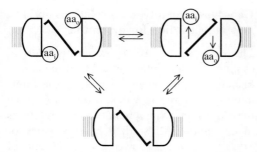

Figure 2 Model of sequential exchange with an asymmetric double-transport pathway with alternating access for the heteromeric amino acid transporters. In this model, the transporter(s) is composed of two transport pathways and is asymmetric (i.e., the efflux and influx pathways are not interchangeable), but a symmetric model may also fit the current data. The scheme is depicted to highlight the formation of a ternary complex aa_o-transporter-aa_i, which is a prerequisite for substrate translocation. The fact that the affinity for the external amino acid is higher than for the internal one suggests a model where the external amino acid binds first to the transporter (Torras-Llort et al. 2001; Reig et al. 2002; Meier et al. 2002).

TM segments (Veenhoff et al. 2002), while LSHATs have 12. If the two-translocation pathway model is confirmed it would be interesting to establish whether the two translocation pores reside in the same LSHAT molecule or if higher oligomers of heterodimeric HATs are required. Clearly, more studies are needed to define the functional and structural unit of the HATs.

CYSTINURIA

Cystinuria (MIM 220100) is an autosomal recessive disorder with an average prevalence of 1 in 7,000 births, ranging between 1 in 2,500 neonates in Israeli Jews of Libyan origin and 1 in 100,000 in Sweden (Segal and Thier 1995). The disease was one of the first inborn errors of metabolism to be described by Sir Archibald Garrod (1908). Cystinuria is caused by the defective influx of cystine and dibasic amino acids (i.e., lysine, arginine, and ornithine)

via system $b^{0,+}$ (see above) across the apical membranes of proximal renal tubular and small intestinal epithelial cells. Because of its poor solubility, cystine precipitates to form calculi in the urinary system that produce obstruction, infection, and ultimately renal insufficiency. Cystinuria represents 1–2% of overall renal lithiasis and 6–8% of renal lithiasis in pediatric patients. Two extensive reviews have dealt with cystinuria before and after the identification of the genes causing the disease (Segal and Thier 1995; Palacín et al. 2001b).

Traditionally, three types of cystinuria have been recognized in humans: type I, type II, and type III (Segal and Thier 1995). This classification now correlates poorly with the findings of molecular analysis and has recently been revised, leading to phenotype I (MIM 220100) and phenotype non-I (MIM 600918) cystinuria (with the latter representing types II and III). Phenotype I and non-I cystinuria are distinguished on the basis of the cystine and dibasic aminoaciduria of the obligate heterozygotes (Segal and Thier 1995; Palacín et al. 2001b): phenotype I heterozygotes are silent, whereas phenotype non-I heterozygotes display a variable degree of urinary hyperexcretion of cystine and dibasic amino acids. Patients with a mixed phenotype, inheriting type I and non-I alleles from either parent, have been also described (Goodyer et al. 1993). Overall, it is believed that phenotypes I and non-I are almost equally distributed, whereas the mixed phenotype is less abundant. Thus, in the cohort of patients of the International Cystinuria Consortium grouping patients mainly from Southern Europe, but also from Central Europe, Israel, and North America, the classified patients were distributed as follows: about 44% of phenotype I, about 42% of phenotype non-I, and about 14% of mixed phenotype (Dello Strologo et al. 2002).

The molecular bases of cystinuria

The characteristics of rBAT (see the section above on the high-affinity reabsorption system of cystine) made human *SLC3A1* on chromosome 2p16 a good candidate for cystinuria. In 1994, it was demonstrated that mutations in *SLC3A1* cause cystinuria phenotype I (Calonge *et al.* 1994). Soon afterwards it was shown that cystinuria phenotype non-I was not due to mutations in *SLC3A1* (Calonge *et al.* 1995; Bisceglia *et al.* 1997; Wartenfeld *et al.* 1997). Over 60 distinct rBAT mutations have been described, including nonsense, missense, splice-site, and frameshift mutations, as well as large deletions (Palacín *et al.* 2000, 2001b). Interestingly, a cystinuria resembling type I due to mutations in canine *SLC3a1* has been reported in the Newfoundland dogs (Henthorn *et al.* 2000). All cystinuria-specific *SLC3A1* mutations analyzed showed trafficking defects (Chillarón *et al.* 1997; Saadi *et al.* 1998) (see the section, the high-affinity reabsorption system of cystine).

The gene causing cystinuria phenotype non-I was assigned by linkage analysis to the 19q12-13.1 region (Bisceglia *et al.* 1997; Wartenfeld *et al.* 1997) and confined to a 1.3 Mb region. Indeed, another study located the subtype II of cystinuria locus in the same chromosomal region (Stoller *et al.* 1999). In 1999 the cystinuria phenotype non-I gene was identified as *SLC7A9* (Feliubadalo *et al.* 1999). This was a positional candidate gene for cystinuria phenotype non-I because it has the proper chromosomal location and tissue expression, and its protein product ($b^{0,+}$AT) induces system $b^{0,+}$ activity when coexpressed with rBAT (Feliubadalo *et al.* 1999). At present, 35 *SLC7A9* mutations causing cystinuria have been described, including nonsense, missense, splice-site, and frameshift mutations (Chillarón *et al.* 2001; Palacín *et al.* 2001b). Recently, a $b^{0,+}$AT-deficient murine model has been obtained that develops cystinuria phenotype non-I. All cystinuria-specific *SLC7A9* mutations analyzed showed loss of transport function (Font *et al.* 2001). Only two mutations have been studied in detail: mutation A354T inactivates the transporter, whereas mutation A182T has a trafficking defect (Reig *et al.* 2002) (see the section above on the high-affinity reabsorption system of cystine).

Genotype–phenotype correlations and a new classification for cystinuria

The study of the urinary excretion levels of cystine and dibasic amino acids in the cohort of heterozygotes (i.e., carriers) of the International Cystinuria Consortium has shown the first genotype–phenotype correlations in cystinuria. All studied heterozygotes of *SLC3A1* mutations showed a urinary profile of amino acid levels within the control range (i.e., within the variability range of control individuals without mutations in the two cystinuria genes), as expected for a gene whose mutation causes cystinuria phenotype I (Dello Strologo *et al.* 2002). In contrast, *SLC7A9* heterozygotes showed a higher variability in the range of cystine and dibasic amino acid excretion in urine: 84% of them hyperexcrete over the control range (i.e., phenotype non-I), and 14% of them excrete within the control range (i.e., phenotype I) (Dello Strologo *et al.* 2002). In addition to a clear individual variability, there are *SLC7A9* mutations that segregate preferentially with one of the two cystinuria phenotypes (Font *et al.* 2001). Thus severe *SLC7A9* mutations (i.e., mutations that show no residual transport activity when coexpressed with rBAT in cultured cells) associated completely or almost completely with phenotype non-I (e.g., 13 out of 13 and 31 out of 32 heterozygotes bearing mutation R333W or G105R, respectively, showed phenotype non-I). In contrast, mild mutations (i.e., mutations with substantial residual transport activity) show a higher tendency towards the phenotype I (e.g., 6 out of 11 A182T heterozygotes showed phenotype I). Interestingly, mutations affecting conserved amino acid residues in the LSHAT family, putatively located within the TM domains and with a short side chain, are severe (Font *et al.* 2001).

The fact that mutations in *SLC3A1* cause cystinuria phenotype I, whereas mutations in *SLC7A9* cause both cystinuria phenotypes (Dello Strologo *et al.* 2002) prompted a new genetic classification of the disease: cystinuria type A, caused by mutations in *SLC3A1*, and cystinuria type B, caused by mutations in *SLC7A9* (Dello Strologo *et al.* 2002). Mutational analysis covering the open reading frame (ORF) of *SLC3A1* and *SLC7A9* of 189 patients of the International Cystinuria Consortium explained about 84% of phenotype I (74% due to cystinuria type A and 10% due to cystinuria type B) and about 84% of phenotype non-I alleles (all due to cystinuria type B). In this cohort of patients, M467T is the most common *SLC3A1* mutation (almost 31% of the type A alleles), and G105R is the more frequent *SLC7A9* mutation (27% of type B alleles). Mutation V170M is the *SLC7A9* mutation explaining almost all the cases of cystinuria among Jews of Libyan origin.

Most probably the 16% of phenotypes I and non-I alleles not yet explained might represent mutations not yet identified in either of the two cystinuria genes. Owing to the extensive mutational analysis performed in the cohort of patients of the International Cystinuria Consortium, only mutations in the promoter region or intronic sequences of *SLC3A1* and *SLC7A9* might explain the remaining alleles. On the other hand, there is no evidence for genetic linkage in loci other than the two identified cystinuria genes.

Pathophysiology of cystinuria

All the evidence discussed above indicates that cystinuria is due to defects in system $b^{0,+}$ (i.e., the rBAT–$b^{0,+}$AT heterodimeric complex), the amino acid transport system responsible for the intestinal absorption and renal reabsorption mechanism shared by cystine and dibasic amino acids. The urinary hyperexcretion of cystine results in most cases in the formation of cystine calculi in the urinary system, mainly in the renal pelvis and calyx, but also in the bladder. Pathological consequences of cystine urolithiasis are those characteristic of urolithasis of any nature: obstruction, infection, and ultimately renal insufficiency.

It is less clear what, besides the increased concentration of cystine in urine, contributes to the formation and growth of cystinuria calculi in the patients. Indeed, there are marked intrafamilial differences in the lithiasic activity between siblings sharing the same mutation (one sibling can have a very aggressive lithiasic phenotype whereas the other does not develop calculi) (Dello Strologo *et al.* 1997). On the other hand, the lithiasic activity of the cystinuria patients does not correlate with the level of hyperexcretion of cystine and dibasic amino acids in urine (Dello Strologo *et al.* 2002). Therefore hyperexcretion of cystine is necessary to develop calculi, but other environmental or genetic factors affect cystine lithiasis. Proteins like bikunin and osteopontin have been proposed to play a role inhibiting calcium oxalate lithiasis (Iida *et al.* 1999; Xie *et al.* 2001). The role of these proteins and other genetic factors in cystine lithiasis is completely unknown.

Apart from cystine urolithaisis, cystinuria would be a metabolic oddity. In this sense treatment of cystinuria is aimed at reducing cystine concentration in urine: increased liquid intake and reduced intake of sodium and methionine, as a precursor of cysteine (conservative treatment). When urolithiasis is persistent, thiol compounds that solubilize cystine by forming adducts with cysteine are prescribed. Finally, surgery or extracorporal wave lithotripsy (for small calculi) are used to remove calculi.

Hyperexcretion of cystine and dibasic amino acids in cystinuria patients is quite significant. In the cohort of the cystinuria patients of the International Cystinuria Consortium (both type A and type B homozygotes) hyperexcretion of cystine and lysine in urine is 30- to 40-fold higher than in controls, and that of ornithine and arginine is about 90-fold and about 170-fold, respectively, higher than in controls (Dello Strologo *et al.* 2002). More than 90% of renal cystine reabsorption occurs in the early parts of the proximal tubule (Völkl and Silbernagl 1982; Silbernagl 1988), where the rBAT-$b^{0,+}$AT heterodimeric complex has its highest expression along the proximal tubule (Fernández *et al.* 2002). Interestingly, cystinuria patients have clearance ratios of cystine/inuline (or creatinine) around unity or even slightly higher, and consequently they reabsorb very little cystine (only occasionally more than 25%) (Frimpter *et al.* 1962; Crawhall *et al.* 1967). All this points to system $b^{0,+}$ (i.e., the rBAT-$b^{0,+}$AT heterodimeric complex) as the main, if not the unique, apical reabsorption system of cystine in kidney. In contrast, for dibasic amino acids, other apical reabsorption systems, not yet identified, should work in kidney in addition to system $b^{0,+}$. Thus cystinuria patients reabsorb substantial amounts of dibasic amino acids (about 50%) (Frimpter *et al.* 1962; Crawhall *et al.* 1967).

There are no pathological consequences of the intestinal phenotype in cystinuria. System $b^{0,+}$ (i.e., the rBAT-$b^{0,+}$AT heterodimeric complex) is expressed in the brush border membranes of the enterocytes of the small intestine and, as expected, absorption of cystine and dibasic amino acids is impaired in patients with cystinuria (Rosenberg *et al.* 1967; de Sanctis *et al.* 2001). The significant loss in urine and the impaired intestinal absorption of cystine and dibasic amino acids would suggest the development of protein malnutrition in cystinuria. In contrast, cystinuria patients do not present with protein malnutrition unless in situations of critical limitation of protein intake (Segal and Thier 1995). Moreover, patients with cystinuria show normal or slightly subnormal plasma levels of amino acids (reviewed by Segal and Thier 1995). Contribution of apical transport of di- and tripeptides via PEPT1 in small intestine (see the section below on the pathophysiology of lysinuric protein intolerance) could be the reason why there is no malnutrition in cystinuria.

As discussed above, $b^{0,+}$AT heterodimerizes with rBAT to form system $b^{0,+}$ in renal brush border membranes (Fernández *et al.* 2002). In addition, part of the rBAT protein heterodimerizes with additional, not yet identified, light subunit(s) (Fernández *et al.* 2002). This suggests that mutations in rBAT (*SLC3A1*) might produce a wider or more severe urinary phenotype than mutations in $b^{0,+}$AT (*SLC7A9*). In contrast with this view, patients with cystinuria type A (due to *SLC3A1* mutations) or type B (due to *SLC7A9* mutations) hyperexcrete similar levels of cystine and dibasic amino acids in urine (Dello Strologo *et al.* 2002). In some patients, hyperexcretion of glycine, methionine, and cystathionine has been described (reviewed by Segal and Thier 1995). Whether hyperexcretion of these amino acids is related to the missing light subunit(s) of rBAT is unknown. Identification of these light subunits will shed new light on our understanding of the cystinuria phenotype.

THE BASOLATERAL AMINO ACID TRANSPORT SYSTEM y^+L

The basolateral system y^+L is required for the absorption of cationic amino acids through the epithelial cells of kidney and intestine. In lysinuric protein intolerance (LPI) there is a reduced cationic amino acid absorption in both the renal tubule and the small intestine owing to a defective cationic amino acid transporter at the basolateral membrane of these epithelial cells (Desjeux *et al.* 1980; Rajantie *et al.* 1980a; Rajantie and Simell 1981). LPI is due to mutations in the gene *SLC7A7* (Borsani *et al.* 1999; Torrents *et al.* 1999) that codes for the light subunit y^+LAT-1 associated with the heavy chain of the cell surface antigen 4F2 (4F2hc) (Torrents *et al.* 1998). This heteromeric complex presents system y^+L amino acid transport activity in *Xenopus* oocytes (Torrents *et al.* 1998; Pfeiffer *et al.* 1999b; Kanai *et al.* 2000).

System y^+L activity

Transport of cationic amino acids has classically been ascribed to the system y^+ described in Ehrlich cells, reticulocytes, and fibroblasts about 40 years ago and then extended to many other cell types (reviewed by Christensen 1984, 1990; White 1985). The cloning of several members of the widely distributed cationic amino acid transporters (CAT) family in the 1990s revealed that these proteins present some, but not all, of the features initially ascribed to system y^+. The CAT proteins show a much weaker interaction with neutral amino acids than expected from previous functional studies (Kim *et al.* 1991; Wang *et al.* 1991; Devés and Boyd 1998). In the 1980s, other transporters handling cationic amino acids were functionally described in blastocysts and named $b^{0,+}$ and $B^{0,+}$ (Van Winkle 1988). More recently, a kinetic study of the partial inhibition of lysine influx by neutral amino acids led to the identification of a novel transport system for cationic amino acids in human erythrocytes designated system y^+L (Devés *et al.* 1992). In that report lysine was shown to enter the erythrocyte through two transporters: (a) a high-affinity low-capacity transporter which recognizes with comparable affinities leucine in the presence of sodium and lysine in either the presence or absence of sodium (system y^+L); (b) a lower-affinity high-capacity transporter which is sodium independent and specific for cationic amino acids (system y^+). The various transporters for cationic amino acids (systems y^+, $B^{0,+}$, $b^{0,+}$, and y^+L) differ mainly in the way they handle neutral amino acids, making this information essential for the functional discrimination of these activities. In addition, the sulfhydryl reagent *N*-ethylmaleimide (NEM) was found to inhibit system y^+ without affecting system y^+L, greatly facilitating the functional study of the latter (Devés *et al.* 1993). In summary, system y^+L transports cationic amino acids with high affinity (K_m in the micromolar range) in a sodium-independent fashion, but requires sodium to transport both small and large neutral amino acids with high affinity. In the absence of sodium, transport of neutral amino acids through system y^+L is of very low affinity (reviewed by Devés and Boyd 1998).

Another remarkable feature of system y^+L is trans-stimulation. System y^+L catalyzes the electroneutral efflux of cationic amino acids in exchange for neutral amino acids plus sodium using the driving force of the sodium concentration gradient (Angelo and Devés 1994; Eleno *et al.* 1994; Chillarón *et al.* 1996). Prior to the identification of system y^+L, several studies described observations that are now easily explained considering the presence of system y^+L in basal membranes. In this sense, a report on vascularly perfused frog small intestine showed that neutral amino acids on the basolateral side stimulated lysine flux through the epithelium due to an increase in the rate of exit through the basolateral membrane (Cheeseman 1983). Other reports presented similar results in different models (reviewed by Devés and Boyd 1998). All these observations are likely to be due to the expression of y^+L transporters not described at that time.

The molecular entity of system y^+L transporters

In the early 1990s, two different groups described expression of a system y^+-like transport activity in *Xenopus* oocytes after injection of 4F2hc cRNA (Bertran *et al.* 1992a; Wells *et al.* 1992). The induced activity corresponded to what we now know as y^+L activity. Those studies were prompted by the discovery of the homologous rBAT protein as a molecule inducing amino acid transport in oocytes. The observation that 4F2hc and rBAT are less hydrophobic than typical transporter proteins and form disulfide-bound complexes with other proteins led to the suggestion that 4F2hc and rBAT were subunits of HATs. Our group showed that y^+L transport activity induced in oocytes by a cysteine-free human 4F2hc was still sensitive to sulfhydryl-specific reagents that modify cysteine residues exposed to the aqueous solvent. These data demonstrated that 4F2hc is associated with another membrane protein for the expression of system y^+L amino acid transport activity at the plasma membrane (Estevez *et al.* 1998).

The molecular identification of the light subunits of heteromeric amino acid transporters revealed the primary structure of two closely related proteins identified by homology screening which, when expressed together with 4F2hc, induce y^+L activity. These proteins, named y^+LAT-1 and y^+LAT-2, are members of the LSHATs (Torrents *et al.* 1998). In contrast with y$^+$LAT-1, which is expressed mainly in tissues affected in LPI like kidney epithelial cells, lung, and small intestine, y$^+$LAT-2 has a wider tissue distribution including brain, heart, testis, kidney, small intestine, and parotid (Torrents *et al.* 1998; Pfeiffer *et al.* 1999b; Bröer *et al.* 2000). The transport activity, tissue distribution, and chromosomal location of y$^+$LAT-1 prompted a mutational analysis that identified mutations in *SLC7A7*, the gene encoding y$^+$LAT-1, in LPI patients (Borsani *et al.* 1999; Torrents *et al.* 1999). The fact that system y$^+$L activity is present in LPI erythrocytes or fibroblasts (Smith *et al.* 1988; Boyd *et al.* 2000; Dall'Asta *et al.* 2000) indicates the expression of a different y$^+$L transporter isoform in these cells, most probably y$^+$LAT-2. In addition, y$^+$LAT-1 mRNA has also been detected in fibroblasts (Shoji *et al.* 2002).

The transport characteristics of 4F2hc-y$^+$LAT-1 have been studied in heterologous expression systems (Torrents *et al.* 1998; Pfeiffer *et al.* 1999b; Kanai *et al.* 2000). The transport mechanism involves a heteroexchange between cationic and neutral amino acids plus sodium and matches the characteristics of system y$^+$L (see above). In a thorough study, Kanai and coworkers found that lowering the pH increased leucine transport through y$^+$LAT-1 without affecting lysine transport. This observation led them to propose that H$^+$, in addition to sodium or lithium, is capable of supporting neutral amino acid transport through system y$^+$L. Na$^+$ and H$^+$ affected leucine transport by decreasing the K_m value without affecting the V_{max} value and the Na$^+$ to leucine coupling ratio was found to be 1:1 (Kanai *et al.* 2000). Functional studies performed with y$^+$LAT-2 showed that it induces y$^+$L activity in association with 4F2hc (Torrents *et al.* 1998; Bröer *et al.* 2000). The transporter functions by exchanging cationic amino acids for neutral amino acids plus Na$^+$. Bröer *et al.* (2000) have reported an unexpected finding for this transporter: glutamate

strongly inhibits L-arginine transport through the carrier.

A recent study has demonstrated the basolateral location of y$^+$LAT-1 in kidney tubules and polarized cellular models (Bauch *et al.* 2003). Immunofluorescence analysis showed that the expression levels for y$^+$LAT-1, LAT-2 and b$^{0,+}$AT follow the same axial gradient along the kidney proximal tubule (higher in the convoluted than in straight proximal tubules). This study also revealed that b$^{0,+}$AT is located at the brush border membrane, whereas both y$^+$LAT1 and LAT2 are found in the basolateral membrane of the same cells (Bauch *et al.* 2003). The coexpression of these transporters in polarized Madin–Darby canine kidney (MDCK) cells corroborated their subcellular localization. In addition, y$^+$L activity has been detected in the basolateral membrane of polarized OK cells, an epithelial proximal tubule derived cell line (Fernández *et al.* 2003).

LYSINURIC PROTEIN INTOLERANCE

LPI (MIM 222700) is a primary inherited aminoaciduria with an autosomal recessive mode of inheritance (reviewed by Simell 2002). LPI is a rare disease, with about 100 patients reported. Nearly half of these patients are from Finland, where the disease was first described, with a prevalence of 1 in 60,000 (Perheentupa and Visakorpi 1965). Two other locations with a relatively high prevalence are southern Italy and Japan (reviewed by Palacín *et al.* 2001a). In LPI there is massive urinary excretion of dibasic amino acids, especially lysine, and intestinal absorption of these amino acids is poor. As a consequence there is a low plasma concentration of dibasic amino acids (Kekomäki *et al.* 1967; Oyanagi *et al.* 1970; Simell *et al.* 1975).

Arginine and ornithine are intermediates of the urea cycle. This is thought to result in a functional deficiency of the urea cycle by reduced availability of intermediates like ornithine, which provides the carbon skeleton to the cycle (Awrich *et al.* 1975; Rajantie *et al.* 1980c). Patients with LPI have periods of hyperammonemia with nausea and vomiting that produce aversion to protein-rich food. Protein malnutrition and deficiency of the essential amino acid lysine contribute to patients' failure to thrive (Simell *et al.* 1975).

Patients with LPI are usually clinically silent while breastfeeding and symptoms appear after weaning (e.g., vomiting, diarrhea, and hyperammonemic coma when force fed high-protein food). After infancy, LPI patients reject high-protein diets and show a delay in bone growth and prominent osteoporosis, bone fractures are common, and they present enlarged liver and spleen, muscle hypotonia, and sparse hair (reviewed by Simell 2002). Most patients have a normal mental development, but some may show moderate retardation. Treatment ameliorates the mental prognosis. Low-protein diet and citrulline, a urea-cycle intermediate, are used to correct the functional deficiency of intermediates of the urea cycle. The final height in treated patients is slightly subnormal or low normal. This treatment does not correct all symptoms like poor growth, hepatosplenomegalia, delayed bone age, and osteoporosis, which are probably due to the lysine deficiency (Awrich *et al.* 1975; Rajantie *et al.* 1980c, 1983; Carpenter *et al.* 1985).

LPI is a multisystemic disease. In addition to the above-mentioned symptoms, about two thirds of the patients have interstitial changes in chest radiographs, and some develop acute or chronic respiratory insufficiency (Parto *et al.* 1993). In a few cases this leads to fatal pulmonary alveolar proteinosis and to a multiple-organ dysfunction syndrome. Some patients also present nephritis, renal insufficiency, and erythrobalstophagia (DiRocco *et al.* 1993). Some of these features (e.g., alveolar proteinosis, glomerulonephritis, and erythroblastophagia) might correspond to alterations of the immune system but their etiology is unknown. Abnormalities of the immune system in LPI have been reported (Nagata *et al.* 1987).

The molecular basis of LPI

The first key step in the identification of the molecular basis of LPI was localization of the gene responsible for LPI to 14q11.2 in a Finnish population (Lauteala *et al.* 1997) and later on in non-Finnish populations (Lauteala *et al.* 1998). The second step was the cloning of the light subunit of the heteromeric amino acid transporters y$^+$LAT-1, encoded by *SLC7A7* (Torrents *et al.* 1998). This subunit heterodimerizes with 4F2hc to express system y$^+$L amino acid transport activity in the basolateral plasma membrane of the epithelial cells of the renal proximal tubule and the small intestine (see the section above on the basolateral amino acid transport system y$^+$L). Moreover, *SLC7A7* maps to the correct location for LPI. This made the gene an excellent candidate for LPI. In the third step, Torrents *et al.* (1999) performed mutational analysis of *SLC7A7* in one Spanish and 31 Finnish LPI patients. A single Finnish mutant allele (1181-2A > T) was found, with an A > T transversion at position -2 of the acceptor splice site in intron 6 of *SLC7A7*. This inactivates the normal splice-site acceptor and activates a cryptic acceptor 10 bp downstream, with the result that 10 bp of the ORF are deleted and the reading frame is shifted. This mutation has been found in all Finnish LPI patients (i.e., "the Finnish mutation") and, with the exception of one patient, with a common haplotype consistent with the expectation of a founder mutation in the Finnish population (Lauteala *et al.* 1998; Torrents *et al.* 1999). The one exception appeared to result from a recombination between the haplotype markers and the LPI locus. The Spanish LPI patient was a genetic compound of two *SLC7A7* mutations (a missense mutation L334R and a 4 bp deletion 1291delCTTT). Expression studies in oocytes showed that L334R inactivated the transporter function (Mykkanen *et al.* 2000). Simultaneously and independently, Borsani *et al.* (1999) surveyed *SLC7A7* in four Finnish and five Italian LPI patients: four of the Italian probands were homozygous for a 4 bp insertion

(1625insATAC) that shifts the ORF at codon 462 with an early STOP codon 13 bp downstream, the remaining Italian proband was homozygous for a 543 bp deletion that removes the first 168 codons of the ORF, and all the Finnish patients were homozygous for the Finnish mutation. These two studies established that mutations in *SLC7A7* cause LPI. A total of 25 *SLC7A7* mutations have been described in 96 LPI patients, where only three alleles have not yet been explained (for a review see Palacín *et al*. 2001a). In contrast, no LPI-associated mutations have been found in *SLC3A2*, coding for the heavy subunit of y$^+$LAT-1 (4F2hc). This strongly suggests that *SLC7A7* is the only gene involved in the primary cause of LPI. It is believed that mutations in *SLC3A2* would be deleterious. 4F2hc serves as the heavy subunit of other heteromeric amino acid transporters (i.e., two isoforms of system L, two isoforms of system y$^+$L, and one isoform of system asc and system x$_c$-; see Introduction). Therefore a defect in 4F2hc will result in six defective amino acid transport activities expressed in many cell types and tissues (Chillarón *et al*. 2001).

Functional studies in oocytes and transfected cells showed that frameshift mutations (e.g., 1291delCTTT, 1548delC, and the Finnish mutations) produce a severe trafficking defect (e.g., the mutated proteins do not localize to the plasma membrane when coexpressed with 4F2hc) (Mykkanen *et al*. 2000; Toivonen *et al*. 2002). In contrast, the missense mutations G54V and L334R inactivate the transporter (e.g., the mutated proteins reach the plasma membrane when coexpressed with 4F2hc but no transport activity is elicited) (Mykkanen *et al*. 2000; Toivonen *et al*. 2002).

Pathophysiology of lysinuric protein intolerance

LPI is a multisystemic disease. Figure 3 summarizes the suggested pathogenesis of some of the symptoms of LPI. Some of these are easily explained by a defect in the basolateral amino acid transport system y$^+$L, like the renal and intestinal phenotypes. In contrast, hyperammonemia and the protein intolerance (e.g., hepatic phenotype) and the immune-disorder-related manifestations (e.g., alveolar proteinosis, erythroblastophagia, glomerulonephritis) are examples of LPI-associated defects that are not so easily explained by mutations in the amino acid transport system y$^+$L. In addition, individual phenotypic variability precluded establishment of genotype–phenotype correlations (Mykkanen *et al*. 2000; Sperandeo *et al*. 2000). Thus Finnish LPI patients, all with the same Finnish mutation in homozygosis, show a wide range of phenotypic severity. This ranges from nearly normal growth with minimal protein intolerance to severe cases with hepatosplenomegalia, osteoporosis, alveolar proteinosis, and severe protein intolerance. In the following sections the mechanisms, or the hypothetical mechanisms, for the pathophysiology of the renal and intestinal phenotypes and the hepatic phenotypes are discussed. Not enough data are available as yet to propose a solid hypothesis for alveolar proteinosis and other immune-related disorders.

The renal and intestinal phenotypes

Perheentupa and Visakorpi (1965) described the first three patients with LPI substantiating two major complications of the disease: "protein intolerance and deficient transport of basic amino acids." Protein intolerance will be discussed in the next section (the hepatic phenotype). The authors already realized that LPI patients have a renal reabsorption defect for dibasic amino acids.

The plasma and urine levels of amino acids in LPI patients have recently been reviewed by Simell (2002). Plasma concentrations of dibasic amino acids (i.e., lysine, arginine, and ornithine) are usually subnormal (one third to half of the normal values) but may occasionally be within the normal range. In contrast, the concentration of the neutral amino acids serine, glycine, citrulline, proline, alanine, and glutamine are increased. The system y$^+$L defective in LPI is highly expressed in white

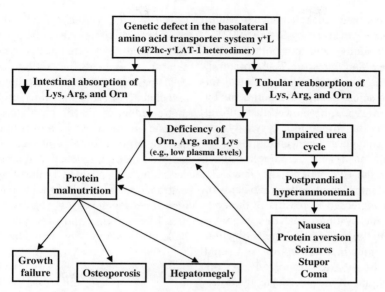

Figure 3 The suggested pathogenesis of lysine, arginine, and ornithine deficiency, hyperammonemia, and aversion to protein in LPI. The 4F2hc-y$^+$LAT-1 heterodimer is expressed on the basolateral membrane of the enterocytes and proximal tubule epithelial cells, where it mediates the efflux of dibasic amino acids from the cell to the blood (in exchange with neutral amino acids plus sodium, y$^+$L system). Mutations in y$^+$LAT1 found in LPI patients decrease this transport activity. This causes malabsorption and reduced reabsorption of arginine, lysine, and ornithine, leading to low plasma levels of these amino acids and protein malnutrition. Malfunctioning of the urea cycle may be the consequence of low intramitochondrial levels of ornithine due to the reduced plasma levels of this amino acid and arginine. The effects of postprandial hyperammonemia reinforce protein malnutrition, which results in growth failure, osteoporosis, and hepatomegaly. Modified from Simell (2002).

blood cells of control individuals (Torrents *et al.* 1998). System y$^+$L mediates the efflux of dibasic amino acids from cells in exchange with neutral amino acids (see the section above on the basolateral amino acid transport system y$^+$L). Then, one could speculate that the LPI-associated defect in this transport system will tend to decrease the concentration of dibasic amino acids and increase that of neutral amino acids in plasma. In addition, the intestinal malabsorption and the deficient renal reabsorption of dibasic amino acids in LPI (see below) should also contribute to a diminished concentration of dibasic amino acids in blood.

Urinary excretion and renal clearance of lysine is massively increased and that of arginine and ornithine is moderately increased in LPI (Simell and Perheentupa 1974). In some older patients who spontaneously restricted their protein intake, showing signs of protein malnutrition, the plasma concentration of dibasic amino acids is extremely low and hyperexcretion of these amino acids in urine is not visible (Simell 2002) even though the clearance of dibasic amino acids is high in these patients. Therefore renal reabsorption of dibasic amino acids is defective in LPI patients. In contrast with cystinuria, there is only a slight increase in renal cystine excretion, which could be explained by the large tubular lysine load (i.e., due to the reabsorption defect of lysine) that competes for absorption through the apical system b$^{0,+}$ (see cystinuria sections above). The increased plasma concentration of serine, glycine, citrulline, proline, alanine, and glutamine explains hyperexcretion of these amino acids, but their renal clearance is within the normal range (reviewed by Simell 2002).

The dibasic amino acid transport defect of LPI in kidney and intestine is basolateral. Indeed, there is a basolateral efflux defect for dibasic amino acids. Beautiful experiments by Rajantie *et al.* (1980a, b) showed that an oral loading with the dipeptide lysyl-glycine increased glycine plasma concentrations properly, but plasma lysine remained almost unchanged in the LPI patients. In clear contrast, the concentration of both amino acids of the dipeptide increased in plasma in the control subjects and cystinuria patients. Figure 4 represents our knowledge of the molecular bases of the intestinal absorption and renal reabsorption of dibasic aminoacids. At the luminal membrane of the enterocyte, the transport of oligopeptides is mediated by PEPT1 (Groneberg *et al.* 2001). This transport is not shared with free amino acids. Dibasic amino acids cross the apical membrane via system $b^{0,+}$. This transporter shares transport of dibasic amino acids and cystine, and it is defective in cystinuria (see cystinuria sections above). The absorbed peptides are hydrolyzed to release amino acids in the cytoplasm of the enterocyte (Adibi 1971; Asatoor *et al.* 1971; Matthews and Adibi 1976), and they are able to cross the basolateral membrane only as free amino acids. The lack of an increase in plasma lysine after the lysyl-glycine load but normal increase in plasma glycine shows that the basolateral efflux of the intracellularly delivered lysine across the basolateral plasma is defective in LPI. In vitro studies demonstrated later the transport defect at the basolateral plasma membrane of the epithelial cell (Desjeux *et al.* 1980). In contrast, in cystinuria patients the cleaved glycine and lysine cross the epithelial cell because the defect is apical (i.e., system $b^{0,+}$).

A defect in the basolateral system y^+L explains straightforwardly the renal and intestinal phenotype in LPI. The protein y^+LAT-1 has a basolateral location in epithelial cells (see the section above on the basolateral amino acid transport system y^+L). System y^+L (i.e., the 4F2hc/y^+LAT-1 heteromeric complex) mediates the efflux of cationic amino acids by exchange with extracellular neutral amino acids and sodium (see above). Thus the loss of transport function due to LPI-associated y^+LAT-1 mutations results in a dramatic reduction of the basolateral efflux of dibasic amino acids in absorptive intestinal and resorptive renal epithelial cells. It is believed that the reduced availability of dibasic amino acids, caused by intestinal and renal malfunction, contributes to the hepatic phenotype (i.e., deficiency of urea-cycle intermediates like arginine and ornithine) and to the bone malformation (i.e., deficiency in lysine).

The hepatic phenotype

Urea-cycle malfunction is a characteristic of patients with LPI after weaning. Patients with LPI have decreased tolerance for nitrogen and present with hyperammonemia after ingestion of even moderate amounts of protein. The severity of the defect of the urea cycle in LPI is less severe than that caused by defects in the enzymes of the cycle (i.e., carbamoyl phosphate synthase, ornithine transcarbamoylase, N-acetylglutamate synthase, arginonosuccinate synthase or lyase) and similar to that of patients with hyperornithinemia–hyperammonemia–homocitrullinuria (HHH) syndrome due to mitochondrial ornithine translocase deficiency (see Chapters 9 and 10, and Simell 2002).

Perheentupa and Visakorpi (1965) noticed that intravenous infusion of ornithine during protein intake or intravenous administration of L-alanine prevented hyperammonemia. Later on, several studies showed that intravenous arginine and oral citrulline produce the same beneficial effect (Awrich *et al.* 1975; Rajantie *et al.* 1980b, c, 1983b). A therapy based on the oral administration of arginine and ornithine to supplement the urea cycle was shown to be minimally effective due to the impaired intestinal absorption of dibasic amino acids in LPI (Desjeux *et al.* 1980; Rajantie *et al.* 1980c, 1983b). Citrulline is a neutral amino acid and

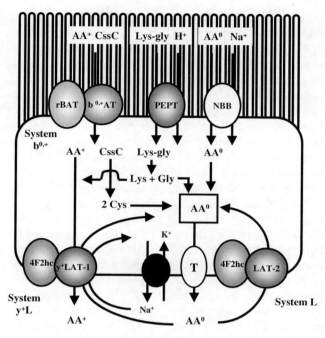

Figure 4 Model for the absorption/reabsorption of different amino acids in an intestinal/proximal tubule epithelial cell. The transepithelial flux of amino acids is ensured by the presence of different transport systems at the apical and basolateral membrane. A tertiary active transport mechanism accounts for the (re)absorption of dibasic amino acids (AA^+) and cystine (CssC); an apical Na^+-dependent neutral amino acid (AA^0) transport system (NBB, not identified at a molecular level) accounts for the high accumulation of neutral amino acids in the cell, which provide the driving force for the entry of cystine and dibasic amino acids through system $b^{0,+}$ (rBAT-$b^{0,+}$AT). Dibasic amino acid and cystine influx are favored by the negative membrane potential and the rapid reduction of cystine to cysteine (CSH), respectively. A defective NBB system may result in a phenotype similar to Hartnup disease. Recently a genetic locus for Hartnup disease has been located on 5p15 (Nozaki et al. 2001). Mutations in either rBAT or $b^{0,+}$AT cause cystinuria, but not protein malnutrition (as would be expected from the localization of rBAT-$b^{0,+}$AT in the enterocyte), most likely because an intestinal peptide transporter (PEPT) is able to maintain the supply of dibasic amino acids. Net efflux of these amino acids is accounted for by exchange with neutral amino acids plus sodium via system y+L (4F2hc-y^+LAT-1) at the basolateral membrane. A defect in the y^+L system in LPI patients causes, besides hyperdibasic aminoaciduria, protein malnutrition because there is no basolateral peptide transporter (and the peptides are hydrolyzed to their amino acid residues in the cell). The pool of intracellular neutral amino acids (including cysteine) can be exchanged with the extracellular pool via the basolateral system L (4F2hc-LAT-2). As long as this exchange is 1:1, the neutral amino acid individual pools, but not the total pool, will change depending on the concentrations of the different amino acids at either side of the basolateral membrane and on the intrinsic asymmetry of the transporter. Therefore a facilitative neutral amino acid transporter (T, not identified at a molecular level) must be present at the basolateral membrane to explain net transport of these amino acids. Recently, it has been shown in the OK cell model that 4F2hc-LAT-2 contributes to the net efflux of cysteine through the basolateral membrane, concomitant with the net influx of alanine, and perhaps serine and threonine (Fernández et al. 2003). Shaded circles represent cloned proteins. ATPase, Na^+, K^+-ATPase. Modified from Chillarón et al. (2001).

an intermediate of the urea cycle. The intestinal absorption of neutral amino acids as citrulline is not affected in LPI. Therefore a therapy based on this amino acid should overcome the problems of the oral administration of dibasic amino acids. Indeed, Awrich *et al.* (1975) showed that oral citrulline prevents hyperammonemia after protein challenge. Several studies have demonstrated the effectiveness and tolerance of this treatment to prevent hyperammonemia in LPI (reviewed by Simell 2002).

As discussed above, the urea-cycle intermediates arginine and ornithine are poorly absorbed in the intestine, are largely lost in the urine, and have low concentrations in plasma. The fact that supplementation with intermediates of the urea cycle prevent hyperammonemia in LPI suggests that the urea-cycle defect associated with the disease is due to a reduced availability of ornithine and/or arginine in liver. In other words, the urea cycle in LPI shows a "functional deficiency" of intermediates. On the other hand, according to northern blot analysis y^+LAT-1 is not expressed in liver (Pfeiffer *et al.* 1999b; Torrents *et al.* 1999). A very faint signal after long exposures of a blot containing liver polyA$^+$ RNA was observed in only one report (Borsani *et al.* 1999). We cannot rule out that the signal was due to cells of hematopoietic origin present in liver. In summary, the "functional deficiency" of intermediates of the cycle seems to be the consequence of the low plasma levels of arginine and ornithine.

A MODEL FOR INTESTINAL/RENAL TRANSEPITHELIAL TRANSPORT OF CYSTINE AND DIBASIC AMINO ACIDS

Over the past few years we have built up an increasingly detailed molecular transport map linking the expression of amino acid carriers in either the apical or the basolateral membrane of epithelial cells to the fluxes of amino acids from the intestinal or tubular lumen to the blood (Calonge *et al.* 1994; Chillarón *et al.* 1996, 2001; Feliubadalo 1999; Torrents *et al.* 1999; Fernández *et al.* 2002, 2003; Bauch

et al. 2003). Their functional characterization in heterologous expression systems and its cell/tissue immunolocalization allowed the proposal of physiological functions for different carriers. Moreover, the involvement of some of them in human aminoacidurias made their particular role in the transepithelial flux of these substrates immediately obvious (Calonge *et al.* 1994; Borsani *et al.* 1999; Torrents *et al.* 1999). The picture that emerges from all these data (Figure 4) is far form complete, however. We have a relatively deep understanding of cystine and dibasic amino acid (re)absorption (see below), but key players of the transepithelial flux of neutral amino acids remain unidentified, particularly NBB (neutral brush border, a putative sodium-dependent apical transport system) and T (postulated to be an efflux carrier for at least one neutral amino acid).

Cystine and dibasic amino acid (re)absorption is ensured by the presence of three different HATs: the apical rBAT-$b^{0,+}$AT and the basolateral 4F2hc-y^+LAT-1 and 4F2hc-LAT-2. As discussed above, both rBAT and $b^{0,+}$AT mutants lead to the cystinuria phenotype, whereas y^+LAT-1 mutations cause LPI (Calonge *et al.* 1994; Borsani *et al.* 1999; Feliubadalo *et al.* 1999; Torrents *et al.* 1999). Their immunolocalization to the apical or basolateral membrane, respectively, in enterocytes and proximal tubule epithelial cells (Furriols *et al.* 1993; Pickel *et al.* 1993; Bauch *et al.* 2003) and the disease phenotype provide the fundamental evidence for the model in Figure 4: the intracellular high concentration of neutral amino acids drives the (re)absorption of dibasic amino acids and cystine through the rBAT-$b^{0,+}$AT heterodimer. The membrane potential favors the influx of dibasic amino acids, whereas the redox potential favors the influx of cystine (i.e., its rapid reduction to cysteine). Although initially questioned because of a lack of complete colocalization of rBAT and $b^{0,+}$AT (Chairoungdua *et al.* 1999; Pfeiffer *et al.* 1999a), recent data demonstrated that this heterodimer mediates most, if not all, cystine (re)absorption in epithelial cells (Fernández *et al.* 2002). The dibasic amino acids are then exported to the blood via the y^+L system (the basolateral heterodimer 4F2hc-y^+LAT-1 defective in LPI) in exchange for neutral amino acids plus sodium (Torrents *et al.* 1999). These transepithelial fluxes have recently been recapitulated by overexpression of rBAT-$b^{0,+}$AT and 4F2hc-y^+LAT-1 in the polarized cell line MDCK (Bauch *et al.* 2003), results which confirm the original model. Moreover, all the implicated proteins colocalize in the same tubule cells (Bauch *et al.* 2003) [i. e., segments S1 and S2, where the bulk of amino acid reabsorption occurs (Silbernagl 1988)]. Interestingly, 4F2hc-LAT-2 was found to colocalize in the same cells (Bauch *et al.* 2003). In an attempt to elucidate the role of LAT-2 in the transepithelial flux of amino acids, Fernández *et al.* (2003) employed an antisense strategy to knock down LAT-2 transport activity in the tubular epithelial OK cell line. Strikingly, they found that, under conditions that roughly mimicked the situation in the early proximal tubule (i.e., equal amino acid concentrations in the tubule lumen and in the blood), the intracellular cysteine concentration increased by up to two- to three-fold, whereas the concentrations of the other amino acids remain unchanged or (particularly for L-alanine, L-threonine, and L-serine) were slightly decreased. This strongly suggests a role for 4F2hc-LAT-2 in mediating the net efflux of cysteine from the epithelia, which makes it a good candidate for a modulating gene in cystinuria (Fernández *et al.* 2003). Therefore it seems that these heteromeric amino acid transporters function in epithelia as exchangers coupled in series (Bauch *et al.* 2003). As long as all of them function as 1 : 1 exchangers (Chillarón *et al.* 1996, 2001; Meier *et al.* 2002; Reig *et al.* 2002), it is clear that the pool of intracellular neutral amino acids serves itself as the driving force for transepithelial flux of cystine and dibasic amino acids. As stated above, unfortunately neither the neutral amino acids entry point (NBB) nor their necessary exit gate (T) are known at a molecular level. Although there

is early physiological data, which suggest the expected transport features of the putative NBB (Silbernagl 1988), there are virtually no clues as to the functional properties that the T efflux carrier should display (with the obvious exception that it cannot be a 1 : 1 exchanger and that it must show neutral amino acid substrate specificity). Another unsolved issue is which other transporters mediate dibasic amino acid (re)absorption through the apical membrane, as cystinuric patients showing almost no reabsorption of cystine still reabsorb significant (even 50% or more) amounts of dibasic amino acids (Crawhall *et al.* 1967; Frimpter *et al.* 1967). Candidates for this function include the $B^{0,+}$ system (Na^+ dependent for neutral and dibasic amino acids) (Van Winkle 1998) and the CAT family of facilitative dibasic amino acid transporters (Devés and Boyd 1998). Clearly, more studies are needed in order to uncover the complexity of transepithelial fluxes of amino acids and other solutes. Knockout mice and model cellular systems should provide new tools for thorough biochemical and physiological analysis. In this sense, the polarized OK cell model has proven to be very useful, at least for the study of heteromeric amino acid transporters. As stated above, a role for basolateral 4F2hc-LAT-2 in net cysteine efflux has been uncovered (Fernández *et al.* 2003). In the same study, the presence of basolateral 4F2hc-y^+LAT-1 was also shown, and in an earlier report the role of rBAT in apical cystine transport was also demonstrated (Mora *et al.* 1996). It is quite possible that OK cells carry most, if not all, proximal tubular amino acid carriers involved in reabsorption of these substrates.

Finally, the fact that whereas LPI provokes malnutrition, cystinuria does not (Segal and Thier 1995; Simell 2002), deserves comment. Physiological studies in the intestine showed that cystinuric patients are still able to absorb dibasic amino acids because of the presence of a proton/peptide symporter in the apical membrane (Silk *et al.* 1975) (PEPT in Figure 4). Most of the dibasic amino acid containing peptides are hydrolyzed inside the cell, and the dibasic amino acids are then exported to the blood through the heterodimer 4F2hc-y$^+$LAT-1. For this reason, and because of the lack of a basolateral peptide transporter, LPI patients, with defective y$^+$LAT-1, suffer from malnutrition. Recently, the intestinal proton/peptide symporter (PEPT-1) has been identified (Boll et al. 1994; Fei et al. 1994). A homologous renal isoform (PEPT-2) has also been identified (Boll et al. 1996), although its physiological role remains uncertain at present.

SUMMARY

Heteromeric amino acid transporters (HATs) are composed of two subunits, a polytopic membrane protein (the light subunit, LSHAT), and a disulfide-linked type II membrane glycoprotein (the heavy subunit, HSHAT). HATs represent several of the classic mammalian amino acid transport systems (e.g., L isoforms, y$^+$L isoforms, asc, xc-, and b$^{0,+}$). The light subunits confer the amino acid transport specificity to the HAT. y$^+$L is formed by the heavy subunit 4F2hc and the light subunit y$^+$LAT-1. Mutations in y$^+$LAT-1 cause y$^+$L transport defects, which are the molecular bases of the lysinuric protein intolerance (LPI) phenotype. The apically expressed heterodimer formed by the heavy subunit rBAT and the light subunit b$^{0,+}$AT is responsible for the bulk of cystine and part of the dibasic amino acid reabsorption in the proximal tubule. Mutations in rBAT or b$^{0,+}$AT lead to cystinuria. A genotype–phenotype correlation has been established for cystinuria, whereas the heterogenous LPI phenotype lacks a consistent explanation. Here we summarize recent advances in our understanding of these two inherited aminoacidurias, from the transport mechanistic level to the clinical manifestation of the diseases.

REFERENCES

Adibi, S.A. (1971). Intestinal Transport of Dipeptides in Man: Relative Importance of Hydrolysis and Intact Absorption. J. Clin. Invest., 50, 2266.

Angelo, S. and Devés, R. (1994). Amino Acid Transport System y$^+$L of Human Erythrocytes: Specificity and Cation Dependence of the Translocation Step. J. Membr. Biol., 141, 183–192.

Asatoor, A.M., Groughman, M.R., Harrison, A.R., Light, F.W., Loughridge, L.W., Milne, M.D., et al. (1971). Intestinal Absorption of Oligopeptides in Cystinuria. Clin. Sci., 41, 23.

Awrich, A.E., Stackhouse, J., Cantrell, J.E., Patterson, J.H., and Rudman, D. (1975). Hyperdibasicaminoaciduria, Hyperammonemia, and Growth Retardation: Treatment with Arginine, Lysine, and Citrulline. J. Pediatr., 87, 731.

Bauch, C., Forster, N., Loffing-Cueni, D., Summa, V., and Verrey, F. (2003). Functional Cooperation of Epithelial Heteromeric Amino Acid Transporters Expressed in Madin–Darby Canine Kidney Cells. J. Biol. Chem., 278, 1316–1322.

Bertran, J., Magagnin, S., Werner, A., Markovich, D., Biber, J., Testar, X., et al. (1992a). Stimulation of System y(+)-Like Amino Acid Transport by the Heavy Chain of Human 4F2 Surface Antigen in Xenopus laevis Oocytes. Proc. Natl Acad. Sci. USA, 89, 5606–5610.

Bertran, J., Werner, A., Moore, M.L., Stange, G., Markovich, D., Biber, J., et al. (1992b). Expression Cloning of a cDNA from Rabbit Kidney Cortex that Induces a Single Transport System for Cystine and Dibasic and Neutral Amino Acids. Proc. Natl Acad. Sci. USA, 89, 5601–5605.

Bertran, J., Werner, A., Chillarón, J., Nunes, V., Biber, J., Testar, X., et al. (1993). Expression Cloning of a Human Renal cDNA that Induces High Affinity Transport of L-Cystine Shared with Dibasic Amino Acids in Xenopus Oocytes. J. Biol. Chem., 268, 14842–14849.

Bisaccia, F., Zara, V., Capobianco, L., Iacobazzi, V., Mazzeo, M., and Palmieri, F. (1996). The Formation of a Disulfide Cross-Link between the Two Subunits Demonstrates the Dimeric Structure of the Mitochondrial Oxoglutarate Carrier. Biochim. Biophys. Acta, 1292, 281–288.

Bisceglia, L., Calonge, M.J., Totaro, A., Feliubadalo, L., Melchionda, S., Garcia, J., et al. (1997). Localization, by Linkage Analysis, of the Cystinuria Type III Gene to Chromosome 19q13.1. Am. J. Hum. Genet., 60, 611–616.

Boll, M., Markovich, D., Weber, W.M., Korte, H., Daniel, H., and Murer, H. (1994). Expression Cloning of a cDNA from Rabbit Small Intestine Related to Proton-Coupled Transport of Peptides, Beta-Lactam Antibiotics and ACE-Inhibitors. Pflügers Arch., 429, 146–149.

Boll, M., Herget, M., Wagener, M., Weber, W.M., Markovich, D., Biber, J., et al. (1996). Expression Cloning and Functional Characterization of the Kidney Cortex High-Affinity Proton-Coupled Peptide Transporter. Proc. Natl Acad. Sci. USA, 93, 284–289.

Borsani, G., Bassi, M.T., Sperandeo, M.P., De Grandi, A., Buoninconti, A., Riboni, M., et al. (1999). SLC7A7, Encoding a Putative Permease-Related Protein, is Mutated in Patients with Lysinuric Protein Intolerance. *Nat. Genet.*, **21**, 297–301.

Boyd, C.A., Devés, R., Laynes, R., Kudo, Y., and Sebastio, G. (2000). Cationic Amino Acid Transport Through System y$^+$L in Erythrocytes of Patients with Lysinuric Protein Intolerance. *Pflügers Arch.*, **439**, 513–516.

Broer, S. and Wagner, C.A. (2002). Structure–Function Relationships of Heterodimeric Amino Acid Transporters. *Cell Biochem. Biophys.*, **36**, 155–168.

Bröer, A., Wagner, C.A., Lang, F., and Bröer, S. (2000). The Heterodimeric Amino Acid Transporter 4F2hc/y$^+$LAT2 Mediates Arginine Efflux in Exchange with Glutamine. *Biochem. J.*, **349**, 787–795.

Busch, A.E., Herzer, T., Waldegger, S., Schmidt, F., Palacín, M., Biber, J., *et al.* (1994). Opposite Directed Currents Induced by the Transport of Dibasic and Neutral Amino Acids in *Xenopus* Oocytes Expressing the Protein rBAT. *J. Biol. Chem.*, **269**, 25581–25586.

Calonge, M.J., Gasparini, P., Chillarón, J., Chillon, M., Gallucci, M., Rousaud, F., *et al.* (1994). Cystinuria Caused by Mutations in rBAT, a Gene involved in the Transport of Cystine. *Nat. Genet.*, **6**, 420–425.

Calonge, M.J., Volpini, V., Bisceglia, L., Rousaud, F., DeSantis, L., Brescia, E., *et al.* (1995). Genetic Heterogeneity in Cystinuria: The rBAT Gene is Linked to Type I but not to Type III Cystinuria. *Proc. Natl Acad. Sci. USA*, **92**, 9667–9671.

Carpenter, T.O., Levy, H.L., Holtrop, M.E., Shih, V.E., and Anast, C.S. (1985). Lysinuric Protein Intolerance Presenting as Childhood Osteoporosis. Clinical and Skeletal Response to Citrulline Therapy. *N. Engl. J. Med.* **312**, 290–294.

Chairoungdua, A., Segawa, H., Kim, J.Y., Miyamoto, K., Haga, H., Fukui, Y., *et al.* (1999). Identification of an Amino Acid Transporter Associated with the Cystinuria-Related Type II Membrane Glycoprotein. *J. Biol. Chem.*, **274**, 28845–28848.

Chairoungdua, A., Kanai, Y., Matsuo, H., Inatomi, J., Kim, D.K., and Endou, H. (2001). Identification and Characterization of a Novel Member of the Heterodimeric Amino Acid Transporter Family Presumed to be Associated with an Unknown Heavy Chain. *J. Biol. Chem.*, **276**, 49390–49399.

Cheeseman, C.I. (1983). Characteristics of Lysine Transport Across the Serosal Pole of the Anuran Small Intestine. *J. Physiol.*, **338**, 87–97.

Chillarón, J., Estévez, R., Mora, C., Wagner, C.A., Suessbrich, H., Lang, F., *et al.* (1996). Obligatory Amino Acid Exchange via Systems b$^{0,+}$-Like and y$^+$L-Like. A Tertiary Active Transport Mechanism for Renal Reabsorption of Cystine and Dibasic Amino Acids. *J. Biol. Chem.*, **271**, 17761–17770.

Chillarón, J., Estévez, R., Samarzija, I., Waldegger, S., Testar, X., Lang, F., *et al.* (1997). An Intracellular Trafficking Defect in Type I Cystinuria rBAT Mutants M467T and M467K. *J. Biol. Chem.*, **272**, 9543–9549.

Chillarón, J., Roca, R., Valencia, A., Zorzano, A., and Palacín, M. (2001). Heteromeric Amino Acid Transporters: Biochemistry, Genetics, and Physiology. *Am. J. Physiol. Renal Physiol.*, **281**, F995–1018.

Christensen, H.N. (1984). Organic Ion Transport during Seven Decades. The Amino Acids. *Biochim. Biophys. Acta*, **779**, 255–269.

Christensen, H.N. (1990). Role of Amino Acid Transport and Countertransport in Nutrition and Metabolism. *Physiol. Rev.*, **70**, 43–77.

Coady, M.J., Chen, X.Z., and Lapointe, J.Y. (1996). rBAT is an Amino Acid Exchanger with Variable Stoichiometry. *J. Membr. Biol.*, **149**, 1–8.

Crawhall, J.C., Scowen, E.F., Thompson, C.J., and Watts, R.W.E. (1967). The Renal Clearance of Amino Acids in Cystinuria. *J. Clin. Invest.*, **46**, 1162–1171.

Dall'Asta, V., Bussolati, O., Sala, R., Rotoli, B.M., Sebastio, G., Sperandeo, M.P., *et al.* (2000). Arginine Transport through System y(+)L in Cultured Human Fibroblasts: Normal Phenotype of Cells from LPI Subjects. *Am. J. Physiol. Cell Physiol.*, **279**, C1829–C1837.

Dello Strologo, L., Carbonari, D., Gallucci, M., Gasparini, P., Bisceglia, L., Zelante, L., *et al.* (1997). Inter and Intrafamilial Clinical Variability in Patients with Cystinuria Type I and Identified Mutation. *J. Am. Soc. Nephrol.*, **8**, 388A.

Dello Strologo, L., Pras, E., Pontesilli, C., Beccia, E., Ricci-Barbini, V., de Sanctis, L., *et al.* (2002). Comparison between SLC3A1 and SLC7A9 Cystinuria Patients and Carriers: A Need for a New Classification. *J. Am. Soc. Nephrol.*, **13**, 2547–2553.

de Sanctis, L., Bonetti, G., Bruno, M., De Luca, F., Bisceglia, L., Palacín, M., *et al.* (2001). Cystinuria Phenotyping by Oral Lysine and Arginine Loading. *Clin. Nephrol.*, **56**, 467–474.

Desjeux, J.-F., Rajantie, J., Simell, O., Dumontier, A.-M., and Perheentupa, J. (1980). Lysine Fluxes across the Jejunal Epithelium in Lysinuric Protein Intolerance. *J. Clin. Invest.*, **65**, 1382–1387.

Devés, R. and Boyd, C.A. (1998). Transporters for Cationic Amino Acids in Animal Cells: Discovery, Structure, and Function. *Physiol. Rev.*, **78**, 487–545.

Devés, R. and Boyd, C.A. (2000). Surface Antigen CD98(4F2): Not a Single Membrane Protein, but a Family of Proteins with Multiple Functions. *J. Membr. Biol.*, **173**, 165–177.

Devés, R., Chavez, P., and Boyd, C.A.R. (1992). Identification of a New Transport System (y$^+$L) in Human Erythrocytes that Recognizes Lysine and Leucine with High Affinity. *J. Physiol.*, **454**, 491–501.

Devés, R., Angelo, S., and Chavez, P. (1993). N-ethylmaleimide Discriminates between Two Lysine Transport Systems in Human Erythrocytes. *J. Physiol.*, **468**, 753–766.

Dierks, T., Riemer, E., and Kramer, R. (1988). Reaction Mechanism of the Reconstituted Aspartate/Glutamate

Carrier from Bovine Heart Mitochondria. *Biochim. Biophys. Acta*, **943**, 231–244.

DiRocco, M., Garibotto, G., Rossi, G.A., Caruso, U., Taccone, A., Picco, P., *et al.* (1993). Role of Haematological, Pulmonary and Renal Complications in the Long-Term Prognosis of Patients with Lysinuric Protein Intolerance. *Eur. J. Pediatr.*, **152**, 437.

Eleno, N., Devés, R., and Boyd, C.A.R. (1994). Membrane Potential Dependence of the Kinetics of Cationic Amino Acid Transport Systems in Human Placenta. *J. Physiol.*, **479**, 291–300.

Estevez, R., Camps, M., Rojas, A.M., Testar, X., Devés, R., Hediger, M.A., *et al.* (1998). The Amino Acid Transport System y⁺L/4F2hc Is a Heteromultimeric Complex. *FASEB J.*, **12**, 1319–1329.

Fei, Y.J., Kanai, Y., Nussberger, S., Ganapathy, V., Leibach, F.H., Romero, M.F., *et al.* (1994). Expression Cloning of a Mammalian Proton-Coupled Oligopeptide Transporter. *Nature*, **368**, 563–566.

Feliubadalo, L., Font, M., Purroy, J., Rousaud, F., Estivill, X., Nunes, V., *et al.* (1999). Non-Type I Cystinuria Caused by Mutations in SLC7A9, Encoding a Subunit (b⁰,⁺AT) of rBAT. *Nat. Genet.*, **23**, 52–57.

Fenczik, C.A., Zent, R., Dellos, M., Calderwood, D.A., Satriano, J., Kelly, C., *et al.* (2001). Distinct domains of CD98hc Regulate Integrins and Amino Acid Transport. *J. Biol. Chem.*, **276**, 8746–8752.

Fernández, E., Carrascal, M., Rousaud, F., Abian, J., Zorzano, A., Palacín, M., *et al.* (2002). rBAT-b(0,+)AT Heterodimer is the Main Apical Reabsorption System for Cystine in the Kidney. *Am. J. Physiol. Renal Physiol.*, **283**, F540–F5488.

Fernández, E., Torrents, D., Chillarón, J., Martín del Río, R., Zorzano, A., and Palacín, M. (2003). Basolateral LAT-2 Has a Major Role in the Transepithelial Flux of L-Cystine in the Renal Proximal Tubule Cell Line, O.K. *J. Am. Soc. Nephrol.*, **14**, 837–847.

Font, M.A., Feliubadalo, L., Estivill, X., Nunes, V., Golomb, E., Kreiss, Y., *et al.* (2001). Functional Analysis of Mutations in SLC7A9, and Genotype/Phenotype Correlation in Non-Type I Cystinuria. *Hum. Mol. Genet.*, **10**, 305–316.

Frimpter, G.W., Horwith, M., Furth, E., Fellows, R.E., and Thompson, D.D. (1962). Inulin and Endogenous Amino Acid Renal Clearances in Cystinuria: Evidence for Tubular Secretion. *J. Clin. Invest.*, **41**, 281–288.

Fukasawa, Y., Segawa, H., Kim, J.Y., Chairoungdua, A., Kim, D.K., Matsuo, H., *et al.* (2000). Identification and Characterization of a Na⁺-Independent Neutral Amino Acid Transporter that Associates with the 4F2 Heavy Chain and Exhibits Substrate Selectivity for Small Neutral D- and L-Amino Acids. *J. Biol. Chem.*, **275**, 9690–9698.

Furriols, M., Chillarón, J., Mora, C., Castello, A., Bertran, J., Camps, M., *et al.* (1993). rBAT, Related to L-Cysteine Transport, is Localized to the Microvilli of Proximal Straight Tubules, and its Expression is Regulated in Kidney by Development. *J. Biol. Chem.*, **268**, 27060–27068.

Garrod, A.E. (1908). Inborn Errors of Metabolism (Lectures I–IV). *Lancet*, **ii**, 1–214.

Goodyer, P.R., Clow, C. Reade, T., and Girardin, C. (1993). Prospective Analysis and Classification of Patients with Cystinuria Identified in a Newborn Screening Program. *J. Pediatr.*, **122**, 568–572.

Groneberg, D.A., Doring, F., Eynott, P.R., Fischer, A., and Daniel, H. (2001). Intestinal Peptide Transport: Ex Vivo Uptake Studies and Localization of Peptide Carrier PEPT1. *Am. J. Physiol. Gastrointest. Liver Physiol.*, **281**, G697–704.

Henthorn, P.S., Liu, J., Gidalevich, T., Fang, J., Casal, M.L., Patterson, D.F., *et al.* (2000). Canine Cystinuria: Polymorphism in the Canine SLC3A1 Gene and Identification of a Nonsense Mutation in Cystinuric Newfoundland Dogs. *Hum. Genet.*, **107**, 295–303.

Iida, S., Peck, A.B., Jonson-Tardieu, J., Moriyama, M., Glenton, P.A., Byer, K.J., *et al.* (1999). Temporal Changes in mRNA Expresión for Bikunin in the Kidneys of Rats during Calcium Oxalate Nephrolithiasis. *J. Am. Soc. Nephrol.*, **10**, 986–996.

Janecek, S., Svensson, B., and Henrissat, B. (1997). Domain Evolution in the Alpha-Amylase Family. *J. Mol. Evol.*, **45**, 322–331.

Kanai, Y., Segawa, H., Miyamoto, K., Uchino, H., Takeda, E., and Endou, H. (1998). Expression Cloning and Characterization of a Transporter for Large Neutral Amino Acids Activated by the Heavy Chain of 4F2 Antigen (CD98). *J. Biol. Chem.*, **273**, 23629–23632.

Kanai, Y., Fukasawa, Y., Cha, S.H., Segawa, H., Chairoungdua, A., Kim, D.K., *et al.* (2000). Transport Properties of a System y⁺L Neutral and Basic Amino Acid Transporter. Insights into the Mechanisms of Substrate Recognition. *J. Biol. Chem.*, **275**, 20787–20793.

Kekomaki, M., Visakorpi, J.K., Perheentupa, J., and Saxen, L. (1967). Familial Protein Intolerance with Deficient Transport of Basic Amino Acids. An Analysis of 10 Patients. *Acta Paediatr. Scand.*, **56**, 617.

Kim, J.W., Closs, E.I., Albritton, L.M., and Cunningham, J.M. (1991). Transport of Cationic Amino Acids by the Mouse Ecotropic Retrovirus Receptor. *Nature*, **352**, 725–728.

Lauteala, T., Sistonen, P., Savontaus, M.-L., Mykkänen, J., Simell, J., Lukkarinen, M., *et al.* (1997). Lysinuric Protein Intolerance (LPI) Gene Maps to the Long Arm of Chromosome 14. *Am. J. Hum. Genet.*, **60**, 1479.

Lauteala, T., Mykkanen, J., Sperandeo, M.P., Gasparini, P., Savontaus, M.L., *et al.* (1998). Genetic Homogeneity of Lysinuric Protein Intolerance. *Eur. J. Hum. Genet.*, **6**, 612.

Lee, W.S., Wells, R.G., Sabbag, R.V., Mohandas, T.K., and Hediger, M.A. (1993). Cloning and Chromosomal Localization of a Human Kidney cDNA Involved in Cystine, Dibasic, and Neutral Amino Acid Transport. *J. Clin. Invest.*, **91**, 1959–1963.

Mannion, B.A., Kolesnikova, T.V., Lin, S.H., Wang, S., Thompson, N.L., and Hemler, M.E. (1998). The Light Chain of CD98 Is Identified as E16/TA1 Protein. *J. Biol. Chem.*, **273**, 33127–33129.

Mastroberardino, L., Spindler, B., Pfeiffer, R., Skelly, P.J., Loffing, J., Shoemaker, C.B., *et al.* (1998). Amino-Acid Transport by Heterodimers of 4F2hc/CD98 and Members of a Permease Family. *Nature*, **395**, 288–291.

Matsuo, H., Kanai, Y., Kim, J.Y., Chairoungdua, A., Kim D.K., Inatomi, J., *et al.* (2002). Identification of a Novel Na$^+$-Independent Acidic Amino Acid Transporter with Structural Similarity to the Member of a Heterodimeric Amino Acid Transporter Family Associated with Unknown Heavy Chains. *J. Biol. Chem.*, **277**, 21017–21026.

Matthews, D.M. and Adibi, S.A. (1976). Peptide Absorption. *Gastroenterology*, **71**, 151.

Meier, C., Ristic, Z., Klauser, S., and Verrey, F. (2002). Activation of System L Heterodimeric Amino Acid Exchangers by Intracellular Substrates. *EMBO J.*, **21**, 580–589.

Mizoguchi, K., Cha, S.H., Chairoungdua, A., Kim, D.K., Shigeta, Y., Matsuo, H., *et al.* (2001). Human Cystinuria-Related Transporter: Localization and Functional Characterization. *Kidney Int.*, **59**, 1821–1833.

Mora, C., Chillarón, J., Calonge, M.J., Forgo, J., Testar, X., Nunes, V., *et al.* (1996). The rBAT Gene Is Responsible for L-Cystine Uptake via the b0,(+)-Like Amino Acid Transport System in a "Renal Proximal Tubular" Cell Line (OK Cells). *J. Biol. Chem.*, **271**, 10569–10576.

Mosckovitz, R., Udenfriend, S., Felix, A., Heimer, E., and Tate, S.S. (1994). Membrane Topology of the Rat Kidney Neutral and Basic Amino Acid Transporter. *FASEB J.*, **8**, 1069–1074.

Mykkanen, J., Torrents, D., Pineda, M., Camps, M., Yoldi, M.E., Horelli-Kuitunen, N., *et al.* (2000). Functional Analysis of Novel Mutations in y(+)LAT-1 Amino Acid Transporter Gene Causing Lysinuric Protein Intolerance (LPI). *Hum. Mol. Genet.*, **9**, 431–438.

Nagata, M., Suzuki, M., Kawamura, G., Kono, S., Koda, N., Yamaguchi, S., *et al.* (1987). Immunological Abnormalities in a Patient with Lysinuric Protein Intolerance. *Eur. J. Pediatr.*, **146**, 427.

Nozaki, J., Dakeishi, M., Ohura, T., Inoue, K., Manabe, M., Wada, Y., *et al.* (2001). Homozygosity Mapping to Chromosome 5p15 of a Gene Responsible for Hartnup Disorder. *Biochem. Biophys. Res. Commun.*, **284**, 255–260.

Oyanagi, K., Miura, R., and Yamanouchi, T. (1970). Congenital Lysinuria: A New Inherited Transport Disorder of Dibasic Amino Acids. *J. Pediatr.*, **77**, 259.

ılacín, M., Estévez, R., Bertran, J., and Zorzano, A. ⁻1998). Molecular Biology of Mammalian Plasma ⁻embrane Amino Acid Transporters. *Physiol. Rev.*, **78**, ⁻–1054.

Palacín, M., Bertran, J., and Zorzano, A. (2000). Heteromeric Amino Acid Transporters Explain Inherited Aminoacidurias. *Curr. Opin. Nephrol. Hypertens.*, **9**, 547–553.

Palacín, M., Borsani, G., and Sebastio, G. (2001a). The Molecular Bases of Cystinuria and Lysinuric Protein Intolerance. *Curr. Opin. Genet. Dev.*, **11**, 328–335.

Palacín, M., Goodyer, P., Nunes, V., and Gasparini, P. (2001b). Cystinuria. In C.R. Scriver, A.L. Beaudet, S.W. Sly, and D. Valle (eds), *Metabolic and Molecular Bases of Inherited Diseases* (8th edn), New York: McGraw-Hill, pp. 4909–4932.

Parto, K., Svedstrom, E., Majurin, M.L., Harkonen, R., and Simell, O. (1993). Pulmonary Manifestations in Lysinuric Protein Intolerance. *Chest*, **104**, 1176.

Perheentupa, J. and Visakorpi, J.K. (1965). Protein Intolerance with Deficient Transport of Basic Amino Acids. *Lancet*, **ii**, 813.

Pfeiffer, R., Spindler, B., Loffing, J., Skelly, P.J., Shoemaker, C.B., and Verrey, F. (1998). Functional Heterodimeric Amino Acid Transporters Lacking Cystine Residues Involved in Disulfide Bond. *FEBS Lett.*, **439**, 157–162.

Pfeiffer, R., Loffing, J., Rossier, G., Bauch, C., Meier, C., Eggermann, T., *et al.* (1999a). Luminal Heterodimeric Amino Acid Transporter Defective in Cystinuria. *Mol. Biol. Cell.*, **10**, 4135–4147.

Pfeiffer, R., Rossier, G., Spindler, B., Meier, C., Kuhn, L., and Verrey F. (1999b). Amino Acid Transport of y$^+$L-Type by Heterodimers of 4F2hc/CD98 and Members of the Glycoprotein-Associated Amino Acid Transporter Family. *EMBO J.*, **18**, 49–57.

Pickel, V.M., Nirenberg, M.J., Chan, J., Mosckovitz, R., Udenfriend, S., and Tate, S.S. (1993). Ultrastructural Localization of a Neutral and Basic Amino Acid Transporter in Rat Kidney and Intestine. *Proc. Natl Acad. Sci. USA*, **90**, 7779–7783.

Pineda, M., Fernández, E., Torrents, D., Estévez, R., Lopez, C., Camps, M., *et al.* (1999). Identification of a Membrane Protein, LAT-2, that Co-expresses with 4F2 Heavy Chain, an L-Type Amino Acid Transport Activity with Broad Specificity for Small and Large Zwitterionic Amino Acids. *J. Biol. Chem.*, **274**, 19738–19744.

Rajantie, J., Simell, O., and Perheentupa, J. (1980a). Basolateral Membrane Transport Defect for Lysine in Lysinuric Protein Intolerance. *Lancet*, **i**, 1219–1221.

Rajantie, J., Simell, O., and Perheentupa, J. (1980b). Intestinal Absorption in Lysinuric Protein Intolerance: Impaired for Diamino Acids, Normal for Citrulline. *Gut*, **21**, 519.

Rajantie, J., Simell, O., Rapola, J., and Perheentupa, J. (1980c). Lysinuric Protein Intolerance: A Two-Year Trial of Dietary Supplementation Therapy with Citrulline and Lysine. *J. Pediatr.*, **97**, 927.

Rajantie, J. and Simell, O. (1981). Lysinuric Protein Intolerance. Basolateral Membrane Transport Defect in Renal Tubuli. *J. Clin. Invest.*, **67**, 1078–1082.

Rajantie, J., Simell, O., and Perheentupa, J. (1983a). Oral administration of ε-N-acetyllysine and Homocitrulline for Lysinuric Protein Intolerance. *J. Pediatr.*, **102**, 388.

Rajantie, J., Simell, O., and Perheentupa, J. (1983b). Oral Administration of Urea Cycle Intermediates in Lysinuric Protein Intolerance: Effect on Plasma and Urinary Arginine and Ornithine. *Metabolism*, **32**, 49.

Reig, N., Chillarón, J., Bartoccioni, P., Fernández, E., Bendahan, A., Zorzano, A., et al. (2002). The Light Subunit of System b$^{o,+}$ is Fully Functional in the Absence of the Heavy Subunit. *EMBO J.*, **21**, 4906–4914.

Rosenberg, L.E., Downing, S., Durant, J.L., and Segal, S. (1967). Cystinuria: Biochemical Evidence of Three Genetically Distinct Diseases. *J. Clin. Invest.*, **46**, 30.

Rossier, G., Meier, C., Bauch, C., Summa, V., Sordat, B., Verrey, F., et al. (1999). LAT2, a New Basolateral 4F2hc/CD98-Associated Amino Acid Transporter of Kidney and Intestine. *J. Biol. Chem.*, **274**, 34948–34954.

Saadi, I., Chen, X.Z., Hediger, M., Ong, P., Pereira, P., Goodyer, P., et al. (1998). Molecular Genetics of Cystinuria: Mutation Analysis of *SLC3A1* and Evidence for Another Gene in Type I (Silent) Phenotype. *Kidney Int.*, **54**, 48–55.

Saier, M.H., Jr, Beatty, J.T., Goffeau, A., Harley, K.T., Heijne, W.H., Huang, S.C., et al. (1999). The Major Facilitator Superfamily. *J. Mol. Microbiol. Biotechnol.*, **1**, 257–279.

Sato, H., Tamba, M., Ishii, T., and Bannai, S. (1999). Cloning and Expression of a Plasma Membrane Cystine/Glutamate Exchange Transporter Composed of Two Distinct Proteins. *J. Biol. Chem.*, **274**, 11455–11448.

Schroers, A., Burkovski, A., Wohlrab, H., and Kramer, R. (1998). The Phosphate Carrier from Yeast Mitochondria. Dimerization is a Prerequisite for Function. *J. Biol. Chem.*, **273**, 14269–14276.

Segal, S. and Thier, S.O. (1995). Cystinuria. In C.H. Scriver, A.L. Beaudet, W.S. Sly, and D. Valle (eds), *Metabolic and Molecular Bases of Inherited Diseases*, New York: McGraw-Hill, pp. 2479–2496.

Shoji, Y., Noguchi, A., Shoji, Y., Matsumori, M., Takasago, Y., Takayanagi, M., et al. (2002). Five Novel SLC7A7 Variants and y$^+$L Gene-Expression Pattern in Cultured Lymphoblasts from Japanese Patients with Lysinuric Protein Intolerance. *Hum. Mutat.*, **20**, 375–381.

Silbernagl, S. (1988). The Renal Handling of Amino Acids and Oligopeptides. *Physiol. Rev.*, **68**, 911–1007.

Silk, D.B., Perrett, D., and Clark, M.L. (1975). Jejunal and Ileal Absorption of Dibasic Amino Acids and an Arginine-Containing Dipeptide in Cystinuria. *Gastroenterology*, **68**, 1426–1432.

Simell, O. (2002). In "The Metabolic and Molecular Bases of Inherited Disease"; Part 21 (Membrane Transport Disorders) Chapter 192 (Lysinuric Protein Intolerance and other Cationic Aminoacidurias). Available at http://genetics.accessmedicine.com.

Simell, O. and Perheentupa, J. (1974). Renal Handling of Diamino Acids in Lysinuric Protein Intolerance. *J. Clin. Invest.*, **54**, 9–17.

Simell, O., Perheentupa, J., Rapola, J., Visakorpi, J.K., and Eskelin, L.-E. (1975). Lysinuric Protein Intolerance. *Am. J. Med.*, **59**, 229.

Smith, D.W., Scriver, C.R., and Simell, O. (1988). Lysinuric Protein Intolerance Mutation is not Expressed in the Plasma Membrane of Erythrocytes. *Hum. Genet.*, **80**, 395–396.

Sperandeo, M.P., Bassi, M.T., Riboni, M., Parenti, G., Buoninconti, A., Manzoni, M., et al. (2000). Structure of the SLC7A7 Gene and Mutational Analysis of Patients Affected by Lysinuric Protein Intolerance. *Am. J. Hum. Genet.*, **66**, 92–99.

Stoller, M.L., Bruce, J.E., Bruce, C.A., Foroud, T., Kirkwood, S.C., and Stambrook, P.J. (1999). Linkage of Type II and Type III Cystinuria to 19q13.1: Codominant Inheritance of Two Cystinuric Alleles at 19q13.1 Produces an Extreme Stone-Forming Phenotype. *Am. J. Med. Genet.*, **86**, 134–139.

Tate, S.S., Yan, N., and Udenfriend, S. (1992). Expression Cloning of a Na$^+$-Independent Neutral Amino Acid Transporter from Rat Kidney. *Proc. Natl Acad. Sci. USA*, **89**, 1–5.

Toivonen, M., Mykkanen, J., Aula, P., Simell, O., Savontaus, M.L., and Huoponen, K. (2002). Expression of Normal and Mutant GFP-Tagged y$^+$L Amino Acid Transporter-1 in Mammalian Cells. *Biochem. Biophys. Res. Commun.*, **291**, 1173–1179.

Torras-Llort, M., Torrents, D., Soriano-Garcia, J.F., Gelpi, J.L., Estévez, R., Ferrer, R., et al. (2001). Sequential Amino Acid Exchange Across b$^{0,+}$-Like System in Chicken Brush Border Jejunum. *J. Membr. Biol.*, **180**, 213–220.

Torrents, D., Estévez, R., Pineda, M., Fernández, E., Lloberas, J., Shi, Y.B., et al. (1998). Identification and Characterization of a Membrane Protein (y$^+$L Amino Acid Transporter-1) that Associates with 4F2hc to Encode the Amino Acid Transport Activity y$^+$L. A Candidate Gene for Lysinuric Protein Intolerance. *J. Biol. Chem.*, **273**, 32437–32445.

Torrents, D., Mykkanen, J., Pineda, M., Feliubadalo, L., Estévez, R., de Cid, R., et al. (1999). Identification of *SLC7A7*, Encoding y$^+$LAT-1, as the Lysinuric Protein Intolerance Gene. *Nat. Genet.*, **21**, 293–296.

Van Winkle, L.J. (1988). Amino Acid Transport in Developing Animal Oocytes and Early Conceptuses. *Biochim. Biophys. Acta*, **947**, 173–208.

Veenhoff, L.M., Heuberger, E.H.M.L., and Poolman, B. (2002). Quaternary Structure and Function of Transport Proteins. *Trends Biochem. Sci.*, **27**, 242–249.

Verrey, F., Jack, D.L., Paulsen, I.T., Saier, M.H., Jr, and Pfeiffer, R. (1999). New Glycoprotein-Associated Amino Acid Transporters. *J. Membr. Biol.*, **172**, 181–192.

Verrey, F., Meier, C., Rossier, G., and Kuhn, L.C. (2000). Glycoprotein-Associated Amino Acid

Exchangers: Broadening the Range of Transport Specificity. *Pflügers Arch.*, **440**, 503–512.

Völkl, H. and Silbernagl, S. (1982). Mutual Inhibition of L-Cystine/L-Cysteine and Other Neutral Amino Acids during Tubular Reabsorption. A Microperfusion Study in Rat Kidney. *Pflügers Arch.*, **395**, 190–195.

Wang, H., Kavanaugh, M.P., North, R.A., and Kabat, D. (1991). Cell-Surface Receptor for Ecotropic Murine Retroviruses is a Basic Amino-Acid Transporter. *Nature*, **352**, 729–731.

Wartenfeld, R., Golomb, E., Katz, G., Bale, S.J., Goldman, B., Pras, M., *et al.* (1997). Molecular Analysis of Cystinuria in Libyan Jews: Exclusion of the *SLC3A1* Gene and Mapping of a New Locus on 19q. *Am. J. Hum. Genet.*, **60**, 617–624.

Watanabe, K., Hata, Y., Kizaki, H., Katsube, Y., and Suzuki, Y. (1997). The Refined Crystal Structure of *Bacillus cereus* Oligo-1,6-Glucosidase at 2.0 Å Resolution: Structural Characterization of Proline-Substitution Sites for Protein Thermostabilization. *J. Mol. Biol.*, **269**, 142–153.

Wells, R.G. and Hediger, M.A. (1992). Cloning of a Rat Kidney cDNA that Stimulates Dibasic and Neutral Amino Acid Transport and has Sequence Similarity to Glucosidases. *Proc. Natl. Acad. Sci. USA*, **89**, 5596–5600.

Wells, R.G., Lee, W.S., Kanai, Y., Leiden, J.M., and Hediger, M.A. (1992). The 4F2 Antigen Heavy Chain Induces Uptake of Neutral and Dibasic Amino Acids in *Xenopus* Oocytes. *J. Biol. Chem.*, **267**, 15285–15288.

White, M.F. (1985). The Transport of Cationic Amino Acids across the Plasma Membrane of Mammalian Cells. *Biochim. Biophys. Acta*, **822**, 355–374.

Xie, Y., Sakatsume, M., Nishi, S., Narita, I., Arakawa, M., and Gejyo, F. (2001). Expression, Roles, Receptors, and Regulation of Osteopontin in the Kidney. *Kidney Int.*, **60**, 1645–1657.

15

FRANS W. VERHEIJEN* AND GRAZIA M.S. MANCINI*

Lysosomal sialic acid transporter sialin (SLC17A5): sialic acid storage disease (SASD)

INTRODUCTION

Lysosomes are intracellular organelles acidified by a vacuolar proton pump. They contain a wide variety of acid hydrolases for degradation of intra- and extracellular macromolecules and are surrounded by a single lipid bilayer membrane. Initially the lysosomal membrane was considered to be only a mechanical border separating the acid lysosomal environment from the neutral environment of the surrounding cytoplasm. Therefore lysosomes were considered the "terminal degradative compartment" of the cell and their function was strictly linked to cellular catabolism. However, two inborn errors of metabolism – cystinosis and sialic acid storage disorders – have contributed to a more thorough understanding of lysosomal membrane transport function. As we know now, the lysosomal membrane contains special transport proteins for both export and import (Mancini *et al.* 2000). A few "importers" are known; these are responsible for the uptake of a variety of small molecules or ions, suggesting a role for the lysosome in the regulation of

certain metabolic processes. Many "exporters" have been characterized; these allow the small degradation products to leave the lysosome to be either reutilized or excreted by the cell. Exporters have been described for a variety of small molecules including amino acids, sugars, and ions. All these transporters have a particular substrate specificity. Two human genetic disorders, cystinosis and sialic acid storage disease, are due to defective export systems. In this chapter we will focus on studies on sialic acid storage diseases (SASDs), recessively inherited neurodegenerative lysosomal storage disorders characterized by an excessive tissue storage and excretion of the carboxylated monosaccharide sialic acid (*N*-acetylneuraminic acid in humans). There are three clinically distinct forms of this disorder: (1) Salla disease, (2) infantile sialic acid storage disease (ISSD), and (3) intermediate phenotypes. These diseases are all caused by mutations in the SLC17A5 gene, with a genotype–phenotype correlation to some extent. The clinical form frequent in northern Finland (Salla disease) produces from early infancy on a slowly progressive neurodegeneration associated with disturbed myelination in the central nervous system (Aula *et al.* 1979). The infantile type of the disease (ISSD) entails generalized storage with visceromegaly,

*Department of Clinical Genetics, Erasmus MC, PO Box 1738, 3000 DR Rotterdam, The Netherlands

cardiomyopathy, nephropathy, dysostosis, fetal hydrops, and outcome within the first years. The symptoms of ISSD start before birth, suggesting the importance of SLC17A5 during fetal development. The existence of intermediate phenotypes between Salla disease and ISSD has also been described (Mancini *et al.* 1992b, 2000). Both Salla disease and ISSD are allelic disorders mapping to chromosome 6q14-q15. Diagnosis of all forms of SASD is based on the demonstration of abnormal excretion of the free (non-oligosaccharide bound) acid monosaccharide sialic acid in urine (sialuria) (Aula *et al.* 1986), and accumulation in cultured fibroblasts, and on microscope evidence of increased and swollen lysosomes, filled with light fibrillo-granular material. This distinguishes SASD from other genetic defects in sialic acid metabolism (discussed later). Prenatal diagnosis is possible, based on sialic acid determination in chorionic villi and cultured amniotic fluid cells (Salomaki *et al.* 2001). However, mutation analysis of SLC17A5 is preferable because the level of sialic acid in prenatal tissue can vary.

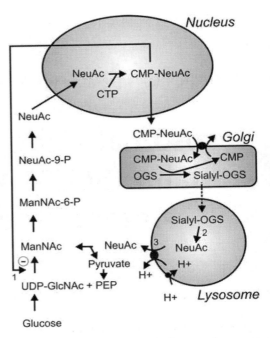

Figure 1   Sialic acid metabolism and genetic defects.

SIALIC ACID METABOLISM AND BIOCHEMISTRY OF TRANSPORT

Sialic acids are essential components of oligosaccharides (OGSs) present in either glycoproteins or gangliosides. OGS-bound sialic acids are involved in main metabolic functions such as immunological processes, hormonal responses, signal transmission in neurons, tumor progression, cell adhesion and migration, and protection from premature degradation (Schauer 2000). The essential steps of intracellular sialic acid metabolism are schematically depicted in Figure 1. Three different genetic metabolic defects in sialic acid metabolism are known, and are indicated in Figure 1: (1) sialuria (OMIM 269921), a feedback inhibition defect in sialic acid biosynthesis (Weiss *et al.* 1989), (2) sialidosis (OMIM 256550), a breakdown defect of sialyloligosaccharides caused by a defect of lysosomal sialidase, and

(3) SASD (OMIM 604369 and 269920), a specific lysosomal membrane transporter defect and the topic of this chapter.

Biochemical studies of lysosomal sialic acid transport demonstrated that an H^+/anionic sugar symporter mechanism is present in the lysosomal membrane (Mancini *et al.* 1989). The lysosomal sialic transport protein is a carrier specific for acid monosaccharides like sialic acid and glucuronic acid, which are degradation products of glycoproteins, glycolipids, or glycosaminoglycans. The existence and properties of the carrier were clarified in vitro using lysosomal membrane vesicles for transport studies, a technique used for reliable kinetic characterization of transport systems. Transport of these acid monosaccharides across the lysosomal membrane is a carrier-mediated process driven by a proton gradient that is maintained by the lysosomal proton pump (Figure 2).

Subsequent studies of the sialic acid transporter in lysosomal membranes from human

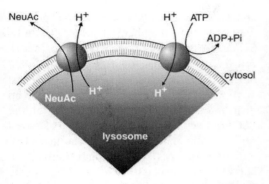

Figure 2 The sialic acid transporter is a proton gradient driven symporter.

fibroblasts demonstrated that the H^+-driven transport of both sialic acid and glucuronic acid is deficient in patients with the different clinical forms of SASD (Mancini *et al.* 1991). Biochemical evidence that the transport defect represents the primary genetic mutation came from the observation of intermediate transport rates in obligate heterozygotes for this autosomal recessive disease. A functional reconstitution system for the sialic acid transporter into proteoliposomes was developed (Mancini *et al.* 1992a). This system provided the tool to start the purification of the lysosomal sialic acid transporter from rat liver lysosomal membranes (Havelaar *et al.* 1998). Functional characterization studies showed that the lysosomal sialic acid carrier transports other nonsugar mono- and dicarboxylated anions, like L-lactate and α-ketoglutarate, as well as structurally different acid monosaccharides. This suggested that the transport protein shared functional similarities with certain anion transporters of well classified transporter families (Havelaar *et al.* 1999).

THE SIALIC ACID TRANSPORTER PROTEIN

The purification of the lysosomal sialic acid transporter initiated a number of detailed studies aimed at understanding its physiological function (Havelaar *et al.* 1998, 1999). The transport activity correlated with a 57 kDa protein isolated from rat liver lysosomes. When the purified 57 kDa protein was reconstituted in artificial proteoliposomes (artificial membrane vesicles suitable for transport studies), it appeared that the same protein was sufficient to reproduce transport function with the same biochemical properties as the carrier in its natural membrane. These studies demonstrate that the anion binding and proton binding sites are present on the same protein. Biochemical studies further suggested that at the anion (sialic acid) binding sites arginine residues were essential for transport function. Characterization of the purified protein revealed that, as well as acidic monosaccharides, other (nonsugar) mono- and dicarboxylated anions are also transported by this protein. In other words, the lysosomal sialic acid transporter has a wide substrate specificity for carboxylated anions. In this respect the sialic acid transporter is unique among the lysosomal transport proteins, since most of these appear to have a very restricted substrate specificity. It would be interesting to know whether the wide substrate specificity of the sialic acid transporter has implications for the pathophysiology of SASD. So far, only a defective transport of sialic acid and glucuronic acid has been established in lysosomal membrane vesicles of fibroblasts from patients with SASD (Renlund *et al.* 1986; Mancini *et al.* 1991). Sialic acid has been demonstrated to accumulate extensively in patient's lysosomes, whereas only a slight lysosomal accumulation of glucuronic acid has been observed. Studies of the contribution of different anions in the pathophysiology of SASD are lacking and might provide new insight into the cause of clinical heterogeneity. The role of glucuronic acid transport in the disease is unclear. Glucuronic acid is a main component of glycosaminoglycans. These are mainly degraded by intralysosomal soluble glycohydrolases. Defects of these hydrolases lead to different forms of mucopolysaccharidoses. It is possible that a defect in the recycling of lysosome-derived glucuronic acid contributes to some of the mucopolysaccharidosis-like signs and symptoms of SASD, mostly as seen in ISSD (visceromegaly, skeletal dysostosis, cardiopathy, and nephropathy). A reduction in the turnover rate of gangliosides, sialioglycoconjugates, and sphingolipids has been observed in SASD fibroblasts as a result of abnormal sialic acid recycling (Pitto *et al.* 1996). It is possible that the disturbance of myelination in severe Salla disease patients, observed with brain imaging (MRI) and pathological studies (Haataja *et al.* 1994), is due to lysosomal accumulation of gangliosides, the main sialic-acid-containing sphingolipids and the major components of myelin. Recent studies have shown the essential role of sialic acid metabolism in progenitor neural cell differentiation (Schwarzkopf *et al.* 2002).

THE SIALIC ACID TRANSPORTER GENE

On the basis of extensive functional characterization, the sialic acid transporter showed similarities with a group of selected anion transporter families with overlapping substrate specificities and transport properties (Havelaar *et al.* 1999). A group of candidate transporter gene families, among which are members of the anion/cation symporter (ACS) family, was considered a good starting point for the identification of the sialic acid tranporter gene. With a genetic linkage approach in Finnish families, the locus for Salla disease was assigned to a region of approximately 200 kb on chromosome 6q14-15 (Schleutker *et al.* 1995a,b; Leppänen *et al.* 1996). Salla disease and ISSD were also shown to be allelic disorders. A search of the Expressed Sequence Tag (EST) database for transporter genes mapping to the known critical SASD region on 6q14-q15 pointed to a strong functional candidate gene belonging to one of the previously selected anion transporter families, the ACS family. Several ESTs mapping near markers in the critical region of SASD proved to belong to the same gene. Such ESTs showed significant homology with mammalian Na^+/phosphate

and bacterial H^+/glucuronate cotransporters, which are both members of the ACS family. Sequence analysis of the ESTs provided a full-length sequence of a new gene, called *SLC17A5* (Verheijen *et al.* 1999). The gene has an open reading frame (ORF) of 1485 bp distributed over 11 exons. Analysis of the *SLC17A5* sequence in SASD patients showed several pathogenic mutations. The gene was initially called AST (anion sugar transporter), but was officially classified as *SLC17A5*, solute carrier number 17A5. The protein encoded by this gene was called sialin. Mutation analysis of this cDNA identified a possible founder mutation in the Finnish patients with Salla disease and a number of deletions and insertions in the ISSD patients (Verheijen *et al.* 1999; Aula *et al.* 2000). Finnish Salla disease patients are homozygous for an R39C mutation of *SLC17A5*. The mutations identified in ISSD patients and in intermediate phenotypes included deletions and insertions predicted to inactivate sialin function (Table 1).

The predicted size of the protein encoded by the *SLC17A5* cDNA is 495 amino acids, which is in agreement with the molecular mass of the purified sialic acid transport protein observed on SDS–PAGE as 57 kDa. Sialin contains seven potential N-glycosylation sites and 12 transmembrane domains, characteristic of a transporter of the major facilitator superfamily (MFS) (Pao *et al.* 1999) to which the ACS family belongs (Figure 3). Functional expression studies (A.C. Havelaar *et al.*, unpublished results) have now shown that the biochemical properties of sialin are identical to the earlier purified lysosomal rat liver sialic acid transporter and that mutations lead to defective sialic acid and glucuronic acid transport in vitro. Therefore it can be concluded that mutations in *SLC17A5* are the primary cause of lysosomal SASD.

The ACS family contains not only anionic sugar transporters, but also Na^+/phosphate symporters (Pao *et al.* 1998). Sialin, the lysosomal sialic acid transporter, has considerable homology with a number of sodium-coupled (phosphate) anion transporters, of which

Table 1 Overview of mutations in SLC17A5 in SASD, Salla disease, intermediate phenotype, and ISSD

Nucleotide change	Mutation in protein	Exon	Predicted consequence in protein
43G > T	E15X	1	Premature truncation
115C > T[a]	R39C	2	Nonfunctional protein
329G > A	W110X	3	Premature truncation
533delC	178, del 1bp	4	Premature truncation
(525–819)del98bp	del98 aa TM3-7	4, 5, 6	Premature truncation
548A > G	H183R	4	His to arg in TM4
719G > A[b]	W240X	6	Premature truncation
753insCCA[c]	252insP	6	Pro ins cytosolic loop between TM6–7
(802–816)del15bp	268,del[SSLRN]	6	Deletion of 5 aa
(978–979)ins500bp	327, ins500bp	7	Premature truncation
991insA[b]	S331I	8	Premature truncation
1001C > G	P334R	8	Pro to arg in TM8
(1112–1259)del148bp	frameshift	9	Premature truncation
(1138–1139)delGT[b]	frameshift	9	Premature truncation

[a] Homozygous form causes classic Salla disease phenotype.
[b] Causes intermediate phenotype in combination with 115C > T.
[c] Causes intermediate phenotype in homozygous state.
All other mutations have been observed in ISSD patients.

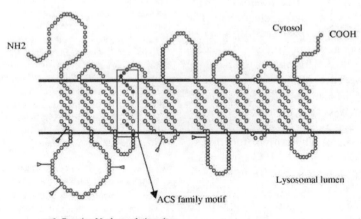

Figure 3 Putative model of sialin, the lysosomal sialic acid transporter. The gene product sialin consists of 495 amino acids with six putative N-glycosylation sites. Hydropathy analysis predicts 12 putative transmembrane domains, with characteristics of the major facilitator superfamily of membrane carriers. Sialin is homologous with the ACS family.

several (BNPI and DNPI) have now been identified as glutamate transporters of synaptic vesicles (VGLUTs) (Bellocchio *et al.* 2000; Hayashi *et al.* 2001). Expression studies indicate that these glutamate transporters are not sodium coupled but are driven by the proton gradient and the membrane potential present across the membrane of synaptic vesicles, with a mechanism similar to sialic acid transport across the lysosomal membrane.

SUMMARY

The lysosomal sialic acid transporter sialin, coded by the *SLC17A5* gene, is defective in sialic acid storage disease (SASD). SASDs are autosomal recessive neurodegenerative disorders that may present as a severe infantile form (ISSD) or a slowly progressive adult form, prevalent in northern Finland, which is referred to as Salla disease because of its geographic distribution. Intermediate forms are also present. The main clinical symptoms are hypotonia, cerebellar ataxia, and mental retardation. In infantile cases visceromegaly and coarse features are also present. A peculiar diffuse developmental defect of the cerebral and cerebellar white matter (hypomyelination and dysmyelination) is observed at brain MRI scans. Biochemically these disorders are characterized by accumulation of the free monosaccharide sialic acid in the lysosomal compartment. The diagnosis is based on the demonstration of abnormal excretion in urine (sialuria) and accumulation in cultured fibroblasts and on the microscope evidence of swollen lysosomes. The acid monosaccharides sialic acid and glucuronic acid are lysosomal breakdown products of sialyloligosaccharides, sialoglycolipids, and glycosaminoglycans. The normal function of sialin, which is present in the lysosomal membrane, is to export these sugars from the lysosome to the cytosol where they can be reutilized for synthesis. Sialin functions as a proton-gradient-dependent transporter. It belongs to the anion/cation symporter family of 12 transmembrane domain transporters and has considerable homology with glutamate transporters of synaptic vesicles.

REFERENCES

Aula, P., Autio, S., Raivio, K.O., Rapola, J., Thodén, C.J., Koskela, S.L., et al. (1979). "Salla disease." A New Lysosomal Storage Disorder. Arch. Neurol., 36, 88–94.

Aula, P., Renlund, M., and Raivio, K. (1986). Screening of Inherited Oligosaccharidurias among Mentally Retarded Patients in Northern Finland. J. Ment. Defic. Res., 30, 365–368.

Aula, N., Salomaki, P., Timonen, R., Verheijen, F., Mancini, G., Mansson, J.E., et al. (2000). The Spectrum of SLC17A5-Gene Mutations Resulting in Free Sialic Acid-Storage Diseases Indicates Some Genotype–Phenotype Correlation. Am. J. Hum. Genet., 67, 832–840.

Bellocchio, E.E., Reimer, R.J., Fremeau, R.T., Jr, and Edwards, R.H. (2000). Uptake of Glutamate into Synaptic Vesicles by an Inorganic Phosphate Transporter. Science, 289, 957–960.

Haataja, L., Parkkola, R., Sonninen, P., Vanhanen, S.L., Schleutker, J., Aarimaa, T., et al. (1994). Phenotypic Variation and Magnetic Resonance Imaging (MRI) in Salla Disease, a Free Sialic Acid Storage Disorder. Neuropediatrics, 25, 238–244.

Havelaar, A.C., Mancini, G.M.S., Beerens, C.E.M.T., Souren, R.M.A., and Verheijen, F.W. (1998). Purification of the Lysosomal Sialic Acid Transporter. Functional Characteristics of a Monocarboxylate Transporter. J. Biol. Chem., 273, 34568–34574.

Havelaar, A.C., Beerens, C.E., Mancini, G.M.S., and Verheijen, F.W. (1999). Transport of Organic Anions by the Lysosomal Sialic Acid Transporter: A Functional Approach towards the Gene for Sialic Acid Storage Disease. FEBS Lett., 446, 65–68.

Hayashi, M., Otsuka, M., Morimoto, R., Hirota, S., Yatsushiro, S., Takeda, J., et al. (2001). Differentiation-Associated Na⁺-Dependent Inorganic Phosphate Cotransporter (DNPI) is a Vesicular Glutamate Transporter in Endocrine Glutamatergic Systems. J. Biol. Chem., 276, 43400–43406.

Leppänen, P., Isosomppi, J., Schleutker, J., Aula, P., and Peltonen, L. (1996). A Physical Map of the 6q14-q15 Region Harboring the Locus for the Lysosomal Membrane Sialic Acid Transport Defect. Genomics, 37, 62–67.

Mancini, G.M.S., de Jonge, H.R., Galjaard, H., and Verheijen, F.W. (1989). Characterization of a Proton-Driven Carrier for Sialic Acid in the Lysosomal Membrane. Evidence for a Group-Specific Transport System for Acidic Monosaccharides. J. Biol. Chem., 264, 15247–15254.

Mancini, G.M.S., Beerens, C.E.M.T., Aula, P.P., and Verheijen, F.W. (1991). Sialic Acid Storage Diseases. A Multiple Lysosomal Transport Defect for Acidic Monosaccharides. J. Clin. Invest., 87, 1329–1335.

Mancini, G.M.S., Beerens, C.E.M.T., Galjaard, H., and Verheijen, F.W. (1992a). Functional Reconstitution of the Lysosomal Sialic Acid Carrier into Proteoliposomes. Proc. Natl Acad. Sci. USA, 89, 6609–6613.

Mancini, G.M.S., Hu, P., Verheijen, F.W., van Diggelen, O.P., Janse, H.C., Kleijer, W.J., et al. (1992b). Salla Disease Variant in a Dutch Patient. Potential Value of

Polymorphonuclear Leucocytes for Heterozygote Detection. *Eur. J. Pediatr.*, **151**, 590–595.

Mancini, G.M.S., Havelaar, A.C., and Verheijen, F.W. (2000). Lysosomal Transport Disorders. *J. Inherit. Metab. Dis.*, **23**, 278–292.

Pao, S.S., Paulsen, I.T., and Saier, M.H., Jr (1998). Major Facilitator Superfamily. *Microbiol. Mol. Biol. Rev.*, **62**, 1–34.

Pitto, M., Chigorno, V., Renlund, M., and Tettamanti, G. (1996). Impairment of Ganglioside Metabolism in Cultured Fibroblasts from Salla Patients. *Clin. Chim. Acta*, **247**, 143–157.

Renlund, M., Tietze, F., and Gahl, W.A. (1986). Defective Sialic Acid Egress from Isolated Fibroblast Lysosomes of Patients with Salla Disease. *Science*, **232**, 759–762.

Salomaki, P., Aula, N., Juvonen, V., Renlund, M., and Aula, P. (2001). Prenatal Detection of Free Sialic Acid Storage Disease: Genetic and Biochemical Studies in Nine Families. *Prenat. Diagn.*, **21**, 354–358.

Schauer, R. (2000). Achievements and Challenges of Sialic Acid Research. *Glycoconj. J.*, **17**, 485–499.

Schleutker, J., Laine, A.P., Haataja, L., Renlund, M., Weissenbach, J., Aula, P., *et al.* (1995a). Linkage Disequilibrium Utilized to Establish a Refined Genetic Position of the Salla Disease Locus on 6q14-q15. *Genomics*, **27**, 286–292.

Schleutker, J., Leppänen, P., Mansson, J.-E., Erikson, A., Weissenbach, J., Peltonen, L., *et al.* (1995b). Lysosomal Free Sialic Acid Storage Disorders with Different Phenotypic Presentations, ISSD and Salla disease, Represent Allelic Disorders on 6q14-15. *Am. J. Hum. Genet.*, **57**, 893–901.

Schwarzkopf, M., Knobeloch, K.P., Rohde, E., Hinderlich, S., Wiechens, N., Lucka, L., *et al.* (2002). Sialylation is Essential for Early Development in Mice. *Proc. Natl Acad. Sci. USA*, **99**, 5267–5270.

Verheijen, F.W., Verbeek, E., Aula, N., Beerens, C.E., Havelaar, A.C., Joosse, M., *et al.* (1999). A New Gene, Encoding an Anion Transporter, is Mutated in Sialic Acid Storage Diseases. *Nat. Genet.*, **23**, 462–465.

Weiss, P., Tietze, F., Gahl, W.A., Seppala, R., and Ashwell, G. (1989). Identification of the Metabolic Defect in Sialuria. *J. Biol. Chem.*, **264**, 17635–17636.

16

JUDITH C. FLEMING* AND ELLIS J. NEUFELD*

Thiamine-responsive megaloblastic anemia (TRMA) syndrome: consequences of defective high-affinity thiamine transport

Thiamine-responsive megaloblastic anemia (TRMA) syndrome (OMIM 249270) comprises a distinctive triad of clinical features: megaloblastic anemia with ringed sideroblasts, diabetes mellitus, and progressive sensorineural deafness. It is an autosomal recessive disorder reported in fewer than 30 families (Raz *et al.* 2000). In addition to the cardinal triad of anemia, deafness, and diabetes, other manifestations, including optic atrophy, cardiomyopathy, and stroke-like episodes, have been described in several kindreds. In all cases, pharmacological doses of thiamine [20–60 fold higher than the 1.5 mg/day suggested as the U.S. recommended dietary allowance (RDA)] improves the anemia substantially. Some improvement in the diabetic state has also been noted (Borgna-Pignatti *et al.* 1989; Rindi *et al.* 1992; Mandel *et al.* 1993; Valerio *et al.* 1998). Biochemical (Neufeld *et al.* 2001) and genetic (Diaz *et al.* 1999; Fleming

et al. 1999; Labay *et al.* 1999) data demonstrate that this rare disease is caused by mutations in a gene encoding a high-affinity thiamine transporter. It is not known how this defect results in the seemingly divergent disorders of megaloblastic anemia, diabetes, and deafness.

CLINICAL FEATURES OF TRMA

TRMA is an inherited disorder with childhood onset. Rogers *et al.* (1969) first described the disease with the case of a child with diabetes mellitus, sensorineural deafness, and megaloblastic anemia that responded to thiamine (vitamin B1) treatment. In all of the more than 20 reported cases, pharmacological doses of thiamine (25–75 mg/day compared with the US RDA of 1.5 mg/day) have ameliorated the anemia but not the macrocytosis, suggesting a persistent erythropoietic abnormality (Haworth *et al.* 1982; Neufeld *et al.* 1977). In addition to megaloblastic erythroid maturation, the marrow often contains ringed sideroblasts. The diabetes is insulin-dependent non-type I, and

*Children's Hospital, Division of Hematology, 300 Longwood Avenue, Boston, MA 02115, USA

the need for insulin is sometimes reduced with thiamine treatment (Neufeld *et al*. 1997; Valero *et al*. 1998). Neither anti-insulin nor anti-islet cell antibodies have been found in TRMA patients (Neufeld *et al*. 1997), but the pancreatic histopathology has not been investigated. The basis of the sensorineural deafness is obscure; it is not known whether the deafness is due to abnormalities of the cochlea or the auditory nervous system. In contrast with the anemia and diabetes, the deafness in TRMA is irreversible and may not be prevented by thiamine treatment. In addition to the cardinal findings for which the syndrome is named, optic atrophy is prevalent and several patients showed cardiovascular abnormalities, including stroke, high output heart failure and congenital heart defects (Abbourd *et al*. 1985; Neufeld *et al*. 1997; Villa *et al*. 2000).

THIAMINE TRANSPORT AND METABOLISM

Thiamine, like other vitamins, is obtained through dietary intake. The pathway of cellular thiamine uptake, conversion, and ultimate role is shown in Figure 1. Normal thiamine homeostasis begins with uptake from the gut, followed by transport in plasma to tissues and into cells. Plasma thiamine concentration is approximately 30 nM. Once inside the cell, cytosolic thiamine pyrophosphokinase (TPK) converts thiamine into thiamine pyrophosphate (TPP), which is the cofactor for four enzymes. These enzymes include transketolase, involved in the pentose phosphate shunt, and three multi-subunit enzyme complexes – pyruvate dehydrogenase, α-ketoglutarate dehydrogenase, and branched-chain ketoacid dehydrogenase – which participate in oxidative decarboxylation. Several human diseases related to thiamine metabolism are known including deficiency disorders *beriberi* (Platt 1967) and *Wernicke–Korsakoff syndrome* (Denny-Brown 1958), as well as defects in the decarboxylase/dehydrogenases. None of these disorders beside TRMA are associated with the unique anemia, progressive deafness, or diabetes mellitus. Blood thiamine levels have been normal in the few untreated TRMA patients tested. Beriberi patients may exhibit some degree of glucose intolerance, but anemia is not prominent.

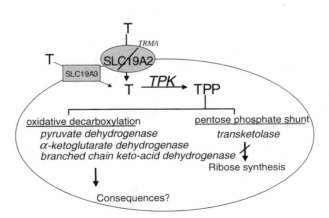

Figure 1   Thiamine metabolic pathways in mammalian cells. Thiamine is taken into cells via the high-affinity thiamine transporters SLC19A2 and SLC19A3. It is then converted to thiamine pyrophosphate (TPP) by thiamine pyrophosphokinase (TPK). Four mammalian enzymes utilize TPP as a cofactor. Thiamine can also enter cells by nonsaturable low-affinity uptake. Whether additional specific carriers exist is not known. Thiamine uptake via SLC19A2 is defective in TRMA syndrome.

THIAMINE TRANSPORT AND TRMA

Defects in thiamine transport in TRMA were postulated by Poggi *et al.* (1984), and subtle red blood cell transport defects at relatively high concentrations were observed by Rindi *et al.* (1992, 1994). Two thiamine uptake processes have been identified in cultured cells. The first mechanism is a saturable high-affinity process that is energy and sodium dependent. The second mechanism is nonsaturable and low affinity, and may represent thiamine entering the cell via diffusion (Rindi *et al.* 1992; Stagg *et al.* 1999). Studies of [^{3}H]thiamine uptake by normal human fibroblasts revealed a saturable process with an apparent K_m of 400–550 nM, while TRMA fibroblasts completely lacked this high-affinity component (Stagg *et al.* 1999). Further, TRMA cells, but not control cells, died of apoptosis in thiamine-depleted medium and accumulated organic acids suggestive of thiamine starvation (Stagg *et al.* 1999).

THIAMINE DEFICIENCY

Chronic thiamine deficiency (e.g., in alcoholics) can cause a neuropsychiatric disorder known as Wernicke–Korsakoff syndrome (Denny-Brown 1958) or beriberi, which is a multisystem disorder associated with a breakdown of cellular energy metabolism (Handin *et al.* 1995). In TRMA patients for which thiamine levels were measured, all were within the normal range (see Box: Pathophysiology of TRMA). In rodents, thiamine deficiency leads to neurodegeneration (Frederikse *et al.* 1999), decreased α-ketoglutarate dehydrogenase activity, and increased branched-chain ketoacid dehydrogenase activity (Blair *et al.* 1999). Interestingly, thiamine deficiency does not result in a TRMA-like syndrome and TRMA patients are not thiamine deficient. This suggests that (1) there is another, possibly intestinal, thiamine transporter and (2) TRMA likely represents a selective deficiency syndrome, perhaps involving intracellular transport.

CLONING OF TRMA GENE

We initially localized the TRMA gene to chromosome 1q23 by homozygosity mapping (Neufeld *et al.* 1997). Analysis of additional families by Raz *et al.* (1998) and Banekazemi *et al.* (1999) shortened the critical region to less than 2 cM. We utilized a candidate gene approach to identify a thiamine transporter gene among the nine predicted genes in this region identified by the Sanger Centre as part of the Human Genome Project (Fleming *et al.* 1999). The annotation for this previously unknown gene referred to its similarity with another vitamin transporter, the reduced folate

---

PATHOPHYSIOLOGY OF TRMA

TRMA is an autosomal recessive disease caused by a defect in thiamine transport via the high-affinity thiamine transporter SLC19A2. The clinical phenotype of TRMA is a thiamine-responsive megaloblastic anemia with ringed sideroblasts in the bone marrow, in combination with non-type I diabetes and progressive sensorineural deafness. Other findings can include progressive optic atrophy and stroke/sudden death. The tissue-restricted expression of the disease phenotype could represent differential requirements for thiamine in various tissues. Relative thiamine deficiency in sensitive tissues may result in (1) impaired nonoxidative pentose phosphate synthesis, leading to reduced de novo RNA/DNA synthesis, and (2) impaired Krebs cycle flux perhaps giving rise to the ringed sideroblasts that are noted in the bone marrow. Indeed, the megaloblastic features of the disease suggest a nucleic acid defect, while the ringed sideroblasts suggest mitochondrial pathology.

carrier (also known as SLC19A1). Using positional cloning approaches, two other groups simultaneously reported mutations in this "reduced folate carrier-like" gene, termed SLC19A2 (also known as THTR-1), in additional TRMA families (Diaz *et al*. 1999; Labay *et al*. 1999). SLC19A2 is predicted to have 12 membrane-spanning domains and is present on the cell surface (Fleming *et al*. 1999, 2001). Eighteen distinct mutations in 25 sibships have

been described as of late 2002 (Table 1). A majority of SLC19A2 mutations known to date are predicted to be null for protein because of nonsense or frameshift mutations. Of the four known missense mutations, all occur within transmembrane domains. Such mutations would likely severely disrupt the folding and membrane targeting of the transporter. Consistently, Balamurugan and Said (2002) showed that introducing several of these

Table 1   Mutations found in TRMA Syndrome

Family	Mutation	Amino acid	Exon/Predicted location	Reference(s)
Pakistan[a]	G196T	E65X	1/extracellular	Haworth *et al*. 1982; Vora and Lilleyman 1993
Denmark	G196T	E65X	1/extracellular	Neufeld, unpublished data
Pakistan	242insA	Ins81fs/ter97	2/transmembrane	Freisinger *et al*. 1999
Iran[a]	242insA	Ins81fs/ter97	2/transmembrane	Vossough *et al*. 1995
France	G277C	D93H	2/transmembrane	Grill *et al*. 1991
Tunisia	287delG	Del96fs/ter117	2/intracellular	Gritli *et al*. 2001
Brazil	C428T	S142F	2/transmembrane	Viana and Carvalho 1978
Iran	429delTT	Del143fs/ter239	2/transmembrane	Vossough *et al*. 1995
Japan	C484T	R162X	2/intracellular	Morimoto *et al*. 1992
Pakistan	C484T	R162X	2/intracellular	Barrett *et al*. 1997
Turkey	C484T	R162X	2/intracellular	Freisinger *et al*. 1999
Italy	G515A	G172D	2/transmembrane	Borgna-Pignatti *et al*. 1989
Italy	G515A/tG1002A	G172D/G335D	2/transmembrane 3/extracellular	Neufeld, unpublished data
Turkey	C697T	Q233X	2/intracellular	Neufeld, unpublished data
Israeli-Arab[a]	724delC	Del242fs/ter259	2/intracellular	Mandel *et al*. 1984; Rindi *et al*. 1994
Lebanon	724delC	Del242fs/ter259	2/intracellular	Bazarbachi *et al*. 1998
India	G750A	W250X	2/intracellular	Raz *et al*. 1998
Alaska	885delT	Del295fs/ter313	3/transmembrane	Fleming *et al*. 1999
Germany	G1074A	W358X	4/transmembrane	Scharfe *et al*. 2000
Turkey	1105delTT	Del369fs/ter372	4/transmembrane	Neufeld, unpublished data
Turkish Kurds	1147delGT	Del383fs/ter385	4/transmembrane	Fleming *et al*. 1999
Lebanon	1223 + 1G>A	408 + 1splice	junction 4&5/	Bazarbachi *et al*. 1998
India	A1315C	T439P	5/transmembrane	Feigenbaum *et al*. 2002

[a] Two distinct families of the same ethnic origin with the same mutation.

mutations into transfected HeLa cells resulted in impaired thiamine uptake. One patient from Italy is a compound heterozygote with distinct mutations from each parent (Fleming, Neufeld, and Fabris, unpublished data).

SLC19A2 AND THIAMINE TRANSPORT

Members of the SLC19A family of transmembrane transporters are involved in the transport of folate and thiamine. When transfected into appropriate host cells, SLC19A2, for example, transports thiamine with nanomolar affinity (Fleming et al. 1999). Defects in these genes result in several genetic disorders and severe phenotypes in mice (Fleming et al. 1999; Zeng et al. 2001; Zhao et al. 2001a), of which the following are examples.

1. Gene targeting studies in mice have shown that a lack of the reduced folate carrier (SLC19A1) leads to intrauterine death prior to embryonic day 9.5 associated with severe impairment of hematopoiesis (Zhao et al. 2001a).
2. Another member of the SLC19A family of vitamin transporters has been identified and named SLC19A3 (Eudy et al. 2000). SLC19A3 functions as a high-affinity thiamine transporter, with missense mutations reported to result in an autosomal recessive disorder known as biotin-responsive basal ganglia disease (BRBGD) (Zeng et al. 2001).
3. As discussed above, mutations in the SLC19A2, a high-affinity thiamine transporter, result in TRMA syndrome.

The name for the TRMA gene is derived from its close homology with the gene for a previously identified reduced folate transporter, SLC19A1. SLC19A2 mRNA is widely expressed, with the highest levels found in skeletal muscle, followed by heart, placenta, kidney, liver, and other tissues (Fleming et al. 1999). Considering the absence of SLC19A2 function in TRMA patients, we speculate that a secondary thiamine transport process (either a second transporter such as SLC19A3 or low-affinity uptake) acts to allow sufficient thiamine uptake to prevent the metabolic condition of severe thiamine deficiency (beriberi). While SLC19A1 cannot transport thiamine, and neither SLC19A2 nor SLC19A3 can transport folate (Dutta et al. 1999; Rajgopal et al. 2001; Fleming and Neufeld unpublished results), a recent report raises the possibility that thiamine pyrophosphate may be carried by SLC19A1 (Zhao et al. 2001a). Homologs or orthologs of the SLC19 family are also present in Caenorhabditis elegans (three family members) and Drosophila, as well as in other mammals.

Unanswered questions about TRMA syndrome abound. For example, why is the deafness in TRMA progressive and irreversible, while the anemia and to some extent the diabetes are reversible? It is possible that the thiamine requirement of cochlea and acoustic nerve cells is significantly higher than that of other cell types. Pharmacologic doses of thiamine may not prevent a local thiamine-deficient environment in cell types that require high-energy usage at all times. This could result in occasional cell death in these sensitive tissues, ultimately progressing to progressive deafness and optic atrophy. Specific localization of SLC19A2 to developing hair cells in the mouse cochlea suggests that lack of cellular thiamine uptake mediated by SLC19A2 by these cells might result in their demise and subsequent deafness (Fleming et al. 2001). In contrast, bone marrow or, more specifically, pluripotent hematopoietic stem cells, might be less sensitive to this effect, with repopulation of the erythroid lineage when patients are thiamine replete. However, the persistent macrocytosis (Neufeld et al. 1997) and poor erythroid colony growth in vitro (Rotoli et al. 1986) suggest that even thiamine-replete erythroid progenitors are sensitive to the thiamine transport defect. The diabetic phenotype in TRMA is reported to be variably sensitive and reversible to thiamine treatment (Neufeld et al. 1997; Valerio et al. 1998; Bappal et al. 2001), suggesting an intermediate sensitivity of pancreatic beta cells or target tissues to cellular thiamine deficiency.

How does thiamine participate in normal hematopoiesis? The combination of megaloblastic changes and ringed sideroblasts in TRMA patients is unique among anemias caused by metabolic or nutritional factors. Among all acquired anemias, this combination is most suggestive of the *myelodysplastic syndromes*, in which sideroblasts and megaloblastosis are frequently observed. One report of a TRMA kindred attaches the sobriquet of "thiamine-responsive myelodysplasia" to the syndrome (Bazarbachi *et al.* 1998). Although it is tempting to postulate that acquired myelodysplastic states might exhibit a defect in thiamine-dependent cellular metabolism, no direct data address this hypothesis yet.

Megaloblastosis reflects dyssynchrony of cellular maturation, with nuclear maturation lagging in relation to the cytoplasm of hematopoietic cells. Why should TRMA have a megaloblastic appearance? Defects in pyrimidine synthesis were considered by Haworth *et al.* (1982), but the deoxyuridine suppression test was normal. In collaboration with Laszlo Boros (Torrance, CA), we have investigated the possibility that TRMA cells cannot adequately synthesize ribose 5-phosphate, the precursor for nucleic acid synthesis. The method used for these studies is mass spectroscopic analysis of stable–isotope-labeled fibroblasts (Lee *et al.* 1998). We have obtained preliminary evidence that ribose synthesis is disordered in TRMA cells (Steinkamp *et al.* 2000). Ribose is synthesized de novo via the pentose phosphate shunt. Synthesis can proceed via oxidative metabolism of glucose, through glucose-6-phosphate dehydrogenase (G6PD), or nonoxidatively via transketolase and transaldolase. The latter pathway requires thiamine pyrophosphate as a cofactor for transketolase. In cultured mammalian cells, ribose synthesis is predominantly through the nonoxidative transketolase pathway, a phenomenon first noted more than 40 years ago (Hiatt 1957). In comparison with normal fibroblasts, TRMA cells exposed to thiamine deprivation are obliged to synthesize ribose-5-P through the G6PD pathway, as transketolase activity must become limiting in the absence of cofactor TPP (Boros *et al.*, in press).

SUMMARY

TRMA is a rare autosomal recessive disease caused by mutations in the high-affinity thiamine transporter SLC19A2. It is not known whether this defect results in the seemingly divergent disorders of megaloblastic anemia, diabetes, and deafness. Mechanisms of thiamine transport in TRMA patients (who lack SLC19A2 altogether) remain to be elucidated. The fact that TRMA patients have neither thiamine deficiency nor widespread tissue involvement, despite the presence of SLC19A2 in intestines (Fleming *et al.* 1999; Reidling *et al.* 2002), illustrates the redundancy in the thiamine uptake system. A second transporter, such as SLC19A3, could be an explanation. Distinctive expression patterns of the two transporters likely result in different disease phenotypes when either transporter is absent. A mouse model of TRMA will be instrumental in the understanding of the disease and the relationship of thiamine to the component disorders. A clearer understanding of the role of high-affinity thiamine transport via SLC19A2 will assist in the understanding of the role of thiamine in the development and maintenance of the erythroid, auditory, and glucose homeostasis systems.

REFERENCES

Abbourd, M.R., Alexander, D., and Najjar, S.S. (1985). Diabetes Mellitus, Thiamine-Dependent Megaloblastic Anemia, and Sensorineural Deafness Associated with Deficient Alpha-Ketoglutarate Dehydrogenase Activity. *J. Pediatr.*, **107**, 537–541.

Balamurugan, K. and Said, H.M. (2002). Functional Role of Specific Amino Acid Residues in Human Thiamine Transporter SLC19A2: Mutational Analysis. *Am. J. Physiol. Gastrointest. Liver Physiol.*, **283**, G37–G43.

Banikazemi, M., Diaz, G.A., Vossough, P., Jalali, M., Desnick, R.J., and Gelb, B.D. (1999). Localization of the Thiamine-Responsive Megaloblastic Anemia Syndrome Locus to a 1.4-cM Region of 1q23. *Mol. Genet. Metab.*, **66**, 193–198.

Bappal, B., Nair, R., Shaikh, H., Al Khusaiby, S.M., and de Silva, V. (2001) Five Years Followup of Diabetes Mellitus in Two Siblings with Thiamine Responsive Megaloblastic Anemia. *Indian Pediatr.*, **38**, 1295–1298.

Barrett, T.G., Poulton, K., Baines, M., and McCowen, C. (1997). Muscle Biochemistry in Thiamine-Responsive Anaemia. *J. Inherit. Metab. Dis.*, **20**, 404–406.

Bazarbachi, A., Muakkit, S., Ayas, M., Taher, A., Salem, Z., Solh, H., *et al.* (1998). Thiamine-Responsive Myelodysplasia. *Br. J. Haematol.*, **102**, 1098–1100.

Blair, P.V., Kobayashi, R., Edwards, H.M.I., Say, N.F., Baker, D.H., and Harris, R.A. (1999). Dietary Thiamin Level Influences Levels of its Diphosphate Form and Thiamin-Dependent Enzymic Activities of Rat Liver. *J. Nutr.*, **129**, 641–648.

Borgna-Pignatti, C., Marradi, P., Pinelli, L., Monetti, N., and Patrini, C. (1989). Thiamine-Responsive Anemia in DIDMOAD Syndrome. *J. Pediatr.*, **114**, 405–410.

Boros, L.G., Steinkamp, M.P., Fleming, J.C., Lee, W.-N.P., Cascante, M., and Newfeld, E.J. (2003). Defective RNA ribose Synthesis in fibroblasts from patients with thiamine-responsive megaloblastic anemia (TRMA): Casual mechanism for the Syndrome. Blood In press.

Denny-Brown, D. (1958). The Neurological Aspects of Thiamine Deficiency. *Fed. Proc.*, **17**(Suppl. 2), 35–39.

Diaz, G.A., Banikazemi, M., Oishi, K., Desnick, R.J., and Gelb, B.D. (1999). Mutations in a New Gene Encoding a Thiamine Transporter Cause Thiamine-Responsive Megaloblastic Anaemia Syndrome. *Nat. Genet.*, **22**, 309–312.

Dutta, B., Huang, W., Molero, M., Kekuda, R., Leibach, F.H., Devoe, L.D., *et al.* (1999). Cloning of the Human Thiamine Transporter, a Member of the Folate Transporter Family. *J. Biol. Chem.*, **274**, 31925–31929.

Eudy, J., Spiegelstein, O., Barber, R., Wlodarczyk, B., Talbot, J., and Finnell, R. (2000). Identification and Characterization of the Human and Mouse SLC19A3 Gene: A Novel Member of the Reduced Folate Family of Micronutrient Transporter Genes. *Mol. Genet. Metab.*, **71**, 581–590.

Feigenbaum, A.S., Hewson, S., Wherrret, D., Doyle, J., Waye, J.S., and Gelb, B.D. (2002). Thiamine-Responsive Megaloblastic Anemia Presenting as Microcytic Anaemia: An Unusual Case with Novel Mutations. *American Society of Human Genetics*, p. 1505.

Fleming, J.C., Tartaglini, E., Steinkamp, M.P., Schorderet, D.F., Cohen, N., and Neufeld, E.J. (1999). The Gene Mutated in Thiamine-Responsive Anaemia with Diabetes and Deafness (TRMA) Encodes a Functional Thiamine Transporter. *Nat. Genet.*, **22**, 305–308.

Fleming, J.C., Steinkamp, M.P., Kawatsuji, R., Tartaglini, E., Pinkus, J.L., Pinkus, G.S., *et al.* (2001). Characterization of a Murine High-Affinity Thiamine Transporter, Slc19a2. *Mol. Genet. Metab.*, **74**, 273–280.

Frederikse, P.H., Farnsworth, P., and Zigler, J.S.J. (1999). Thiamine Deficiency In Vivo Produces Fiber Cell Degeneration in Mouse Lenses. *Biochem. Biophys. Res. Commun.*, **258**, 703–707.

Freisinger, P., Lapp, B., Baumeister, F., Muller-Weihrich, S., Jaksch, M., and Rabl, W. (1999). Thiamine-Responsive Diabetes Mellitus, Megaloblastic Anemia, and Sensorineural Deafness: Clinical Variability and Response to High-Dose Thiamine Therapy in Two Siblings. *Horm. Res.*, **51**(Suppl 2): 102 p352.

Grill, J., Leblanc, T., Baruchel, A., Daniel, M.T., Dresch, C., and Schaison, G. (1991). Thiamine Responsive Anemia: Report of a New Case Associated with a Thiamine Pyrophosphokinase Deficiency. *Nouv. Rev. Fr. Hematol.*, **33**, 543–544.

Gritli, S., Omar, S., Tartaglini, E., Guannouni, S., Fleming, J.C., Steinkamp, M.P., *et al.* (2001). A Novel Mutation in the SLC19A2 Gene in a Tunisian Family with Thiamine-Responsive Megaloblastic Anaemia, Diabetes and Deafness Syndrome. *Br. J. Haematol.*, **113**, 508–513.

Handin, R.I., Lux, S.E., and Stossel, T.P. (1995). *Blood: Principles and Practice of Hematology*, Philadelphia, Saunders.

Haworth, C., Evans, D.I.K., Mitra, J., and Wickramasinghe, S.N. (1982). Thiamine Responsive Anaemia: A Study of Two Further Cases. *Br. J. Haematol.*, **50**, 549–561.

Hiatt, H.H. (1957). Studies of Ribose Metabolism. I. The Pathway of Nucleic Acid Synthesis in a Human Carcinoma Cell in Tissue Culture. *J. Clin. Invest.*, **37**, 1408–1415.

Labay, V., Raz, T., Baron, D., Mandel, H., Williams, H., Barrett, T., *et al.* (1999). Mutations in SLC19A2 Cause Thiamine-Responsive Megaloblastic Anaemia Associated with Diabetes Mellitus and Deafness. *Nat. Genet.*, **22**, 300–304.

Lee, W.N., Boros, L.G., Puigjaner, J., Bassilian, S., Lim, S., and Cascante, M. (1998). Mass Isotopomer Study of the Nonoxidative Pathways of the Pentose Cycle with [1,2-13C2]Glucose. *Am. J. Physiol.*, **274**, E843–E851.

Mandel, H., Berant, M., Hazani, A., and Naveh, Y. (1984). Thiamine-Dependent Beriberi in the "Thiamine-Responsive Anemia Syndrome." *N. Engl. J. Med.*, **311**, 836–838.

Mandel, H., Vardi, P., and Berant, M. (1993). Thiamine-Responsive Diabetes: A Novel Category of Diabetes Mellitus. *Pediatr. Res.*, Suppl. 33, 193A.

Morimoto, A., Kizaki, Z., Konishi, K., Sato, N., Kataoka, T., Hayashi, R., *et al.* (1992). A Case of Thiamine-Responsive Megaloblastic Anemia Syndrome with Nystagmus, Cerebral Infarction, and Retinal Degeneration. *J. Jpn Pediatr. Soc.*, **96**, 2137–2145.

Neufeld, E.J., Mandel, H., Raz, T., Szargel, R., Yandava, C.N., Stagg, A., *et al.* (1997). Localization of the Gene for Thiamine-Responsive Anemia Syndrome on the Long Arm of Chromosome 1 by Homozygosity Mapping. *Am. J. Hum. Genet.*, **61**, 1335–1341.

Neufeld, E.J., Fleming, J.C., Tartaglini, E., and Steinkamp, M.P. (2001). Thiamine-Responsive Megaloblastic Anemia Syndrome: A Disorder of High-Affinity Thiamine Transport. *Blood Cells Mol. Dis.*, **27**, 135–138.

Platt, B.S. (1967). Thiamine Deficiency in Human Beriberi and in Wernicke's Encephalopathy. In G.E.W. Wolstenholme and M. O'Connor (eds), *Thiamine Deficiency*, Boston, pp. 135–145.

Poggi, V., Longo, G., DeVizia, B., Andria, G., Rindi, G., Patrini, C., et al. (1984). Thiamin-Responsive Megaloblastic Anemia: A Disorder of Thiamin Transport? *J. Inherit. Metab. Dis.*, **7**, 153–154.

Rajgopal, A., Edmondson, A., Goldman, I.D., and Zhao, R. (2001). SLC19A3 Encodes a Second Thiamine Transporter ThTr2. *Biochim. Biophys. Acta*, **1537**, 175–178.

Raz, T., Barrett, T., Szargel, R., Mandel, H., Neufeld, E.J., Nosaka, K., et al. (1998). Refined Mapping of the Gene for Thiamine-Responsive Megaloblastic Anemia Syndrome and Evidence for Genetic Homogeneity. *Hum. Genet.*, **103**, 455–461.

Raz, T., Labay, V., Baron, D., Szargel, R., Anbinder, Y., Barrett, T., et al. (2000). The Spectrum of Mutations, Including Four Novel Ones, in the Thiamine-Responsive Megaloblastic Anemia Gene SLC19A2 of Eight Families. *Hum. Mutat.*, **16**, 37–42.

Reidling, J.C., Subramanian, V.S., Dudeja, P.K., and Said, H.M. (2002). Expression and Promoter Analysis of SLC19A2 in the Human Intestine. *Biochim. Biophys. Acta*, **1561**, 180–187.

Rindi, G., Casirola, D., Poggi, V., De Vizia, B., Patrini, C., and Laforenza, U. (1992). Thiamine Transport by Erythrocytes and Ghosts in Thiamine-Responsive Megaloblastic Anemia. *J. Inherit. Metab. Dis.*, **15**, 231–242.

Rindi, G., Patrini, C., Laforenza, U., Mandel, H., Berant, M., et al. (1994). Further Studies on Erythrocyte Thiamin Transport and Phosphorylation in Seven Patients with Thiamine-Responsive Megaloblastic Anaemia. *J. Inherit. Metab. Dis.*, **17**, 667–677.

Rogers, L.E., Porter, F.S., and Sidbury, J.B.J. (1969). Thiamine-Responsive Megaloblastic Anemia. *J. Pediatr.*, **74**, 494–504.

Rotoli, B., Poggi, V., De Renzo, A., and Robledo, R. (1986). In Vitro Addition of Thiamin does not Restore BFU-E Growth in Thiamin-Responsive Anemia Syndrome. *Haematologica*, **71**, 441–443.

Scharfe, C., Hauschild, M., Klopstock, T., Janssen, A.J., Heidemann, P.H., Meitinger, T., et al. (2000). A Novel Mutation in the Thiamine Responsive Megaloblastic Anaemia Gene SLC19A2 in a Patient with Deficiency of Respiratory Chain Complex I. *J. Med. Genet.*, **37**, 669–673.

Stagg, A.R., Fleming, J.C., Baker, M.A., Sakamoto, M., Cohen, N., and Neufeld, E.J. (1999). Defective High-Affinity Thiamine Transporter Leads to Cell Death in Thiamine-Responsive Megaloblastic Anemia Syndrome Fibroblasts. *J. Clin. Invest.*, **103**, 723–729.

Steinkamp, M.P., Fleming, J.C., Boros, L.G., and Neufeld, E.J. (2000). Thiamine Depletion in Thiamine-Responsive Megaloblastic Anemia (TRMA) Mutant Fibroblasts Leads to a Reduction in Non-Oxidative Ribose Synthesis. *FASEB J.*, **14**, A1547.

Valerio, G., Franzese, A., Poggi, V., and Tenore, A. (1998). Long-Term Follow-up of Diabetes in Two Patients with Thiamine-Responsive Megaloblastic Anemia Syndrome. *Diabetes Care*, **21**, 38–41.

Viana, M.B. and Carvalho, R.I. (1978). Thiamine-Responsive Megaloblastic Anemia, Sensorineural Deafness, and Diabetes Mellitus: A New Syndrome? *J. Pediatr.*, **93**, 235–238.

Villa, V., Rivellese, A., Di Salle, F., Iovine, C., Poggi, V., and Capaldo, B. (2000). Acute Ischemic Stroke in a Young Woman with the Thiamine-Responsive Megaloblastic Anemia Syndrome. *J. Clin. Endocrinol. Metab.*, **85**, 947–949.

Vora, A.J. and Lilleyman, J.S. (1993). Wolfram Syndrome: Mitochondrial Disorder. *Lancet*, **342**, 1059.

Vossough, P., Jalai, M., and Alebouyeh, M. (1995). Thiamine responsive megaloblastic anemia (Abstract), *Eur. J. Pediatr.*, **154**, 782.

Zeng, W., Al-Yamani, E., Acierno, J.S., Ozand, P., and Gusella, J.F. (2001). Mutations in SLC19A3 Encoding a Novel Transporter cause Biotin-Responsive Basal Ganglia Disease. *American Society of Human Genetics*, p. 101.

Zhao, R., Gao, F., Wang, Y., Diaz, G.A., Gelb, B.D., and Goldman, I.D. (2001a). Impact of the Reduced Folate Carrier on the Accumulation of Active Thiamin Metabolites in Murine Leukemia Cells. *J. Biol. Chem.*, **276**, 1114–1118.

Zhao, R., Russell, R.G., Wang, Y., Liu, L., Gao, F., Kneitz, B., et al. (2001b). Rescue of Embryonic Lethality in Reduced Folate Carrier-Deficient Mice by Maternal Folic Acid Supplementation Reveals Early Neonatal Failure of Hematopoietic Organs. *J. Biol. Chem.*, **276**, 10224–10228.

ABC TRANSPORTERS

17

STEFAN BRÖER*, WOLFGANG E. KAMINSKI**, AND
GERD SCHMITZ***

Introduction

OVERVIEW

ABC proteins are a diverse group of proteins, including some soluble proteins not considered here, that are defined by the presence of a binding motif for ATP – the *A*TP-*B*inding *C*assette (ABC) (Higgins 1992). ABC transporters are found in all species and serve a wide variety of functions. Most ABC proteins move substances across the membrane; however, some ABC proteins only regulate other transport proteins or ion channels. Studies of bacterial ABC transporters established that the free energy of ATP hydrolysis is used to drive uphill transport of valuable nutrients and ions into cells. Thus ABC transporters are primary active transporters. This statement also applies to mammalian ABC transporters, and in some cases substrate-induced ATP hydrolysis has been demonstrated (e.g., Chapter 20, Figure 4). Mammalian ABC transporters, in contrast to most bacterial ABC transporters, move solutes from the cytosol either into the extracellular space or into an intracellular organelle, such as the endoplasmic reticulum or peroxisomes. Exceptions are mitochondrial

ABC transporters, which are thought to export Fe–S clusters into the cytosol. Thus mammalian ABC transporters can be considered as efflux pumps. The best described member of the mammalian ABC protein family is the multiple drug resistance protein MDR1, also named P-glycoprotein, which confers resistance of malignant cells to cytostatic drugs (Borst and Elferink 2002). Another well-described protein of this family is the cystic fibrosis transmembrane conductance regulator protein (CFTR), the protein that is mutated in patients suffering from cystic fibrosis (Riordan *et al.* 1989). The CFTR protein is not known to be an efflux pump but functions as a chloride channel, and will only be mentioned briefly. No additional transport function has been assigned to CFTR; however, it appears to regulate a number of other ion channels and transporters. More details of the CFTR protein and further references can be found in monographs describing ion-channel diseases and ABC proteins in general (Ashcroft 2000; Holland *et al.* 2003).

Another nontransporter ABC protein that is briefly mentioned here is the ion-channel regulating protein SUR (sulfonylurea receptor). This protein regulates ATP-dependent potassium channels in pancreatic β cells and is the target of sulfonylurea compounds used in the treatment of type II diabetes (Aguilar-Bryan *et al.* 1995). Potassium channels regulated by SUR play a key role in the homeostasis of

---

*School of Biochemistry and Molecular Biology, Faculty of Science, Australian National University, Canberra ACT 0200, Australia

**

***Institute for Clinical Chemistry and Laboratory Medicine, University Hospital Regensburg, Franz-Josef-Straus Allee II, D-93053 Regensburg, Germany

plasma glucose levels (see Chapter 9). For further information, the reader is referred to monographs describing ion-channel diseases (Ashcroft 2000).

Discovery of many more ABC proteins over recent years has revealed a number of subfamilies and resulted in a confusing multiplicity of names. This issue has recently been resolved by a new nomenclature, which is based on sequence similarity and which will be used in this book (Chapter 1, Table 2). There are now seven subfamilies of ABC proteins, designated ABCA to ABCG. The MDR1 protein, for example, has now been renamed ABCB1. CFTR belongs to a different subfamily and has been given the alias ABCC7. The ion-channel-regulating sulfonylurea receptors SUR1 and SUR2 have been named ABCC8 and ABCC9, respectively. Members of families ABCD and ABCE are cytosolic proteins which have an ATP-binding cassette but do not catalyze any transport function.

STRUCTURE OF ABC TRANSPORTERS

Hydropathy plots of ABC transporter sequences indicate the presence of multiple transmembrane helices. As ABC proteins are defined by the presence of an ATP-binding cassette, which resides in a cytosolic domain, the design of the transmembrane region can vary significantly. Subfamilies have been arranged by sequence similarity and also display different arrangements of transmembrane helices (Figure 1). Another feature of ABC proteins is their modular design. In some members a single polypeptide chain can have several domains, whereas in others these domains are constituted separate proteins that form a larger protein complex (Borst and Elferink 2002).

The prototypical mammalian ABC transporter is the multiple drug resistance protein (MDR1 or P-glycoprotein). According to hydropathy analysis, the ABCB1 protein has a tandem arrangement of 2×6 transmembrane helices on a single polypeptide chain (Figure 1). The nucleotide binding domain, which contains

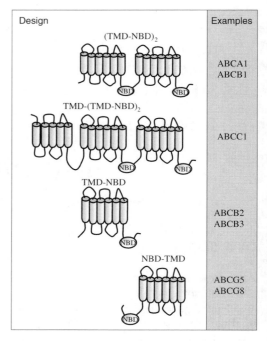

Figure 1 Topology of ABC transporters. ABC transporters are comprised of transmembrane domains (TMDs) and nucleotide binding domains (NBDs). A functional transporter requires at least two transmembrane domains and two nucleotide binding domains. Half-size transporters, such as ABCB2/3 and ABCG5/8, form heterodimers as a functional unit. The ABCC family is characterized by an additional transmembrane domain that precedes a typical ABC transporter structure.

the ATP binding cassette, is located at the carboxyterminal end of each group of six transmembrane helices. Each ATP binding site comprises three submotifs called the Walker-A and Walker-B motifs and the signature sequence. The design is often referred to as an ABC transporter with four domains (transmembrane domain 1 – nucleotide binding domain 1 – transmembrane domain 2 – nucleotide binding domain 2 or TMD1–NBD1–TMD2–NBD2 for short) (Figure 1).

Members of the ABCA group (ABCA1, Chapter 19; ABCA4, Chapter 20) also have the TMD1–NBD1–TMD2–NBD2 design, but the arrangement of loops between the helices differs from that in the ABCB1 protein.

The arrangement of helices in the ABCA family resembles that of the bacterial multiple drug resistance protein MsbA (see below). The ABCC group includes larger proteins, such as ABCC1/MRP1 and ABCC2 (Chapter 18), which have an additional five transmembrane helices. The structure is TMD0–L0–TMD1–NBD1–TMD2–NBD2. L0 is a linking loop between transmembrane domain 0 and the rest of the protein, which looks like the ABCB1 design (Figure 1). A number of ABC proteins have primary structures that look like half of an ABC transporter and are thought to act as homo- or heterodimers, such as TAP1(ABCB2)/TAP2 (ABCB3) (Chapter 21) or ABCG5/ABCG8 (Figure 1).

The fine structure of ABC transporters became much clearer after the crystallization of the *Escherichia coli* multidrug resistance protein MsbA (Chang and Roth 2001) (Figure 2). Several pieces of evidence suggest that mammalian ABC transporters are likely to have a similar design.

1. MsbA is a phospholipid transporter (Karow and Georgopoulos 1993), as are many mammalian ABC transporters (see below and Chapter 19).

2. MsbA is related to the LmrA protein of *Lactococcus lactis*, which has a similar substrate specificity to ABCB1 and has significant sequence similarity to it (Poelarends *et al.* 2002).

3. Expression of LmrA in mammalian cells confers drug resistance.

4. The structure is immediately suggestive of a flippase type transport mechanism, which appears to be the common transport mechanism of most, if not all, ABC transporters (Figure 3).

The structure shows two large transmembrane domains that are strongly tilted towards each other, like the handles of a nutcracker,

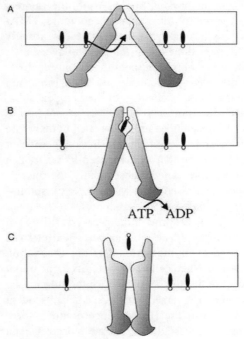

Figure 3 Transport mechanism of ABC transporters. ABC transporter substrates are amphiphilic molecules (black ovals with white head groups) that insert spontaneously into the membrane. (A) Substrates have access to the transporter from the inner leaflet of lipid bilayer. (B) Hydrolysis of ATP occludes the substrate and flips it 180°. (C) After ATP hydrolysis the substrate is directly extruded into the extracellular fluid.

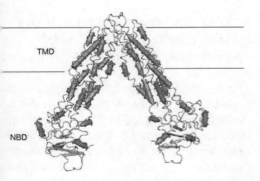

Figure 2 Molecular structure of the *E. coli* MsbA protein. The *E. coli* MsbA protein is an ABC transporter that acts as a lipid flippase. The structure of this protein is thought to be similar to that of eukaryotic ABC proteins. The location of the membrane–water interphase is depicted by a black line; α-helices are shown as tubes. The nucleotide-binding domains are indicated.

which merge in the plane of the membrane. The tip of each handle is the nucleotide binding domain. One particular striking feature of this structure is the accessibility of the translocation domain from the inner leaflet of the lipid bilayer (Figure 2).

TRANSPORT MECHANISM OF ABC TRANSPORTERS

Most ABC transporters mediate the translocation of amphiphilic molecules. These can be amphiphilic drugs, phospholipids, sterols, fatty acids, etc. Common to all these molecules is that they partition rapidly into the plasma membrane. Studies using the fluorescent amphiphilic reporter molecule TMA-diphenylhexatriene, which fluoresces strongly when partitioned into the lipid bilayer but is essentially nonfluorescent in aqueous environments, gave particularly useful insights into the transport mechanism of the *L. lactis* LmrA ABC transporter (Poelarends *et al.* 2002). TMA-diphenylhexatriene inserted quickly into the outer leaflet of the plasma membrane of deenergized cells, as evidenced by a fast rise of the fluorescence. Subsequently, the fluorescence increased slowly because of spontaneous flipping of the compound to the inner leaflet of the membrane and refilling of the "empty space" in the outer leaflet from TMA-diphenylhexatriene in the incubation solution. When cells were energized, the initial increase in the fluorescence could not be reduced or reversed. However, the subsequent slow rise of the fluorescence could be prevented or reversed if energization occurred at a later stage. These experiments demonstrate two main features of the transport mechanism: first, the amphiphilic compound is picked up by the transporter from the inner leaflet; secondly, it is extruded directly into the extracellular space. Occupation of the outer leaflet cannot be prevented.

Drug extrusion requires ATP hydrolysis and is prevented by vanadate. Vanadate is a phosphate analog that inhibits ABC transporters by forming a complex with Mg^{2+} and ADP. The complex traps the transporter in a conformation that occludes the substrate (Urbatsch *et al.* 1995). Thus it appears that the transporter is stopped halfway through its conformational change that moves the substrate across the membrane. Vanadate experiments have demonstrated the existence of two substrate binding sites in the LmrA ABC transporter: a high-affinity cytosolic substrate binding site, which is not accessible after vanadate treatment, and a low-affinity binding site, which is accessible from the extracellular medium. Taken together with the structural information, the transport mechanism can be envisaged as follows (Figure 3). The substrate is picked up from the inner leaflet of the membrane and binds to a high-affinity binding site. ATP hydrolysis occurs and results in a closing movement of the two nutcracker handles, resulting in the occlusion of the high-affinity binding site. The binding site is subsequently exposed to the extracellular medium. The conformational change decreases the affinity for the substrate and causes release of the substrate (Figure 3).

ROLE OF ABC TRANSPORTERS IN BILE FORMATION AND SECRETION OF AMPHIPHILIC COMPOUNDS

Lipids are a heterogeneous group of compounds related more by their physical than their chemical properties. They are relatively insoluble in water and as a result form higher-order structures in aqueous environments such as membranes, vesicles, and micelles. Lipids are energy metabolites and building blocks of membranes at the same time. Simple lipids, such as triacylglycerols (fat), form a major energy store in the body. Hydrolysis of triacylglycerols releases fatty acids, the metabolism of which is introduced in more detail in Chapters 5–8. Complex lipids and cholesterol, on the other hand, are the building blocks of cell membranes. Lipids are synthesized by mammalian cells but also form a significant part of our diet.

The digestion, metabolism, and transport of lipids poses physical problems for a cellular system that is largely based on metabolic reactions in aqueous systems. Three different principles have evolved to facilitate the metabolism of lipids. First, lipids can be solubilized with the help of detergents (bile acids). Secondly, lipids can be bound to proteins which act as a scaffold for lipid metabolism or can transport them to other locations. Thirdly, reactions take place at the membrane–water interface. It appears that ABC transporters have evolved to extrude lipophilic and amphiphilic molecules from the cytosol.

In the following, a short overview of lipid metabolism with a particular emphasis on transport processes will be presented. For a more detailed review the reader is referred to biochemistry textbooks. Lipids and triacylglycerols are hydrolyzed by lipases in the stomach and intestine to form free fatty acids and monoacylglycerols. Hydrolysis of lipids is facilitated by the secretion of bile from the liver. Bile is composed of three compounds: *bile acids, cholesterol*, and *phosphatidylcholine*. It appears that phosphatidylcholine owes its presence in bile to the fact that it greatly enhances the solubility of cholesterol in bile. Otherwise, cholesterol is almost insoluble in water or solutions of bile acids. All three components together form a stable micellar liquid which allows fats to be hydrolyzed by lipases in the intestine. Phosphatidylcholine and cholesterol, in addition to their role in the formation of micelles, are thought to protect cell membranes from the detergent action of bile acids.

To generate bile in the liver all three compounds – bile acids, cholesterol, and phosphatidylcholine – have to be secreted across the canalicular membrane. The secretion is carried out by three different ABC transporters (Figure 4).

Phosphatidylcholine is secreted into the bile by the ABCB4 transporter (Small 2003),

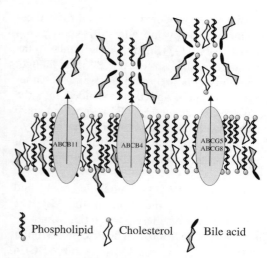

Phospholipid   Cholesterol   Bile acid

Figure 4 Bile production by ABC transporters. Three ABC transporters are involved in bile formation, namely ABCB4, ABCB11, and ABCG5/8. Bile is composed of bile salts, phospholipids, and cholesterol. Bile salts and phospholipids form mixed micelles that act as an acceptor for cholesterol.

also known as MDR-3 or, in mice, mdr-2. It is thought that ABCB4 flips phosphatidylcholine from the inner leaflet of the membrane and extrudes it into the extracellular space. Disruption of ABCB4 in mice prevents secretion of both phosphatidylcholine and cholesterol into the bile, confirming that phosphatidylcholine acts as a cholesterol acceptor in the bile. Defects in human ABCB4 cause *progressive familial intrahepatic cholestasis type III* (PFIC3) (de Vree *et al.* 1998). Heterozygous ABCB4 mutations have also been found in association with cholestasis of pregnancy (Jacquemin *et al.* 1999). The pathophysiology of PFIC3 is thought to reside in the lack of phospholipid protection against the detergent effect of bile salts, resulting in damage to the biliary epithelium, bile ductular proliferation, and progressive portal fibrosis. One of the typical diagnostic findings in individuals with PFIC3 is elevated levels of γ-glutamyl-transferase, a marker enzyme indicating the death of liver cells.

Bile acids are pumped across the canalicular membrane by the bile salt export pump (BSEP, ABCB11) (Small 2003). Similar to the transport mechanism of ABCB4, it is thought that bile acids are removed from the inner leaflet, flipped, and then extruded into the extracellular space. ABCB11 knockout mice have a twofold increase in bile phosphatidylcholine content and a sevenfold increase of bile cholesterol, whereas bile acids are largely missing.

In both ABCB11 knockout mice and humans carrying mutations in the ABCB11 gene, the lack of bile acid secretion results in *intrahepatic cholestasis*. The disease is similar to that caused by mutations in the ABCB4 transporter and thus is classified as *PFIC2* (Strautnieks *et al.* 1998).

Cholesterol is released into the bile by an ABC transporter which comprises two half-transporters encoded by the ABCG5 and ABCG8 genes (Small 2003). Both genes are located on chromosome 2p21 in head-to-head orientation. Thus it is likely that they share a bidirectional promoter and are expressed in a coordinated fashion. In line with this view, both ABCG5 and ABCG8 are expressed predominantly in hepatocytes and enterocytes of the proximal small intestine of humans and mice. The transport mechanism of the heterodimeric transporter is less clear. Cholesterol flips spontaneously between the two leaflets of a membrane at a high rate. Thus a likely function for the ABCG5/8 heterodimer could be to extrude cholesterol into the extracellular medium, where it is picked up by mixed phosphatidylcholine–bile acid micelles in the bile (Figure 4). ABCG5/8 is not only involved in the secretion of cholesterol into the bile but also appears to be directly involved in the prevention of hyperabsorption of dietary sterols by pumping them back into the lumen of the intestine. Mutations in both genes cause β-*sitosterolemia*, a rare autosomal recessive disorder which is characterized by hyperabsorption and decreased biliary excretion of all dietary neutral sterols (Berge *et al.* 2000). This leads to dramatically elevated plasma levels of sitosterol and other plant sterols. Importantly, individuals with defective ABCG5/ABCG8 characteristically present with hypercholesterolemia. β-Sitosterolemia phenotypically resembles homozygous familial hypercholesterolemia in that both diseases manifest with xanthomas (lipid deposits in the skin) in childhood and premature coronary atherosclerosis.

Further support for the function of this heterodimeric ABC transporter in sterol secretion is provided by the observation that increased expression of ABCG5 and ABCG8 in transgenic mice promotes biliary neutral sterol secretion and reduces intestinal cholesterol absorption (Yu *et al.* 2002).

The secretion of bile, apart from its role in the digestion of lipids, is also a major pathway for the secretion of lipophilic compounds that are either toxic or metabolic end-products that need to be secreted. To increase the solubility of lipophilic compounds they are frequently coupled to hydrophilic compounds. Three compounds are used in human metabolism: glucuronic acid, sulfate, and glutathione. As a result xenobiotics are frequently secreted as glucuronides, glutathione–S conjugates, or sulfates. Coupling requires the presence of hydroxyl groups on the lipophilic compound/drug. As this is not always the case, hydroxyl groups are created by cytochrome P450 mediated oxidation. Oxidation of lipophilic compounds is known as phase I metabolism; the subsequent coupling to hydrophilic groups is known as phase II metabolism (see Chapter 18, Figure 3). To allow secretion of the large variety of drugs, metabolites, and their conjugates, the canalicular membrane contains multispecific drug/conjugate transporters. Drug conjugates may include those of cytostatic drugs. ABC transporters that extrude conjugated drugs can confer multiple drug resistance because they transport a variety of anionic compounds, and thus they have been named multiple drug resistance proteins (MRPs). The ABCC2 (MRP2) protein exports a wide range of organic anions such as bilirubin glucuronides (end-products of heme degradation), glutathione–S conjugates, and dianionic xenobiotics. Mutations in the ABCC2 gene cause *Dubin–Johnson syndrome*, a liver disorder characterized by chronically increased blood levels of conjugated bilirubin as described in Chapter 18.

The bile forms mixed micelles with the products of lipid digestion, such as fatty acids, 2-monoacylglycerols, and 1-monoacylglycerols, to allow their absorption. Monoacylglycerols are further hydrolyzed by membrane-associated lipases to form fatty acids and glycerol. These are efficiently absorbed in the small intestine (see Chapters 9–16 for a description of fatty acid transport), whereas bile acids are later absorbed in the jejunum to recapture these valuable metabolites. The recycling pathway of bile is known as the enterohepatic circulation.

Bile acids are reabsorbed in the intestine by the apical sodium-dependent bile acid transporter ASBT (SLC10A2), which is an electrogenic $2Na^+$/bile acid cotransporter. Both primary and secondary conjugated and unconjugated bile acids are substrates for ASBT. Mutations in the ASBT gene result in *primary bile acid malabsorption* (Oelkers *et al.* 1997). The clinical phenotype includes severe diarrhea, malabsorption of fat, and malnutrition. In parallel with the Na^+-dependent uptake of bile acids, Na^+-independent absorption is mediated by the organic anion transporting polypeptide 3 (OATP3, SLC21A7). However, the significance of this transporter compared with ASBT has not yet been established, particularly because uptake of bile acids occurs in exchange for valuable intracellular substrates, such as glutathione. Release of bile acids on the basolateral side is similarly carried out by anion exchangers, possibly by an alternatively spliced ASBT transporter.

After resorption in the intestine, bile acids return to the liver via the portal vein which delivers all metabolites that are resorbed in the intestine (Meier and Steiger 2002). The portal vein faces the basolateral membrane of hepatocytes. Similar to the uptake across the apical membrane of enterocytes, bile acids are transported across the basolateral membrane of hepatocytes by Na^+-dependent and Na^+-independent mechanisms. Na^+-dependent transport is mediated by the Na^+-taurocholate cotransporting polypeptide (NTCP, SLC10A1), which is related to ASBT. Several members of the organic anion transporting peptide family mediate Na^+-independent antiport of bile acids against intracellular anions, possibly HCO_3^- or glutathione.

A comparison of bile acid transport in the canalicular membrane with the transport process in epithelial cells or the basolateral membrane of hepatocytes illustrates that transmembrane passage of amphiphilic compounds can be equally well catalyzed by primary active and secondary active transporters. ABC transporters are generally found in membranes where lipophilic compounds are secreted.

ROLE OF ABC TRANSPORTERS IN LIPID METABOLISM

After resorption into the enterocytes, glycerol and fatty acids are used to resynthesize triacylglycerols, which are exported from the cell to form chylomicrons that are secreted into the lymphatic vessels (not into the portal vein). The chylomicrons are subsequently passed into the circulation via the thoracic duct. The reason for this separate pathway may lie in the size of chylomicrons, which are too large to pass through the gaps between the endothelial cells lining the blood vessels.

Chylomicrons are a form of lipoprotein. Lipoproteins are the major carriers of lipids in the blood and are classified by their density, which is determined by the lipid-to-protein ratio. In order of increasing density, lipoproteins are named chylomicrons, VLDL (very-low-density lipoprotein), LDL (low-density lipoprotein), and HDL (high-density lipoprotein). All lipoproteins contain apolipoproteins, and in different proportions triacylglycerols, phospholipids, cholesteryl esters, cholesterol, and very small amounts of free fatty acids. Free fatty acids, instead, are bound to albumin in the circulation. Apolipoproteins play several roles: (1) they are enzyme cofactors, (2) they act as a scaffold for lipid binding, and (3) they are recognized by receptors to allow resorption of the lipid content of lipoproteins by specific cell types. For example, apolipoprotein A-I (Apo A-I) is an activator of lecithin:cholesterol acyltransferase (LCAT), an enzyme of the blood plasma that synthesizes cholesterol esters that are found in lipoproteins. In addition, ApoA-I also acts as a ligand for the HDL receptor and the ABCA1 transporter (see Chapter 19).

Chylomicrons and VLDLs are the major carriers of triacylglycerols from the intestine and liver, respectively, to peripheral tissues. The ability of peripheral tissues to consume triacylglycerols from lipoproteins relies on the expression of lipoprotein lipase in the endothelium of blood vessels in these tissues. The enzyme hydrolyzes triacylglycerol to generate free fatty acids and glycerol. Fatty acids can then be taken up into cells by fatty acid transporters (see Chapters 9–16). Removal of triacylglycerols from VLDL particles ultimately results in the generation of LDL particles, which are cleared from the blood by endocytosis into hepatocytes or peripheral cells. LDL particles are the major carrier of cholesterol from the liver to the extrahepatic tissues.

The HDL protein, which has the highest protein-to-lipid ratio, is the only carrier that transports cholesterol and phospholipids from the periphery to the liver (reverse cholesterol transport) (Figure 5). Initially, ApoA-I is synthesized in the liver or intestine and secreted into the extracellular space by the secretory pathway. Subsequently, phospholipids are loaded onto ApoA-I, converting it into a discoidal HDL particle. The loading of phospholipids onto the nascent ApoA-I requires the ABCA1 phospholipid transporter. Similar to the situation in bile, phospholipids are required to act as cholesterol acceptors. Moreover, the acyl residues are needed for cholesteryl ester formation. The discoidal HDL subsequently picks up cholesterol from extrahepatic tissues. The plasma enzyme

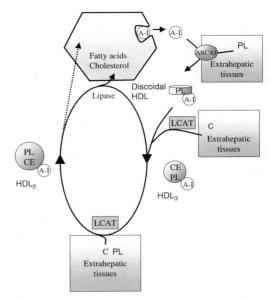

Figure 5 The role of HDL in lipid metabolism. HDL transports choletserol from extrahepatic tissues back to the liver. Apolipoprotein A is produced in hepatocytes and the intestine and subsequently phospholipids (PL) are loaded onto it by the ABCA1 transporter to form discoidal HDL. Discoidal HDL is converted into HDL_3 by the recruitment of cholesteryl esters, formed by LCAT from cholesterol. Increased uptake of cholesterol and cholesteryl esters in extrahepatic tissues converts HDL_3 into HDL_2. Cholesterol esters and phospholipids are subsequently unloaded in the liver. To some extent, the whole HDL particle is taken up by endocytosis.

LCAT converts cholesterol into cholesteryl esters, which are included in the hydrophobic core of HDL_3, thereby decreasing its density and forming HDL_2 (Figure 5). The high-density HDL_3 is restored by removal of lipids and cholesteryl ester from HDL in the liver or by endocytosis of the whole lipoprotein. Mutations in the ABCA1 protein result in *Tangier disease*, an HDL-deficiency syndrome. As HDL is involved in the removal of cholesterol from cells of the periphery, this disease results in high levels of plasma cholesterol as described in Chapter 19. As a result, patients suffering Tangier disease develop arteriosclerosis early in life.

Plant lipids are valuable metabolites in our nutrition but frequently have an unusual structure. Very long fatty acids (>C22), for example, cannot immediately be metabolized by mitochondrial β oxidation (see Chapters 9–16). Instead, they are first transported into peroxisomes, where they are truncated by the peroxisomal β oxidation until octanoyl-CoA is formed. Transport of very long fatty acids into peroxisomes is mediated by the ABCD1 protein. Like ABCG5 and ABCG8, ABCD1 is a half-size transporter. The functional unit is likely to be a homodimer of the protein. Again, it appears that the hydrophobic part of long-chain fatty acids inserts into the lipid bilayer of the peroxisome, where they are picked up by the ABC transporter. After flipping they are extruded into the lumen of the peroxisome where they can be processed by the β oxidation machinery. The final product, octanoyl-CoA, is thought to be transported across the peroxisomal membrane by a carnitine-dependent transport mechanism similar to the one described in Chapters 9–16.

The importance of peroxisomal ABC transporters for oxidation of very-long-chain fatty acids is evidenced by mutations in the ABCD1 gene, which have been shown to cause the neurodegenerative disorder *X-linked adrenoleukodystrophy* (X-ALD) (Mosser *et al.* 1993). This lipid storage disease is characterized by a striking and unpredictable variation in phenotypic expression. Clinical manifestations include a rapidly progressive childhood cerebral form, a milder adult form, adrenomyeloneuopathy, and variants without neurological involvement. The neurological symptoms are thought to be a consequence of the physical properties of very-long-chain fatty acids, which are believed to destroy the myelin ensheathing axons of the central nervous system.

ABC TRANSPORTERS INVOLVED IN OTHER DISEASES

ABCA4 encodes a 2273 amino acid full-size ABC transporter protein localized to the rims of the rod and cone outer-segment disks (Chapter 20). Available evidence suggests that ABCA4 mediates the transport of *N*-retinylidene-phosphatidylethanolamine from the luminal to the cytosolic face of the photoreceptor disks. *N*-retinylidene-phosphatidylethanolamine is a metabolic by-product which is generated during regeneration of the photopigment retinal. Mutations in the ABCA4 gene have been linked to a wide spectrum of degenerative retinal diseases, including *Stargardt disease* (STGD), *cone–rod dystrophy* (CRD), *atypical retinitis pigmentosa* (RP), and *age-related macular degeneration* (AMD), as described in Chapter 20.

ABCC6 and pseudoxanthoma elasticum

Three independent studies have demonstrated that mutations in ABCC6 cause the autosomal recessive disorder *pseudoxanthoma elasticum* (Bergen *et al.* 2000; LeSaux *et al.* 2000; Ringpfiel *et al.* 2000). This rare disorder is characterized by calcification of elastic fibers in the skin, arteries, and retina. It results in dermal lesions, arterial insufficiency, and retinal hemorrhages, ultimately leading to macular degeneration. In pseudoxanthoma elasticum, characteristic changes are found in the skin of the neck, axilla, and other flexural areas, resulting in angioid streaks in the retina and arteries. This produces hemorrhage of gastrointestinal and other blood vessels, early calcification, and occlusive vascular changes.

The ABCC6 gene encodes a 1503 amino acid protein with a molecular weight of 165 kDa. Its primary structure predicts a full-size transporter with 17 membrane spanning helices grouped into three transmembrane domains, typical of ABCC subfamily transporters. ABCC6 is located in the plasma membrane and predominantly expressed in kidney and liver; however, ABCC6 mRNA expression is also detectable in the retina, skin, and vasculature.

Recent work by Ilias *et al.* (2002) suggests that ABCC6 functions as an organic anion transporter. Overexpression in Sf9 insect cells demonstrated transport activity for glutathione conjugates, including leukotriene C4 and *N*-ethylmaleimide-*S*-glutathione. Organic anions such as probenecid, benzbromarone, and indomethacin, which are known to interfere with transport of glutathione conjugates in ABCC1 and ABCC2, also inhibited ABCC6 transport activity. The normal function of ABCC6, including its physiological substrate, remains to be established.

An important aspect of the biology of ABCC6 is that pseudoxanthoma elasticum is typically complicated by cardiovascular disease, which points to its implication in the development of atherosclerosis. Importantly, the involvement of ABCC6 in atherogenesis is supported by a recent report that has revealed an association between the frequent R1141X mutation in the ABCC6 gene and the prevalence of premature cardiovascular disease. The prevalence of the R1141X mutation was 4.2-fold higher among cardiovascular disease patients than among controls (Trip *et al.* 2002).

ABCB2/ABCB3 (TAP) and its involvement in immunological responses

The ABCB2 and ABCB3 proteins are half transporters that have been shown to act as a heterodimer in the translocation of immunogenic peptides from the cytosol into the endoplasmic reticulum. Antigen processing and the role of this transporter in *bare lymphocyte syndrome type I* and its role in virus persistence are discussed in detail in Chapter 21.

REFERENCES

Aguilar-Bryan, L., Nichols, C.G., Wechsler, S.W., Clement, J.P., Boyd, A.E., 3rd, Gonzalez, G., et al. (1995). Cloning of the Beta Cell High-Affinity Sulfonylurea Receptor: A Regulator of Insulin Secretion. *Science*, **268**, 423–426.

Ashcroft, F.M. (2000). *Ion Channels and Disease*, San Diego, CA: Academic Press.

Berge, K.E., Tian, H., Graf, G.A., Yu, L., Grishin, N.V., Schultz, J., et al. (2000). Accumulation of Dietary Cholesterol in Sitosterolemia Caused by Mutations in Adjacent ABC Transporters. *Science*, **290**, 1771–1775.

Bergen, A.A., Plomp, A.S., Schuurman, E.J., Terry, S., Breuning, M., Dauwerse, H., et al. (2000). Mutations in ABCC6 Cause Pseudoxanthoma Elasticum. *Nat. Genet.*, **25**, 228–231.

Borst, P. and Elferink, R.O. (2002). Mammalian ABC Transporters in Health and Disease. *Annu. Rev. Biochem.*, **71**, 537–592.

Chang, G. and Roth, C.B. (2001). Structure of MsbA from *E. coli*: A Homolog of the Multidrug Resistance ATP Binding Cassette (ABC) Transporters. *Science*, **293**, 1793–1800.

de Vree, J.M., Jacquemin, E., Sturm, E., Cresteil, D., Bosma, P.J., Aten, J., et al. (1998). Mutations in the MDR3 Gene Cause Progressive Familial Intrahepatic Cholestasis. *Proc. Natl Acad. Sci. USA*, **95**, 282–287.

Higgins, C.F. (1992). ABC Transporters: From Microorganisms to Man. *Annu. Rev. Cell Biol.*, **8**, 67–113.

Holland, I.B., Cole, SPC., Kuchler, K., and Higgins, C.F. (2003). *ABC Proteins: From Bacteria to Man*. San Diego; CA: Academic Press.

Ilias, A., Urban, Z., Seidl, T.L., Le Saux, O., Sinko, E. Boyd, C.D., et al. (2002). Loss of ATP-Dependent Transport Activity in Pseudoxanthoma Elasticum-Associated Mutants of Human ABCC6 (MRP6). *J. Biol. Chem.*, **277**, 16860–16867.

Jacquemin, E., Cresteil, D., Manouvrier, S., Boute, O., and Hadchouel, M. (1999). Heterozygous Non-Sense Mutation of the MDR3 Gene in Familial Intrahepatic Cholestasis of Pregnancy. *Lancet*, **353**, 210–211.

Karow, M. and Georgopoulos, C. (1993). The Essential *Escherichia coli* msbA Gene, a Multicopy Suppressor of Null Mutations in the htrB Gene, is Related to the Universally Conserved Family of ATP-Dependent Translocators. *Mol. Microbiol.*, **7**, 69–79.

Le Saux, O., Urban, Z., Tschuch, C., Csiszar, K., Bacchelli, B., Quaglino, D., et al. (2000). Mutations in a Gene Encoding an ABC Transporter Cause Pseudoxanthoma Elasticum. *Nat. Genet.*, **25**, 223–227.

Meier, P.J. and Stieger, B. (2002). Bile Salt Transporters. *Annu. Rev. Physiol.*, **64**, 635–661.

Mosser, J., Douar, A.M., Sarde, C.O., Kioschis, P., Feil, R., Moser, H., et al. (1993). Putative X-Linked Adrenoleukodystrophy Gene Shares Unexpected Homology with ABC Transporters. *Nature*, **361**, 726–730.

Oelkers, P., Kirby, L.C., Heubi, J.E., and Dawson, P.A. (1997). Primary Bile Acid Malabsorption Caused by Mutations in the Ileal Sodium-Dependent Bile Acid Transporter Gene (SLC10A2). *J. Clin. Invest.*, **99**, 1880–1887.

Poelarends, G.J., Mazurkiewicz, P., and Konings, W.N. (2002). Multidrug Transporters and Antibiotic Resistance in *Lactococcus lactis*. *Biochim. Biophys. Acta*, **1555**, 1–7.

Ringpfeil, F., Lebwohl, M.G., Christiano, A.M., and Uitto, J. (2000). Pseudoxanthoma Elasticum: Mutations in the MRP6 Gene Encoding a Transmembrane ATP-Binding Cassette (ABC) Transporter. *Proc. Natl Acad. Sci. USA*, **97**, 6001–6006.

Riordan, J.R., Rommens, J.M., Kerem, B., Alon, N., Rozmahel, R., Grzelczak, Z., *et al.* (1989). Identification of the Cystic Fibrosis Gene: Cloning and Characterization of Complementary DNA. *Science*, **245**, 1066–1073.

Small, D.M. (2003). Role of ABC Transporters in Secretion of Cholesterol from Liver into Bile. *Proc. Natl Acad. Sci. USA*, **100**, 4–6.

Strautnieks, S.S., Bull, L.N., Knisely, A.S., Kocoshis, S.A., Dahl, N., Arnell, H., *et al.* (1998). A Gene Encoding a Liver-Specific ABC Transporter is Mutated in Progressive Familial Intrahepatic Cholestasis. *Nat. Genet.*, **20**, 233–238.

Trip, M.D., Smulders, Y.M., Wegman, J.J., Hu, X., Boer, J.M., ten Brink, J.B., *et al.* (2002). Frequent Mutation in the ABCC6 Gene (R1141X) is Associated with a Strong Increase in the Prevalence of Coronary Artery Disease. *Circulation*, **106**, 773–775.

Urbatsch, I.L., Sankaran, B., Weber, J., and Senior, A.E. (1995). P-Glycoprotein is Stably Inhibited by Vanadate-Induced Trapping of Nucleotide at a Single Catalytic Site. *J. Biol. Chem.*, **270**, 19383–19390.

Yu, L., Hammer, R.E., Li-Hawkins, J., Von Bergmann, K., Lutjohann, D., Cohen, J.C., *et al.* (2002). Disruption of Abcg5 and Abcg8 in Mice Reveals Their Crucial Role in Biliary Cholesterol Secretion. *Proc. Natl Acad. Sci. USA*, **99**, 16237–16242.

18

MORIMASA WADA*, TAKESHI UCHIUMI*, AND
MICHIHIKO KUWANO*,**

Canalicular multispecific organic anion transporter ABCC2

ABCC2 PROTEIN AND ABCC SUBFAMILY

The ABCC2 protein, also designated the canalicular multispecific organic anion transporter (cMOAT) or multidrug resistance protein 2 (MRP2), belongs to the subfamily C of ABC transporter proteins (http://nutrigene.4t.com/humanabc.htm). It is now known that this subfamily comprises 12 members (Figure 1), including six MRP-related proteins (ABCC1–6, Cole *et al.* 1992; Paulusma *et al.* 1996; Taniguchi *et al.* 1996; Kool *et al.* 1997; Suguki *et al.* 1997; Lee *et al.* 1998; Kool *et al.* 1999; McAleer *et al.* 1999; Schuetz *et al.* 1999), the cystic fibrosis transmembrane conductance regulator (CFTR), and the sulfonylurea receptors SUR1 and SUR2. One of the major characteristics of subfamily C members is the relatively low sequence homology between their NH_2- and COOH-terminal nucleotide binding domains (NBDs). Four of the MRP-related proteins, ABCC1, 2, 3, and 6, have an atypical structure displaying an additional transmembrane domain at the NH_2-terminus. This fifth domain

[the membrane spanning domain 1 (MSD1) or transmembrane domain 0 (TMD0)], of approximately 200 amino acids, has an extracytosolic NH_2-terminus and consists of five transmembrane segments connected to the typical four-domain transporter core by an intracellular loop [the cytoplasmic loop 3 (CL3) or linker 0 (L0)] of approximately 130 amino acids (Figure 2) (Cole *et al.* 1992; Bakos *et al.* 1996; Stride *et al.* 1996; Hipfner *et al.* 1997). The ABCC1 protein was the first identified MRP-related gene from a multidrug resistant human small-cell lung cancer cell line and was followed by the identification of ABCC2 as described below. Because ABCC1 is closely related to ABCC2, the topology and biochemical features of which have been studied most extensively, some of the knowledge obtained about ABCC1 will be presented and discussed together with ABCC2.

IDENTIFICATION OF THE *ABCC2* GENE

The *ABCC2* cDNA was isolated by two independent approaches. In one case, ABCC2 was identified as the transporter responsible for multidrug resistance in human cancer cells (Taniguchi *et al.* 1996); the second approach identified ABCC2 as a multispecific organic anion excreting transporter across the hepatocyte canalicular membrane (Buchler

*Department of Medical Biochemistry, Graduate School of Medical Sciences, Kyushu University, Fukuoka 812–8582, Japan
** Present address: Research Center for Innnovative Cancer Therapy, Kurume University, Kurume 830 0011, Japan

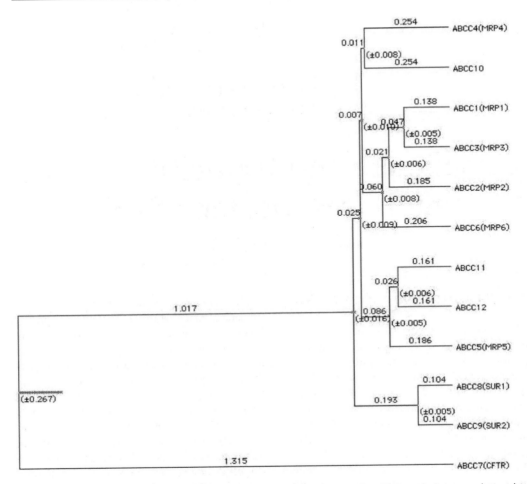

Figure 1   Evolutionary relationship between ABCC subfamily proteins. The evolutionary relationship between ABCC subfamily proteins was calculated by the unweighted pair group method with arithmetic mean (UPGMA) algorithm (Nei 1987). The length of the horizontal lines connecting one sequence to another is proportional to the estimated genetic distance between the sequences. The thick gray lines at branch points in the tree are error bars showing the standard error of the branch position. The score was calculated from the number of mismatches dividing by the length of the sequence using GeneWorks software.

et al. 1996; Paulusma *et al.* 1996; Ito *et al.* 1997). Multidrug resistance in cancer cells is often associated with overexpression of two different ABC transporters, such as ABCB1 [also called multidrug resistance 1 (MDR1) or P-glycoprotein (P-gp)] (Kartner *et al.* 1983, 1985; Chen *et al.* 1986; Alvarez *et al.* 1995) or ABCC1 (Cole *et al.* 1992; Nooter and Stoter 1996). Owing to the fact that the

chemotherapeutic drug cisplatin can be conjugated to glutathione (Ishikawa and Ali-Osman 1993) and resistant cells actively excrete the conjugate (Ishikawa 1992), it has been expected that ABCC1 and/or ABCB1 may be involved in resistance to cisplatin. However, cells transfected with cDNA of either of the two transporters do not show resistance to cisplatin, and cell lines selected for cisplatin

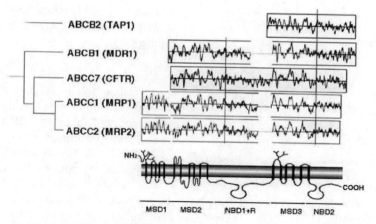

Figure 2   Hydropathy analysis of ABC family proteins and expected topology of ABCC1 Kyte–Doolittle hydrophobicity (Kyte and Doolittle 1982) of representative ABC proteins are presented in the upper panel. The lower panel shows the expected topology of ABCC1 predicted by the hydrophobicity and experimental data including inhibition of glycosylation and partial digestion by peptidases. MSD, membrane spanning domain; NBD, nucleotide binding domain; R, regulatory domain carried by CFTR.

resistance do not show overexpression of ABCB1 or ABCC1. Consequently, we hypothesized that the transporter responsible for cisplatin resistance may be a novel type of ABC transporter. To isolate the cDNA of the putative transporter we designed highly degenerate PCR primer mixtures that were complementary to the conserved ATP binding domain of the three ABC proteins, namely ABCB1, ABCC1, and CFTR (Higgins 1992) (Figure 2). This approach resulted in the isolation of a novel human ABC protein gene from a cisplatin-resistant derivative of an epidamoid cancer cell line. The gene was subsequently mapped to chromosome 10q24 (Taniguchi *et al.* 1996). The amino acid sequence predicted from the nucleotide sequence of the gene showed 47%, 30%, and 18% homology to that of *ABCC1, CFTR,* and *ABCB1,* respectively, and a hydrophobicity analysis of the protein showed a structure highly similar to ABCC1 (Figure 2). Hence the protein has been named ABCC2 to indicate the similarity to other members of the ABCC subfamily.

   The second approach took advantage of biochemical and genetic studies using rat models of Dubin–Johnson syndrome (DJS) (TR⁻, Groningen yellow (GY), and Eisai

hyperbilirubinemic (EHBR) rats), which indicated the presence of a multispecific organic anion transporter in the hepatocyte canalicular membranes, transporting substrates such as glutathione and glucuronide conjugates (Figure 3) (Kartenbeck *et al.* 1996; Paulusma and Oude Elferink 1997; Suzuki and Sugiyama 1998; Konig *et al.* 1999b). The putative transporter was designated cmoat (canalicular multispecific organic anion transporter) (Paulusma and Oude Elferink 1997). Although Abcc1, the rat ortholog of ABCC1, transports glutathione and glucuronide conjugates, it was an unlikely candidate for this activity (Cole *et al.* 1992; Loe *et al.* 1996a, b) because expression of Abcc1 is very low in the liver (Cole *et al.* 1992). Thus it was hypothesized that a canalicular isoform of Abcc1 could be the putative cmoat. The absence of canalicular isoforms of Abcc1 and ABCC1 (which were later identified as Abcc2 and ABCC2, respectively) in the mutant rats and DJS patients, respectively, was subsequently demonstrated by immunohistochemical and other analysis, further supporting this hypothesis (Buchler *et al.* 1996; Kartenbeck *et al.* 1996; Keppler and Kartenbeck 1996). Finally, Abcc2 and ABCC2 cDNAs were

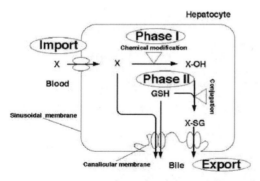

Figure 3 Three steps of the detoxification process and excretion pathways for xenobiotics in liver. The metabolism of most drugs in humans is thought to occur in three phases. In phase I, drugs are oxidized by enzymes such as the cytochrome P450 isoforms. In phase II, drugs are coupled to glutathione, glucuronic acid, or sulfate to improve solubility. Conjugated drugs are then exported out of the cells. X, xenobiotics.

cloned by RT-PCR using primer pairs complementary to ABCC1 (Bucher *et al.* 1996; Paulusma *et al.* 1996).

The success of the two cloning strategies suggests that the ABCC2 protein has a dual function. On the one hand, it is responsible for the secretion of drug conjugates across the canalicular membrane; on the other hand, it can provide resistance against anionic drugs such as cisplatin.

Isolation of genomic clones covering the human *ABCC2* gene, sequencing, and comparison with the *ABCC2* cDNA sequence (Taniguchi *et al.* 1996; Toh *et al.* 1999; Tsujii *et al.* 1999) indicated the presence of 32 exons. The complete *ABCC2* gene spans a region of 68.82 kb on chromosome 10q24. The exon splitting pattern of the *ABCC2* gene, particularly in the region adjacent to the nucleotide binding domains and in the posterior half of the gene, was similar to the human *ABCC1* gene (Grant *et al.* 1997), suggesting a close evolutionary relationship between the two ABC transporter genes.

TISSUE DISTRIBUTION AND SUBCELLULAR LOCALIZATION OF ABCC2

Despite the close structural similarity among MRP-related proteins (Figures 1 and 2), their tissue distribution and subcellular localization differ widely (Table 1). ABCC1 is ubiquitously expressed throughout the body, whereas ABCC2 and ABCC3 are mainly expressed in the liver, kidney, and small intestine (Borst *et al.* 1999). Expression of ABCC2 in other tissues including colon (Konig *et al.* 2003), gallbladder (Rost *et al.* 2001), lung (Konig *et al.* 2003), and placenta (St. Pierre *et al.* 2000), has also been reported.

At the subcellular level, ABCC1 and ABCC3 are present in the basolateral membrane, while ABCC2 is localized to the apical membrane of polarized cells (Buchler *et al.* 1996; Evers *et al.* 1996, 1998; Borst *et al.* 1999; Konig *et al.* 1999a, b). Accordingly, ABCC2 is expressed in the canalicular membrane, which is the apical domain of hepatocytes. Similarly, ABCC2 is expressed in the apical domain of proximal tubule epithelial cells of human and rat kidney, where it presumably mediates the export of conjugated organic anions into the urine (Schaub *et al.* 1999). Studies of the expression of rat and rabbit ABCC2 homologues (Abcc2) in the small intestine further revealed that the protein is found in the apical brush border membrane of villi (Mottino *et al.* 2000; Van Aubel *et al.* 2000).

The mechanisms involved in apical targeting have not been fully resolved. Comparison of the primary structure of ABCC2 with those of other ABC subfamily members, reveals a seven-amino-acid extension at its COOH terminus, the last three amino acids (TKF) of which comprise a PDZ-interacting motif. It has been reported that interaction of this motif with the PDZ domain containing protein is important for apical localization of membrane proteins (Songyang *et al.* 1997; Fanning and Anderson 1999). The PDZ-interacting motif, for example, is reported to be a determinant for

Table 1 Comparison of chromosome location, expression in normal tissues and tumors, and substrates of six ABC transporter family genes

Gene symbol	ABCB1	ABCC1	ABCC2	ABCC3	ABCC4	ABCG2
Alias	PGY1/MDR1	MRP1	MRP2/cMOAT	MRP3	MRP4	MXR1/BCRP
Protein	P-glycoprotein (P-gp)	Multidrug resistance protein (MRP)	Canalicular multispecific organic anion transporter	MRP3	MRP4	MXR1/BCRP
Amino acids	1280	1531	1545	1527	1325	655
Chr location	7q21	16p13.1	10q24	17q22	13q32	4q22
Tissue expression	Adrenal gland, kidney, liver (canalicular membrane)[a], colon, brain (endothelial cell), uterus (pregnancy)	Placenta, testis, lung, skeletal muscle, heart, monocyte, liver (sinusoidal membrane)	Liver (canalicular membrane), duodenum, kidney	Liver, duodenum, colon, adrenal gland, pancreas, kidney	(Pancreas, skeletal muscle, lung, kidney, bladder)[b]	Placenta, breast, liver, intestine
Altered expression in human cancer	Neuroblastoma, leukemia, lymphoma, breast cancer, pancreas cancer, colon cancer, adrenal cancer, liver cancer, renal cancer, ovarian cancer, non-small-cell lung carcinoma	Neuroblastoma, lymphoma, leukemia, non-small-cell lung cancer, anaplasticthyroid cancer, esophageal cancer, gastric cancer	Leukemia, colon cancer; hepatocellular carcinoma, renal clear cell carcinoma	Colon cancer	NE	Breast cancer
Substrates	Neutral and cationic organic compounds	Glutathione and glucuronate conjugates, glutathione	Glucuronate and glutathione conjugates, glutathione	Glucuronate and sulfite conjugates, bile salts	Nucleoside monophosphate analogs	NE
Anticancer drugs	Vinca alkaloids, Vincristine, Vinblastine, Anthracyclines, Doxorubicin, Daunorubicin, Epipodophylotoxins, Etoposide, Teniposide, Others, Pacritaxel, Colchicine, Actionmycin D	Vinca alkaloids, Vincristine, Vinblastine, Anthracyclines, Doxorubicin, Daunorubicin, Epipodophylotoxins, Etoposide, Teniposide, Antimetabolites, Methotrexate	Vinca alkaloids, Vincristine, Vinblastine, Camptothesins, Cpt-11, Sn-38, Antimetabolites, Methotrexate, Others, Cisplatin	Anthracyclines, Doxorubicin, Daunorubicin, Epipodophylotoxins, Etoposide, Teniposide, Antimetabolites, Methotrexate	Antimetabolites, Methotrexate, Thiopurines	Anthracyclines, Doxorubicin, Daunorubicin, Camptothesins, Cpt-11, Sn-38, Topotecan, Antimetabolites, Mitoxantrone
References	Kartner et al. 1985; Chen et al. 1986; Ueda et al. 1987; Alvarez et al. 1995; Ambudkar et al. 1999	Cole et al. 1992; Muller et al. 1994; Zaman et al. 1994; Noter and Stoter 1996; Jedlitschky et al. 1997; Loe et al. 1998; Hooijberg et al. 1999	Kartenbeck et al. 1996; Paulusma et al. 1996; Taniguchi et al. 1996; Koike et al. 1997; Kool et al. 1997; Narasaki et al. 1997; Evers et al. 1998; Cui et al. 1999; Hooijberg et al. 1999; Kawabe et al. 1999; Schaub et al. 1999	Uchiumi et al. 1998; Belinsky et al. 1999; Kubitz et al. 1999	Schuetz et al. 1999; Chen et al. 2001, 2002	Allikmets et al. 1998; Doyle et al. 1998; Miyake et al. 1999; Robey et al. 2001

NE, not established.

[a] Parentheses indicate precise subcellular location or cell type which express the each mRNA in organs ("normal tissue" line) or more precise cancer type in each human cancer ("human cancer" line).

[b] Low expression.

apical localization of CFTR (Milewski et al. 2001). Furthermore, ABCC2 was isolated by an yeast two-hybrid system screen using a PDZ domain containing protein as the bait (Kocher et al. 1999). These results suggest that the COOH terminus, particularly the PDZ-interacting motif, might be important for apical sorting of ABCC2. However, the experimental results examining the role of this motif in apical targeting are controversial. Harris et al. (2001) showed that deletion of the TKF motif resulted in basolateral targeting of ABCC2. In contrast, using an NH_2-terminal fusion of ABCC2 to GFP in transiently transfected cells, Nies et al. (2002) showed that deletion of the COOH-terminal 11 amino acids, including the PDZ-interacting motif, did not abolish apical targeting, but that deletion of 15 or more amino acids of the COOH terminus impaired the localization of ABCC2 to the apical membrane of human HepG2 cells. On the other hand, the N-terminal MSD1 and/or linker region 3 (CL3) has been shown to be necessary for membrane routing by deletion experiment of these regions in ABCC1 and ABCC2, respectively (Bakos et al. 2000a; Fernandez et al. 2002). It appears likely that this region of the ABCC2 protein is involved in the integration into the membrane of the endoplasmic reticulum, intracellular trafficking, or recycling by endocytosis. However, additional motifs are required to determine either apical or basolateral localization in polarized cells. We took advantage of the close structural similarity of ABCC1 and ABCC2 and the distinct localization of the two proteins to generate sequential chimeric proteins and to analyze their localization. We found that substitution of L0-MSD2 (Figure 2) but not that of the COOH-terminal 65 amino acids of ABCC2, by the corresponding region of ABCC1 resulted in alteration of protein targeting from apical to basolateral membrane, suggesting that at least L0-MSD2 is involved in distinct localization of ABCC1 and ABCC2 and that the COOH-terminus may be necessary but is exchangeable between ABCC1 and ABCC2 (Konno et al. 2003). Further analysis is necessary to identify the responsible amino acid sequences or motifs and precise mechanism for proper cellular localization of ABCC2 and related proteins.

SUBSTRATE SPECIFICITY OF ABCC2

Among the ABCC family members, substrate specificity has been most extensively characterized in ABCC1 cDNA transfected cells. ABCC1 has a broad substrate specificity covering anticancer drugs (Jedlitschky et al. 1996; Loe et al. 1998; Priebe et al. 1998), organic anions derived from phase I and II metabolism of xenobiotics (Ishikawa et al. 1992; Loe et al. 1996a, b, 1997; Bakos et al. 2000b), and endogenous compounds

Table 2   Substrate affinities of ABCC1 and ABCC2

Substrate	K_m (μM)		Reference(s)
	ABCC1	ABCC2	
Leukotriene C_4	0.1	1.0	Leier et al. 1996; Loe et al. 1996a; Stride et al. 1997; Cui et al. 1999; Konno et al. 2003
S-Glutathionyl-2,4-dinitrobenzene (DNP-SG)	3.6	6.5	Jedlitschky et al. 1996; Evers et al. 1998
17β-Glucuronosyl estradiol	1.5–4.8	7.2	Jedlitschky et al. 1996; Loe et al. 1996b; Stride et al. 1997; Cui et al. 1999
p-Aminohippurate	372	880	Leier et al. 2000
Methotrexate	2,200–3,500	250–3,000	Bakos et al. 2000b; Zeng et al. 2001; Konno et al. 2003

like cysteinyl leukotriene LTC4 (Ishikawa *et al.* 1990), bilirubin-glucuronides, sulfate-conjugated bile salts, glutathione disulfide (GSSG), prostaglandin A2, estradiol-gluconuride ($E_2$17βG), and others (Jedlitschky *et al.* 1996, 1997; Evers *et al.* 1997; Heijn *et al.* 1997; Hooijberg *et al.* 1997). The substrate specificity of ABCC2 has been characterized not only using transfected cells but also by comparing canalicular membrane preparations of wild-type rats with those of rat strains lacking hepatobiliary excretion activity as described above. Although the substrate specificity or affinity of some substrates could differ between human ABCC2 and rat Abcc2, as seen in studies using human ABCC1 and rat Abcc1 (Stride *et al.* 1997, 1999; Leslie *et al.* 2001a; Zhang 2001), they nevertheless provide valuable data about the general substrate range of ABCC2. ABCC2 translocates a similar spectrum of compounds as ABCC1, but with different affinities (Table 2). A comprehensive list of ABCC2 substrates can be found in previous reviews (Konig *et al.* 1999b; Suzuki and Sugiyama 2002).

MECHANISM OF TRANSPORT BY ABCC2

Transport processes including substrate binding, ATP binding, ATP hydrolysis, substrate release, and cooperation among those processes in the ABCC2 protein during the catalytic cycle remain to be clarified. It is likely that conformational changes are necessary to transit from one process to the next. Recent success in determination of the crystal structures of two bacterial ABC proteins, MsbA (Chang and Roth 2001) and BtuCD (Locher *et al.* 2002), should facilitate the general understanding of the transport processes, but knowledge about those processes in mammalian ABC proteins, including ABCC2, is still very limited.

Several indispensable regions or amino acid residues have been identified by substitution or deletion experiment of amino acid residues. By this type of approach, it has been shown that the linker region connecting MSD1 and MSD2 is involved in membrane integration of

ABCC1 and in LTC$_4$ transport (Bakos *et al.* 1998, 2000a; Gao *et al.* 1998), whereas most of the linker connecting NBD1 and MSD3 is dispensable (Gao *et al.* 1998). MSD1 has also been shown to be dispensable for LTC$_4$ transport (Bakos *et al.* 1998), but could be important for other substrates. Actually, mutation of a cysteine located in TM1 of MSD1 changes the NH$_2$-terminal conformation and LTC$_4$ transport activity (Yang *et al.* 2002). We have also shown that MSD1 is involved in determination of transport character. Despite the close similarity of substrate specificity between ABCC1 and ABCC2, their affinity to LTC$_4$ differs 10-fold. By examining the kinetic properties of ABCC1/ABCC2 chimeric proteins for LTC$_4$ transport in inside-out membrane vesicles, we revealed the following:

1. When the NH$_2$-proximal 108 amino acids of ABCC2 including transmembrane helices (TM) 1–3 were exchanged with the corresponding region of ABCC1, affinity for LTC$_4$ increased fivefold relative to wild-type ABCC2, suggesting that TM 1–3 are involved in substrate affinity.

2. When the NH$_2$-terminal two thirds of ABCC2 was exchanged with the corresponding ABCC1 region, the chimeric protein transported LTC$_4$ with an efficiency comparable to that of wild-type ABCC1, suggesting that the COOH-terminal half is exchangeable between two proteins (Konno *et al.* 2003).

It has also been reported that basic residues in TM6, 9, 16, and 17 of ABCC2 that are conserved in ABCC1 are important for transport of glutathione–methylfluorescein (Ryu *et al.* 2000), charged amino acid residues in TM6 are important for specificity and/or binding of LTC$_4$ and other substrates (Haimeur *et al.* 2002), and a tryptophan residue in TM17 is important for transport of E217βG by ABCC1 and transport of methotrexate, LTC$_4$, and E217βG by ABCC2 (K. Ito *et al.* 2001a, b). Similarly, it has been reported that charged residues in TM6, 11 and 16, TM11 and 14, and TM16 of rat Abcc2 are involved in transport of glutathione conjugates, monovalent bile salts, and glucuronide conjugates, respectively (K. Ito *et al*. 2001c, d).

Recent photolabeling studies using *N*-(hydrocinchonidin-8′-yl)-4-azido-2-hydroxybenzamide (Daoud *et al*. 2001), iodoaryl azidorhodamine 123 (Daoud *et al*. 2000), LTC_4 (Qian *et al*. 2001a), agosterol A (Ren *et al*. 2001), and LY475776 (Mao *et al*. 2002) have suggested that TM10/11 and/or TM16/17 are the substrate binding sites for these substrates. One important character is that ABCC1 and ABCC2 transport anticancer drugs that are not conjugated require glutathione for this function (Zaman *et al*. 1995; Loe *et al*. 1996a; Lorico *et al*. 1996; Rappa *et al*. 1997). A cotransport mechanism of the substrates with reduced glutathione (GSH) rather than that of substrate–GSH conjugates has been proposed (Cole and Deeley 1998; Loe *et al*. 1998). Altogether, it is expected that multiple amino acid residues in MSD2 and MSD3 are involved in substrate binding including GSH through forming a substrate binding pocket. On the other hand, MSD1 could modulate the substrate recognition and/or subsequent processes.

ABCC1, ABCC2, and probably other ABCC subfamily members show low sequence homology between their NH_2- and COOH-terminal NBDs, and lack 13 amino acids between the Walker A and Walker B motifs in NBD1 that are present in NBD2 and in both NBDs of most other eukaryotic ABC transporters (Cole *et al*. 1992; Hipfner *et al*. 1999). Consistent with this structural character, the two NBDs in ABCC1, ABCC2, and their relatives are functionally nonequivalent (Matsuo *et al*. 1999; Nagel *et al*. 1999; Gao *et al*. 2000; Hou *et al*. 2000, 2002; Nagata *et al*. 2000).

PHYSIOLOGICAL AND CLINICAL RELEVANCE OF ABCC2 AND ITS GENETIC DEFECT

Hyperbilirubinemia II/Dubin–Johnson syndrome (DJS; MIM237500) is a hereditary disease transmitted as an autosomal recessive trait. It is characterized by increased plasma levels of conjugated bilirubin and an increase in the urinary excretion of coproporphyrin isomer I, both of which are metabolites of heme degradation. In addition, it is characterized by deposition of a melanin-like pigment in hepatocytes and prolonged retention of sulfobromophthalein, a marker for secretion of anionic drugs. Otherwise, liver function is normal (Dubin and Johnson 1954; Sprinz and Nelson 1954; Shani *et al*. 1970; Kondo *et al*. 1974). These characteristics suggested defects in excretion rather than excessive import or conjugation of bilirubin. Paulusma *et al*. (1996) reported that TR^- rats, animal models of DJS, lack Abcc2 protein, and found a single nucleotide deletion at position 1179 in the gene, resulting in reduced mRNA abundance and absence of the protein. Ito *et al*. (1997) independently reported that expression of *cmoat* (*Abcc2*) is absent in Eisai hyperbilirubinemic (EHB) rats, another animal model of DJS, and found a transition mutation ($G \rightarrow A$ at nucleotide 2564) that creates a premature stop codon.

Our group and others have analyzed genomic DNA of DJS patients to provide genetic evidence that *ABCC2* is the responsible gene for DJS. The mutations identified so far are summarized in Table 3. Strong genetic evidence supporting the proposal that *ABCC2* is responsible for DJS was obtained from the analysis of family pedigrees of affected subjects (Wada *et al*. 1998; Toh *et al*. 1999; Mor-chohen *et al*. 2001). The cosegregation profile of mutations and the DJS trait in one family is presented in Figure 4. The chromosomal localization on 10q24 (Allikmets *et al*. 1996; Taniguchi *et al*. 1996; Van Kuijck *et al*. 1997) is also consistent with the autosomal recessive inheritance of the syndrome.

Interestingly, most of the mutations observed in DJS patients are localized in the NBDs or its adjacent regions. In the *CFTR* gene, about 80% of mutations in patients are identified within the NBD region, and almost all of them cause a severe form of the disease (Welsh and Smith 1993), suggesting that alteration in the NBDs impairs the transporter

Table 3 Summary of mutations identified in DJS

Mutation	Exon	IVS	Amino acid alteration	Reference
298C > T	3		R100X	Wada and Adachi, unpublished data
1815+2T > A		13	Exon13 skip	Toh et al. 1999
1967+2T > C		15	Exon15 skip	Kajihara et al. 1998
2302C > T	18		R768W	Wada et al. 1998
2439+2T > C		18	Exon18 skip	Toh et al. 1999
3196C > T	23		R1066X	Paulusma et al. 1997
3517A > T	25		I1173F	Mor-Cohen et al. 2001
3449G > A	25		R1150H	Mor-Cohen et al. 2001
3928C > T	28		R1310X	Tate et al. 2002
4145A > G	29		Q1382R	Toh et al. 1999
4175−4180del	30		RM1392−1393del	Tsujii et al. 1999

IVS, intervening sequence.

activity. Site-directed mutagenesis of ABCB1 showed that the transmembrane domain is important for substrate specificity, but not for the transport activity itself (Gottesman 1995; Taguchi et al. 1997; Loo and Clarke 1999). Thus mutations localized around the NBDs might disrupt the function of the transporter completely, not only in CFTR and ABCB1 but also in ABCC2. In contrast, mutations in the transmembrane domain are more likely to alter substrate specificity. No naturally occurring inherited base change has been reported in the transmembrane domain so far (Table 3).

Table 4 summarizes the genetic and laboratory information on DJS patients analyzed by our group. According to Kondo et al. (1974), the serum bilirubin concentration of 40 DJS patients varied from 1.3 to 6.9 mg/dl. We identified the homozygous mutation 2302C > T in patients DJ1 and DJ8 whose bilirubin concentrations were 5.0 and 4.8, respectively. This concentration is one of the highest levels among the range reported by Kondo et al. (1974). This mutation causes the amino acid substitution R768W in the C motif (also called the ABC signature), which is a highly conserved ATP binding domain found not just in the ABC transporter family. The mutation is likely to cause severe disruption of the transporter activity and results in relatively high serum bilirubin concentration in DJS patients.

A similarly high level of bilirubin to that observed in DJ1 and DJ8 was observed in patient DJ7, whose mutation in *ABCC2* was 1815 + 2T > A. This mutation causes abnormal splicing and consequently a 147 bp deletion in *ABCC2* mRNA, but the remaining exons are linked in frame after abnormal splicing. Thus the mutation does not generate an immature stop codon. This deletion of the NBD1 is likely to reduce the levels of normal ABCC2 protein and its activity. The absence of ABCC2 in the liver of patient DJ7 was confirmed by immunohistochemical analysis.

As described above, the level of conjugated bilirubin in blood is increased in DJS patients. Bilirubin is conjugated in hepatocytes and is expected to back-flux into sinusoidal direction because of the ABCC2 defect. This in turn suggests the presence of putative transporters on the sinusoidal membrane (Figure 3). ABCC3 could be responsible for this role because it can transport conjugated bilirubin in inside-out vesicle experiments and is localized at sinusoidal/basolateral membrane of hepatocytes (Konig et al. 1999a; Kool et al. 1999). In agreement with this notion, expression levels of ABCC3 are strongly increased in the liver of cholestatic rats (Hirohashi et al. 1998; Ortiz et al., 1999) and humans (Kool et al. 1999; Konig et al. 1999b). ABCC3 may act as an emergency export pump, when ABCC2 function fails. In support of this

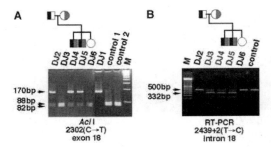

Figure 4 Familial study of two mutations in human *ABCC2* gene in DJS patients. (A) Restriction digestion analysis of the genomic region around the mutation 2302(C>T) reveals that the mutation is homozygous in patient DJ1 and heterozygous in patients DJ4 and DJ5 and their father DJ2. The mutation 2302(C>T) destroys a naturally occuring *Aci*I restriction site, whereas wild-type products are digested into 88 bp and 82 bp when PCR-amplified genomic DNA (170 bp) is digested with *Aci*I. (B) On the other hand, the splicing mutation 2439 + 2(T>C) is also detected in DJ4 and DJ5 and their mother DJ3. The splicing mutation is assessed by RT-PCR analysis, where a single RT-PCR product of 500 bp is amplified from wild-type allele, whereas a smaller fragment with 332 bp is mainly amplified from the mutant allele. Closed and shaded symbols in the pedigree show the 2302(C>T) and the 2439 + 2(T>C) mutation, respectively. The 100 bp ladder marker was run on lane M. The analysis showed perfect cosegregation of the mutations with the DJS trait, indicating that patients DJ4 and DJ5 are compound heterozygote for both mutations, whereas parents DJ2 and DJ3, who have been both diagnosed as carriers by urinary excretion of coproporphyrin I, are heterozygous states of one of each mutation, respectively.

view, ABCC3 expression is up-regulated in the basolateral hepatocyte membrane in livers from patients with DJS, where ABCC2 has lost its function (Guhlmann *et al.* 1995; see Box: Clinical and Pathological Characteristics of Dublin–Johnson Syndrome).

BIOCHEMICAL CONSEQUENCES OF GENETIC DEFECT IN DJS PATIENTS

Absence of the ABCC2 protein in DJS patients was first reported by Kartenbeck and colleagues (Kartenbeck *et al.* 1996; Keppler and Kartenbeck 1996). Subsequently, absence of ABCC2 was reported in hepatocytes of patients carrying nonsense mutations or deletion mutations (Paulusma *et al.* 1997; Tsujii *et al.* 1999). Splicing mutations that cause exon skipping and shortened mRNA molecules have also been observed (Wada *et al.* 1998; Toh *et al.* 1999; Tate *et al.* 2002). Besides the possible instability of the mutant protein, the instability and degradation of mRNA may also contribute to the decrease of the ABCC2 protein level in patients carrying nonsense, deletion, or splicing mutations, as seen in mutant rats (Paulusma *et al.* 1996). Among these mutations, the biochemical consequences of the deletion mutation, RM1392–1393del, was analyzed in detail in vitro (Keitel *et al.* 2000). The mutation is associated with absence of the ABCC2 protein from the apical membrane of hepatocytes. Transfection of mutated *ABCC2* cDNA led to a mutant protein that was only core glycosylated, sensitive to endoglycosidase H digestion, and located in the endoplasmic reticulum (ER) of transfected HEK293 and HepG2 cells, suggesting that the deletion leads to impaired maturation and trafficking of the protein from the ER to the Golgi complex. Inhibition of proteasome function resulted in a paranuclear accumulation of the mutant protein, suggesting the involvement of proteasomes in the degradation.

Analysis of the missense mutations as they occur in patients suffering DJS not only provides information about the molecular events leading to the pathogenesis of DJS, but also give hints about the functional roles of certain amino acid residues in the ABCC2 protein. Mor-Cohen *et al.* (2001) analyzed in vitro two novel missense mutations identified in exon 25 in DJS (Table 3). Continuous measurement of probenecid-sensitive carboxyfluorescein efflux, a measure of ABCC2 activity, revealed that both mutations impaired the transport activity. Immunoblot analysis and immunocytochemistry showed that one mutant ABCC2 (R1150H) matured properly and was localized at the plasma membrane of transfected cells. In contrast, expression of another mutant

CLINICAL AND PATHOLOGICAL CHARACTERISTICS OF DUBIN–JOHNSON SYNDROME

Hereditary hyperbilirubinemia is a syndrome associated with increased level of bilirubin that are caused by inborn errors in the metabolism of bilirubin. These can be divided into two groups. In Crigler–Njjar syndrome and Gilbert's syndrome unconjugated hyperbilirubinemia is observed (normal range, 0.1–0.6 mg/dl) caused by defects in the glucuronosyl transferase system in the hepatocytes. In the second group, comprising of Dubin–Johnson and Rotor's syndrome, conjugation of bilirubin occurs but secretion is blocked. Dubin–Johnson syndrome is characterized by a chronic or intermittent nonhemolytic type of jaundice in which an increase of both conjugated (normal range, 0.3–1.0 mg/dl) and unconjugated bilirubin occurs. However, conjugated hyperbilirubinemia prevails. Similarly, the secretion of sulfobromophthalein (BSP) and iopanoic acid is affected, so that there is BSP retention with a late rise after 90 mins and the gallbladder cannot be viewed radiographically. Most helpful in diagnosis is the abnormal excretion of coproporphyrin isomers in the urine. Isomer I is increased and isomer III is decreased, resulting in an increased ratio of urinary isomer I. The parenchymal cells, especially in the perivenous zone 3, contain large granules of lipochrome pigment, often in such amounts that the liver is black. The disease is inherited as an autosomal recessive trait with a high frequency of consanguinity of parents. The disease is rare and more often found in Iranian Jews with a minimal frequency of 1 in 1,300 and in Japanese at 1 in 1 million.

ABCC2 (I1173F) was low and found to be mislocated to the ER of the transfected cells.

ABCC2 (R768W), which is mutated in the C motif of NBD1, was localized in the cytoplasm with an ER-like distribution (Hashimoto *et al.* 2002). This result is consistent with immunohistochemical data showing that there was no apparent expression of ABCC2 protein in the canalicular membrane of hepatocytes in a DJS patient carrying the same mutation. Thus the mutation in the NBD1 appeared to block the maturation process of ABCC2 protein during membrane sorting, plausibly from the ER to the Golgi apparatus. The mutations in the NBD1 might induce an inadequate conformational change, resulting in a defective sorting of the ABCC2 protein and a failure to reach the plasma membrane. MG132, an inhibitor of the cytosolic proteasome, blocked the degradation of the precursor form of ABCC2 (R768W), suggesting that ABCC2 (R768W) is degraded by the proteasome pathway which is involved in the degradation of newly synthesized, misfolded, and unassembled proteins in the ER (Schwartz and Ciechanover 1999).

Another missense mutation, Q1382R in NBD2, was found in one DJS patient with compound heterozygous mutations (Table 4) (Toh *et al.* 1999). The precursor form of the ABCC2 (Q1382R) was rapidly converted to the mature form, which was resistant to Endo H, and sorted to the apical membrane of the LLC-PK$_1$ cells as the wild-type ABCC2. These results suggested that, unlike the R768W mutation, the Q1382R mutation does not affect either the maturation process or the subcellular localization of ABCC2. However, efflux of glutathione conjugate of monochlorobimane (GS-MCLB) and ATP-dependent LTC$_4$ uptake into plasma membrane vesicles derived from HEK293 cells expressing ABCC2 (Q1382R) was markedly reduced compared with that from cells expressing wild-type ABCC2. This indicated that ABCC2 (Q1382R), although localized on the apical membrane, was nonfunctional. The ATP binding site of ABCC1 can be specifically labeled when crude membranes containing ABCC1 are incubated with light-sensitive 8-azido-[α-^{32}P]ATP in the presence of excess vanadate, as established by Senior and colleagues (Senior *et al.* 1995;

Table 4 Genetic and laboratory profiles of DJS patients and their family members

Pedigree	Patient or family member[a]	Alteration in ABCC2 gene	Exon	Putative consequence	T-bilirubin (0.3–1.0 mg/dl)[b]	D-bilirubin (0.1–0.6 mg/dl)[b]	Urinary coproporphyrin I fraction (<27%)[b, c]
1	DJ1[d]	2302C>T/ 2302C>T	18	R768W/ R768W	5.0	3.8	NT
2	DJ2	2302C>T/wt	18	R768W/wt	NT	NT	42.1
	DJ3	2439+2T>C/wt	18	Splice donor/wt	NT	NT	43.5
	DJ4[d]	2302C>T/ 2439+2T>C	18 18	R768W/ Splice donor	1.3	0.8	94.5
	DJ5[d]	2302C>T/ 2439+2T>C	18 18	R768W/ Splice donor	1.3	0.8	93.6
	DJ6	wt/wt	–	wt/wt	NT	NT	NT
3	DJ7[d]	1815+2T>A/ 1815+2T>A	13	Splice donor/ Splice donor	5.2	3.8	NT
4	DJ8[d]	2302C>T/ 2302C>T	18	R768W/ R768W	4.8	3.2	NT
5	DJ9[d]	2439+2T>C/ 4145A>G	18 29	Splice donor/ Q1382R	2.5	1.6	80.0
6	DJ10[d]	2439+2T>C/ 2439+2T>C	18	Splice donor/ Splice donor	2.1	1.6	85.7
	DJ11	2439+2T>C/wt	18	Splice donor/wt	0.9	0.4	48.0
	DJ12	2439+2T>C/wt	18	Splice donor/wt	0.5	0.2	36.9

NT, not tested.
a DJ4 and DJ5 are brothers, and DJ2, DJ3, and DJ6 are their father, mother, and sister, respectively (Kimura *et al.* 1995). DJ 11 and DJ12 are the father and the mother, respectively, of DJ10.
b Normal ranges for each value are in parentheses.
c Frank *et al.* 1990.
d DJS patients.

Urbatsch *et al.* 1995). Labeling occurs because a stable inhibitory complex of ABCC1, MgADP, and vanadate is formed, which is an analog of the ABCC1·MgADP and Pi transition state complex formed after ATP hydrolysis (Gao *et al.* 2000; Nagata *et al.* 2000). Vanadate-induced nucleotide trapping in wild-type ABCC2 was stimulated by the transporter substrate estradiol-glucuronide (E$_2$17βG), but that in ABCC2 (Q1382R) was not. In the absence of vanadate, E$_2$17βG did not enhance photoaffinity labeling of the wild type ABCC2. ABCC2 was barely photoaffinity labeled by 8-azido-[γ-^{32}P]ATP. These results suggest that 8-azido-[α-^{32}P]ATP is trapped together with vanadate after hydrolysis, and that the Q1382R mutation impaired substrate-induced ATP hydrolysis.

Glutamine located between the Walker A and Walker C motifs is highly conserved among the ABC superfamily proteins. Crystal structure analysis of the ATP binding domain of the bacterial histidine permease (HisP), an ABC transporter, suggests that the corresponding glutamine in the HisP molecule (Q100) is likely to form hydrogen bonds with a water molecule which interacts with the γ-phosphate of ATP (Hung *et al.* 1998). This water molecule is the most likely candidate for attacking water during ATP hydrolysis (Hung *et al.* 1998). The comparable amino acid substitution, Q1291R, in CFTR was observed in patients suffering from cystic fibrosis (Dork *et al.* 1994). It has also been reported that the CFTR (Q1291R) shows no chloride-channel function, although it reaches the plasma membrane as a fully glycosylated mature protein. However, the role of the glutamine in the Q-loop has been controversial, since the correspondent glutamine residue in the bacterial maltose permease was suggested to be located too far away from the nucleotide to coordinate Mg^{2+} and the water molecule that attacks the γ-phosphate bond (Diederichs *et al.* 2000). In our study, the lack of substrate-induced vanadate trapping in the ABCC2 (Q1382R) may suggest that Q1382 is directly involved in ATP hydrolysis (Hashimoto *et al.* 2002).

FURTHER PHYSIOLOGICAL RELEVANCE OF ABCC2

Protection against xenobiotics

Besides identification of *ABCC2* as the responsible gene that is defective in DJS, the broad substrate specificity including xenobiotics as described above and apical localization of the ABCC2 in hepatocytes suggest a protective role as a xenobiotics export pump in vivo. Such a role has already been proposed for the ABCC1 protein (Leslie *et al.* 2001a). Compounds of toxicological relevance that are presently known to be substrates and modulators of ABCC1 include carcinogenic mycotoxin aflatoxin B1-epoxide–glutathione conjugate (Loe *et al.* 1997), metabolites of the tobacco-specific carcinogens [4-(methylnitrosamino)-1-(3-pyridyl)-1-butyl]-β-*O*-D-glucosiduronic acid (NNAL-*O*-glucuronide) (Leslie *et al.* 2001b), 4-hydroxynonenol–glutathione conjugate (Renes *et al.* 2000), 2,4-dinitrophenyl–glutathione conjugate (Jedlitschky *et al.* 1966), a herbicide metolachlor–glutathione conjugate which is a potential carcinogen (Leslie *et al.* 2001b), oxidized glutathione (Leier *et al.* 1996), estrone 3-sulfate (Qian *et al.* 2001b), and the antitumor agents described below. These compounds could also be substrates of ABCC2, and some of them, including NNAL-*O*-glucuronide (Leslie *et al.* 2001b), dinitrophenyl–glutathione conjugate (Evers *et al.* 1998), and antitumor agents (Table 1), have actually been demonstrated to be such substrates. The food-derived carcinogen 2-amino-1-methyl-6-phenylimidazo[4,5-b]pyridine (PhIP) is also a substrate of ABCC2 (Dietrich *et al.* 2001). Studies in Abcc2-deficient TR$^-$/GY or EHB rats have implicated rat Abcc2 in biliary excretion of zinc, copper, manganese, and arsenic (Paulusma and Oude Elferink *et al.* 1997; Kala *et al.* 2000). Metal excretion by mouse Abcc2 is also GSH dependent, and recently arsenic triglutathione and methylarsenic glutathione have been isolated from bile, providing the first in vivo evidence that these GSH complexes are formed and then effluxed (Kala *et al.* 2000).

The substrates of ABCC2 include carcinogens, suggesting that ABCC2 may be involved in certain carcinogenesis. Several have reported that hepatocellular carcinoma (HCC) and adenoma occurred in DJS patients (Okamura *et al.* 1980; Roth *et al.* 1982; Sakamoto *et al.* 1987; Bernard *et al.* 2000) at a frequency of 2 in 57 (Adachi *et al.* 1992); six cases were reported between 1980 and 1993 (Ueno *et al.* 1998). Because the number of DJS patients in the Japanese population is estimated as 120 (Takino *et al.* 1977), it is estimated that 2–6 in 120 cases of DJS are complicated by HCC. We also observed that two of seven DJS patients analyzed developed HCC (Toh *et al.* 1999). It is impossible to assess whether or not the frequency of HCC in DJS is significant because the disease is very rare. However, it should be noted that the occurrence of HCC is 4 in 100 in the Japanese population and almost all cases carry infection of C- or B-type hepatitis virus, whereas two cases analyzed by us were free of both types of virus. In this regards, it is interesting that DNA damage was significantly increased in mice disrupted in *Abcbl (mdrla)*, ortholog of human *ABCBI* which is another member of ABC transporters, compared with wild-type mice (Mochida *et al.* 2003). Surprisingly, others and we also found that statistically smaller members of polyps was generated in *Abcbl*-disrupted mice compared with wild-type mice under *Apc*^Min^ background (Mochida *et al.* 2003; Yamada *et al.* 2003). Any contribution of interindividual variations of ABCC2 expression and activity in certain carcinogenesis remains to be analyzed further (see below).

Homeostasis of physiological substances

ABCC2 could also participate in hepatobiliary elimination of endogenous substrates, resulting in contribution to homeostasis of those substrates. The inflammatory mediator LTC_4 is rapidly degraded by enzymes to yield the cysteinyl–glycine derivative LTD_4 and the cysteinyl derivative LTE_4 after being released

into the blood circulation. Those LTC_4-related substances are then rapidly eliminated from the blood circulation. The predominant organ responsible for the elimination is the liver, as demonstrated in the rat (Denzlinger *et al.* 1985; Guhlmann *et al.* 1995) and human (Huber *et al.* 1990) (for review see Jedlitschky and Keppler 2002). The transport of LTC_4-related substances in the hepatocyte canalicular membrane of Abcc2 deficient rats is hereditarily defective (Huber *et al.* 1987), suggesting its involvement in this transport function. The transport defect of LTC_4-related substances and inflammatory alteration have not reported in DJS so far, and the pathophysiological significance of ABCC2 in inflammatory responses remains to be clarified.

GSH has many important roles in the protection of cells from oxidative stress. GSH is a reducing agent of hydrogen and lipid peroxides and also inactivates electrophilic compounds through spontaneous or glutathione transferase catalyzed formation of conjugates. The ratio of reduced GSH to the oxidized form, glutathione disulfide (GSSG), is a reflection of cellular redox status. During cellular oxidative stress, GSH oxidation is more rapid than GSSG reduction by GSH reductases. Maintenance of low cellular GSSG concentrations and high GSH levels is important for cellular homeostasis. The antisense suppression of ABCC2 levels in HepG2 human hepatoma cells elevates the intracellular concentration of GSH (Koike *et al.* 1997). This observation, together with earlier studies demonstrating that biliary GSH is markedly decreased in Abcc2 deficient rats (Lu *et al.* 1996), suggests that ABCC2, like ABCC1, is capable of effluxing GSH either alone or together with other substrates (Paulusma *et al.* 1999).

PARTICIPATION OF ABCC2 AND RELATED PROTEINS IN MULTIDRUG RESISTANCE TO ANTICANCER AGENTS

The acquisition of multidrug resistance in cancer cells is often associated with increased expression of various cell surface ABC transporters. ABCB1 (MDR1/P-gp) and several members of the ABCC subfamily including ABCC2, ABCG2 (BCRP), and others, have been shown to be involved in the multidrug resistance mainly by studies using transfected cells (Kartner *et al.* 1985; C.J. Chen *et al.* 1986; Ueda *et al.* 1987; Cole *et al.* 1992; Zaman *et al.* 1994; Koike *et al.* 1997; Allikmets *et al.* 1998; Doyle *et al.* 1998; Evers *et al.* 1998; Cui *et al.* 1999; Hooijberg *et al.* 1999; Kawabe *et al.* 1999; Kool *et al.* 1999; Miyake *et al.* 1999; Schuetz *et al.* 1999; Z.S. Chen *et al.* 2001, 2002; reviewed by Borst *et al.* 2000; Gottesman *et al.* 2002; Kuwano *et al.* 2003). Representative anticancer agents recognized by these ABC transporters are listed in Table 1. The spectrum of drug resistance induced in cells by ABCC2 appears to be similar to that shown for ABCC1 with one exception: ABCC2 induces cisplatin resistance (Kool *et al.* 1997; Cui *et al.* 1999; Kawabe *et al.* 1999), which has never been seen in cells overexpressing ABCC1. As for other ABC transporters (Kartner *et al.* 1985; Alvarez *et al.* 1995; Nooter and Stoker 1996; Robey *et al.* 2001), overexpression of the ABCC2 protein or ABCC2 mRNA has been reported in multidrug resistant cell lines (Taniguchi *et al.* 1996; Kool *et al.* 1997; Minemura *et al.* 1999) and tumors (Narasaki *et al.* 1997; Schaub *et al.* 1999). Furthermore, a significant correlation has been observed between ABCC2 mRNA levels and cisplatin resistance in colorectal carcinoma (Hinoshita *et al.* 2000). Although ABCB1 and ABCC1 are expressed in various hematopoietic cancers and solid tumors, their clinical significance in limiting drug sensitivity/resistance is often discussed (Table 1). Further studies are required to determine whether other ABC transporters, including ABCC2, play clinical roles in the limitation of drug sensitivities to anticancer treatment in human malignancies.

PHARMACOLOGICAL AND CLINICAL IMPACT OF GENETIC ALTERATIONS AND POLYMORPHISMS IN THE *ABCC2* GENE

Most drugs are detoxified and conjugated in hepatocytes, and are subsequently disposed by active extrusion into the bile or the urine. The activity of the detoxification system affects the pharmacokinetics of drugs. Many studies have recently reported that some polymorphisms of detoxification-related genes are correlated with efficacy and the side effects on drugs (Evans 1999; Roden and George 2002). Metabolism of most drugs in humans is thought to occur in three phases (Figure 3). In phase I, drugs are oxidized by enzymes such as the cytochrome P450 isoforms (CYPs). In phase II, drugs are coupled to glutathione, glucuronic acid, or sulfate to improve solubility. Conjugated drugs are then exported out of the cells. ABCC2 is responsible for this export step as mentioned in this chapter. Another member of ABC transporters, ABCB1 (P-gp), also affects the pharmacokinetics as a rate-limiting factor in drug absorption (Schinkel *et al.* 1997; Schinkel 1998). Thus interindividual variations in both activity and expression level of ABC transporters might be a critical factor for pharmacokinetics, and clarifying the molecular basis for the interindividual variation is necessary for the optimization of medication.

Possible relevance of ABCC2 mutations in the pharmacokinetics of DJS carriers may be suggested by the laboratory data (Table 4). The urinary coproporphyrin I is an undesired by-product of heme biosynthesis. All DJS patients showed more than 80% of urinary coproporphyrine I fraction, while normal controls showed less than 27% (Frank *et al.* 1990). Interestingly, all family members examined (DJ2, DJ3, DJ11, and DJ12) who carry a heterozygous mutation in the *ABCC2* gene showed normal levels of T- and D-bilirubin, whereas they showed slightly higher than normal levels of urinary coproporphyrine I fraction (Table 4). The mechanisms for abnormal coproporphyrine I fraction in the urine is unknown, but a correlation may exist between the urinary coproporphyrine I level and the homozygous/heterozygous status of mutation in the *ABCC2* gene. In Japan, the expected number of people carrying a heterozygous mutation in *ABCC2* is at least 200,000, which is calculated by frequency of DJS patients, that is, 121 patients out of 100 million people from a nationwide survey in Japan (Takino et al. 1977). The transport and/or pharmacokinetics of some substrates might be affected by a heterozygous mutation in the ABCC2 gene as observed in coproporphyrine level, and a putative differential responsiveness to some drugs and their side effects in these carriers might be also an important clinical factor.

Genetic polymorphisms also have potential significance in drug disposition and pharmacokinetics in the same sense. Although numerous single-nucleotide polymorphisms (SNPs) in ABCC2 have been collected in several databases and publications (S. Ito et al. 2001; Saito et al. 2002), their association with clinical phenotype including drug disposition remains to be clarified. Naturally occurring base substitutions of the ABCC2 gene accompanied by amino acid substitutions reported so far are shown in Table 5.

TRANSCRIPTIONAL REGULATION OF ABCC2 GENE

The regulation of ABCC2 gene expression has been investigated under various conditions associated with changes in mRNA and protein levels. ABCC2 mRNA levels are reduced in some kinds of disease states. ABCC2 mRNA is reduced in lipopolysaccharide-treated animals (Nakamura et al. 1999) and in human liver with hepatitis C virus infection (Hinoshita et al. 2001). Under these pathological conditions, cytokines such as IL-1β, TNF-α and IL-6 might be involved in reducing the expression level of ABCC2 mRNA and protein (Nakamura et al. 1999; Hinoshita et al. 2001). Using normal tissues from patients undergoing surgical resection of liver metastasis, the ABCC2 mRNA level was determined by semiquantitative RT-PCR (Zollner et al. 2001). It was found that there is an approximately 15-fold difference in the expression of ABCC2 among 13 samples (Zollner et al. 2001). An even larger variation was observed in ABCC2 mRNA levels in tissues from hepatitis patients (Hinoshita et al. 2001).

The promoter region of the human ABCC2 and rat Abcc2 genes have been cloned and characterized. Sequence analysis of the 5'-flanking region of the human ABCC2 gene identified a variety of consensus binding sites for both ubiquitous and liver-specific transcription factors (Tanaka et al. 1999; Stockel et al. 2000). Using luciferase and gel shift assays, it has been suggested that the CCAAT-enhancer binding protein (C/EBP) binding sequence located between −356 and −343 is required for the basal expression of ABCC2 (Tanaka et al. 1999). Moreover, the hepatic nuclear factor 1 (HNF1) and upstream stimulatory factor (USF) like elements located between +81 and +248 might also be important in the basal expression of human ABCC2 (Tanaka et al. 1999).

ABCC2 expression in liver cells is increased in response to various drugs, such as the phenobarbital, the chemopreventive agent oltipraz, and the anticancer drug cisplatin. Most of the chemical inducers are well known to upregulate some hepatic-drug-metabolizing enzymes. This induction is thought to be mediated by nuclear receptors, such as the pregnane-X receptor (PXR; also known as the steroid and xenobiotics activated receptor) or the constitutive androsterone receptor (CAR). These results suggested a coordinate regulation of liver-detoxifying proteins in response to these compounds.

Regulation by drugs and other chemical compounds

The barbiturate phenobarbital has been shown to alter ABCC2 expression. Human hepatoma cells exposed to phenobarbital increased ABCC2 mRNA and protein levels. ABCC2 promoter activity was also enhanced (Kauffmann and Schrenk 1998). However, in vivo treatment by phenobarbital failed to induce ABCC2 expression in rat liver (Ogawa et al. 2000). These data suggest differential regulation of ABCC2 by phenobarbital in in vitro and in vivo situations. The mechanism for the induction of ABCC2 has been studied

Table 5 Naturally occurring base-change in *ABCC2* gene accompanied by amino acid substitution

Location (exon)	Nucleic acid substitution	Amino acid substitution	Domain	Pathogenetic consequence (biochemical defect)	Frequency (%) Jews	Frequency (%) Japanese	Reference(s)
7	842G>A	S281N	Linker	Unkown	2.4	0	Mor-Cohen *et al.* 2001
10	1249G>A	V417I	MSD2	Unkown	22.7	10.9	Ito *et al.* 2001; Mor-Cohen *et al.* 2001; Saito *et al.* 2002; Ebihara, Kuwano, and Wada, unpublished data
18	2302C>T	R768W	NBD1	DJS (protein maturation)	Not reported	0.4	Hashimoto *et al.* 2002; Ito *et al.* 2001
18	2366C>T	S789F	NBD1	(transport activity)	Not reported	0.9	Ito *et al.* 2001; Ebihara, Kuwano, and Wada, unpublished data
25	3449G>A	R1150H	MSD3	DJS (transport activity)	0.3	0	Mor-Cohen *et al.* 2001
25	3517A>T	I1173F	MSD3	DJS (protein maturation)	1.4	0	Mor-Cohen *et al.* 2001
28	3895A>C	K1299Q	NBD2	Unkown	Not reported	1?	Saito *et al.* 2002
29	4145A>G	Q1382R	NBD2	DJS (ATP hydrolysis)	Not reported	0	Hashimoto *et al.* 2002
31	4348G>A	A1450T	NBD2	(transport activity)	Not reported	0.4	Ito *et al.* 2001; Ebihara, Kuwano, and Wada, unpublished data

in animal models and human cell lines. The response of ABCC2 induction to increasing doses of phenobarbital in primary cultured human hepatocytes was different from induction of CYP2B1 and CYP2B2 via the CAR (Courtois *et al.* 2002). It is hypothesized that the prolonged cell survival and/or inhibition of cellular proliferation by phenobarbital might be related to the increase in human ABCC2 mRNA (Courtois *et al.* 2002).

Several ligands, such as the HIV protease inhibitor ritonavir, bind to the steroid and xenobiotic receptor (SXR) and activate target genes including the human *ABCC2* gene. The ability of SXR to activate *ABCC2* suggests that it may regulate the biliary excretion of xenobiotic compounds. Downregulation of human ABCC2 mRNA was found in livers from patients with primary sclerosing cholangitis, as demonstrated by quantitative RT-PCR (Oswald *et al.* 2001).

It was demonstrated that rifampicin, a PXR ligand, induced ABCC2 mRNA in human hepatocytes (Kast *et al.* 2002). This induction was in parallel with that of CYP3A4 (Kast *et al.* 2002). Moreover, in PXR −/− mice, no Abcc2 induction by pregnenolone-16-carbonitrile, a PXR ligand, was observed, suggesting that PXR is involved in the induction of Abcc2 (Kast *et al.* 2002). The fact that PXR, CAR, and farnesoid X-activated receptor (FXR) bind to an everted repeat element located at 2440 bp of the rat *Abcc2* gene further supports these ideas (Kast *et al.* 2002). It is possible that a similar pathway also induces transcription of human *ABCC2*. Using human duodenal biopsies obtained after oral rifampin treatment, it was found that an induction of ABCC2 mRNA and protein occurred (Fromm *et al.* 2000). Following oral administration of rifampin, 14 and 10 subjects out of 16 exhibited increased ABCC2 mRNA and protein levels, respectively (Fromm *et al.* 2000).

Glucocorticoids such as dexamethasone have been shown to have a differential effect on the expression of biliary drug efflux pumps (Courtois *et al.* 1999). Dexamethasone induces expression of Abcc2 and of the bile salt export pump (Bsep) in primary rat hepatocytes.

Treatment of rats with dexamethasone induced hepatic transcript levels of Abcc2. High dexamethasone treatment induced the expression of Abcc2 mRNA levels in primary hepatocytes; moreover, RU486, a glucocorticoid receptor (GR) antagonist, inhibits *Abcc2* gene regulation in dexamethasone-treated primary rat hepatocyte (Courtois *et al.* 1999). These results do not support a major role of the glucocorticoid receptor in Abcc2 regulation by dexamethasone and are in favor of an involvement of the PXR that has been shown to mediate glucocorticoid effects on some isoforms of the cytochrome P450 family.

Treatment of rat liver cells with the cholestatic compound ethinylestradiol led to an increase in Abcc2 mRNA and Abcc2 promoter activity. However, Abcc2 protein levels were decreased by this compound through a post-transcriptional mechanism. This downregulation of Abcc2 protein may contribute to the decrease of bile-salt-independent bile flow occurring in response to ethinylestradiol (Kauffmann and Schrenk 1998).

P-glycoprotein expression has also been shown to be regulated by numerous chemical carcinogens in human hepatocytes. In addition to P-glycoprotein, Abcc2 and ABCC2 have also been found to be upregulated by 2-acetylaminofluorene (2-AAF) in primary rat hepatocytes and in human and rat hepatoma cells, respectively (Kauffmann and Schrenk 1998). However, this carcinogen decreased human ABCC2 promoter activity, suggesting that its effects on ABCC2 may involve post-transcriptional events such as enhanced mRNA stability (Stockel *et al.* 2000).

Abcc2 and P-glycoprotein expression have been shown to be regulated by various structurally unrelated anticancer drugs in rat (Schenk *et al.* 2001). Acute treatment with anthracyclines, such as doxorubicin and daurorubicin, or by mitoxanthrone increased P-glycoprotein levels in rat hepatocytes through activation of the *mdr1b* gene (Schrenk *et al.* 1996). However, primary hepatocytes exposed to cisplatin displayed enhanced Abcc2 expression (Kauffmann *et al.* 1997). These regulations might involve DNA damage or production of reactive oxygen species triggered by the anticancer compounds used.

The analysis of rat *Abcc2* gene expression has been performed in cholestasis models including endotoxin treatment, ethinylestradiol treatment, and common bile duct ligation (Trauner *et al.* 1997; Vos *et al.* 1998; Kubitz *et al.* 1999; Paulusmn *et al.* 2000). Endotoxin-mediated cholestasis stems from impaired hepatobiliary transport of bile acids and organic anions due to altered expression and activity of transporters. Lipopolysaccharide, IL-1, and IL-6 administration suppressed Abcc2, Oatp1, Oatp2, and Bsep mRNA level. Bile acid treatment increased the in vivo expression of the bile acid export pump but not Abcc2 (Hartmann *et al.* 2002).

SUMMARY

The ABCC2 protein belongs to the subfamily C of the ABC proteins and was identified as a putative transporter responsible for multidrug resistance in human cancer cells and as an activity excreting multispecific organic anions across the canalicular membrane of hepatocytes. Human *ABCC2* has been identified as the responsible gene for Dubin–Johnson syndrome (DJS), a hereditary disease characterized by conjugated hyperbilirubinemia. Biochemical and genetic analysis of ABCC2 led to the conclusion that ABCC2 has a role in hepatobiliary extrusion of a wide variety of endogenous and exogenous substances. In addition to the pathogenetic importance of ABCC2 in the DJS, the data support important roles of ABCC2 and related proteins in protection against a wide variety of xenobiotics, including pharmaceutical substances and carcinogens, and in homeostasis of endogenous regulatory substances including hormones, glutathione, and cysteinyl leukotrienes. Thus further analysis of participation of ABCC2 in physiological and clinical aspect such as tumor progression, interindividual variety of drug response, and homeostatic regulation is an important future direction.

REFERENCES

Adachi, Y., Nanno, T., and Yamamoto, T. (1992). Japanese Clinical Statistical Data of Patients with Constitutional Jaundice. *Nippon Rinsho (Jpn. J. Clin. Medi.)*, **50**, 677–685 (in Japanese).

Allikmets, R., Gerrard, B., Hutchinson, A., and Dean, M. (1996). Characterization of the Human ABC Superfamily: Isolation and Mapping of 21 New Genes Using the Expressed Sequence Tags Database. *Hum. Mol. Genet.*, **5**, 1649–1655.

Allikmets, R., Schriml, L.M., Hutchinson, A., Romano-Spica, V., and Dean, M. (1998). A Human Placenta-Specific ATP-Binding Cassette Gene (ABCP) on Chromosome 4q22 that is Involved in Multidrug Resistance. *Cancer Res.*, **58**, 5337–5339.

Alvarez, M., Paull, K., Monks, A., Hose, C., Lee, J.S., Weinstein, J., *et al.* (1995). Generation of a Drug Resistance Profile by Quantitation of mdr-1/P-glycoprotein in the Cell Lines of the National Cancer Institute Anticancer Drug Screen. *J. Clin. Invest.*, **95**, 2205–2214.

Ambudkar, S.V., Dey, S., Hrycyna, C.A., Ramachandra, M., Pastan, I., and Gottesman, M.M. (1999). Biochemical, Cellular, and Pharmacological Aspects of the Multidrug Transporter. *Ann. Rev. Pharm. Toxicol.*, **39**, 361–398.

Bakos, E., Hegedus, T., Hollo, Z., Welker, E., Tusnady, G.E., Zaman, G.J.R., *et al.* (1996). Membrane Topology and Glycosylation of the Human Multidrug Resistance-Associated Protein. *J. Biol. Chem.*, **271**, 12322–12326.

Bakos, E., Evers, R., Szakacs, G., Tusnady, G.E., Welker, E., Szabo, K., *et al.* (1998). Functional Multidrug Resistance Protein (MRP1) Lacking the N-Terminal Transmembrane Domain. *J. Biol. Chem.*, **273**, 32167–32175.

Bakos, E., Evers, R., Calenda, G., Tusnady, G.E., Szakacs, G., Varadi, A., *et al.* (2000a). Characterization of the Amino-Terminal Regions in the Human Multidrug Resistance Protein (MRP1). *J. Cell. Sci.*, **113**, 4451–4461.

Bakos, E., Evers, R., Sinkule, J.A., Varadi, A., Borst, P., and Sarkadi, B. (2000b). Interactions of the Human Multidrug Resistance Proteins MRP1 and MRP2 with Organic Anions. *Mol. Pharmacol.*, **57**, 760–768.

Belinsky, M.G. and Kruh, G.D. (1999). MOAT-E (ARA) Is a Full-Length MRP/cMOAT Subfamily Transporter Expressed in Kidney and Liver. *Br. J. Cancer.*, **80**, 1342–1349.

Bernard, P.H., Blanc, J.F., Paulusma, C., Le Bail, B., Carles, J., Balabaud, C., *et al.* (2000). Multiple Black Hepatocellular Adenomas in a Male Patient. *Eur. J. Gastroenterol. Hepatol.*, **12**, 1253–1257.

Borst, P., Evers, R., Kool, M., and Wijnholds, J. (1999). The Multidrug Resistance Protein Family. *Biochim. Biophys. Acta*, **1461**, 347–357.

Borst, P., Evers, R., Kool, M., and Wijnholds, J. (2000). A Family of Drug Transporters: The Multidrug

Resistance-Associated Proteins. *J. Natl. Cancer. Inst.*, **92**, 1295–302.

Buchler, M., Konig, J., Brom, M., Kartenbeck, J., Spring, H., Horie, T., *et al.* (1996). cDNA Cloning of the Hepatocyte Canalicular Isoform of the Multidrug Resistance Protein, cMrp, Reveals a Novel Conjugate Export Pump Deficient in Hyperbilirubinemic Mutant Rats. *J. Biol. Chem.*, **271**, 15091–15098.

Chang, G. and Roth, C.B. (2001). Structure of MsbA from *E. coli*: A Homolog of the Multidrug Resistance ATP Binding Cassette (ABC) Transporters. *Science*, **293**, 1793–1800.

Chen, C.J., Chin, J.E., Ueda, K., Clark, D.P., Pastan, I., Gottesman, M.M., *et al.* (1986). Internal Duplication and Homology with Bacterial Transport Proteins in the mdr1 (P-glycoprotein) Gene from Multidrug-Resistant Human Cells. *Cell*, **47**, 381–389.

Chen, Z.S., Lee, K., and Kruh, G.D. (2001). Transport of Cyclic Nucleotides and Estradiol 17-beta-D-Glucuronide by Multidrug Resistance Protein 4. Resistance to 6-Mercaptopurine and 6-Thioguanine. *J. Biol. Chem.*, **276**, 33747–33754.

Chen, Z.S., Lee, K., Walther, S., Raftogianis, R.B., Kuwano, M., Zeng, H., *et al.* (2002). Analysis of Methotrexate and Folate Transport by Multidrug Resistance Protein 4 (ABCC4): MRP4 is a Component of the Methotrexate Efflux System. *Cancer Res.*, **62**, 3144–3150.

Cole, S.P. and Deeley, R.G. (1998). Multidrug Resistance Mediated by the ATP-Binding Cassette Transporter Protein MRP. *Bioessays*, **20**, 931–940.

Cole, S.P.C., Bhardwaj, G., Gerlach, J.H., Mackie, J.E., Grant, C.E., Almquist, K.C., *et al.* (1992). Overexpression of a Transporter Gene in a Multidrug-Resistant Human Lung Cancer Cell Line. *Science*, **258**, 1650–1654.

Courtois, A., Payen, L., Guillouzo, A., and Fardel, O. (1999). Up-Regulation of Multidrug Resistance-Associated Protein 2 (MRP2) Expression in Rat Hepatocytes by Dexamethasone. *FEBS Lett.*, **459**, 381–385.

Courtois, A., Payen, L., Le Ferrec, E., Scheffer, G.L., Trinquart, Y., Guillouzo, A., *et al.* (2002). Differential Regulation of Multidrug Resistance-Associated Protein 2 (MRP2) and Cytochromes P450 2B1/2 and 3A1/2 in Phenobarbital-Treated Hepatocytes. *Biochem. Pharmacol.*, **63**, 333–341.

Cui, Y., Konig, J., Buchholz, J.K., Spring, H., Leier, I., and Keppler, D. (1999). Drug Resistance and ATP-Dependent Conjugate Transport Mediated by the Apical Multidrug Resistance Protein, MRP2, Permanently Expressed in Human and Canine Cells. *Mol. Pharmacol.*, **55**, 929–937.

Daoud, R., Kast, C., Gros, P., and Georges, E. (2000). Rhodamine 123 Binds to Multiple Sites in the Multidrug Resistance Protein (MRP1). *Biochemistry*, **39**, 15344–15352.

Daoud, R., Julien, M., Gros, P., and Georges, E. (2001). Major Photoaffinity Drug Binding Sites in Multidrug Resistance Protein 1 (MRP1) Are within Transmembrane Domains 10–11 and 16–17. *J. Biol. Chem.*, **276**, 12324–12330.

Denzlinger, C., Rapp, S., Hagmann, W., and Keppler, D. (1985). Leukotrienes as Mediators in Tissue Trauma. *Science*, **230**, 330–332.

Diederichs, K., Diez, J., Greller, G., Muller, C., Breed, J., Schnell, C., *et al.* (2000). Crystal Structure of MalK, the ATPase Subunit of the Trehalose/Maltose ABC Transporter of the Archaeon *Thermococcus litoralis*. *EMBO J.*, **19**, 5951–5961.

Dietrich, C.G., de Waart, D.R., Ottenhoff, R., Schoots, I.G., and Elferink, R.P. (2001). Increased Bioavailability of the Food-Derived Carcinogen 2-amino-1-methyl-6-phenylimidazo[4,5-b]pyridine in MRP2-Deficient Rats. *Mol. Pharmacol.*, **59**, 974–980.

Dork, T., Mekus, F., Schmidt, K., Bosshammer, J., Fislage, R., Heuer, T., *et al.* (1994). Detection of More than 50 Different CFTR Mutations in a Large Group of German Cystic Fibrosis Patients. *Hum. Genet.*, **94**, 533–542.

Doyle, L.A., Yang, W., Abruzzo, L.V., Krogmann, T., Gao, Y., Rishi, A.K., *et al.* (1998). A Multidrug Resistance Transporter from Human MCF-7 Breast Cancer Cells. *Proc. Natl Acad. Sci.*, **95**, 15665–15670.

Dubin, I.N. and Johnson, F.B. (1954). Chronic Idiopathic Jaundice with Unidentified Pigment in Liver Cells: New Clinico-Pathologic Entity with Report of 12 Cases. *Medicine*, **33**, 155–197.

Evans, W.E. and Relling, M.V. (1999). Pharmacogenomics: Translating Functional Genomics into Rational Therapeutics. *Science*, **286**, 487–491.

Evers, R., Zaman, G.J.R., van Deemter, L., Jansen, H., Calafat, J., Oomen, L.J.M, *et al.* (1996). Basolateral Localization and Export Activity of the Human Multidrug Resistance-Associated Protein in Polarized Pig Kidney Cells. *J. Clin. Invest.*, **97**, 1211–1218.

Evers, R., Cnubben, N.H., Wijnholds, J., van Deemter, L., van Bladeren, P.J., and Borst, P. (1997). Transport of Glutathione Prostaglandin A Conjugates by the Multidrug Resistance Protein 1. *FEBS Lett.*, **419**, 112–116.

Evers, R., Kool, M., van Deemter, L., Jansen, H., Calafat, J., Oomen, L.C., *et al.* (1998). Drug Export Activity of the Human Canalicular Multispecific Organic Anion Transporter in Polarized Kidney MDCK Cells Expressing cMOAT (MRP2) cDNA. *J. Clin. Invest.*, **101**, 1310–1319.

Fanning, A.S. and Anderson, J.M. (1999). PDZ Domains: Fundamental Building Blocks in the Organization of Protein Complexes at the Plasma Membrane. *J. Clin. Invest.*, **103**, 767–772.

Fernandez, S.B., Hollo, Z., Kern, A., Bakos, E., Fischer, P.A., Borst, P., *et al.* (2002). Role of the N-Terminal Transmembrane Region of the Multidrug Resistance Protein MRP2 in Routing to the Apical

Membrane in MDCKII Cells. *J. Biol. Chem.*, **277**, 31048–31055.

Frank, M., Doss, M., and de Carvalho, D.G. (1990). Diagnostic and Pathogenetic Implications of Urinary Coproporphyrin Excretion in the Dubin–Johnson Syndrome. *Hepato Gastroenterol.*, **37**, 147–151.

Fromm, M.F., Kauffmann, H.M., Fritz, P., Burk, O., Kroemer, H.K., Warzok, R.W., *et al.* (2000). The Effect of Rifampin Treatment on Intestinal Expression of Human MRP Transporters. *Am. J. Pathol.*, **157**, 1575–1580.

Gao, M., Yamazaki, M., Loe, D.W., Westlake, C.J., Grant, C.E., Cole, S.P., *et al.* (1998). Multidrug Resistance Protein. Identification of Regions Required for Active Transport of Leukotriene C4. *J. Biol. Chem.*, **273**, 10733–10740.

Gao, M., Cui, H.R., Loe, D.W., Grant, C.E., Almquist, K.C., Cole, S.P., *et al.* (2000). Comparison of the Functional Characteristics of the Nucleotide Binding Domains of Multidrug Resistance Protein 1. *J. Biol. Chem.*, **275**, 13098–13108.

Gottesman, M.M., Hrycyna, C.A., Schoenlein, P.V., Germann, U.A., and Pastan, I. (1995). Genetic Analysis of the Multidrug Transporter. *Annu. Rev. Genet.*, **29**, 607–649.

Gottesman, M.M., Fojo, T., and Bates, S.E. (2002). Multidrug Resistance in Cancer: Role of ATP-Dependent Transporters. *Nat. Rev. Cancer*, **2**, 48–58.

Grant, C.E., Kurz, E.U., Cole, S.P., and Deeley, R.G. (1997). Analysis of the Intron–Exon Organization of the Human Multidrug-Resistance Protein Gene (*MRP*) and Alternative Splicing of its mRNA. *Genomics*, **45**, 368–378.

Guhlmann, A., Krauss, K., Oberdorfer, F., Siegel, T., Scheuber, P.H., Muller, J., *et al.* (1995). Noninvasive Assessment of Hepatobiliary and Renal Elimination of Cysteinyl Leukotrienes by Positron Emission Tomography. *Hepatology*, **21**, 1568–1575.

Haimeur, A., Deeley, R.G., and Cole, S.P.C. (2002). Charged Amino Acids in the Sixth Transmembrane Helix of Multidrug Resistance Protein 1 (MRP1) Are Critical Determinants of Transport Activity. *J. Biol. Chem.*, **277**, 41326–41333.

Harris, M., Kuwano, M., Webb, M., and Board, P.G. (2001). Identification of the Apical Membrane-Targeting Signal of the Multidrug Resistance-Associated Protein 2 (MRP2/MOAT). *J. Biol. Chem.*, **276**, 20876–20881.

Hartmann, G., Cheung, A.K., and Piquette-Miller, M. (2002). Inflammatory Cytokines, but not Bile Acids, Regulate Expression of Murine Hepatic Anion Transporters in Endotoxemia. *J. Pharmacol. Exp. Ther.*, **303**, 273–281.

Hashimoto, K., Uchiumi, T., Konno, T., Ebihara, T., Nakamura, T., Wada, M., *et al.* (2002). Trafficking-Defect and Functional–Defect by Mutations of the ATP-Binding Domains in Multidrug Resistance Protein

2 in Patients with Dubin-Johnson Syndrome. *Hepatology*, **36**, 1236–1245.

Heijn, M., Hooijberg, J.H., Scheffer, G.L., Szabo, G., Westerhoff, H.V., and Lankelma, J. (1997). Anthracyclines Modulate Multidrug Resistance Protein (MRP) Mediated Organic Anion Transport. *Biochim. Biophys. Acta*, **1326**, 12–22.

Higgins, C.F. (1992). ABC Transporters: From Microorganisms to Man. *Annu Rev. Cell Biol.*, **8**, 67–113.

Hinoshita, E., Uchiumi, T., Taguchi, K., Kinukawa, N., Tsuneyoshi, M., Maehara, Y., *et al.* (2000). Increased Expression of an ATP-Binding Cassette Superfamily Transporter, Multidrug Resistance Protein 2, in Human Colorectal Carcinomas. *Clin. Cancer Res.*, **6**, 2401–2407.

Hinoshita, E., Taguchi, K., Inokuchi, A., Uchiumi, T., Kinukawa, N., Shimada, M., *et al.* (2001). Decreased Expression of an ATP-Binding Cassette Transporter, MRP2, in Human Livers with Hepatitis C Virus Infection. *J. Hepatol.*, **35**, 765–773.

Hipfner, D.R., Almquist, K.C., Leslie, E.M., Gerlach, J.H., Grant, C.E., Deeley, R.G., *et al.* (1997). Membrane Topology of the Multidrug Resistance Protein, MRP: A Study of Glycosylation-Site Mutants Reveals an Extracytosolic NH$_2$-Terminus. *J. Biol. Chem.*, **272**, 23623–23630.

Hipfner, D.R., Deeley, R.G., and Cole, S.P. (1999). Structural, Mechanistic and Clinical Aspects of MRP1. *Biochim. Biophys. Acta*, **1461**, 359–376.

Hirohashi, T., Suzuki, H., Ito, K., Ogawa, K., Kume, K., Shimizu, T., *et al.* (1998). Hepatic Expression of Multidrug Resistance-Associated Protein-Like Proteins Maintained in Eisai Hyperbilirubinemic Rats. *Mol. Pharmacol.*, **53**, 1068–1075.

Hooijberg, J.H., Broxterman, H.J., Heijn, M., Fles, D.L., Lankelma, J., and Pinedo, H.M. (1997). Modulation by (Iso)flavonoids of the ATPase Activity of the Multidrug Resistance Protein. *FEBS Lett.*, **413**, 344–348.

Hooijberg, J.H., Broxterman, H.J., Kool, M., Assaraf, Y.G., Peters, G.J., Noordhuis, P., *et al.* (1999). Antifolate Resistance Mediated by the Multidrug Resistance Proteins MRP1 and MRP2. *Cancer Res.*, **59**, 2532–2535.

Hou, Y., Cui, L., Riordan, J.R., and Chang, X. (2000). Allosteric Interactions Between the Two Non-Equivalent Nucleotide Binding Domains of Multidrug Resistance Protein MRP1. *J. Biol. Chem.*, **275**, 20280–20287.

Hou, Y.X., Cui, L., Riordan, J.R., and Chang, X.B. (2002). ATP Binding to the First Nucleotide-Binding Domain of Multidrug Resistance Protein MRP1 Increases Binding and Hydrolysis of ATP and Trapping of ADP at the Second Domain. *J. Biol. Chem.*, **277**, 5110–5119.

Huber, M., Guhlmann, A., Jansen, P.L., and Keppler, D. (1987). Hereditary Defect of Hepatobiliary Cysteinyl

Leukotriene Elimination in Mutant Rats with Defective Hepatic Anion Excretion. *Hepatology*, **7**, 224–228.

Huber, M., Muller, J., Leier, I., Jedlitschky, G., Ball, H.A., Moore, K.P., *et al.* (1990). Metabolism of Cysteinyl Leukotrienes in Monkey and Man. *Eur. J. Biochem.*, **194**, 309–315.

Hung, L.W., Wang, I.X., Nikaido, K., Liu, P.Q., Ames, G.F., and Kim, S.H. (1998). Crystal Structure of the ATP-Binding Subunit of an ABC Transporter. *Nature*, **396**, 703–707.

Ishikawa, T., (1992). The ATP-Dependent Glutathione S-Conjugate Export Pump. *Trends Biochemi. Sci.*, **17**, 463–468.

Ishikawa, T and Ali-Osman, F. (1993) Glutathione-Associated *cis*-Diamminedichloroplatinum(II) Metabolism and ATP-Dependent Efflux from Leukemia Cells. Molecular Characterization of Glutathione-Platinum Complex and its Biological Significance. *J. Biol. Chem.*, **268**, 20116–20125.

Ishikawa, T., Muller, M., Klunemann, C., Schaub, T., and Keppler, D. (1990). ATP-Dependent Primary Active Transport of Cysteinyl Leukotrienes Across Liver Canalicular Membrane. Role of the ATP-Dependent Transport System for Glutathione S-Conjugates. *J. Biol. Chem.*, **265**, 19279–19286.

Ito, K., Suzuki, H., Hirohashi, T., Kume, K., Shimizu, T., and Sugiyama, Y. (1997). Molecular Cloning of Canalicular Multispecific Organic Anion Transporter Defective in EHBR. *Am. J. Physiol.*, **272**, G16–G22.

Ito, K., Olsen, S.L., Qiu, W., Deeley, R.G., and Cole, S.P. (2001a). Mutation of a Single Conserved Tryptophan in Multidrug Resistance Protein 1 (MRP1/ABCC1) Results in Loss of Drug Resistance and Selective Loss of Organic Anion Transport. *J. Biol. Chem.*, **276**, 15616–15624.

Ito, K., Oleschuk, C.J., Westlake, C., Vasa, M.Z., Deeley, R.G., and Cole, S.P. (2001b). Mutation of Trp1254 in the Multispecific Organic Anion Transporter, Multidrug Resistance Protein 2 (MRP2) (ABCC2), Alters Substrate Specificity and Results in Loss of Methotrexate Transport Activity. *J. Biol. Chem.*, **276**, 38108–38114.

Ito, K., Suzuki, H., and Sugiyama, Y. (2001c). Charged Amino Acids in the Transmembrane Domains Are Involved in the Determination of the Substrate Specificity of Rat MRP2. *Mol. Pharmacol.*, **59**, 1077–1085.

Ito, K., Suzuki, H., and Sugiyama, Y. (2001d). Single Amino Acid Substitution of Rat MRP2 Results in Acquired Transport Activity for Taurocholate. *Am. J. Physiol.*, **281**, G1034–G1043.

Ito, S., Ieiri, I., Tanabe, M., Suzuki, A., Higuchi, S., and Otsubo, K. (2001). Polymorphism of the ABC Transporter Genes, MDR1, MRP1 and MRP2/cMOAT, in Healthy Japanese Subjects. *Pharmacogenetics*, **11**, 175–184.

Jedlitschky, G. and Keppler, D. (2002). Transport of Leukotriene C4 and Structurally Related Conjugates. *Vitam. Horm.*, **64**, 153–184.

Jedlitschky, G., Leier, I., Buchholz, U., Barnouin, K., Kurz, G., and Keppler, D. (1996). Transport of Glutathione, Glucuronate, and Sulfate Conjugates by the MRP Gene-Encoded Conjugate Export Pump. *Cancer Res.*, **56**, 988–994.

Jedlitschky, G., Leier, I., Buchholz, U., Hummel-Eisenbeiss, J., Burchell, B., and Keppler, D. (1997). ATP-Dependent Transport of Bilirubin Glucuronides by the Multidrug Resistance Protein MRP1 and its Hepatocyte Canalicular Isoform MRP2. *Biochem. J.*, **327**, 305–310.

Kajihara, S., Hisatomi, A., Mizuta, T., Hara, T., Ozaki, I., Wada, I., *et al.* (1998). A Splice Mutation in the Human Canalicular Multispecific Organic Anion Transporter Gene Causes Dubin–Johnson Syndrome. *Biochem. Biophys. Res. Commun.*, **253**, 454–457.

Kala, S.V., Neely, M.W., Kala, G., Prater, C.I., Atwood, D.W., Rice, J.S., *et al.* (2000). The MRP2/cMOAT Transporter and Arsenic–Glutathione Complex Formation Are Required for Biliary Excretion of Arsenic. *J. Biol. Chem.*, **275**, 33404–33408.

Kartenbeck, J., Leuschner, U., Mayer, R., and Keppler, D. (1996). Absence of the Canalicular Isoform of the MRP Gene-Encoded Conjugate Export Pump from the Hepatocytes in Dubin–Johnson Syndrome. *Hepatology*, **23**, 1061–1066.

Kartner, N., Riordan, J.R., and Ling, V. (1983). Cell Surface P-glycoprotein Associated with Multidrug Resistance in Mammalian Cell Lines. *Science*, **221**, 1285–1288.

Kartner, N., Evernden-Porelle, D., Bradley, G., and Ling, V. (1985). Detection of P-glycoprotein in Multidrug-Resistant Cell Lines by Monoclonal Antibodies. *Nature*, **316**, 820–823.

Kast, H.R., Goodwin, B., Tarr, P.T., Jones, S.A., Anisfeld, A.M., Stoltz, C.M., *et al.* (2002). Regulation of Multidrug Resistance-Associated Protein 2 (ABCC2) by the Nuclear Receptors Pregnane X Receptor, Farnesoid X-Activated Receptor, and Constitutive Androstane Receptor. *J. Biol. Chem.*, **277**, 2908–2915.

Kauffmann, H.M. and Schrenk, D. (1998). Sequence Analysis and Functional Characterization of the 5′-Flanking Region of the Rat Multidrug Resistance Protein 2 (mrp2) Gene. *Biochem. Biophys. Res. Commun.*, **245**, 325–331.

Kauffmann, H.M., Keppler, D., Kartenbeck, J., and Schrenk, D. (1997). Induction of cMrp/cMoat Gene Expression by Cisplatin, 2-Acetylaminofluorene, or Cycloheximide in Rat Hepatocytes. *Hepatology*, **26**, 980–985.

Kawabe, T., Chen, Z.S., Wada, M., Uchiumi, T., Ono, M., Akiyama, S., *et al.* (1999). Enhanced Transport of Anticancer Agents and Leukotriene C4 by the Human Canalicular Multispecific Organic Anion Transporter (cMOAT/MRP2). *FEBS Lett.*, **456**, 327–331.

Keitel, V., Kartenbeck, J., Nies, A.T., Spring, H., Brom, M., and Keppler, D. (2000). Impaired Protein Maturation of the Conjugate Export Pump Multidrug Resistance Protein 2 as a Consequence of a Deletion Mutation in Dubin–Johnson Syndrome. *Hepatology*, **32**, 1317–1328.

Keppler, D. and Kartenbeck, J. (1996). The Canalicular Conjugate Export Pump Encoded by the *cmrp/cmoat* Gene. *Prog. Liver Dis.*, **14**, 55–67.

Kimura, A., Yuge, K., Kosai, K.I., Kage, M., Fujisawa, T., Inoue, T., *et al.* (1995). Neonatal Cholestasis in Two Siblings: A Variant of Dubin–Johnson Syndrome? *J. Paediatr. Child Health*, **31**, 557–560.

Kocher, O., Comella, N., Gilchrist, A., Pal, R., Tognazzi, K., Brown, L.F., *et al.* (1999). PDZK1, a Novel PDZ Domain-Containing Protein Up-Regulated in Carcinomas and Mapped to Chromosome 1q21, Interacts with cMOAT (MRP2), the Multidrug Resistance-Associated Protein. *Lab. Invest.*, **79**, 1161–1170.

Koike, K., Kawabe, T., Tanaka, T., Toh, S., Uchiumi, T., Wada, M., *et al.* (1997). A Canalicular Multispecific Organic Anion Transporter (cMOAT) Antisense cDNA Enhances Drug Sensitivity in Human Hepatic Cancer Cells. *Cancer Res.*, **57**, 5475–5479.

Kondo, T., Kazuo, K., Ohtsuka, Y., Yanagisawa, W., and Shiomura, T. (1974). Clinical and Genetic Studies on Dubin–Johnson Syndrome in a Cluster Area in Japan. *Jpn. J. Hum. Genet.*, **18**, 378–392.

Konig, J., Rost, D., Cui, Y., and Keppler, D. (1999a). Characterization of the Human Multidrug Resistance Protein Isoform MRP3 Localized to the Basolateral Hepatocyte Membrane. *Hepatology*, **29**, 1156–1163.

Konig, J., Nies, A.T., Cui, Y., Leier, I., and Keppler, D. (1999b). Conjugate Export Pumps of the Multidrug Resistance Protein (MRP) Family: Localization, Substrate Specificity and MRP2-Mediated Drug Resistance. *Biochim. Biophys. Acta.*, **1461**: 377–394.

Konig, J., Nies, A.T., Cui, Y., and Keppler, D. (2003). In ABC proteins, from bacteria to man. Edited by Holland, I.B., Cole, S.P.C., Kuchler, K., Higgins, C.F. UK: Academic Press, pp. 423–443.

Konno, T., Ebihara, T., Hisaeda, K., Uchiumi, T., Nakamura, T., Shirakusa, T., *et al.* (2003). Identification of Domains Participating in the Substrate Specificity and Subcellular Localization of the Multidrug Resistance Proteins, MRP1 and MRP2. *J. Biol. Chem.*, **278**, 22908–22917.

Kool, M., de Haas, M., Scheffer, G.L., Scheper, R.J., van Eijk, M.J., Juijn, J.A., *et al.* (1997). Analysis of Expression of cMOAT (MRP2), MRP3, MRP4, and MRP5, Homologues of the Multidrug Resistance-Associated Protein Gene (MRP1), in Human Cancer Cell Lines. *Cancer Res.*, **57**, 3537–3547.

Kool, M., van der Linden, M., de Haas, M., Scheffer, G.L., de Vree, J.M., Smith, A.J., *et al.* (1999). MRP3, an

Organic Anion Transporter Able to Transport Anti-cancer Drugs. *Proc. Natl Acad. Sci. USA*, **96**, 6914–6919.

Kubitz, R., Warskulat, U., Schmitt, M., and Haussinger, D. (1999). Dexamethasone- and Osmolarity-dependent Expression of the Multidrug-Resistance Protein 2 in Cultured Rat Hepatocytes. *Biochem. J.*, **340**, 585–591.

Kuwano, M., Uchiumi, M., Hayakawa, H., Ono, M., Wada, M., Izumi, H., *et al.* (2003). The Basic and Clinical Implication of ABC transporters, YB-1 and Angiogenesis in Human Malignancies. *Cancer Sci.*, **94**, 9–14.

Kyte, J. and Doolittle, R.F. (1982). A simple method for displaying the hydropathic character of a protein. *J. Mol. Biol.*, **157**, 105–132.

Lee, K., Belinsky, M.G., Bell, D.W., Testa, J.R., and Kruh, G.D. (1998). Isolation of *MOAT-B*, a Widely Expressed Multidrug Resistance-Associated Protein/ Canalicular Multispecific Organic Anion Transporter-Related Transporter. *Cancer Res.*, **58**, 2741–2747.

Leier, I., Jedlitschky, G., Buchholz, U., Center, M., Cole, S.P.C., Deeley, R.G., *et al.* (1996). ATP-Dependent Glutathione Disulfide Transport Mediated by the *MRP* Gene Encoded Conjugate Export Pump. *Biochem. J.*, **314**, 433–437.

Leier, I., Hummel-Eisenbeiss, J., Cui, Y., and Keppler, D. (2000). ATP-Dependent Paraaminohippurate Transport by Apical Multidrug Resistance Protein MRP2. *Kidney Int.*, **57**, 1636–1642.

Leslie, E.M., Deeley, R.G., and Cole, S.P. (2001a). Toxicological Relevance of the Multidrug Resistance Protein 1, MRP1 (ABCC1) and Related Transporters. *Toxicology*, **167**, 3–23.

Leslie, E.M., Ito, K.-I., Upadhyaya, P., Hecht, S.S., Deeley, R.G., Cole, S.P.C. (2001b). Transport of the β-*O*-Glucuronide Conjugate of the Tobacco-Specific Carcinogen 4- (methylnitrosamino)-1-(3-pyridyl)-1-butanol (NNAL) by the Multidrug Resistance Protein 1 (MRP1). Requirement for Glutathione or a Non-Sulfur-Containing Analog. *J. Biol. Chem.*, **276**, 27846–27854.

Locher, K.P., Lee, A.T., and Rees, D.C. (2002). The *E. coli* BtuCD Structure: A Framework for ABC Transporter Architecture and Mechanism. *Science*, **296**, 1091–1098.

Loe, D.W., Almquist, K.C., Deeley, R.G., and Cole, S.P.C. (1996a). Multidrug Resistance Protein (MRP)-mediated Transport of Leukotriene C4 and Chemotherapeutic Agents in Membrane Vesicles: Demonstration of Glutathione-Dependent Vincristine Transport. *J. Biol. Chem.*, **271**, 9675–9682.

Loe, D.W., Almquist, K.C., Cole, S.P.C., and Deeley, R.G. (1996b). ATP-Dependent 17 β-estradiol 17-(β-D-glucuronide) Transport by Multidrug Resistance Protein: Inhibition by Cholestatic Steroids. *J. Biol. Chem.*, **271**, 9683–9689.

Loe, D.W., Stewart, R.K., Massey, T.E., Deeley, R.G., and Cole, S.P.C. (1997). ATP-Dependent Transport of Aflatoxin B1 and its Glutathione Conjugates by the

Product of the MRP Gene. *Mol. Pharmacol.*, **51**, 1034–1041.

Loe, D.W., Deeley, R.G., and Cole, S.P. (1998). Characterization of Vincristine Transport by the M(r) 190,000 Multidrug Resistance Protein (MRP): Evidence for Cotransport With Reduced Glutathione. *Cancer Res.*, **58**, 5130–5136.

Loo, T.W. and Clarke, D.M. (1999). Determining the Structure and Mechanism of the Human Multidrug Resistance P-glycoprotein using Cysteine-Scanning Mutagenesis and Thiol-Modification Techniques. *Biochim. Biophy. Acta*, **1461**, 315–325.

Lorico, A., Rappa, G., Flavell, R.A., and Sartorelli, A.C. (1996). Double Knockout of the MRP Gene Leads to Increased Drug Sensitivity In Vitro. *Cancer Res.*, **56**, 5351–5355.

Lu, S.C., Cai, J., Kuhlenkamp, J., Sun, W.-M., Takikawa, H., Takenaka, O., *et al.* (1996). Alterations in Glutathione Homeostasis in Mutant Eisai Hyperbilirubinemic Rats. *Hepatology*, **24**, 253–258.

Mao, Q., Qiu, W., Weigl, K.E., Lander, P.A., Tabas, L.B., Shepard, R.L., *et al.* (2002). GSH-Dependent Photolabeling of Multidrug Resistance Protein MRP1 (ABCC1) by [^{125}I]LY475776. Evidence of a Major Binding Site in the Cooh-Proximal Membrane Spanning Domain. *J. Biol. Chem.*, **277**, 28690–28699.

Matsuo, M., Kioka, N., Amachi, T., and Ueda, K. (1999). ATP Binding Properties of the Nucleotide-Binding Folds of SUR1. *J. Biol. Chem.*, **274**, 37479–37482.

McAleer, C., Breen, M.A., White, N.L., and Matthews, N. (1999). pABC11 (Also Known as MOAT-C and MRP5), a Member of the ABC Family of Proteins, Has Anion Transporter Activity but Does Not Confer Multidrug Resistance When Overexpressed in Human Embryonic Kidney 293 Cells. *J. Biol. Chem*, **274**, 23541–23548.

Milewski, M.I., Mickle, J.E., Forrest, J.K., Stafford, D.M., Moyer, B.D., Cheng, J., *et al.* (2001). A PDZ-Binding Motif is Essential but not Sufficient to Localize the C Terminus of CFTR to the Apical Membrane. *J. Cell. Sci*, **114**, 719–726.

Minemura, M., Tanimura, H., and Tabor, E. (1999). Overexpression of Multidrug Resistance Genes MDR1 and cMOAT in Human Hepatocellular Carcinoma and Hepatoblastoma Cell Lines. *Int. J. Oncol.*, **15**, 559–563.

Miyake, K., Mickley, L., Litman, T., Zhan, Z., Robey, R., Cristensen, B., *et al.* (1999). Molecular Cloning of cDNAs Which are Highly Overexpressed in Mitoxantrone-Resistant Cells: Demonstration of Homology to ABC Transport Genes. *Cancer Res.*, **59**, 8–13.

Mochida, Y., Taguchi, K., Tanigachi, S., Tsuhcyoshi, M., Kuwano, H., Tsazaki, T., *et al.* (2003). The role of P-glycoprotein in intestinal tamorigenesis: disruption of *indula* suppresses polyp formation in Apc^{Min,4} mice. *Carcinogenesis*, **24**, 1219–1224.

Mor-Cohen, R., Zivelin, A., Rosenberg, N., Shani, M., Muallem, S., and Seligsohn, U. (2001). Identification and Functional Analysis of Two Novel Mutations in the Multidrug Resistance Protein 2 Gene in Israeli Patients with Dubin-Johnson Syndrome. *J. Biol. Chem.*, **276**, 36923–36930.

Mottino, A.D., Hoffman, T., Jennes, L., and Vore, M. (2002). Expression and Localization of Multidrug Resistant Protein MRP2 in Rat Small Intestine. *J. Pharmacol. Exp. Ther.*, **293**, 717–723.

Muller, M., Meijer, C., Zaman, G.J., Borst, P., Scheper, R.J., Mulder, N.H., *et al.* (1994). Overexpression of the Gene Encoding the Multidrug Resistance-Associated Protein Results in Increased ATP-Dependent Glutathione S-Conjugate Transport. *Proc. Natl Acad. Sci. USA*, **91**, 13033–13037.

Nagata, K., Nishitani, M., Matsuo, M., Kioka, N., Amachi, T., and Ueda, K. (2000). Nonequivalent Nucleotide Trapping in the Two Nucleotide Binding Folds of the Human Multidrug Resistance Protein MRP1. *J. Biol. Chem.*, **275**, 17626–17630.

Nagel, G. (1999). Differential Function of the Two Nucleotide Binding Domains On Cystic Fibrosis Transmembrane Conductance Regulator. *Biochim. Biophys. Acta*, **1461**, 263–274.

Nakamura, J., Nishida, T., Hayashi, K., Kawada, N., Ueshima, S., Sugiyama, Y., *et al.* (1999). Kupffer Cell-Mediated Down Regulation of Rat Hepatic CMOAT/MRP2 Gene Expression. *Biochem. Biophys. Res. Commun.*, **255**, 143–149.

Narasaki, F., Oka, M., Nakano, R., Ikeda, K., Fukuda, M., Nakamura, T., *et al.* (1997). Human Canalicular Multispecific Organic Anion Transporter (cMOAT) Is Expressed in Human Lung, Gastric, and Colorectal Cancer Cells. *Biochem. Biophys. Res. Commun.*, **240**, 606–611.

Nei, M. (1987). *Molecular Evolutionary Genetics*. Columbia University Press, New York, pp. 293–298.

Nies, A.T., Konig, J., Cui, Y., Brom, M., Spring, H., and Keppler, D. (2002). Structural Requirements for the Apical Sorting of Human Multidrug Resistance Protein 2 (ABCC2). *Eur. J. Biochem.*, **269**, 1866–1876.

Nooter, K. and Stoter, G. (1996). Molecular Mechanisms of Multidrug Resistance in Cancer Chemotherapy. *Pathol. Res. Pract.*, **192**, 768–780.

Ogawa, K., Suzuki, H., Hirohashi, T., Ishikawa, T., Meier, P.J., Hirose, K., *et al.* (2000). Characterization of Inducible Nature of MRP3 in Rat Liver. *Am. J. Physiol.*, **278**, G438–G446.

Okamura, J., Monden, M., Horikawa, S., Sikujara, O., Kosaki, G., Seki, K., *et al.* (1980). Hepatocellular Carcinoma in a Case of Dubin–Johnson Syndrome Treated Successfully with Right Extended Lobectomy. *Jpn. J. Surg.*, **10**, 343–347.

Ortiz, D.F., Li, S., Iyer, R., Zhang, X., Novikoff, P., and Arias, I.M. (1999). MRP3, a New ATP-Binding Cassette Protein Localized to the Canalicular Domain of the Hepatocyte. *Am. J. Physiol.*, **276**, G1493–G5100.

Oswald, M., Kullak-Ublick, G.A., Paumgartner, G., and Beuers, U. (2001). Expression of Hepatic Transporters

OATP-C and MRP2 in Primary Sclerosing Cholangitis. *Liver*, **21**, 247–253.

Paulusma, C.C. and Oude Elferink, R.P.J. (1997). The Canalicular Multispecific Organic Anion Transporter and Conjugated Hyperbilirubinemia in Rat and Man. *J. Mol. Med.*, **75**, 420–428.

Paulusma, C.C., Bosma, P.J., Zaman, G.J., Bakker, C.T., Otter, M., Scheffer, G.L., *et al.* (1996). Congenital Jaundice in Rats With a Mutation in a Multidrug Resistance Associated Protein Gene. *Science*, **271**, 1126–1128.

Paulusma, C.C., Kool, M., Bosma, P.J., Scheffer, G.L., Ter Borg, F., Scheper, R.J., *et al.* (1997). A Mutation in the Human Canalicular Multispecific Organic Anion Transporter Gene Causes the Dubin–Johnson Syndrome. *Hepatology*, **25**, 1539–1542.

Paulusma, C.C., van Geer, M.A., Evers, R., Heijn, M., Ottenhoff, R., Borst, P., *et al.* (1999). Canalicular Multispecific Organic Anion Transporter/Multidrug Resistance Protein 2 Mediates Low-Affinity Transport of Reduced Glutathione. *Biochem. J.*, **338**, 393–401.

Paulusma, C.C., Kothe, M.J., Bakker, C.T., Bosma, P.J., van Bokhoven, I., van Marle, J., *et al.* (2000). Zonal Down-Regulation and Redistribution of the Multidrug Resistance Protein 2 During Bile Duct Ligation in Rat Liver. *Hepatology*, **31**, 684–693.

Priebe, W., Krawczyk, M., Kuo, M.T., Yamane, Y., Savaraj, N., and Ishikawa, T. (1998). Doxorubicin– and Daunorubicin–Glutathione Conjugates, but not Unconjugated Drugs, Competitively Inhibit Leukotriene C4 Transport Mediated by MRP/GS-X Pump. *Biochem. Biophys. Res. Commun.*, **247**, 859–863.

Qian, Y.M., Qiu, W., Gao, M., Westlake, C.J., Cole, S.P., and Deeley, R.G. (2001a). Characterization of Binding of Leukotriene C4 by Human Multidrug Resistance Protein 1: Evidence of Differential Interactions with NH_2- and COOH-Proximal Halves of the Protein. *J. Biol. Chem.*, **276**, 38636–38644.

Qian, Y.M., Song, W.C., Cui, H.-R., Cole, S.P.C., and Deeley, R.G. (2001b). Glutathione Stimulates Sulfated Estrogen Transport by Multidrug Resistance Protein 1. *J. Biol. Chem.*, **276**, 6404–6411.

Rappa, G., Lorico, A., Flavell, R.A., and Sartorelli, A.C. (1997). Evidence that the Multidrug Resistance Protein (MRP) Functions as a Co-transporter of Glutathione and Natural Product Toxins. *Cancer Res.*, **57**, 5232–5237.

Ren, X.Q., Furukawa, T., Aoki, S., Nakajima, T., Sumizawa, T., Haraguchi, M., *et al.* (2001). Glutathione-Dependent Binding of a Photoaffinity Analog of Agosterol A to the C-Terminal Half of Human Multidrug Resistance Protein. *J. Biol. Chem.*, **276**, 23197–23206.

Renes, J., de Vries, E.G.E., Hooiveld, G.J.E.J., Krikken, I., Jansen, P.L.M., and Muller, M. (2000). Multidrug Resistance Protein MRP1 Protects against the Toxicity of the Major Lipid Peroxidation Product 4-Hydroxy-nonenal. *Biochem. J.*, **350**, 555–561.

Robey, R.W., Medina-Perez, W.Y., Nishiyama, K., Lahusen, T., Miyake, K., Litman, T., *et al.* (2001). Overexpression of the ATP-Binding Cassette Half-transporter, ABCG2 (Mxr/BCrp/ABCP1), in Flavopiridol-Resistant Human Breast Cancer Cells. *Clin. Cancer Res.*, **7**, 145–152.

Roden, D.M. and George, A.L., Jr (2002). The Genetic Basis of Variability in Drug Responses. *Nat. Rev. Drug Discov.*, **1**, 37–44.

Rost, D., Konig, J., Weiss, G., Klar, E., Stremmel, W., and Keppler, D. (2001). Expression and Localization of the Multidrug Resistance Proteins MRP2 and MRP3 in Human Gallbladder Epithelia. *Gastroenterology*, **121**, 1203–1208.

Roth, J.A., Berman, E., Befeler, D., and Johnson, F.B. (1982). A Black Hepatocellular Carcinoma with Dubin–Johnson-Like Pigment and Mallory Bodies: A Histochemical and Ultrastructural Study. *Am. J. Surg. Pathol.*, **6**, 375–382.

Ryu, S., Kawabe, T., Nada, S., and Yamaguchi, A. (2000). Identification of Basic Residues Involved in Drug Export Function of Human Multidrug Resistance-Associated Protein 2. *J. Biol. Chem.*, **275**, 39617–39624.

Saito, S., Iida, A., Sekine, A., Miura, Y., Ogawa, C., Kawauchi, S., *et al.* (2002). Identification of 779 Genetic Variations in Eight Genes Encoding Members of the ATP-Binding cassette, Subfamily C (ABCC/MRP/CFTR). *J. Hum. Genet.*, **47**, 147–171.

Sakamoto, A., Mori, I., Kawai, K., and Tsuchiyama, H. (1987). Dubin–Johnson Syndrome Associated with Hepatocellular Carcinoma – Report of an Autopsy Case. *Gan No Rinsho* (*Jpn. J. Cancer. Clin.*), **33**, 1361–1367.

Schaub, T.P., Kartenbeck, J., Konig, J., Spring, H., Dorsam, J., Staehler, G., *et al.* (1999). Expression of the MRP2 Gene-Encoded Conjugate Export Pump in Human Kidney Proximal Tubules and in Renal Cell Carcinoma. *J. Am. Soc. Nephrol.*, **10**, 1159–1169.

Schinkel, A.H. (1998). Pharmacological Insights from P-glycoprotein Knockout Mice. *Int. J. Clin. Pharmacol. Ther.*, **36**, 9–13.

Schinkel, A.H., Mayer, U., Wagenaar, E., Mol, C.A., van Deemter, L., Smit, J.J., *et al.* (1997). Normal Viability and Altered Pharmacokinetics in Mice Lacking MDR1-Type (Drug-Transporting) P-glycoproteins. *Proc. Natl. Acad. Sci. USA*, **94**, 4028–4033.

Schrenk, D., Michalke, A., Gant, T.W., Brown, P.C., Silverman, J.A., and Thorgeirsson, S.S. (1996). Multidrug Resistance Gene Expression in Rodents and Rodent Hepatocytes Treated with Mitoxantrone. *Biochem. Pharmacol.*, **52**, 1453–1460.

Schrenk, D., Baus, P.R., Ermel, N., Klein, C., Vorderstemann, B., and Kauffmann, H.M. (2001). Up-Regulation of Transporters of the MRP Family by Drugs and Toxins. *Toxicol. Lett.*, **120**, 51–57.

Schuetz, J.D., Connelly, M.C., Sun, D., Paibir, S.G., Flynn, P.M., Srinivas, R.V., *et al.* (1999) MRP4: A Previously

Unidentified Factor in Resistance to Nucleoside-Based Antiviral Drugs. *Nat. Med.*, **5**, 1048–1051.

Schwartz, A.L. and Ciechanover, A. (1999). The Ubiquitin–Proteasome Pathway and Pathogenesis of Human Diseases. *Annu. Rev. Med.*, **50**, 57–74.

Senior, A.E., al-Shawi, M.K., and Urbatsch, I.L. (1995). The Catalytic Cycle of P-glycoprotein. *FEBS Lett.*, **377**, 285–289.

Shani, M., Seligsohn, U., Gilon, E., Sheba, C., and Adam, A. (1970). Dubin–Johnson Syndrome in Israel. I. Clinical, Laboratory, and Genetic Aspects of 101 Cases. *Q. J. Med.*, **39**, 549–567.

Songyang, Z., Fanning, A.S., Fu, C., Xu, J., Marfatia, S.M., Chishti, A.H., *et al.* (1997). Recognition of Unique Carboxyl-Terminal Motifs by Distinct PDZ Domains. *Science*, **275**, 73–77.

Sprinz, H. and Nelson, R.S. (1954). Persistent Nonhemolytic Hyperbilirubinemia Associated with Lipochrome-Like Pigment in Liver Cells: Report of Four Cases. *Ann. Int. Med.*, **41**, 952–962.

Stockel, B., Konig, J., Nies, A.T., Cui, Y., Brom, M., and Keppler, D. (2000). Characterization of the 5′-Flanking Region of the Human Multidrug Resistance Protein 2 (MRP2) Gene and its Regulation in Comparison with the Multidrug Resistance Protein 3 (MRP3) Gene. *Eur. J. Biochem.*, **267**, 1347–1358.

St. Pierre, M.V., Serrano, M.E., Macias, R.I.R., Dubs, U., Hoechli, M., Lauper, U., *et al.* (2000). Expression of Members of the Multidrug Resistance Protein Family in Human Term Placenta. *Am. J. Physiol.*, **279**, R1495–R1503.

Stride, B.D., Valdimarsson, G., Gerlach, J.H., Wilson, G.M., Cole, S.P.C., and Deeley, R.G. (1996). Structure and Expression of the mRNA Encoding the Murine Multidrug Resistance Protein (MRP), an ATP-Binding Cassette Transporter. *Mol. Pharmacol.*, **49**, 962–971.

Stride, B.D., Grant, C.E., Loe, D.W., Hipfner, D.R., Cole, S.P., and Deeley, R.G. (1997). Pharmacological Characterization of the Murine and Human Orthologs of Multidrug-Resistance Protein in Transfected Human Embryonic Kidney Cells. *Mol. Pharmacol.*, **52**, 344–353.

Stride, B.D., Cole, S.P., and Deeley, R.G. (1999). Localization of a Substrate Specificity Domain in the Multidrug Resistance Protein. *J. Biol. Chem.*, **274**, 22877–22883.

Suzuki, H. and Sugiyama, Y. (1998). Excretion of GSSG and Glutathione Conjugates Mediated by MRP1 and cMOAT/MRP2. *Semin. Liver Dis.*, **18**, 359–376.

Suzuki, H. and Sugiyama, Y. (2002). Single Nucleotide Polymorphisms in Multidrug Resistance Associated Protein 2 (MRP2/ABCC2): Its Impact on Drug Disposition. *Adv. Drug Deliver. Rev.*, **54**, 1311–1331.

Suzuki, T., Nishio, K., Sasaki, H., Kurokawa, H., Saito-Ohara, F., Ikeuchi, T., *et al.* (1997). cDNA Cloning of a Short Type of Multidrug Resistance Protein Homologue, SMRP, from a Human Lung Cancer Cell Line. *Biochem. Biophys. Res. Commun.*, **238**, 790–794.

Takino, T., Takahashi, T., and Okuno, T. (1977). Clinical Study of the Constitutional Hyperbilirubinemia in Japan. A Nationwide Survey between 1970 and 1974. *Jpn. J. Gastroenterol.*, **74**, 1518–1528.

Taguchi, Y., Kino, K., Morishima, M., Komano, T., Kane, S.E., and Ueda, K. (1997). Alteration of Substrate Specificity by Mutations at the His61 Position in Predicted Transmembrane Domain 1 of Human MDR1/P-glycoprotein. *Biochemistry*, **36**, 8883–8889.

Tanaka, T., Uchiumi, T., Hinoshita, E., Inokuchi, A., Toh, S., Wada, M., *et al.* (1999). The Human Multidrug Resistance Protein 2 Gene: Functional Characterization of the 5′-Flanking Region and Expression in Hepatic Cells. *Hepatology*, **30**, 1507–1512.

Taniguchi, K., Wada, M., Kohno, K., Nakamura, T., Kawabe, T., Kawakami, M., *et al.* (1996). A Human Canalicular Multispecific Organic Anion Transporter (cMOAT) Gene is Overexpressed in Cisplatin-Resistant Human Cancer Cell Lines with Decreased Drug Accumulation. *Cancer Res.*, **56**, 4124–4129.

Tate, G., Li, M., Suzuki, T., and Mitsuya, T. (2002). A New Mutation of the ATP-Binding Cassette, Sub-Family, C, Member 2 (ABCC2) Gene in a Japanese Patient with Dubin–Johnson Syndrome. *Genes Genet. Syst.*, **77**, 117–121.

Toh, S., Wada, M., Uchiumi, T., Inokuchi, A., Makino, Y., Horie, Y., *et al.* (1999). Genomic Structure of the Canalicular Multispecific Organic Anion-Transporter Gene (MRP2/cMOAT) and Mutations in the ATP-Binding-Cassette Region in Dubin–Johnson Syndrome. *Am. J. Hum. Genet.*, **64**, 739–746.

Trauner, M., Arrese, M., Soroka, C.J., Ananthanarayanan, M., Koeppel, T.A., Schlosser, S.F., *et al.* (1997). The Rat Canalicular Conjugate Export Pump (Mrp2) is Down-Regulated in Intrahepatic and Obstructive Cholestasis. *Gastroenterology*, **113**, 255–264.

Tsujii, H., Konig, J., Rost, D., Stockel, B., Leuschner, U., and Keppler, D. (1999). Exon–Intron Organization of the Human Multidrug-Resistance Protein 2 (MRP2) Gene Mutated in Dubin–Johnson Syndrome. *Gastroenterology*, **117**, 653–660.

Uchiumi, T., Hinoshita, E., Haga, S., Nakamura, T., Tanaka, T., Toh, S., *et al.* (1998). Isolation of a Novel Human Canalicular Multispecific Organic Anion Transporter, cMOAT2/MRP3, and its Expression in Cisplatin-Resistant Cancer Cells with Decreased ATP-Dependent Drug Transport. *Biochem. Biophys. Res. Commun.*, **252**, 103–110.

Ueda, K., Cardarelli, C., Gottesman, M.M., and Pastan, I. (1987). Expression of a Full-Length cDNA for the Human "MDR1" Gene Confers Resistance to Colchicine, Doxorubicin, and Vinblastine. *Proc. Natl Acad. Sci. USA*, **84**, 3004–3008.

Ueno, S., Tanabe, G., Hanazono, K., Ogawa, H., Yoshidome, S., Aikou, T., *et al.* (1998). Postoperative Management Following Massive Hepatectomy in a

Patient with Dubin–Johnson Syndrome: Report of a Case. *Surg. Today*, **28**, 1274–1278.

Urbatsch, I.L., Sankaran, B., Weber, J., and Senior, A.E. (1995). P-glycoprotein Is Stably Inhibited by Vanadate-Induced Trapping of Nucleotide at a Single Catalytic Site. *J. Biol. Chem.*, **270**, 19383–19390.

van Aubel, R.A.M.H., Hartog, A., Bindels, R.J.M., van Os, C.H., and Russel, F.G.M. (2000). Expression and Immunolocalization of Multidrug Resistance Protein 2 in Rabbit Small Intestine. *Eur. J. Pharmacol.*, **400**, 195–198.

van Kuijck, M.A., Kool, M., Merkx, G.F.M., Geurts van Kessel, A., Bindels, R.J.M., Deen, P.M.T., *et al.* (1997). Assignment of the Canalicular Multispecific Organic Anion Transporter Gene (CMOAT) to Human Chromosome 10q24 and Mouse Chromosome 19D2 by Fluorescent In Situ Hybridization. *Cytogenet. Cell Genet.*, **77**, 285–287.

Vos, T.A., Hooiveld, G.J., Koning, H., Childs, S., Meijer, D.K., Moshage, H., *et al.* (1998). Up-Regulation of the Multidrug Resistance Genes, Mrp1 and Mdr1b, and Down-regulation of the Organic Anion Transporter, Mrp2, and the Bile Salt Transporter, Spgp, in Endotoxemic Rat Liver. *Hepatology*, **28**, 1637–1644.

Wada, M., Toh, S., Taniguchi, K., Nakamura, T., Uchiumi, T., Kohno, K., *et al.* (1998). Mutations in the Canilicular Multispecific Organic Anion Transporter (cMOAT) Gene, a Novel ABC Transporter, in Patients with Hyperbilirubinemia II/Dubin–Johnson Syndrome. *Hum. Mol. Genet.*, **7**, 203–207.

Welsh, M.J. and Smith, A.E. (1993). Molecular Mechanisms of CFTR Chloride Channel Dysfunction in Cystic Fibrosis. *Cell*, **73**, 1251–1254.

Yamada, T., Mari, Y., Hayashi, R., Takada, M., Ino, Y., Naishiao, Y., *et al.* (2003). Suppression of intestinal polyposis in *Mdnl*-deficient $Apc^{Min/+}$ mice. *Cancer Res.*, **63**, 895–901.

Yang, Y., Chen, Q., and Zhang, J.T. (2002). Structural and Functional Consequences of Mutating Cysteine Residues in the Amino Terminus of Human Multidrug Resistance-Associated Protein 1. *J. Biol. Chem.*, **277**, 44268–44277.

Zaman, G.J., Flens, M.J., van Leusden, M.R., de Haas, M., Mulder, H.S., Lankelma, J., *et al.* (1994). The Human Multidrug Resistance-Associated Protein MRP is a Plasma Membrane Drug-Efflux Pump. *Proc. Natl Acad. Sci. USA*, **91**, 8822–8826.

Zaman, G.J., Lankelma, J., van Tellingen, O., Beijnen, J., Dekker, H., Paulusma, C., *et al.* (1995). Role of Glutathione in the Export of Compounds from Cells by the Multidrug-Resistance-Associated Protein. *Proc. Natl Acad. Sci. USA*, **92**, 7690–7694.

Zeng, H., Chen, E.-S., Belinsky, M.G., Rea, P.A. and Kruh, G.D. (2001). Transport of methotrexate (MTX) and folates by multidrug resistance protein (MRP) 3 and MRP1: effect of polyglutamylation on MTX transport. *Cancer Res.*, **61**, 7225–7232.

Zhang, D.W., Cole, S.P., and Deeley, R.G. (2001). Identification of an Amino Acid Residue in Multidrug Resistance Protein 1 Critical for Conferring Resistance to Anthracyclines. *J. Biol. Chem.*, **276**, 13231–13239.

Zollner, G., Fickert, P., Zenz, R., Fuchsbichler, A., Stumptner, C., Kenner, L., *et al.* (2001). Hepatobiliary Transporter Expression in Percutaneous Liver Biopsies of Patients with Cholestatic Liver Diseases. *Hepatology*, **33**, 633–646.

19

GERD SCHMITZ* AND
WOLFGANG E. KAMINSKI**

Phospholipid transporters ABCA1 and ABCA7

ABCA1 – FAMILIAL HDL-DEFICIENCY SYNDROMES AND ATHEROGENESIS

So far, a total of 12 genes have been assigned to the ABCA subclass of ABC transporters (Dean *et al.* 2001). The complete human coding regions and genomic structures have been determined for seven transporters of this subfamily. In addition to the prototypic members ABCA1 (Langman *et al.* 1999) and ABCA2 (Kaminski *et al.* 2001a), these include ABCA3 (Connors *et al.* 1997), ABCA4 (ABCR) (Chapter 20), ABCA6 (Kaminski *et al.* 2001b), ABCA7 (Kaminski *et al.* 2000a), and ABCA9 (Piehler *et al.* 2002). Among the ABCA proteins, two members, ABCA1 and ABCA4, have been directly linked to human disease. Evidence has accumulated to suggest that ABCA1 functions as a facilitator of cellular cholesterol and phospholipid transport (Orsó *et al.* 2000) and ABCA4 plays a pivotal role in retinaldehyde processing (Chapter 20). In addition, ABCA3 has recently been identified as a constituent of lamellar bodies that may play a role in lung surfactant processing (Yamano *et al.* 2001), and ABCA7 has been identified as a modulator of phosphatidylserine and ceramide transport (D. Kielar *et al.*, unpublished observations).

ABCA1 (formerly ABC1), the defining member of a subclass of ABC proteins, is expressed in a multitude of human organs with highest expression levels in placenta, liver, lung, adrenal glands and fetal tissues (Langmann *et al.* 1999). A characteristic feature of ABCA1 is that it is upregulated by cholesterol influx in macrophages and suppressed by HDL_3-mediated cholesterol efflux (Langmann *et al.* 1999). The human cDNA encodes a 2261 amino acid polypeptide with a molecular weight of 220 kDa (Langmann *et al.* 1999). Structurally, ABCA1 exhibits all the characteristics of a full-size transporter, that is, it consists of two tandemly oriented transmembrane domain-ABC subunits (See Chapter 17, Figure 1). ABCA1 shows highest amino acid homology (54%) to the recently cloned leukocyte transporter ABCA7 (Kaminski *et al.* 2000a) and the retina-specific ABCA4 (52%) (Chapter 20). The human gene has been localized to chromosome 9q31 (Luciani *et al.* 1994) and shown to be composed of 50 exons spanning a region of app. 149 kb (Singaraja *et al.* 2002).

---
*Institute for Clinical Chemistry and Laboratory Medicine, University Hospital Regensburg, Franz-Josef-Strauss Allee 11, D-93053, Regensburg, Germany
**

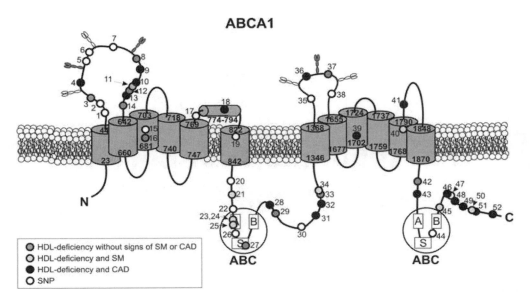

Figure 1 Molecular structure of human ABCA1 and localization of known mutations and single-nucleotide polymorphisms. Allelic variants (1–52) are shown as circles and numbered relative to the ABCA1 N-terminus. The gray code indicates the association of HDL deficiency and splenomegaly (SM) or cardiovascular disease (CAD). ABC, ATP binding cassette; SNP, single nucleotide polymorphism. Numbers in barrels designate the flanking amino acid positions of the respective transmembrane membrane domains.

Although ABCA1 had already been identified in the mouse in 1994 (Luciani *et al.* 1994), little was known about its biologic function until recently. The major breakthrough came 5 years later when it was discovered that mutations in the human ABCA1 gene are the underlying molecular defect in familial high-density lipoprotein (HDL) deficiency syndromes such as Tangier disease (TD), which identified ABCA1 as a major regulator of HDL metabolism (Bodzioch *et al.* 1999) (Figure 1). TD is an autosomal recessive disorder of lipid metabolism characterized by almost complete absence of plasma HDL and the accumulation of cholesteryl esters in the cells of the reticuloendothelial system leading to splenomegaly and enlargement of tonsils or lymph nodes (Schmitz *et al.* 2000). In agreement with the finding that mutations in ABCA1 cause an HDL-deficiency phenotype in humans, a recent study by our group revealed that mice lacking functional ABCA1 exhibit plasma lipid alterations that are concordant with those found in TD (Orsó *et al.* 2000).

PHYSIOLOGICAL ROLE OF ABCA1

Initial studies of the biological role of ABCA1 supported the view that ABCA1, like MDR1 and MDR3 (van Helvoort *et al.* 1996), functions as a translocator of lipids between the innner and outer leaflet of the plasma membrane (Lawn *et al.* 1999). This was based on experiments showing an increase in cholesterol and phospholipid export under conditions of forced expression of ABCA1 in ABCA1 null mutant cells from patients with genetic HDL deficiency which characteristically display an increased cholesterol and phospholipid uptake (Orsó *et al.* 2000). Moreover, ABCA1 appears to be localized on the plasma membrane and surface expression of ABCA1 is upregulated in macrophages by cholesterol loading (Orsó *et al.* 2000). However, recent work from our laboratories indicates that the ATP turnover of ABCA1 occurs at a very low rate, whereas nucleotide binding induces conformational changes. As a result it is likely that ABCA1 acts as a facilitator of cholesterol/ choline-phospholipid export within the cellular lipid export machinery rather than exerting a genuine translocator function (Lawn *et al.* 1999). It will be exciting to elucidate the exact molecular mechanisms by which ABCA1 mediates the export of lipid compounds from the cell. In this context, the most critical aspect is the question as to which molecular partners interact with and thus potentially modulate ABCA1 function. Work from our laboratory has provided potential clues by demonstrating that ABCA1 interacts with a β_2-syntrophin–utrophin complex that may couple ABCA1 to the F-actin cytoskeleton (Büchler *et al.* 2002a). Moreover, we could demonstrate that ABCA1 function depends on binding to Fas-associated death domain protein, an adapter molecule involved in cell death receptor signal transduction (Büchler *et al.* 2002b) (Figure 2). The association of Fas-associated death

domain protein with ABCA1 provides a surprising link between HDL metabolism and the cell death receptor signaling machinery. ABCA1 has been proposed to be a phosphatidylserine translocase that facilitates phosphatidylserine exofacial flipping. The transient local exposure of anionic phospholipids in the outer-membrane leaflet enhances the engulfment of apoptotic cells, endocytosis, and binding of apoprotein A-I (apoA-I). Importantly, redistribution of phosphatidylserine at the cell surface is one of the early characteristics of cells undergoing apoptosis. Thus the physical linkage of Fas-associated death domain protein and ABCA1 supports the notion that ABCA1 may act as an anti-apoptotic molecule.

Recent evidence indicates that ABCA1 and the cell division cycle 42 (Cdc42) protein are associated with a raft subfraction resistant to the detergent Lubrol, whereas ABCA1 is not detectable in Triton-resistant rafts (Drobnik *et al.* 2002). Moreover, the fact that ABCA1 is detectable in the cytosol and Golgi compartment of unstimulated fibroblasts also raises the intriguing possibility that it is a mobile molecule that may shuttle between the plasma

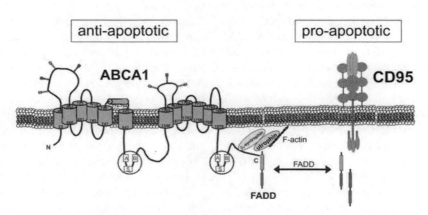

Figure 2   ABCA1 and its known C-terminal interaction partners β_2-syntrophin, utrophin, and the Fas-associated death domain protein (FADD). Note that the β_2-syntrophin–utrophin establishes a physical link between ABCA1 and the cytoskeleton (F-actin). Moreover, ABCA1 binding to FADD provides a novel interlink between this transporter and the CD95- dependent apoptosis pathway. CD95, Fas antigen.

membrane and the Golgi as constituent of a vesicular transport route.

ABCA1 CONTROLS SUSCEPTIBILITY TO ATHEROSCLEROSIS

Macrophages play a key role in the initiation and progression of atherosclerotic lesions. In the nascent lesion, they transform into foam cells through the excessive accumulation of cholesteryl esters. Dysfunctional lipid homeostasis in macrophages and foam cells ultimately results in the breakdown of membrane integrity and cell death. An interesting clue as to how ABCA1 may be implicated in the control of monocyte/macrophage targeting on the cellular level comes from the observation that apoA-I mediated lipid efflux in ABCA1 deficient cells is paralleled by the downregulation of the protein Cdc42 and filopodia formation (Diederich et al. 2001). Cdc42, the ras homolog protein (rho), and the ras-related C3 botulinum toxin substrate (rac) are members of the family of small GTP-binding proteins. Importantly, all three proteins are sequentially activated by extracellular stimuli in mammalian cells. Cdc42 controls formation of filopodia and rho proteins are known to induce the formation of stress fibers and focal adhesions; rac proteins regulate formation of lamellipodia and membrane ruffles. Thus it is conceivable that ABCA1 modulates cellular mobility of monocytes/macrophages through this mechanism and hence may affect recruitment of monocytes into the vessel wall. Within the vascular wall, this regulator function in filopodia formation may even extend to platelets, vascular smooth muscle cells, and endothelial cells since these cells have been shown to express ABCA1 (Bellincampi et al. 2001).

The findings that ABCA1 is upregulated in human macrophages during sustained uptake of cholesterol (Langmann et al. 1999; Orsó et al. 2000) and that it facilitates cholesterol export from the cell render it a likely player in foam cell development and the formation of lesions of atherosclerosis.

REGULATION OF ABCA1 TRANSCRIPTION

Since the cloning of the ABCA1 promoter region, a number of transcriptional control elements have been characterized. In particular, the zinc-finger protein ZNF202 appears to function as a major repressor of transcriptional activation of ABCA1 which is capable of suppressing its induction by 20(S)OH-cholesterol and 9-cis-retinoic acid (Porsch-Ozcurumez et al. 2001). Other factors that regulate the transcription of ABCA1 include the pleiotropic growth-inducing factor oncostatin M, a member of the IL-6 family of cytokines (Langmann et al. 2002), and geranylgeranyl pyrophosphate, one of the major products of the mevalonate pathway (Gan et al. 2001). Recent in vitro data demonstrated that the nuclear peroxisome proliferator-activated receptor PPAR-γ induces ABCA1 expression in macrophage-derived foam cells (Chawla et al. 2001). Moreover, chimeric LDL-receptor$-/-$ mice deficient in PPAR-γ in their hematopoietic cells show a significant increase in atherosclerosis, supporting the concept that PPAR-γ-dependent activation of macrophage ABCA1 expression is atheroprotective in vivo (Chawla et al. 2001). Based on this, considerable attention has recently been paid to nuclear liver X receptors (LXRs) as inducers of ABCA1 expression in response to lipid loading, and LXR agonists have been proposed to be promising candidates for therapeutic activation of ABCA1 (Chawla et al. 2001). However, because ATP is a factor that critically controls the activity of ABC transporters, it should be noted that sustained activation of significant energy consumers such as ABCA1 may lead to excessive mitochondrial energy production and induce mitochondrial exhaustment.

Importantly, among patients with defective ABCA1 one subgroup develops premature atherosclerosis, whereas another presents predominantly with splenomegaly (Schmitz et al. 2000), indicating differences in macrophage targeting to tissues in ABCA1 deficiency which may be the reflection of the nature of the mutation in the ABCA1 gene. Based on this view, we hypothesized that ABCA1 acts as a factor that determines monocyte targeting into tissues. In recent experiments using chimeric low-density lipoprotein receptor (LDLR)−/− mice lacking ABCA1 in their circulating cells, we tested the hypothesis that the selective disruption of ABCA1 in circulating cells, in particular macrophages, affects macrophage targeting into the vascular wall and lesion formation in vivo. Our results demonstrated that the absence of ABCA1 from leukocytes only is sufficient to induce aberrant monocyte recruitment into the spleen, identifying ABCA1 as a critical leukocyte factor in the control of monocyte targeting (van Eck *et al.* 2002). Moreover, we found that LDLR−/− chimeras deficient in ABCA1 develop significantly larger (60%) and more advanced atherosclerotic lesions compared with chimeric LDLR−/− mice with functional ABCA1 in hematopoietic cells (van Eck *et al.* 2002). Importantly, targeted disruption of leukocyte ABCA1 function in this study did not affect plasma HDL cholesterol levels. These results provided direct evidence that leukocyte ABCA1 exerts significant antiatherosclerotic activity which is clearly independent of plasma HDL. Because macrophages are by far the predominant leukocyte cell type within lesions of atherosclerosis and lymphocytes, which occur in lesions, if any, only in limited numbers, express only minute amounts of ABCA1 mRNA (unpublished observation), these data strongly suggest that the observed increased susceptibility to atherosclerosis is largely attributable to the absence of ABCA1 from lesional monocytes/macrophages. The view that macrophage ABCA1 functions as an important antiatherosclerotic factor has also been supported by work from others. Another recent study reported that targeted inactivation of ABCA1 in macrophages leads to increased lesion formation and foam cell accumulation in apolipoprotein E (ApoE) null mutant mice (Aiello *et al.* 2002). Furthermore, in a reciprocal approach using ABCA1 overexpressing transgenic ApoE null mice it has been shown

that forced expression of human ABCA1 results in the formation of significantly smaller and less complex lesions in ApoE knockout mice. Interestingly, we detected high expression of ABCA1 in differentiating megakaryocytes and circulating platelets, indicating that monocytes/macrophages are not the only source of ABCA1 within the hematopoietic system (unpublished data). Thus it is possible that altered platelet function due to the lack of ABCA1 may contribute to the observed antiatherogenic effect. In agreement with a role for ABCA1 in platelets, we observed previously that ABCA1 null mice have abnormalities in their platelet morphology and function including reduced platelet counts, increased platelet mass and volume, and impaired ADP-induced leukocyte aggregation (Orsó *et al.* 2000). Taken together, these findings point to a likely involvement of platelet ABCA1 in atherosclerosis and its clinical sequelae such as myocardial infarction and stroke.

ASSOCIATION BETWEEN MUTATION TOPOLOGY AND CLINICAL PHENOTYPES IN INDIVIDUALS WITH DEFECTIVE ABCA1

As already mentioned, HDL-deficiency syndromes caused by dysfunctional ABCA1 typically manifest with splenomegaly or cardiovascular disease or a combination thereof (Schmitz *et al.* 2000). More than 50 mutations within the coding region of the human ABCA1 gene have been identified so far (Figure 1). When looking at the topological distribution of known mutations within the ABCA1 protein product, two findings are worth noting. First, mutations within the ABCA1 gene appear to occur in clusters. Major mutation clusters can be found in the C-terminal moiety of the first extracellular domain, the N-terminal ATP binding cassette, and the C-terminus (Figure 1). Secondly, mutations in the first extracellular domain and the C-terminus mutations are frequently associated with a cardiovascular phenotype. In contrast, amino acid exchanges in or near the N-terminal ATP binding cassette appear to coincide with splenomegaly. These observations raise the intriguing possibility that amino acid substitutions within a few selected domains of the ABCA1 protein may to a significant degree determine the type of clinical presentation. In light of this, one may postulate a critical function for the C-terminal moiety of the first extracellular domain and the C-terminus in the development of the cardiovascular phenotype.

Recent experiments from our laboratory revealed that human macrophages express a variety of ABC transporters which are regulated by cholesterol uptake and/or HDL-dependent cholesterol efflux (Klucken et al. 2000). Intriguingly, we found the vast majority of the currently known human 48 ABC genes to be expressed in macrophages (Klucken et al. 2000). A significant portion of these genes are upregulated during monocyte differentiation into macrophages, and numerous ABC transporters showed cholesterol-influx- or efflux-dependent gene regulation. Based on this survey one can hypothesize the potential involvement of at least some of these cholesterol-responsive ABC molecules in atherogenesis (Schmitz and Kaminski 2002).

ABCA7/HA-1 – POTENTIAL MODULATORS OF IMMUNE FUNCTION?

The human A subfamily transporter ABCA7 was recently cloned in our laboratory (Kaminski et al. 2000a). Its cDNA is of size 6.8 kb and contains an open reading frame encoding a polypeptide of 2146 amino acids with a calculated molecular weight of 220 kDa. The predicted protein product is composed of two transmembrane domains and two nucleotide binding folds, indicating that it pertains to the group of full-size ABC transporters. Interestingly, this novel transporter displays highest protein sequence homology with the cholesterol and phospholipid export facilitator ABCA1 (54%) and the retinal transporter ABCR (49%), both members of the ABC transporter subfamily A. ABCA7 is predominantly expressed in myelolymphatic tissues, with highest expression in peripheral leukocytes, thymus, spleen, and bone marrow (Kaminski et al. 2000a). As observed for other members of the ABCA subclass (Langmann et al. 1999; Kaminski et al. 2001a, b; Piehler et al. 2002), ABCA7 expression is responsive to cholesterol flux and regulated during differentiation of human monocytes into macrophages. Identical regulatory profiles during monocyte differentiation and cholesterol flux in macrophages and the finding that ABCA7 is the closest relative of ABCA1 suggest a role for this transporter in monocyte/macrophage lipid homeostasis. Recent overexpression experiments in HeLa cells suggest that ABCA7 facilitates the export of phophatidylserine and ceramide metabolites from the cell.

The biological role of ABCA7 is currently unknown. Of particular interest, however, is the observation that the ABCA7 gene is intimately physically linked to the gene for the minor histocompatibility antigen HA-1 on chromosome 19p13.3, a genomic region between the CDC42 gene and the marker D19S342 (Kaminski et al. 2000b). Both genes are arranged in a head-to-tail array with only 1.7 kb distance between the terminal exon of ABCA7 and the initial exon of the HA-1 gene. Thus it is likely that part of the ABCA7 – HA-1 intergene region contains the 5′ UTR of the HA-1 gene. This raises the intriguing possibility that there may be a functional and regulatory interlink between ABCA7 and HA-1. The ABCA7 gene consists of 46 small exons, which are separated by relatively small intron sequences (Kaminski et al. 2000b). The length of the coding region and the number of exons of ABCA7 is similar to that of ABCA1 and ABCR, two other human ABCA subfamily members that have been associated with inherited diseases. Another striking feature of the ABCA7 gene is the presence of a series of small introns whose lengths consist of multiples of 3, which may represent potential candidate sequences for alternative splicing.

HA-1 is a cytosolic protein composed of 1165 amino acids (Nagase *et al.* 1996) and with unknown function that bears significant homology to members of the chimerin family of proteins. A database homology search revealed the presence of a rhoGAP domain and a diacylglycerol binding domain (C1-domain). It has recently been implicated in the survival of cytotoxic T cells in graft versus host disease (GVHD) (den Haan *et al.* 1998). GVHD is a major cause of mortality and morbidity after hematopoietic cell transplantation. Recipient disparity for major histocompatibility antigens is the most important risk factor for GVHD (Hansen *et al.* 1998). Interestingly, among the class of minor histocompatibility antigens, only HA-1 has been causatively linked to the pathogenesis of GVHD (Goulmy *et al.* 1996). A polymorphism at position 504 of the reported HA-1 cDNA sequence, which encodes either His or Arg within a nonapeptidic epitope of the polypeptide sequence, has been identified as the immunologic target for HA-1-specific cytotoxic T cells (den Haan *et al.* 1998). Work from our laboratory has established additional single-nucleotide polymorphisms (SNPs) within the coding region of the ABCA1/HA-1 locus. These allelic variants may provide important information on the potential association of both genes with candidate diseases.

Several findings point to the notion that ABCA1/HA-1 may be involved in events associated with autoimmune disease and programmed cell death. First, both genes are expressed in macrophages and lymphocytes, major effector cells in chronic inflammation. Second a database search revealed that a polypeptide of length approximately 150 amino acids within the first extracellular domain of ABCA7 is recognized by antisera from patients with Sjögren's syndrome, raising the possibility that ABCA7 is associated with the pathogenesis of this autoimmune disease (unpublished data). Third we found that ABCA7 mRNA expression is significantly upregulated during keratinocyte differentiation, which is paralleled by an increase in intracellular and cell surface content

of the pro-apoptotic sphingolipid ceramide. Finally, overexpression of ABCA7 increases the cellular phosphatidylserine (PS) content (unpublished data). The exposure of PS at the cell surface is a hallmark of the apoptotic execution phase. It is possible that the increase in intracellular PS levels, due to enhanced keratinocyte ABCA7 expression, may promote the cell surface exposure of PS during apoptosis and thus contribute to the proapoptotic effects and the increased cell death we observed during ABCA7 overexpression. The notion that ABCA7 has proapoptotic activities is also consistent with our finding that ABCA7-overexpressing cells exhibit a cell cycle arrest in the G_2/M phase (unpublished data).

Based on the still limited available information, it is likely that ABCA7 and HA-1 serve functions as modulators of chronic inflammatory processes. However, more work is required to define the potential role of ABCA7 and HA-1 in autoimmune disease and apoptosis.

SUMMARY

Human ABCA1 is a polypeptide of 2261 amino acids and consists of two tandemly oriented transmembrane domain ABC subunits. Mutations in the human ABCA1 gene are the underlying molecular defect in the familial high-density lipoprotein (HDL) deficiency syndrome Tangier disease (TD). TD is an autosomal recessive disorder of lipid metabolism characterized by an almost complete absence of plasma HDL and the accumulation of cholesteryl esters in the cells of the reticuloendothelial system leading to splenomegaly, enlargement of tonsils or lymph nodes, and increased susceptibility to atherosclerosis. ABCA1, like MDR1 and MDR3, functions as an active translocator of lipids between the inner and outer leaflet of the plasma membrane. As part of this function, it has been suggested to be a phosphatidylserine translocase that facilitates phosphatidylserine exofacial flipping. The transient local exposure of anionic phospholipids in the outer-membrane leaflet enhances the engulfment of apoptotic cells, endocytosis, and binding of apolipoprotein A-I.

Macrophages play a key role in the initiation and progression of atherosclerotic lesions. In the nascent lesion, they transform into foam cells through the excessive accumulation of cholesteryl esters. ABCA1 is likely to modulate cellular mobility of monocytes/ macrophages and thus may affect recruitment of monocytes into the vessel wall. The findings that ABCA1 is upregulated in human macrophages during sustained uptake of cholesterol and that it facilitates cholesterol export from the cell render it a likely player in foam cell development and the formation of lesions of atherosclerosis. Current evidence strongly suggest that the observed increased susceptibility to atherosclerosis is largely attributable to the absence of ABCA1 from lesional monocytes/macrophages.

The human ABCA7 transporter is a polypeptide of 2146 amino acids with a similar structure to ABCA1. ABCA7 expression is responsive to cholesterol flux and regulated during differentiation of human monocytes into macrophages. Identical regulatory profiles during monocyte differentiation and cholesterol flux in macrophages and the finding that ABCA7 is the closest relative of ABCA1 suggest a role for this transporter in monocyte/ macrophage lipid homeostasis.

REFERENCES

Aiello, R.J., Brees, D., Bourassa, P.A., Royer, L., Lindsey, S., Coskran, T., et al. (2002). Increased Atherosclerosis in Hyperlipidemic Mice with Inactivation of ABCA1 in Macrophages. Arterioscler. Thromb. Vasc. Biol., 22, 630–637.

Bellincampi, L., Simone, M.L., Motti, C., Cortese, C., Bernardini, S., Bertolini, S., et al. (2001). Identification of an Alternative Transcript of ABCA1 Gene in Different Human Cell Types. Biochem. Biophys. Res. Communi., 283, 590–597.

Bodzioch, M., Orsó, E., Klucken, J., Böttcher, A., Diederich, W., Drobnik, W., et al. (1992). The ATP Binding Cassette Transporter-1 Gene (ABCA1) Is Mutated in Familial HDL Deficiency (Tangier Disease). Nat. Genet., 22, 347–351.

Büchler, C., Böttcher, A., Bared, S.M., Probst, M.C., and Schmitz, G. (2002a). The Carboxyterminus of the ATP-Binding Cassette Transporter A1 Interacts with a β2-Syntrophin/Utrophin Complex. Biochem. Biophys. Res. Commun., 293, 759–765.

Büchler, C., MaaBared, S., Aslanidis, C., Ritter, M., Drobnik, W., and Schmitz, G. (2002b). Molecular and Functional Interaction of the ATP Binding Cassette Transporter A1 with Fas-Associated Death Domain Protein. J. Biol. Chem., 277, 41307–41310.

Chawla, A., Boisvert, W.A., Lee, C.H., Laffitte, B.A., Barak, Y., Joseph, S.B., et al. (2001). A PPAR Gamma-LXR-ABCA1 Pathway in Macrophages is Involved in Cholesterol Efflux and Atherogenesis. Mol. Cell, 7, 161–171.

Connors, T.D., Van Raay, T.J., Petry, L.R., Klinger, K.W., Landes, G.M., and Burn, T.C. (1997). The Cloning of a Human ABC Gene (ABC3) Mapping to Chromosome 16p13.3. Genomics, 39, 231–234.

Dean, M., Hamon, Y., and Chimini, G. (2001). The Human ATP-Binding Cassette (ABC) Transporter Superfamily. J. Lipid Res., 42, 1007–1017.

den Haan, J.M., Meadows, L.M., Wang, W., Pool, J., Blokland, E., Bishop, T.L., et al. (1998). The Minor Histocompatibility Antigen HA-1: A Diallelic Gene with a Single Amino Acid Polymorphism. Science, 279, 1054–1057.

Diederich, W., Orsó, E., Drobnik, W., and Schmitz, G. (2001). Apolipoprotein AI and HDL(3) Inhibit Spreading of Primary Human Monocytes through a Mechanism that Involves Cholesterol Depletion and Regulation of CDC42. Atherosclerosis, 159, 313–324.

Drobnik, W., Borsukova, H., Böttcher, A., Pfeiffer, A., Liebisch, G., Schutz, G.J., et al. (2002). Apo AI/ABCA1-Dependent and HDL3-Mediated Lipid Efflux from Compositionally Distinct Cholesterol-Based Microdomains. Traffic, 3, 268–278.

Gan, X., Kaplan, R., Menke, J.G., MacNaul, K., Chen, Y., Sparrow, C.P., et al. (2001). Dual Mechanisms of ABCA1 Regulation by Geranylgeranyl Pyrophosphate. J. Biol. Chem., 276, 48702–48708.

Goulmy, E., Schipper, R., Pool, J., Blokland, E., Falkenburg, J.H., Vossen, J., et al. (1996). Mismatches of Minor Histocompatibility Antigens between HLA-Identical Donors and Recipients and the Development of Graft-versus-Host Disease after Bone Marrow Transplantation. N. Engl. J. Med., 334, 281–285.

Hansen, J.A., Gooley, T.A, Martin, P.J., Appelbaum, F., Chauncey, T.R., Clift, R.A., et al. (1998). Bone Marrow Transplants from Unrelated Donors for Patients with Chronic Myeloid Leukemia. N. Engl. J. Medi., 338, 962–968.

Kaminski, W.E., Orsó E., Diederich, W., Klucken, J., Drobnik, W., and Schmitz, G. (2000a). Identification of a Novel Human Sterol-Sensitive ATP-Binding Cassette Transporter (ABCA7). Biochem. Biophys. Res Commun., 273, 532–538.

Kaminski, W.E., Piehler, A., and Schmitz, G. (2000b). Genomic Organization of the Human Cholesterol-Responsive ABC Transporter ABCA7: Tandem Linkage with the Minor Histocompatibility Antigen HA-1 Gene. *Biochem. Biophys. Res. Commun.*, **278**, 782–789.

Kaminski, W.E., Piehler, A., Püllmann, K., Porsch-Ozcurumez, M., Duong, C., Maa Bared, G., et al. (2001a). Complete Coding Sequence, Promoter Region, and Genomic Structure of the Human ABCA2 Gene and Evidence for Sterol-Dependent Regulation in Macrophages. *Biochem. Biophys. Res. Commun.*, **281**, 249–258.

Kaminski, W.E., Wenzel, J.J., Piehler, A., Langmann, T., and Schmitz, G. (2001b). ABCA6, a Novel a Subclass ABC Transporter. *Biochem. Biophys. Res. Commun.*, **285**, 1295–1301.

Klucken, J., Büchler, C., Orsó, E., Kaminski, W.E., Porsch-Özcürümez, M., Liebisch, G., et al. (2000). ABCG1 (ABC8), the Human Homolog of the *Drosophila* White Gene, Is a Regulator of Macrophage Cholesterol and Phospholipid Transport. *Proc. Nat Acad. Sci. USA*, **97**, 817–822.

Langmann, T., Klucken, J., Reil, M., Liebisch, G., Luciani, M.-F., Chimini, G., et al. Molecular Cloning of the Human ATP-Binding Cassette Transporter 1 (hABC1): Evidence for Sterol-Dependent Regulation in Macrophages. *Biochem. Biophys. Res. Commun.*, **257**, 29–33.

Langmann, T., Porsch-Ozcurumez, M., Heimerl, S., Probst, M., Moehle, C., Taher, M., et al. (2002). Identification of Sterol-Independent Regulatory Elements in the Human ATP-Binding Cassette Transporter A1 Promoter: Role of Sp1/3, E-Box Binding Factors, and an Oncostatin M-Responsive Element. *J. Biol. Chem.*, **277**, 14443–14450.

Lawn, R.M., Wade, D.P., Garvin, M.R., Wang, X., Schwartz, K., Porter, J.G., et al. (1999). The Tangier Disease Gene Product ABC1 Controls the Cellular Apolipoprotein-Mediated Lipid Removal Pathway. *J. Clin. Invest.*, **104**, R25–31.

Luciani, M.-F., Denizot, F., Savary, S., Mattei, M.G., and Chimini, G. (1994). Cloning of Two Novel ABC Transporters Mapping on Human Chromosome 9. *Genomics*, **21**, 150–159.

Nagase, T., Seki, N., Ishikawa, K., Ohira, M., Kawarabayasi, Y., Ohara, O., et al. (1996). Prediction of the Coding Sequences of Unidentified Human Genes. VI. The Coding Sequences of 80 New Genes (KIAA0201-KIAA0280) Deduced by Analysis of cDNA Clones from Cell Line KG-1 and Brain. *DNA Res.*, **3**, 321–329.

Orsó, E., Broccardo, C., Kaminski, W.E., Böttcher A., Liebisch, G., Drobnik, W., et al. (2000). Transport of Lipids from Golgi to Plasma Membrane is Defective in Tangier Disease Patients and *Abc1* Deficient Mice. *Nat. Genet.*, **24**, 192–196.

Piehler, A., Kaminski, W.E., Wenzel, J.J., Langmann, T., and Schmitz, G. (2002). Molecular Structure of a Novel Cholesterol-Responsive A Subclass ABC Transporter, ABCA9. *Biochem. Biophys. Res. Commun.*, **295**, 408–416.

Porsch-Ozcurumez, M., Langmann, T., Heimerl, S., Borsukova, H., Kaminski, W.E., Drobnik, W., et al. (2001). The Zinc Finger Protein 202 (ZNF202) Is a Transcriptional Repressor of ATP Binding Cassette Transporter A1 (ABCA1) and ABCG1 Gene Expression and a Modulator of Cellular Lipid Efflux. *J. Biol. Chem.*, **276**, 12427–12433.

Raybould, M.C., Birley, A.J., Moss, C., Hulten, M., and McKeown, C.M.E. (1994). Exclusion of an Elastin Gene (ELN) Mutation as the Cause of Pseudoxanthoma Elasticum (PXE) in One Family. *Clin. Genet.*, **45**, 48–51.

Santamarina-Fojo, S., Peterson, K., Knapper, C., Qiu, Y., Freeman, L., Cheng, J.F., et al. (2000). Complete Genomic Sequence of the Human ABCA1 Gene: Analysis of the Human and Mouse ATP-Binding Cassette A Promoter. *Proc Nat Acad. Sci. USA*, 97, 7987–7992.

Schmitz, G. and Kaminski, W.E. (2002). ATP-Binding Cassette (ABC) Transporters in Atherosclerosis. *Curr. Atheroscler. Rep.*, **4**, 243–251.

Schmitz, G., Kaminski, W.E., and Orsó, E. (2000). ABC Transporters in Cellular Lipid Trafficking. *Curr. Opin. Lipidol.*, **11**, 493–501.

Singaraja, R.R., Fievet, C., Castro, G., James, E.R., Hennuyer, N., Clee, S.M., et al. (2002). Increased ABCA1 Activity Protects against Atherosclerosis. *J. Clin. Invest.*, **110**, 35–42.

van Eck, M., Bos, I.S., Kaminski, W.E., Orsó, E., Rothe, G., Twisk, J., et al. (2002). Leukocyte ABCA1 Controls Susceptibility to Atherosclerosis and Macrophage Recruitment into Tissues. *Proc. Natl Acad. Sci. USA*, **99**, 6298–6303.

van Helvoort, A., Smith, A.J., Sprong, H., Fritzsche, I., Schinkel, A.H., Borst, P., et al. (1996). MDR1 P-glycoprotein is a Lipid Translocase of Broad Specificity, while MDR3 P-Glycoprotein Specifically Translocates Phosphatidylcholine. *Cell*, **87**, 507–517.

Yamano, G., Funahashi, H., Kawanami, O., Zhao, L.X., Ban, N., Uchida, Y., et al. (2001). ABCA3 is a Lamellar Body Membrane Protein in Human Lung Alveolar Type II Cells. *FEBS Lett.*, **508**, 221–225.

20

ROBERT S. MOLDAY* AND JINHI AHN*

The role of ABCR (ABCA4) in photoreceptor cells and Stargardt macular degeneration

INTRODUCTION

ATP binding cassette (ABC) transporters are a superfamily of membrane proteins found in virtually all living organisms. They generally function in the active transport of a wide variety of compounds across cell membranes (Higgins 1992; Borst and Elferink 2002). These include amino acids, peptides, ions, metabolites, vitamins, fatty acid derivatives, steroids, organic anions, phospholipids, drugs, and other compounds. ABC transporters typically consist of two membrane domains that provide a pathway for the translocation of a substrate across the membrane and two ATP binding cassettes or nucleotide binding domains that provide the energy for the substrate transport. These functional domains can exist as individual or multidomain subunits that assemble into an oligomeric complex, or alternatively they all can reside on one large single polypeptide chain (see Chapter 17, Figure 1). The detailed mechanism by which ABC transporters actively translocate substrates across membranes is not known, although it is generally thought that the energy derived from the hydrolysis of ATP is coupled to the transport of a substrate across cell membranes via a series of allosteric protein conformational changes.

To date, 49 human genes are known to encode ABC transporters (Dean and Allikmets 2001; Borst and Elferink 2002). Recently, these proteins have been organized into seven subfamilies (ABCA – ABCG), each having one or more members (http://nutrigene.4t.com/humanabc.htm). Compounds that are transported by these proteins are known for only some transporters (Borst and Elferink 2002). Multidrug resistance proteins, which have been most extensively studied, actively extrude toxic compounds and drugs or drug conjugates from cells and in some instances flip phospholipids across cell membranes (Juranka et al. 1989; Hipfner et al. 1999; Borst et al. 2000). The adrenoleukodystrophy protein (ALDP or ABCD1) has been implicated in the transport of very-long-chain saturated fatty acid derivatives across the peroxisomal membrane (Aubourg et al. 1993; Hettema and Tabak 2000), while TAP proteins translocate peptides across the endoplasmic reticulum membrane as part of the antigen-processing reactions (Monaco et al. 1990; See Chapter 21). The cystic fibrosis

---

*Biochemistry and Molecular Biology, Faculty of Medicine, University of British Columbia, 2146 Health Sciences Mall, Vancouver, BC V6T 1Z3, Canada

transmembrane regulator CFTR is somewhat unusual inasmuch as it acts as a chloride channel instead of a transporter (Anderson *et al.* 1991).

Many ABC transporters are associated with human diseases (Gottesman and Ambudkar 2001). Multidrug resistance proteins have been implicated in the resistance of cancer to drug therapy. Genetic defects in genes for many ABC transporters are responsible for inherited human diseases including cystic fibrosis (CFTR or ABCC7), adrenomyeloneuropathy (ALDP or ABCD1), Zellweger syndrome 2 (peroxisomal membrane protein or ABCD3), sitosterolemia (ABCG5 and ABCG8), persistent hyperinsulinemic hypoglycemia (sulfonylurea receptor or ABCC8), Tangier disease (ABCA1) (see Chapter 19), and others.

Stargardt disease is an inherited eye disorder that has been associated with mutations in a photoreceptor-specific ABC transporter (Stargardt 1909; Weleber 1994; Allikmets *et al.* 1997b). It is the most common autosomal recessive juvenile macular dystrophy, with an estimated incidence of 1 in 7,000. Affected individuals experience impaired central vision with a marked decrease in visual acuity in their first or second decade of life. This is associated with degeneration of photoreceptor and retinal pigment epithelial cells in the central region of the retina known as the macula. The retinas of patients with Stargardt disease often display yellow-orange flecks that extend outward from the macula when viewed under an ophthalmoscope. A later onset and slower progressing disease called fundus flavimaculatus exhibits similar clinical features to Stargardt disease (Franceschetti 1965; Klein and Krill 1967). Clinical analysis and more recent genotype–phenotype studies indicate that Stargardt disease and fundus flavimaculatus are genetic variations of the same disease (Weleber 1994; Rozet *et al.* 1999).

The gene mutated in Stargardt disease was first identified by Allikmets *et al.* (1997b) and shown to encode a photoreceptor-specific ABC transporter which they named ABCR. At the same time, the cDNA for the rim protein, an abundant high-molecular-weight photoreceptor membrane protein, first described by Papermaster *et al.* (1978), was cloned and found to encode a novel member of the superfamily of ABC transporters (Azarian and Travis 1997; Illing *et al.* 1997). Comparison of the ABCR and rim protein sequences firmly established that these are one and the same protein (Azarian and Travis 1997; Nasonkin *et al.* 1998).

Since the initial report showing that mutations in ABCR are responsible for Stargardt disease, an international effort has been undertaken to screen large populations of individuals with various types of inherited retinal degenerative diseases for mutations in the *ABCA4* gene (Rozet *et al.* 1998; Lewis *et al.* 1999; Allikmets 2000; Maugeri *et al.* 2000; Papaioannou *et al.* 2000; Rivera *et al.* 2000; Briggs *et al.* 2001; Webster *et al.* 2001). Over 300 different mutations have now been implicated in autosomal recessive Stargardt disease (Allikmets 2000). Selected mutations in the *ABCR* gene are also responsible for most forms of autosomal recessive cone-rod dystrophy (Cremers *et al.* 1998; Maugeri *et al.* 2000) and some cases of autosomal recessive retinitis pigmentosa (Cremers *et al.* 1998; Martinez-Mir *et al.* 1998), two more aggressive retinal dystrophies that affect both central and peripheral vision. It has also been reported that individuals heterozygous for specific Stargardt mutations in ABCR are at increased risk for age-related macular degeneration (AMD), a late-onset disease that causes mild to severe loss in central vision (Allikmets *et al.* 1997a). A genotype–phenotype correlation model has been proposed to explain the spectrum of clinical diseases associated with ABCR mutations (Van Driel *et al.* 1998; Maugeri *et al.* 1999, 2000; Rozet *et al.* 1999). This model suggests that the severity of the disease is correlated with the extent to which mutations in both alleles affect ABCR function. Thus two null alleles result in the most severe phenotype, retinitis pigmentosa, one null and one mutant allele that leads to minimal ABCR activity cause cone–rod dystrophy, missense mutations in one or both alleles, which partially reduce ABCR function, produce Stargardt or fundus flavimaculatus, and one wild-type and one mutated allele, which result in close to normal ABCR activity, can increase the risk that an individual will develop AMD.

Over the past several years, biochemical and immunocytochemical studies have been carried out in an effort to determine the role of ABCR in vision and to understand the pathogenesis of Stargardt diseases. These studies, together with the analysis of mice harboring the disrupted *abcr* gene and genetic and clinical analysis of Stargardt patients, have led to a unifying model for how the loss in the putative transport function of ABCR can produce many of the characteristic features of Stargardt disease. In this chapter, we briefly review the molecular and cellular properties of ABCR and discuss key studies that have led to the current model for the role of ABCR as a retinal transporter in photoreceptor cells. We also discuss the possible mechanism by which mutations in ABCR can lead to Stargardt disease. Additional information can be found in several excellent papers discussing the genetic and mechanistic properties of ABCR (Rozet *et al.* 1999; Allikmets 2000; Sun and Nathans 2001).

STRUCTURAL FEATURES OF ABCR (ABCA4)

ABCR, also known as ABCA4, is a member of the ABCA subclass of ABC transporters (Broccardo *et al.* 1999). Members of this group, now totalling 13, are distinguished from other ABC transporters on the basis of their overall amino acid sequence identity, typically in the range of 30–50%, and the presence of two isolated hydrophobic segments, one close to the N-terminus and another downstream from the first nucleotide binding domain (NBD1). ABCR is most similar to ABCA1, an ABC transporter that mediates the transport of cholesterol and phospholipids across the plasma membrane of cells (see Chapter 19).

Both proteins are 50% identical in sequence and have a similar topological organization (Bungert *et al.* 2001; Fitzgerald *et al.* 2002). Mutations in ABCA1 are responsible for Tangier disease and familial hypoalphalipoproteinemia, autosomal recessive disorders associated with low levels of plasma high-density lipoprotein, and the accumulation of cholesterol and cholesterol esters in macrophage foam cells (Brooks-Wilson *et al.* 1999; Wang *et al.* 2001; Oram 2002). The tissue expression of many other ABCA subfamily members have been examined, but little is known about their structure, function, or involvement in human disease.

Human ABCR is a large membrane glycoprotein consisting of 2273 amino acids (Allikmets *et al.* 1997b; Nasonkin *et al.* 1998). Like P-glycoprotein and CFTR, it consists of two structurally related tandem-arranged halves, each having a nucleotide binding domain (NBD) and a membrane spanning domain (MSD). A topological model for ABCR has been deduced from hydropathy plots and biochemical studies, and is shown in Figure 1 (Illing *et al.* 1997; Bungert *et al.* 2001). A relatively short positively charged N-terminus is predicted to reside on the cytoplasmic side of the membrane. This region is followed by a stretch of predominantly hydrophobic amino acids sufficient to cross the lipid bilayer in an α-helix. A large exocytoplasmic (extracellular or lumen) domain (ECD) separates the first transmembrane segment (H1) from MSD, predicted to contain five additional transmembrane segments (H2–H6). The MSD is followed by the first NBD. This domain organization is repeated in the carboxyl-terminal half of ABCR. ECD1 and ECD2 each contain four N-linked glycosylation sites and numerous conserved cysteine residues predicted to form intrachain disulfide bonds. These structural features are consistent with the exposure of this domain on the lumen side of the disk membrane. Both NBD1 and NBD2 of ABCR consist of approximately 140 amino acids. Within these domains, the characteristic active transport signature sequence is flanked by two nucleotide binding

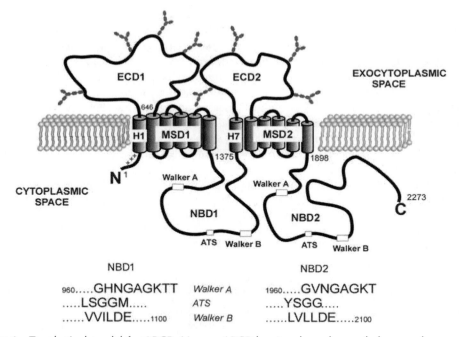

Figure 1 Topological model for ABCR. Human ABCR has two homologous halves, each comprising a large exocytoplasmic domain (ECD), a membrane spanning domain (MSD), and a cytoplasmic nucleotide binding domain (NBD). There are 12 predicted membrane spanning segments; two segments H1 in the N-terminal half and H7 in the C-terminal half are separated from the remaining transmembrane segments by the ECDs. Each ECD has four N-linked oligosaccharide chains, indicated by hexagons. The Walker A, active transporter signature (ATS) and Walker B motifs in the NBD1 and NBD2 are shown. Adapted from Bungert et al. (2001).

consensus sequences known as the Walker A and Walker B motifs. The identity of the transmembrane segments within the MSDs is not clearly resolved. Computer programs predict between five and six membrane spanning segments in each MSD. Further experiments are required to resolve the number and identity of the transmembrane segments within the MSD.

Similarity between the amino acid sequence of ABCR and various members of the ABCA subfamily strongly suggests that these transporters exhibit a similar membrane topology (Bungert et al. 2001). Indeed, a recent mutagenesis study has provided strong experimental evidence that ABCA1 has a similar membrane organization to ABCR (Fitzgerald et al. 2002). This study has also shown that the ECDs of ABCA1 participate in the binding of

apolipoprotein A-I, a protein implicated as an acceptor for cholesterol and phospholipids.

PHOTORECEPTOR CELLS AND THE LOCALIZATION OF ABCR

Rod and cone photoreceptor cells are specialized neurons of the vertebrate retina that function in the capture of light, the conversion of light into electrical signals, and the transmission of these signals to secondary neurons in the retina as the initial steps in vision. Rod cells are highly sensitive to light and function under dim lighting conditions; cone cells are less sensitive and operate under normal light and in color vision. In the retina of most mammals, including humans, rods are the most abundant photoreceptor cell type, comprising over 95% of the photoreceptors. Cone cells are sparsely dispersed in a mosaic pattern throughout the peripheral retina. In humans, a high density of cone cells is found in the fovea, a specialized region of the central retina or macula responsible for high visual acuity.

Rod and cone photoreceptors are highly differentiated and polarized cells (Figure 2). They contain a specialized compartment called the outer segment where photons are captured and converted into electrical signals in the process called phototransduction. The outer segment is attached by a thin connecting cilium to the inner segment, which contains the mitochondria, endoplasmic reticulum, Golgi apparatus, and other common subcellular organelles

involved in metabolic and biosynthetic reactions. The inner segment is joined to the nuclear region. At the opposite end of the photoreceptor lies the synaptic region containing synaptic vesicles filled with the neurotransmitter glutamate.

The outer segment of rod cells consists of a plasma membrane that encloses a stack of over 500 closed disks. Each disk is composed of two flattened membranes circumscribed by a curved rim region interrupted by one or more incisures (Figure 2). Cone outer segments also contain disk-like membranes, but these structures remain continuous with the plasma membrane.

Photoreceptor outer segments are constantly renewed (Bok 1993). New disks are added at the base of the outer segment, while

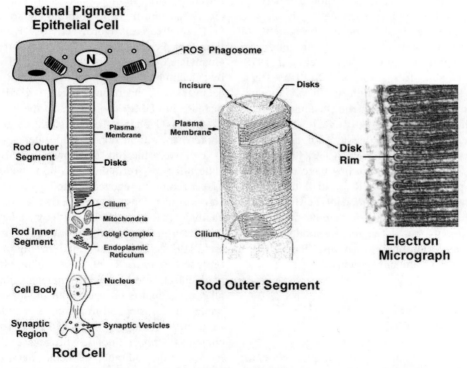

Figure 2 A diagram of a rod photoreceptor cell and retinal pigment epithelial (RPE) cell. Left: a rod photoreceptor situated against the apical side of the retinal pigment epithelial cell. Center: a schematic representation of the specialized outer-segment compartment consisting of a stack of disks enclosed by a separate plasma membrane and a single incisure. Right: an electron micrograph of a frog rod outer segment showing the rim region which harbors the ABCR transporter.

packets of aged disks are shed and ingested by the adjacent retinal pigment epithelium (RPE). A tenth of each outer segment is removed by phagocytosis every 24 h. Between 20 and 30 photoreceptor outer segments are in contact with each RPE cell. Thus RPE cells have a remarkable capacity to ingest and degrade large numbers of photoreceptor outer segment disks on a daily basis. In addition, the RPE cells play other essential roles in photoreceptor cell biology. One involves the recycling of retinal following the bleaching of rhodopsin, a process known as the visual cycle (Saari 2000). RPE cells also provide metabolites, ions, antioxidants, and trophic factors important for the survival of photoreceptors. Finally, RPE also contribute to the interphotoreceptor matrix, an extracellular scaffold, which plays a role in the attachment of the retina to the RPE (Bok 1993).

ABCR is primarily, if not exclusively, expressed in photoreceptor cells. The protein has been localized to the outer segments of human rod and cone cells by immunocytochemical techniques (Papermaster *et al.* 1978; Illing *et al.* 1997; Sun and Nathans 1997; Molday *et al.* 2000). High-resolution immuno-electron microscope labeling techniques have further revealed that ABCR is distributed along the rims and incisures of disk membranes (Papermaster *et al.* 1982; Illing *et al.* 1997). Biochemical studies have confirmed the presence of ABCR in disk membrane preparations (Illing *et al.* 1997). ABCR comprises 1–3% of the disk membrane protein, making it one of the most abundant outer-segment membrane proteins (Papermaster *et al.* 1978; Sun and Nathans 1997).

PHOTOTRANSDUCTION AND THE VISUAL CYCLE

The principal function of the photoreceptor outer segment is to carry out phototransduction. The basic biochemical reactions involved in this process are shown in Figure 3 for rods. Similar reactions occur in cones. Briefly, a photon of light isomerizes the 11-*cis* retinal

chromophore of rhodopsin to its all-*trans* isomer, resulting in a conformational change in rhodopsin and an activation of the visual cascade. Photoactivated rhodopsin or R* catalyzes the exchange of GDP for GTP on the trimeric G-protein, transducin (T$\alpha\beta\gamma$), and a dissociation of the Tα from T$\beta\gamma$. Tα binds to and activates phosphodiesterase which catalyzes the hydrolysis of cGMP to 5′GMP. The decrease in cGMP levels causes the cyclic nucleotide-gated (CNG) channels in the plasma membrane to close, resulting in a reduction in inward current carried by Na$^+$ and Ca^{2+} and a hyperpolarization of the rod cell.

Following photoexcitation, the rod cell returns to its dark state through inactivation of the visual cascade system and resynthesis of cGMP. Rhodopsin is inactivated by a rhodopsin kinase catalyzed phosphorylation reaction followed by the binding of arrestin. Phosphodiesterase returns to its inhibited state when GTP bound to Tα is hydrolyzed to GDP, a reaction that is facilitated by RGS-9 protein. Guanylate cyclase, an enzyme that is activated by calcium binding proteins known as GCAPs, catalyzes the conversion of GTP to cGMP. The increase in cGMP levels leads to the opening of the CNG channel and restoration of the photoreceptor to its depolarized state.

To restore the rod to its preactivated dark state, all-*trans* retinal has to be recycled to 11-*cis* retinal to regenerate rhodopsin. This is carried out through a complex series of biochemical reactions, known as the visual cycle, that occur in the photoreceptor outer segments and RPE cells as illustrated in Figure 3B (Palczewski *et al.* 1999; Saari 2000; McBee *et al.* 2001). Briefly, following photoexcitation, all-*trans* retinal is released from rhodopsin. In a rate-limiting step, all-*trans* retinal is reduced to all-*trans* retinol by the NADPH-dependent enzyme all-*trans* retinol dehydrogenase (RDH) located in the rod outer segments (Saari *et al.* 1998). All-*trans* retinol is then translocated from the rod outer segment across the subretinal space to the RPE cell, where a series of enzyme-catalyzed reactions convert all-*trans* retinol to 11-*cis* retinal. The latter is transported

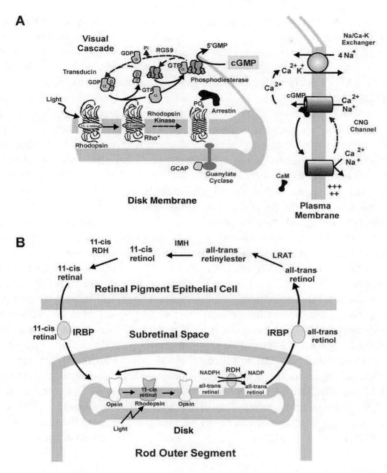

Figure 3 Phototransduction and the visual cycle in rod outer segments and RPE cells. (A) Basic scheme for phototransduction. Light converts 11-*cis*-retinal of rhodopsin to its all-*trans* isomer. The resulting activated form of rhodopsin (Rho*) catalyzes the exchange of bound GDP for GTP on the α-subunit of transducin (Tα). Upon dissociation, Tα binds and activates phosphodiesterase, which catalyzes the hydrolysis of cGMP. A decrease in cGMP levels leads to the closure of the cyclic nucleotide-gated (CNG) channel in the plasma membrane to Ca^{2+} and Na^+ causing a hyperpolarization of the cell. A decrease in Ca^{2+} levels due to continued extrusion of Ca^{2+} by the Na/Ca–K exchanger results in the activation of guanylate cyclase via the guanylate cyclase activating protein GCAP and causes calmodulin (CaM) to dissociate from the channel. The resynthesis of cGMP, together with the inactivation of rhodopsin and the visual cascade, leads to the recovery of the cell to its depolarized state. Adapted from Molday (1998). (B) Basic reactions involved in the visual cycle. Following the photoactivation of rhodopsin, all-*trans* retinal dissociates from rhodopsin. All-*trans* retinol dehydrogenase (RDH) catalyzes the NADPH-dependent reduction of all-*trans* retinal to all-*trans* retinol. This retinoid is transported to the RPE cell, possibly by interaction with the interphotoreceptor retinoid binding protein (IRBP). In the RPE cell, it is first converted to all-*trans* retinyl ester by the enzyme lecithin:retinol acetyltransferase (LRAT) and subsequently isomerized to 11-*cis* retinol by isomerohydrolase (IMH) and finally oxidized to 11-*cis* retinal by 11-*cis* retinal dehydrogenase. The 11-*cis* retinal is then transported back to the outer segment disk where it recombines with opsin to regenerate rhodopsin. The role of ABCR in this process is not shown. Adapted from Saari (2000).

back to the rod outer segment where it combines with opsin to regenerate rhodopsin.

ABCR AS A PUTATIVE RETINAL TRANSPORTER

The finding that ABCR is expressed in photo-receptor outer segments and involved in a pho-toreceptor degenerative disease led to the initial suggestion that this protein may function in the transport of a key substrate across the disk membrane (Allikmets *et al.* 1997b; Illing *et al.* 1997). All-*trans* retinal derived from the photo-bleaching of rhodopsin was considered as a prime candidate despite earlier studies indi-cating that retinal derivatives can diffuse freely across membranes.

To begin to examine the possible function of ABCR as a transporter, an immunoaffinity purification procedure was developed utilizing a monoclonal antibody directed against an epi-tope near the C-terminus of ABCR (Illing *et al.* 1997). Initial studies demonstrated that purified ABCR and ABCR in rod disk mem-branes could be photoaffinity labeled with 8-azido ATP, confirming that at least one of the NBDs binds ATP (Illing *et al.* 1997).

Purified ABCR reconstituted into brain lipid vesicles was used to examine the ATPase activ-ity of ABCR and screen for possible substrate that are transported by ABCR. ATPase assay are straightforward and highly sensitive Furthermore, substrates that are transported by other ABC proteins, such as P-glycoprotein are known to stimulate the ATP hydrolysis (Ambudkar *et al.* 1992; Shapiro and Ling 1994). This reflects a direct coupling between substrate-stimulated ATP hydrolysis and sub strate transport. Sun *et al.* (1999) first showed that reconstituted ABCR has basal ATPase activity that is stimulated three- to fourfold by the addition of either all-*trans* or 11-*cis* retinal The dependence of velocity on ATP concentra-tions obeys Michaelis–Menten kinetics with retinal showing uncompetitive activation kinet-ics as shown in Figure 4 (Sun *et al.* 1999; Ahn *et al.* 2000). ATPase stimulation was specific for retinal derivatives since other retinoids including retinoic acid, retinol, and retinal esters, showed little if any activation. A few unrelated compounds among a large number of chemicals tested also increased the ATPase activity of ABCR, but the kinetic profile and synergistic effect of these compounds with retinal indicated that they bound ABCR a

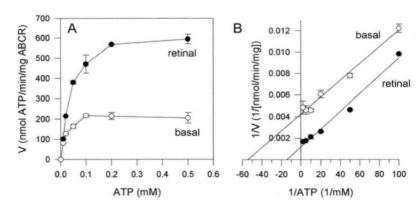

Figure 4   Kinetics of ATP hydrolysis for ABCR reconstituted in ROS lipids. (A) Basal (○) and all-*trans* retinal-stimulated (●) ATPase activities were measured as a function of ATP concentration. (B) The double-reciprocal plot was used to determine K_m and V_{max}. Each data point is the mean ± SD of triplicate values from a typical experiment. Basal $K_m = 19\,\mu M$ and $V_{max} = 230\,nmol/min/mg$; retinal stimulated $K_m = 71\,\mu M$ and $V_{max} = 750\,nmol/min/mg$. Adapted from Ahn *et al.* (2000).

a different site than retinal and stimulated the ATPase activity by a different mechanism (Sun *et al.* 1999; Sun and Nathans 2001).

The basal and retinal-stimulated ATPase activity is also dependent on the lipid composition of the vesicles (Ahn *et al.* 2000). The highest basal and retinal-stimulated ATPase activity is present in ABCR preparations reconstituted into rod outer-segment lipids that have a high (>40%) content of phosphatidylethanolamine (PE), a result that could be reproduced by the addition of exogenous PE to brain lipids. The importance of PE was further demonstrated by the finding that ABCR reconstituted in vesicles lacking PE is devoid of ATPase activity (Sun *et al.* 1999; Ahn *et al.* 2000).

Although the ATPase activity of ABCR is stimulated by the addition of retinal, it is not known whether retinal or retinyl–PE mediates this effect. Early studies have shown that all-*trans* retinal released after the photobleaching of rhodopsin reacts with PE in the disk membrane to form the Schiff base adduct retinyl–PE (Poincelot *et al.* 1969). This reaction also occurs upon the addition of all-*trans* retinal to lipid vesicles containing PE (Ahn *et al.* 2000). Approximately 50% of the retinal exists as free retinal and 50% as a retinyl–PE adduct. Thus it remains to be directly determined whether free retinal or retinyl–PE activates the ATPase activity and therefore is the possible substrate which is transported by ABCR.

The role of ABCR in photoreceptor biology was further examined by Weng *et al.* (1999) using an ABCR knockout mouse. Following light exposure, the retinas of the ABCR −/− mice show elevated levels of all-*trans* retinal, PE, and protonated retinyl–PE. These results, together with the observation that ABCR −/− mice exhibit delayed dark adaptation presumably due to the presence of all-*trans* retinal–opsin complexes, support the role of ABCR in facilitating the removal of all-*trans* retinal following the photobleaching of rhodopsin. Their results further suggest that ABCR may function to actively translocate retinyl–PE from the lumen to the cytoplasmic side of the disk membrane (Weng *et al.* 1999).

Biochemical and cell biology studies have led to a conceptual model for the role of ABCR as a retinoid transporter in the visual cycle as shown in Figure 5. Following the photobleaching of rhodopsin, all-*trans* retinal dissociates from opsin. In the disk membrane all-*trans* retinal reacts with PE to form a mixture of all-*trans* retinyl–PE and free all-*trans* retinal. A fraction of the all-*trans* retinal most likely diffuses to the cytoplasmic side of the disk membrane where it is directly reduced to all-*trans* retinol by RDH. All-*trans* N-retinyl–PE (retinyl–PE) trapped on the lumen side of the disk membrane is inaccessible to RDH. ABCR is envisioned to translocate or flip retinyl–PE from the lumen to the cytoplasmic side of the disk membrane, thereby facilitating the reduction of all-*trans* retinal to retinol by RDH. Thus ABCR insures that all of the retinal produced from the photobleaching of rhodopsin is made accessible to RDH for reduction to retinol. ABCR in cone photoreceptors is believed to play a similar role in the visual cycle (Molday *et al.* 2000)

Demonstration of ATP-dependent transport of retinal or translocation of retinyl-PE across the disc membrane is crucial to rigorously prove that ABCR functions as an active retinoid transporter. Such assays are difficult to develop particularly when the putative substrate is a hydrophobic compound that has a tendency to interact nonspecifically with membranes. At the present time direct transport assays for ABCR have yet to be developed.

MODEL FOR THE ROLE OF ABCR DYSFUNCTION IN STARGARDT DISEASE

Over 300 missense and in-frame deletions in ABCR have been linked to Stargardt disease (Allikmets 2000). These mutations, which are distributed throughout the protein, are predicted to diminish the ability of ABCR to actively transport all-*trans* retinal or retinyl–PE to the cytoplasmic surface of the disk membrane owing to partial misfolding of the protein and/or inefficient protein

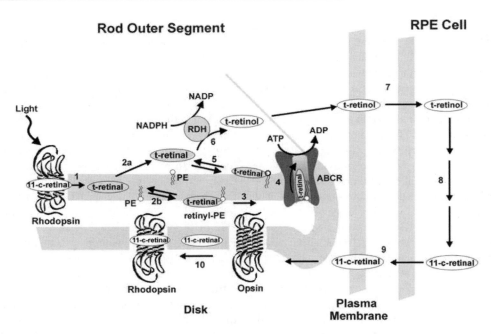

Figure 5 Role of ABCR as a retinoid transporter. All-*trans* retinal (t-retinal) generated from the photobleaching of rhodopsin (1) can diffuse through the disk lipid bilayer (2a) where it is directly reduced to all-*trans* retinol (t-retinol) by RDH on the cytoplasmic side of the disk membrane (6). A fraction of all-*trans* retinal, possibly released into the disk lumen, reacts with PE to form retinyl–PE (2b). This adduct diffuses to the rim region of the disk (3) where it is actively translocated to the cytoplasmic surface of the disk membrane by ABCR (4). N-retinyl–PE dissociates (5), allowing all-*trans* retinal to be reduced to all-*trans* retinol by RDH (6). After translocation to the RPE cell (7), all-*trans* retinol is converted to 11-*cis* retinal (11-c-retinal) in a series of reactions (8) (also see Figure 3B). After transport back to the outer segment (9), 11-*cis* retinal recombines with opsin to regenerate rhodopsin (10). ABCR insures that all-*trans* retinal does not accumulate in disk membranes.

targeting to outer-segment disk membranes. The diminished transport function of ABCR will result in a build-up of all-*trans* retinal and retinyl–PE in the disk membranes following the photobleaching of rhodopsin. The accumulation of retinal and ensuing side reactions can explain many characteristic features found in individuals with Stargardt disease. These include delayed dark adaptation of rod cells, the presence of orange-yellow fluorescent deposits, termed lipofuscin, in RPE cells, and the degeneration of photoreceptor cells and RPE cells. A generalized model of the involvement of ABCR in Stargardt disease is illustrated in Figure 6A.

A delay in dark adaptation can arise from the accumulation of all-*trans* retinal in the disk membrane following photoexcitation of rhodopsin. Several studies have shown that all-*trans* retinal can reassociate with opsin to form a complex that activates the visual cascade, although less efficiently than the photoactivated form of rhodopsin (Buczylko *et al.* 1996; Surya and Knox 1998). This low level of activity can contribute to background noise and a delay in the recovery of rod photoreceptors to their dark-adapted state.

Lipofuscin consists of autofluorescent pigments that accumulate in the lysosomal compartment of aged retinal pigment epithelial

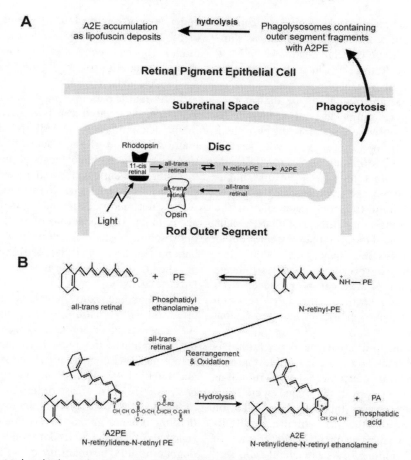

Figure 6 Molecular basis for Stargardt disease. (A) Sequence of events leading to a delay in dark adaptation in rod cells and accumulation of A2E-containing lipofuscin in retinal pigment epithelial cells. All-*trans* retinal released after the photobleaching of rhodopsin reacts with PE to form an equilibrium mixture of free all-*trans* retinal and *N*-retinyl–PE. Decreased or absent ABCR retinoid transport activity will enable these compounds to accumulate in disks. All-*trans* retinal can recombine with opsin to produce an all-*trans* retinal–opsin complex that can activate the visual cascade resulting in background noise and a delay in the recovery of the rod to its dark-adapted state. *N*-retinyl–PE can react with another molecule of all-*trans* retinal to generate the diretinal compound A2PE. Upon phagocytosis of outer segments, A2PE is converted to A2E in the phagolysosomal compartment of the retinal pigment epithelial cell. Since A2E is not degraded, it accumulates as lipofuscin deposits in individuals with Stargardt disease. (B) Key chemical reactions responsible for the formation of A2E from all-*trans* retinal and PE. Modified from Sakai *et al.* (1996) and Mata *et al.* (2000).

cells as a by-product of photoreceptor outer-segment phagocytosis (Feeney-Burns *et al.* 1984; Dorey *et al.* 1989; Kennedy *et al.* 1995). Individuals with Stargardt disease and some forms of age-related macular degeneration have abnormally high levels of lipofuscin that appear as yellow-orange deposits in the central region of the retina at the level of the RPE cells (Delori *et al.* 1995, 2000). A major component of lipofuscin is the fluorescent compound A2E, a pyridinium bis-retinoid (Eldered and Lasky 1993; Sakai *et al.* 1996; Parish *et al.*

1998). This retinoid derivative is formed through a series of chemical reactions as summarized in Figure 6B (Sakai *et al.* 1996; Mata *et al.* 2000). Retinyl–PE in disk membranes condenses with a second molecule of all-*trans* retinal. Rearrangement and oxidation produces a pyridinium derivative known as A2PE. In a subsequent reaction, the phosphate ester group of the PE moiety is hydrolyzed to form A2E. The formation of *N*-retinyl–PE and A2PE takes place mainly in the disk membranes, whereas the hydrolysis of A2PE to A2E is thought to occur primarily in the phagolysosomal compartment of the RPE cells after ingestion of the shed photoreceptor outer-segment disks (Ben-Shabat *et al.* 2000; Mata *et al.* 2000). However, recent studies indicate that a portion of A2E can be produced from A2PE in outer segments through a reaction catalyzed by phosphodiesterase (Ben-Shabat *et al.* 2000).

In normal individuals all-*trans* retinal is efficiently transferred to the cytoplasmic surface where it is reduced to all-*trans* retinol as a rate-limiting step in the visual cycle. This most likely occurs through a combination of diffusion of all-*trans* retinal and ABCR-mediated transport of all-*trans* retinal or retinyl–PE (Figure 5). As a result all-*trans* retinal and retinyl–PE derivatives do not accumulate in disk membranes, and side reactions that lead to the production of A2E do not occur under normal circumstances. Small amounts of all-*trans* retinal may evade reduction to retinol. This can account for the slow progressive accumulation of A2E found in RPE of aged individuals.

In Stargardt patients, mutations in ABCR that decrease retinoid transport activity result in the accumulation of retinyl–PE in disk membranes and the production of A2PE (Figure 6). Upon phagocytosis of photoreceptor outer segments, A2PE undergoes hydrolysis in the phagolysosomes to form A2E . Since A2E is not readily degraded by RPE cells, it accumulates over a period of time, forming lipofuscin deposits characteristic of Stargardt patients (Delori *et al.* 1995; see Box Summary of the Clinical Appearance of Stargadt Disease).

A2E has been shown to have a negative effect on RPE function. Studies suggest that A2E can act as a detergent, photosensitizer for the generation of free radicals and an inhibitor of normal RPE degradative functions (Eldred and Lasky 1993; Holz *et al.* 1999; Sparrow *et al.* 2000). The decrease or loss in the ability of RPE cells to carry out phagocytosis of disk membranes as part of the outer-segment renewal process can lead to photoreceptor degeneration. The inability of RPE cells to provide factors required for photoreceptor survival may also contribute to photoreceptor cell death and the loss in vision.

SUMMARY OF THE CLINICAL APPEARANCE OF STARGADT DISEASE

Stargardt disease is the most common hereditary juvenile macular degeneration, affecting as many as 1 in 7,000 individuals in the developed world. It is characterized by a significant loss in central visual acuity in the first or second decade of life, the appearance of yellow-orange flecks in the central retina or macula at the level of the retinal pigment epithelium, and progressive bilateral atrophy of rod and cone photoreceptors and underlying retinal pigment epithelium in and around the fovea. Fluorescein angiography often shows marked hypofluorescence of the choroid, a condition referred to as a dark choroid. Electrophysiological recordings reveal a significant decrease in the dark adaptation of rod photoreceptors, and electroretinograms exhibit variable loss in signal amplitudes arising from rod and/or cone photoreceptor photoactivation. Fundus flavimaculatus is a later-onset genetic variant of Stargardt disease.

BIOCHEMICAL ANALYSIS OF ABCR DISEASE MUTANTS AND *ABCR* KNOCKOUT MICE

Heterologous cell expression has been used to examine the effect of mutations causing Stargardt disease on the expression levels and properties of ABCR (Sun *et al.* 2000). Deletions and missense mutations in the putative transmembrane segments of ABCR were found to express poorly, presumably due to misfolding and degradation in the endoplasmic reticulum. Mutants that expressed at levels comparable to wild-type ABCR were examined for ATP binding and hydrolysis (Sun *et al.* 2000). A number of these disease-causing mutants exhibited basal ATPase activity comparable to wild-type ABCR, but showed a significant decrease in retinal-stimulated activity. A comprehensive study of 34 mutants has been reported by Sun *et al.* (2000). Similar results have been obtained in our laboratory as shown in Figure 7 for several disease-causing mutations. At the present time, it is not known whether one or both NBDs actively hydrolyze ATP, but analysis of mutations in the NBDs has led to a model in which ATP hydrolysis in the absence of retinal occurs at NBD-1, whereas retinal-stimulated ATPase activity coupled to the transport of retinal derivatives across the membrane requires ATPase hydrolysis at both NBDs (Sun *et al.* 2000; Sun and Nathans 2001). By analogy with other well-studied ABC transporters, substrate transport most likely occurs through allosteric coupling between individual NBDs and between NBDs and MSDs of ABCR.

Mice harboring the disrupted *abcr* gene have proven to be a valuable model for understanding the biochemical basis for Stargardt disease (Weng *et al.* 1999; Mata *et al.* 2001). Abcr −/− mice exhibit delayed rod dark adaptation similar to that observed in Stargardt patients. Moreover, after light exposure abcr −/− mice show an elevated level of all-*trans* retinal, PE, and retinyl–PE consistent with the inability of all-*trans* retinal to be efficiently

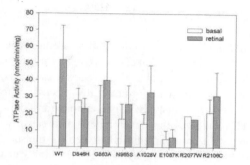

Figure 7 ATPase activity of ABCRs with mutations linked to Stargardt disease. ABCR was immunopurified from COS-1 cells transfected with the ABCR variants indicated on the *x*-axis and reconstituted in phospholipid. ATPase activity was determined in the absence or presence of 50 μM all-*trans* retinal. Error bars were calculated from three or more experiments. Values without error bars are averages of two experiments. While the ATPase activity of the D846H and R2077W variants was not stimulated by all-*trans* retinal, the E1087K variant reduced both basal and retinal-stimulated activity.

reduced to all-*trans* retinol. These mice also accumulate A2E-containing lipofuscin deposits in RPE cells in a light-dependent manner (Weng *et al.* 1999). These results provide important evidence for the role of ABCR in retinoid processing in photoreceptors and suggest that ABCR functions in the translocation of retinyl–PE derivatives across the disk membrane. Photoreceptors from 6-week-old abcr −/− mice appeared normal, indicating that ABCR, unlike other disk rim proteins, does not play a role in the morphogenesis or structure of photoreceptor outer segments. However, older mice exhibit mild photoreceptor cell degeneration. These studies are consistent with a model in which the absence of ABCR does not affect phototransduction, photoreceptor structure, or short-term survival. Instead, the gradual accumulation of A2E compromises the ability of RPE cells to support photoreceptors leading to slow progressive photoreceptor degeneration and the loss in vision.

UNRESOLVED ISSUES

Although studies to date lead to a conceptual model for how defects in ABCR can cause Stargardt disease, several key issues remain to be resolved. Most importantly, it is necessary to show directly that ABCR is a retinoid transporter. This involves the development of a sensitive assay that can distinguish between the transport of free retinal and retinyl–PE. The possibility that ABCR can flip PE across disk membranes also needs to be addressed. Another issue that requires clarification is the directionality of transport. The model is most consistent with the transport of retinal or retinyl–PE from the exocytoplasmic or lumen side to the cytoplasmic surface of disk membranes. However, most eukaryotic ABC transporters transport substrates in the opposite direction, namely from the cytoplasmic side of the plasma membrane to the extracellular side. Likewise, cellular studies of ABCA1 suggest that this transporter also translocates phospholipids to the extracellular surface. If ABCR transports its substrate from the lumen to the cytoplasmic side of the disk membrane then, it will be necessary to understand how two related transporters ABCR and ABCAI move substrates in opposite directions. Detailed analysis of the mechanism by which ATP hydrolysis at one or both NBDs is coupled to transport would also provide important information on how mutations in ABCR can lead to partial or complete loss in transport activity. The effect of mutations on the trafficking of ABCR to outer-segment disk membranes also needs to be examined and analyzed in terms of how ABCR variants may cause Stargardt disease. Finally, it will be important to determine how the activity of ABCR is regulated, and more specifically, whether ABCR interacts with other proteins in disk membranes. Although significant progress has been made over the last 5 years, there is still much to be learned about ABCR and its relation to other ABC transporters.

SUMMARY

ABCR is a photoreceptor-specific ABC transporter that has been linked to Stargardt disease, a common juvenile macular dystrophy characterized by a loss in central vision, delayed dark adaptation, the presence of yellow flecks in the macular region of the retina, and degeneration of the central photoreceptors and underlying retinal pigment epithelial cells. The 2273 amino acid protein is organized in two tandem-arranged halves, each containing a large exocytoplasmic domain, a multispanning membrane domain, and a nucleotide binding domain with an active transport signature and Walker A and B motifs characteristic of ABC transporters. Biochemical studies of purified and reconstituted ABCR and analysis of ABCR knockout mice, together with genetic and clinical analyses of Stargardt patients, have led to a general model for the role of ABCR in photoreceptor cell function and a mechanism for the pathogenesis of Stargardt disease. In this model ABCR functions to actively transport retinal derivatives across the photoreceptor disk membrane following the photobleaching of rhodopsin in rod cells and cone opsin in cone cells. Although the exact mechanism and identity of the substrate remain to be determined experimentally, there is convincing evidence that ABCR functions as a retinoid transporter, possibly flipping retinyl–PE from the lumen side of the disk membrane to the cytoplasmic side, where all-*trans* retinal can be reduced to all-*trans* retinol by retinol dehydrogenase. This insures that all the retinal is accessible to retinol dehydrogenase for reduction to retinol as a key reaction in the visual cycle. A decrease or absence of ABCR transport activity due to disease-linked mutations results in an accumulation of retinal and retinyl–phosphatidylethanolamine in disk membranes, causing delayed dark adaptation and the production of diretinal compounds. The latter is converted into a toxic diretinal pyridinium ethanolamine derivative, known as A2E, in retinal pigment epithelial cells as a by-product of outer segment phagocytosis. A2E, which accumulates as fluorescent lipofuscin deposits, compromises the function and viability of retinal pigment epithelial cells and photoreceptors, resulting in a marked loss of vision.

REFERENCES

Ahn, J., Wong, J.T., and Molday, R.S. (2000). The Effect of Lipid Environment and Retinoids on the ATPase Activity of ABCR, the Photoreceptor ABC Transporter Responsible for Stargardt Macular Dystrophy. *J. Biol. Chem.*, **275**, 20399–20405.

Allikmets, R. (2000). Simple and Complex ABCR: Genetic Predisposition to Retinal Disease. *Am. J. Hum. Genet.*, **67**, 793–799.

Allikmets, R., Shroyer, N.F., Singh, N., Seddon, J.M., Lewis, R.A., Bernstein, P.S., *et al.* (1997a). Mutation of the Stargardt Disease Gene (ABCR) in Age-Related Macular Degeneration. *Science*, **277**, 1805–1807.

Allikmets, R., Singh, N., Sun, H., Shroyer, N.F., Hutchinson, A., Chidambaram, A., *et al.* (1997b). A Photoreceptor Cell-Specific ATP-Binding Transporter Gene (ABCR) Is Mutated in Recessive Stargardt Macular Dystrophy. (See comments) *Nat. Genet.*, **15**, 236–246.

Ambudkar, S.V., Lelong, I.H., Zhang, J., Cardarelli, C.O., Gottesman, M.M., and Pastan, I. (1992). Partial Purification and Reconstitution of the Human Multidrug- Resistance Pump: Characterization of the Drug-Stimulatable ATP Hydrolysis. *Proc. Natl Acad. Sci. USA*, **89**, 8472–8476.

Anderson, M.P., Gregory, R.J., Thompson, S., Souza, D.W., Paul, S., Mulligan, R.C., *et al.* (1991). Demonstration that CFTR Is a Chloride Channel by Alteration of its Anion Selectivity. *Science*, **253**, 202–205.

Aubourg, P., Mosser, J., Douar, A.M., Sarde, C.O., Lopez, J., and Mandel, J.L. (1993). Adrenoleuko-dystrophy Gene: Uexpected Homology to a Protein Involved in Peroxisome Biogenesis. *Biochimie*, **75**, 293–302.

Azarian, S.M. and Travis, G.H. (1997). The Photoreceptor Rim Protein is an ABC Transporter Encoded by the Gene for Recessive Stargardt's Disease (ABCR). *FEBS Lett.*, **409**, 247–252.

Ben-Shabat, S., Parish, C.A., Vollmer, H.R., Itagaki, Y., Fishkin, N., Nakanishi, K., and Sparrow, J.R. (2002). Biosynthetic Studies of A2E, a Major Fluorophore of Retinal Pigment Epithelial Lipofuscin. *J. Biol. Chem.*, **277**, 7183–7190.

Bok, D. (1993). The Retinal Pigment epithelium: A Versatile Partner in Vision. *J. Cell. Sci. Suppl.*, **17**, 189–195.

Borst, P. and Elferink, R.O. (2002). Mammalian ABC Transporters in Health and Disease. *Annu. Rev. Biochem.*, **71**, 537–592.

Borst, P., Zelcer, N., and van Helvoort, A. (2000). ABC Transporters in Lipid Transport. *Biochim. Biophys. Acta*, **1486**, 128–144.

Briggs, C.E., Rucinski, D., Rosenfeld, P.J., Hirose, T., Berson, E.L., and Dryja, T.P. (2001). Mutations in ABCR (ABCA4) in Patients with Stargardt Macular Degeneration or Cone–Rod Degeneration. *Invest. Ophthalmol. Vis. Sci.*, **42**, 2229–2236.

Broccardo, C., Luciani, M., and Chimini, G. (1999). The ABCA Subclass of Mammalian Transporters. *Biochim. Biophys. Acta*, **1461**, 395–404.

Brooks-Wilson, A., Marcil, M., Clee, S.M., Zhang, L.H., Roomp, K., van Dam, M., *et al.* (1999). Mutations in ABC1 in Tangier Disease and Familial High-Density Lipoprotein Deficiency. *Nat. Genet.*, **22**, 336–345.

Buczylko, J., Saari, J.C., Crouch, R.K., and Palczewski, K. (1996). Mechanisms of Opsin Activation. *J. Biol. Chem.*, **271**, 20621–20630.

Bungert, S., Molday, L.L., and Molday, R.S. (2001). Membrane Topology of the ATP Binding Cassette Transporter ABCR and its Relationship to ABC1 and Related ABCA Transporters: Identification of N-Linked Glycosylation Sites. *J. Biol. Chem.*, **276**, 23539–23546.

Cremers, F.P., van de Pol, D.J., van Driel, M., den Hollander, A.I., van Haren, F.J., Knoers, N.V., *et al.* (1998). Autosomal Recessive Retinitis Pigmentosa and Cone–Rod Dystrophy Caused by Splice Site Mutations in the Stargardt's Disease Gene ABCR. *Hum. Mol. Genet.*, **7**, 355–362.

Dean, M. and Allikmets, R. (2001). Complete Characterization of the Human ABC Gene Family. *J. Bioenerg. Biomembr.*, **33**, 475–479.

Delori, F.C., Staurenghi, G., Arend, O., Dorey, C.K., Goger, D.G., and Weiter, J.J. (1995). In Vivo Measurement of Lipofuscin in Stargardt's Disease – Fundus Flavimaculatus. *Invest. Ophthalmol. Vis. Sci.*, **36**, 2327–2331.

Delori, F.C., Fleckner, M.R., Goger, D.G., Weiter, J.J., and Dorey, C.K. (2000). Autofluorescence Distribution Associated with Drusen in Age-Related Macular Degeneration. *Invest. Ophthalmol. Vis. Sci.*, **41**, 496–504.

Dorey, C.K., Wu, G., Ebenstein, D., Garsd, A., and Weiter, J.J. (1989). Cell Loss in the Aging Retina. Relationship to Lipofuscin Accumulation and Macular Degeneration. *Invest. Ophthalmol. Vis. Sci.*, **30**, 1691–1699.

Eldred, G.E. and Lasky, M.R. (1993). Retinal Age Pigments Generated by Self-Assembling Lysosomotropic Detergents. *Nature*, **361**, 724–726.

Feeney-Burns, L., Hilderbrand, E.S., and Eldridge, S. (1984). Aging Human RPE: Morphometric Analysis of Macular, Equatorial, and Peripheral Cells. *Invest. Ophthalmol. Vis. Sci.*, **25**, 195–200.

Fitzgerald, M.L., Morris, A.L., Rhee, J.S., Andersson, L.P., Mendez, A.J., and Freeman, M.W. (2002). Naturally Occurring Mutations in the Largest Extracellular Loops of ABCA1 Can Disrupt its Direct Interaction with Apolipoprotein A-I. *J. Biol. Chem.*, **277**, 33178–33187.

Franceschetti, A. and Francois, J. (1965). Fundus Flavimaculatus. *Arch. Ophthalmol.*, **25**, 505–530.

Gottesman, M.M. and Ambudkar, S.V. (2001). Overview: ABC Transporters and Human Disease. *J. Bioenerg. Biomembr.*, **33**, 453–458.

Hettema, E.H. and Tabak, H.F. (2000). Transport of Fatty Acids and Metabolites Across the Peroxisomal Membrane. *Biochim. Biophys. Acta*, **1486**, 18–27.

Higgins, C.F. (1992). ABC Transporters, From Microorganisms to Man. *Annu. Rev. Cell. Biol.*, **8**, 67–113.

Hipfner, D.R., Deeley, R.G., and Cole, S.P. (1999). Structural, Mechanistic, and Clinical Aspects of MRP1. *Biochim. Biophys. Acta*, **1461**, 359–376.

Holz, F.G., Schutt, F., Kopitz, J., Eldred, G.E., Kruse, F.E., Volcker, H.E., *et al.* (1999). Inhibition of Lysosomal Degradative Functions in RPE Cells by a Retinoid Component of Lipofuscin. *Invest. Ophthalmol. Vis. Sci.*, **40**, 737–743.

Illing, M., Molday, L.L., and Molday, R.S. (1997). The 220-kDa Rim Protein of Retinal Rod Outer Segments Is a Member of the ABC Transporter Superfamily. *J. Biol. Chem.*, **272**, 10303–10310.

Juranka, P.F., Zastawny, R.L., and Ling, V. (1989). P-glycoprotein: Multidrug-Resistance and a Superfamily of Membrane-Associated Transport Proteins. *FASEB J.*, **3**, 2583–2592.

Kennedy, C.J., Rakoczy, P.E., and Constable, I.J. (1995). Lipofuscin of the Retinal Pigment Epithelium: A Review. *Eye*, **9**, 763–771.

Klein, B.A. and Krill, A.E. (1967). Fundus Flavimaculatus: Clinical, Functional and Histopathologic Observations. *Am. J. Ophthalmol.*, **64**, 3–23.

Lewis, R.A., Shroyer, N.F., Singh, N., Allikmets, R., Hutchinson, A., Li, Y., *et al.* (1999). Genotype/Phenotype Analysis of a Photoreceptor-Specific ATP-Binding Cassette Transporter Gene, ABCR, in Stargardt Disease. *Am. J. Hum. Genet.*, **64**, 422–434.

Martinez-Mir, A., Paloma, E., Allikmets, R., Ayuso, C., del Rio, T., Dean, M., *et al.* (1998). Retinitis Pigmentosa Caused by a Homozygous Mutation in the Stargardt Disease Gene ABCR. *Nat. Genet.*, **18**, 11–12.

Mata, N.L., Weng, J., and Travis, G.H. (2000). Biosynthesis of a Major Lipofuscin Fluorophore in Mice and Humans with ABCR-Mediated Retinal and Macular Degeneration. *Proc. Natl Acad. Sci. USA*, **97**, 7154–7159.

Mata, N.L., Tzekov, R.T., Liu, X., Weng, J., Birch, D.G., and Travis, G.H. (2001). Delayed Dark-Adaptation and Lipofuscin Accumulation in abcr+/− Mice: Implications for Involvement of ABCR in Age-Related Macular Degeneration. *Invest. Ophthalmol. Vis. Sci.*, **42**, 1685–1690.

Maugeri, A., van Driel, M.A., van de Pol, D.J., Klevering, B.J., van Haren, F.J., Tijmes, N., *et al.* (1999). The 2588G>C Mutation in the ABCR Gene is a Mild Frequent Founder Mutation in the Western European Population and Allows the Classification of ABCR Mutations in Patients with Stargardt Disease. *Am. J. Hum. Genet.*, **64**, 1024–1035.

Maugeri, A., Klevering, B.J., Rohrschneider, K., Blankenagel, A., Brunner, H.G., Deutman, A.F., *et al.* (2000). Mutations in the ABCA4 (ABCR) Gene Are the Major Cause of Autosomal Recessive Cone–Rod Dystrophy. *Am. J. Hum. Genet.*, **67**, 960–966.

McBee, J.K., Palczewski, K., Baehr, W., and Pepperberg, D.R. (2001). Confronting Complexity: The Interlink of Phototransduction and Retinoid Metabolism in the Vertebrate Retina. *Prog. Retin. Eye. Res.*, **20**, 469–529.

Molday, R.S. (1998). Photoreceptor Membrane Proteins, Phototransduction, and Retinal Degenerative Diseases. The Friedenwald Lecture. *Invest. Ophthalmol. Vis. Sci.*, **39**, 2491–2513.

Molday, L.L., Rabin, A.R., and Molday, R.S. (2000). ABCR expression in Foveal Cone Photoreceptors and its Role in Stargardt Macular Dystrophy. *Nat. Genet.*, **25**, 257–258.

Monaco, J.J., Cho, S., and Attaya, M. (1990). Transport Protein Genes in the Murine MHC: Possible Implications for Antigen Processing. *Science*, **250**, 1723–1726

Nasonkin, I., Illing, M., Koehler, M.R., Schmid, M., Molday, R.S., and Weber, B.H. (1998). Mapping of the Rod Photoreceptor ABC Transporter (ABCR) to 1p21-p22.1 and Identification of Novel Mutations in Stargardt's Disease. *Hum. Genet.*, **102**, 21–26.

Oram, J.F. (2002). ATP-Binding Cassette Transporter A1 and Cholesterol Trafficking. *Curr. Opin. Lipidol.*, **13**, 373–381.

Palczewski, K., Van Hooser, J.P., Garwin, G.G., Chen, J., Liou, G.I., and Saari, J.C. (1999). Kinetics of Visual Pigment Regeneration in Excised Mouse Eyes and in Mice with a Targeted Disruption of the Gene Encoding Interphotoreceptor Retinoid-Binding Protein or Arrestin. *Biochemistry*, **38**, 12012–12019.

Papaioannou, M., Ocaka, L., Bessant, D., Lois, N., Bird, A., Payne, A., *et al.* (2000). An Analysis of ABCR Mutations in British Patients with Recessive Retinal Dystrophies. *Invest. Ophthalmol. Vis. Sci.*, **41**, 16–19.

Papermaster, D.S., Schneider, B.G., Zorn, M.A., and Kraehenbuhl, J.P. (1978). Immunocytochemical Localization of a Large Intrinsic Membrane Protein to the Incisures and Margins of Frog Rod Outer Segment Disks. *J .Cell Biol.*, **78**, 415–425.

Papermaster, D.S., Reilly, P., and Schneider, B.G. (1982). Cone Lamellae and Red and Green Rod Outer Segment Disks Contain a Large Intrinsic Membrane Protein on their Margins: An Ultrastructural Immunocytochemical Study of Frog Retinas. *Vision Res.*, **22**, 1417–1428.

Parish, C.A., Hashimoto, M., Nakanishi, K., Dillon, J., and Sparrow, J. (1998). Isolation and One-Step Preparation of A2E and Iso-A2E, Fluorophores from Human Retinal Pigment Epithelium. *Proc. Natl Acad. Sci. USA*, **95**, 14609–14613.

Poincelot, R.P., Millar, P.G., Kimbel, R.L., and Abrahamson, E.W. (1969). Lipid to Protein Chromophore Transfer in the Photolysis of Visual Pigments. *Nature*, **221**, 256–257.

Rivera, A., White, K., Stohr, H., Steiner, K., Hemmrich, N., Grimm, T., *et al.* (2000). A Comprehensive Survey of Sequence Variation in the *ABCA4* (ABCR) Gene in

Stargardt Disease and Age-Related Macular Degeneration. *Am. J. Hum. Genet.*, **67**, 800–813.

Rozet, J.M., Gerber, S., Souied, E., Perrault, I., Chatelin, S., Ghazi, I., *et al.* (1998). Spectrum of ABCR Gene Mutations in Autosomal Recessive Macular Dystrophies. *Eur. J. Hum. Genet.*, **6**, 291–295.

Rozet, J.M., Gerber, S., Souied, E., Ducroq, D., Perrault, I., Ghazi, I., *et al.* (1999). The ABCR Gene: A Major Disease Gene in Macular and Peripheral Retinal Degenerations with Onset from Early Childhood to the Elderly. *Mol. Genet. Metab.*, **68**, 310–315.

Saari, J.C. (2000)Biochemistry of Visual Pigment Regeneration: The Friedenwald Lecture. *Invest. Ophthalmol. Vis. Sci.*, **41**, 337–348.

Saari, J.C., Garwin, G.G., van Hooser, J.P., and Palczewski, K. (1998). Reduction of all-*trans* Retinal Limits Regeneration of Visual Pigment in Mice. *Vision Res.*, **38**, 1325–1333.

Sakai, N., Decatur, J., Nakanishi, K., and Eldred, G.E. (1996). Ocular Age Pigment "A2-E": An Unprecented Pyridinium Bisretinoid. *J. Am. Chem. Soc.*, **118**, 1559–1560.

Shapiro, A.B. and Ling, V. (1994). ATPase Activity of Purified and Reconstituted P-Glycoprotein from Chinese Hamster Ovary Cells. *J. Biol. Chem.*, **269**, 3745–3754.

Sparrow, J.R., Nakanishi, K., and Parish, C.A. (2000). The Lipofuscin Fluorophore A2E Mediates Blue Light-Induced Damage to Retinal Pigmented Epithelial Cells. *Invest. Ophthalmol. Vis. Sci.*, **41**, 1981–1989.

Stargardt, K. (1909). Über Familiare, Progressiv Degeneration under Makulagegend des Augen. *Albrecht von Graefes Arch. Klin Exp Ophthalmol.*, **71**, 534–550.

Sun, H. and Nathans, J. (2001). Mechanistic Studies of ABCR, the ABC Transporter in Photoreceptor Outer Segments Responsible for Autosomal Recessive Stargardt Disease. *J. Bioenerg. Biomembr.*, **33**, 523–530.

Sun, H. and Nathans, J. (1997). Stargardt's ABCR is Localized to the Disc Membrane of Retinal Rod Outer Segments (Letter). *Nat. Genet.*, **17**, 15–16.

Sun, H., Molday, R.S., and Nathans, J. (1999). Retinal Stimulates ATP Hydrolysis by Purified and Reconstituted ABCR, the Photoreceptor-Specific ATP-Binding Cassette Transporter Responsible for Stargardt Disease. *J. Biol. Chem.*, **274**, 8269–8281.

Sun, H., Smallwood, P.M., and Nathans, J. (2000). Biochemical Defects in ABCR Protein Variants Associated with Human Retinopathies. *Nat. Genet.*, **26**, 242–246.

Surya, A. and Knox, B.E. (1998). Enhancement of Opsin Activity by all-*trans* Retinal. *Exp. Eye Res.*, **66**, 599–603.

van Driel, M.A., Maugeri, A., Klevering, B.J., Hoyng, C.B., and Cremers, F.P. (1998). ABCR Unites what Ophthalmologists Divide(s). *Ophthalmic Genet.*, **19**, 117–122.

Wang, N., Silver, D.L., Thiele, C., and Tall, A.R. (2001). ATP-Binding Cassette Transporter A1 (ABCA1) Functions as a Cholesterol Efflux Regulatory Protein. *J. Biol. Chem.*, **276**, 23742–23747.

Webster, A.R., Heon, E., Lotery, A.J., Vandenburgh, K., Casavant, T.L., Oh, K.T., *et al.* (2001). An Analysis of Allelic Variation in the *ABCA4* Gene. *Invest. Ophthalmol. Vis. Sci.*, **42**, 1179–1189.

Weleber, R.G. (1994). Stargardt's Macular Dystrophy. *Arch. Ophthalmol.*, **112**, 752–754.

Weng, J., Mata, N.L., Azarian, S.M., Tzekov, R.T., Birch, D.G., and Travis, G.H. (1999). Insights into the Function of Rim Protein in Photoreceptors and Etiology of Stargardt's Disease from the Phenotype in *abcr* Knockout Mice. *Cell*, **98**, 13–23.

21

SILKE BEISMANN-DRIEMEYER[*] AND ROBERT TAMPÉ[*]

Function of the transporter associated with antigen processing (TAP) in cellular immunity, tumor escape, and virus persistence

INTRODUCTION

The transporter associated with antigen processing (TAP) plays a pivotal role in the cellular immune response. The heterodimeric transporter complex translocates peptides derived from endogenous proteins from the cytosol into the endoplasmic reticulum (ER). This transport is a requirement for subsequent peptide presentation on the cell surface. Viral and tumor-specific peptides can be recognized by cytotoxic T lymphocytes (CTL), which subsequently kill the infected or tumorigenic cells. Furthermore, surface-exposed major histocompatibility complex (MHC) I molecules may stimulate the activity of natural killer (NK) cells and $\gamma\delta$ T cells, which are also involved in the elimination of virus-infected and malignant cells. On the other hand, viruses have evolved sophisticated mechanisms to escape immune recognition by impairing antigen presentation, for example, on the level of TAP-mediated peptide transport. Tumors can downregulate MHC surface expression, which in several cases is due to reduced or abolished TAP expression. In addition, mutations in the *TAP* genes, which lead to nonfunctional proteins, have a severe impact on the cellular immune system and lead to the rare bare lymphocyte syndrome.

Since the loss of TAP function is associated with severe disturbance of the immune system, it is important to understand TAP function and the mechanisms leading to TAP malfunction in great detail.

OVERVIEW OF THE ANTIGEN PROCESSING PATHWAY

Antigens are recognized by the immune system by either B or T lymphocytes. There are two types of T lymphocytes, CD4[+] T cells ("helper" cells and inflammatory cells), which

* Institute of Biochemistry, Biocenter Frankfurt, Goethe-Univeristy Frankfurt, Marie-Curie-Strasse 9, D-60439 Frankfurt, Germany

promote the maturation of antibody-producing B cells, and CD8$^+$ T cells (CTLs), which kill virus-infected and malignant cells. Antigen recognition involves the MHC I or the MHC II pathway. MHC I molecules are presented on nearly all nucleated cells, whereas MHC II are found primarily on professional antigen presenting cells (e.g., macrophages, dendritic cells, and B cells). Peptides bound to MHC I are recognized by CD8$^+$ T cells; those bound to MHC II are recognized by CD4$^+$ T cells. The professional antigen presenting cells (APCs) present MHC II bound peptides to T helper cells, but they also play an important role in the presentation of antigens bound to MHC I and in the activation of CTLs.

Dendritic cells are especially abundant in the skin, mucosa, and lymphoid organs. Immature dendritic cells take up and process antigens and then migrate in the now activated state into lymphoid organs, where they initiate the primary immune response. Activated dendritic cells are essential for the subsequent activation of naive and activated T cells. In addition, they release cytokines, which are important for the recruitment or activation of effector cells like NK cells, macrophages, and B cells. Since this chapter deals with the transporter associated with antigen processing (TAP, ABCB2/ABCB3), only the MHC I pathway, in which TAP is involved, will be discussed (for detailed reviews of MHC-I-restricted antigen processing, see York and Rock 1996; Pamer and Cresswell 1998).

MHC genes are polygenic: there are three MHC I genes, called HLA-A, HLA-B, and HLA-C (for human leukocyte antigen, which is used in parallel with MHC). In addition, there are "non-classical" MHC I genes (HLA-E, HLA-F, HLA-G, HLA-H, HLA-J, and HLA-X) which are grouped as class IB molecules. The MHC genes show strong polymorphism, and the large number of alleles is expressed codominantly. Almost all people are heterozygous in the MHC loci. Allelic polymorphism in the heavy chain occurs primarily at the peptide binding site and thus alters the substrate specificity of the MHC I molecules. In humans, there are six MHC I alleles which together are able to bind a vast pool of different peptides.

MHC I molecules have the dual role of presenting intracellular antigenic peptides to CTLs and modulating the activity of cells bearing MHC I binding receptors, such as NK and γδ T cells. NK cells are lymphatic non-B non-T cells. They selectively kill cells which present no or only a few MHC molecules at their surface, because these cells are not protected through the killer cell receptor–MHC interaction ("missing self" hypothesis) (Ljunggren and Karre 1990).

The ligand of a T-cell receptor is always an MHC molecule. The specificity of T-cell receptors is determined by the peptide bound to a MHC molecule as well as by the MHC molecule itself. The T-cell receptor resembles the antibody Fab fragment in that it also consists of an α and a β chain which are linked by a disulfide bridge (αβ T-cell receptor in αβ T cells). There also exists a specific type of T-cell receptor called the γδ receptor, which is present in γδ T cells, whose function is not completely understood. The ligands of γδ cell receptors are mainly unknown, but it is supposed that they can bind some MHC IB molecules.

Presentation of self-peptides is critical for the selection of cytotoxic T cells in the thymus. Only a small fraction of the double-positive (CD4$^+$ CD8$^+$) immature T cells with low affinity for self-peptide–MHC complexes receive a signal from the T-cell receptor bound to a cortical epithelial cell that allows them to survive and differentiate into either CD4$^+$ or CD8$^+$ cells (positive selection). When double-positive T cells bind with high affinity to bone-marrow-derived macrophages or dendritic cells expressing MHC–self peptide complexes, they receive an apoptosis signal and die (negative selection). By this mechanism the organism avoids reactivity against self-peptides (reviewed by Sprent *et al.* 1988; Ohashi 1996; Sprent and Kishimoto 2001).

Antigen processing

Antigenic peptides presented by MHC I molecules derive from cytosolic proteins or protein fragments from organellar proteins (e.g., from mitochondria). Non-self-peptides are cleavage products of viral proteins (Townsend et al. 1986, 1989; Rock and Goldberg 1999) or they are generated from proteins of certain bacteria, which replicate in the cytosol (e.g., *Listeria* spp., *Shigella* spp., *Salmonella* spp.) (Kaufmann 1993; Kurty and Bevan 1999). The peptides are produced mainly by the multicatalytic protease complex called proteasome (for recent reviews, see Baumeister et al. 1998; Früh and Yang 1999) and to a lesser extent also by other cytosolic and ER-resident proteases.

The proteasome contains a 20S core particle that degrades proteins in an ATP-independent manner, but most peptides are derived from the 26S proteasome in which additional ATP-dependent regulatory complexes are recruited to the 20S core complex (Kloetzel 2001). The 26S proteasome cleaves ubiquitinated and some nonubiquitinated proteins into peptides of 3 to 30 residues with an optimum of 8 to 11 residues (Ehring et al. 1996; Kisselev et al. 1999; Toes et al. 2001). The size distribution of peptides generated by the proteasome overlaps the size distribution of peptides bound by MHC I.

The cytokine interferon-γ (IFN-γ) leads to the exchange of specific constitutively expressed proteasomal subunits by inducible catalytic β-type subunits, namely low-molecular-weight protein 2 (LMP2), LMP7 (Belich et al. 1994), and multicatalytic endopeptidase complex-like protein 1 (MECL-1, also called LMP10) (Groettrup et al. 1996; Nandi et al. 1996). The two former subunits are encoded within in MHC locus in close proximity to TAP, whereas the other proteasomal β-subunits are encoded elsewhere in the genome (Goldberg and Rock 1992). The proteasome containing the inducible subunits is called "immunoproteasome" because of the switch to altered substrate specificity. The immunoproteasome generates preferentially peptides with hydrophobic and basic carboxy-termini (Driscoll et al. 1993), which are favored by both TAP and MHC I molecules. The peptides may be amino-terminally trimmed before they are accepted by MHC (Beninga et al. 1998).

TAP translocates peptides from the cytosol into the lumen of the ER where they are loaded onto newly synthesized and assembled MHC I molecules. Any defect that causes reduced peptide translocation into the ER results in reduced surface expression of MHC I molecules. This is important to note for understanding the effects of TAP mutations or viral inhibition of TAP (see below). Nevertheless, viral antigens can in some instances also be presented in a TAP-independent manner, as shown, for example, for the late lytic cycle protein BCRF1 from Epstein–Barr virus (Saulquin et al. 2001).

Antigen presentation

Efficient antigen presentation by MHC I requires the assembly of a macromolecular loading complex. To date, both the precise role of some of the components and the series of events leading to the formation of peptide-loaded MHC I molecules are a matter of debate (Pamer and Cresswell 1998; Deidrich et al. 2001; Antoniou et al. 2002; Gao et al. 2002). One recent model suggests the following order of events. Newly synthesized MHC I heavy chains assemble in the ER with β₂-microglobulin (β₂m) to MHC I heterodimers. This step is assisted by the membrane-associated chaperone calnexin, which is subsequently replaced by the soluble chaperone calreticulin and the thiol oxido reductase ERpS7 (Wilson et al. 2000; Gao et al. 2002). In the calreticulin-containing complex, the MHC I molecules can be loaded transiently with suboptimal peptides. Newly synthesized tapasin, a 48 kDa ER-resident type I membrane glycoprotein, binds to TAP (Diedrich et al. 2001). A macromolecular loading complex is formed by association of MHC I molecules TAP and tapasin. The peptides which are

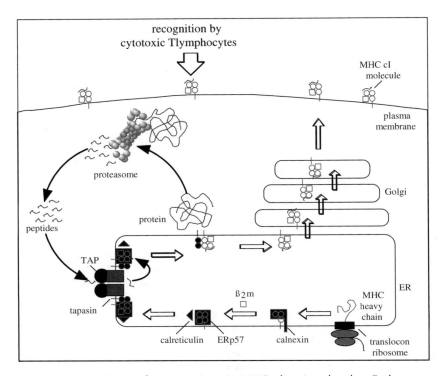

Figure 1   Antigen processing and presentation via MHC class I molecules. Endogenous proteins including viral or tumor-associated proteins are degraded by the proteasome, and peptides are transported into the lumen of the ER by the TAP complex. Several molecules are involved in folding, assembly, and loading of the MHC I molecules, including calnexin, calreticulin, tapasin, and ERp57. MHC complexes loaded optimally with peptides can leave the ER via the Golgi compartment to the cell surface for recognition by cytotoxic T lymphocytes.

translocated into the ER lumen by TAP are loaded onto MHC I. The MHC I–peptide assembly is translocated to the cell surface for recognition by $CD8^+$ T cells. The steps of MHC I-dependent antigen processing and presentation are summarized in Figure 1.

When the T-cell receptor of a cytotoxic T cell recognizes a MHC I molecule loaded with a non-self (or altered) peptide, the CTL is triggered to eliminate the target cell. This is an essential protective mechanism of the T-cell dependent ("cellular") immune system against viral infection or malignant cell growth. The antigen processing pathway by MHC I is constitutively active, but is upregulated by inflammatory cytokines like IFN-γ, which are secreted by activated CTLs.

GENOMIC ORGANIZATION AND REGULATION OF TAP GENES

Genomic organization and TAP polymorphism

TAP1 and *TAP2* are located in the *MHC II* locus on chromosome 6 band p21.3 (Trowsdale *et al.* 1991). The genes are 8–12 kb in size and consist of 11 exons (Hanson and Trowsdale 1991). The TAP1 cDNA contains 2,244 bp, and the TAP2 cDNA contains 2,058 bp. A splicing variant for *TAP2* has been described (*TAP2iso*) which lacks exon 11 but contains in addition exon 12 and a different 3′ untranslated region. Since this isoform has been described only in a single publication (Yan *et al.* 1999), its existence and possible function requires further evidence.

Human *TAP1* and *TAP2* show limited polymorphism (Colonna *et al.* 1992; Powis *et al.* 1992; Jackson and Capra 1993). In contrast to the *TAP2* splicing variant described by Yan *et al.* (1999), *TAP* polymorphism seems not to influence the substrate specificity, since no differences were observed in various *TAP1* and *TAP2* alleles (Obst *et al.* 1995; Daniel *et al.* 1997). Furthermore, different alleles could not be linked to specific diseases as was suspected, for example, for rheumatoid arthritis (Vejbaesya *et al.* 2000), atopic dermatitis (Lee *et al.* 2001), or ankylosing spondylitis (Konno *et al.* 1998).

Expression of TAP

The expression of *TAP1* and *TAP2* is controlled by transcriptional and post-transcriptional mechanisms (Wright *et al.* 1995; Cramer and Klemsz 1997). On the transcriptional level, stimulation with IFN-γ accelerates the rate of gene expression. On the other hand, the *TAP* gene expression is downregulated after 24 h of IFN-γ stimulation by destabilizing the corresponding mRNAs. Costimulation with IFN-γ and lipopolysaccharide (LPS) alters the kinetics of *TAP* gene transcription as well as protein expression (Cramer and Klemsz 1997).

Antigen presentation can be regulated on different levels: transcription rate of *TAP* and *MHC I* genes; ratio of TAP1 to TAP2 and thus amount of stable TAP complex; the nature of a pathogen, the presence of other immune cells, and the level of produced cytokines.

IFN-γ stimulates not only the expression of both *TAP* genes, but also that of proteasomal *LMP2* and *LMP7* genes (Epperson *et al.* 1992). For *TAP2* expression, there are three promotors which can act independently, but in a physiological context probably act in concert (Arons *et al.* 2001; Guo *et al.* 2002). The coordinated expression of *TAP1* and *LMP2* is facilitated by their common bidirectional intergenic promotor, which is regulated by at least three sequences located in the *TAP1*

proximal region. Importantly, different *TAP1/LMP2* promotor alleles respond similar to IFN-γ stimulation and could not be linked to the observed downregulation of *TAP1* and *LMP2* expression in different tumors (Seliger *et al.* 2002a). In addition to IFN-γ, *TAP1* is strongly induced by the "tumor marker protein" p53 (Zhu *et al.* 1999). The physiological role of this effect has to be clarified.

STRUCTURE AND FUNCTION OF THE TAP COMPLEX

Structural organization of TAP

TAP is the only known ABC transporter permanently resident in the ER membrane; however, an ER retention signal has not been identified. Human TAP consists of 748 amino acids, while human TAP2 consists of 686 amino acids. The TAP complex consists of two modules of a membrane spanning domain (MSD) fused to a cytosolic nucleotide binding domain (NBD). The NBDs contain approximately 210 amino acids and are well conserved in all ABC transporters, while the MSDs are much more diverse in sequence and length. The identity between the TAP1 and TAP2 NBDs is around 60%, while that of the MSD is around 30%.

Depending on the algorithm used, different numbers of transmembrane helices were predicted for the TAP1 and TAP2 MSDs. A current model derived from hydrophobicity analysis and sequence alignments of TAP proteins predicts the typical six helix core of ABC transporters for both TAP subunits plus four additional N-terminal helices for TAP1 and three additional N-terminal helices for TAP2, leading to a sum of 19 membrane helices in the TAP1–TAP2 complex (Figure 2) (Lankat-Buttgereit and Tampé 2002). The function of the amino-terminal membrane domain extension is unknown.

The two subunits, TAP1 and TAP2, form the functional heterodimeric TAP complex by rapid association *in vivo* (Russ *et al.* 1995). In

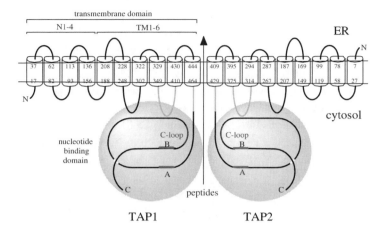

Figure 2 Proposed membrane topology of the human TAP complex. The transmembrane helices are predicted from hydrophobicity plots and sequence alignments with other ABC transporters. The hydrophobic N-terminal domain (N) of TAP1 and TAP2 comprises three or four predicted transmembrane helices (gray symbols). The nucleotide binding domain consists of the Walker A and B motifs (A, B) and the C loop (C). The orange lines illustrate binding regions for peptides, as identified by photo-cross-linking and deletion mutagenesis experiments.

contrast to two other ABC transporters of medical relevance, MDR1 and CFTR, only a minor, probable misfolded, fraction of TAP seems to be glycosylated (Russ *et al.* 1995).

ATP binding and hydrolysis

Peptide transport requires ATP binding and hydrolysis, and thus the peptide binding MSDs are functionally coupled to the NBDs. Peptide binding to the membrane domains, on the other hand, leads to stimulation of the ATPase activity (Gorbulev *et al.* 2001). Both MSDs must be present and functional to allow peptide binding and transport (Alberts *et al.* 2001; Daumke and Knittler 2001), but there is some uncertainty about the requirement of both the TAP1 and the TAP2 NBD (Alberts *et al.* 2001; Arora *et al.* 2001). As in all ABC transporters, TAP binds ATP using the highly conserved motifs, namely the Walker A (GX_4GKS/T) and Walker B ($RX_{6-8}\Phi_4D$) motifs (X, any amino acid; Φ, hydrophobic amino acid) and the C loop with the consensus sequence LSGGQ (in TAP2 LAAGQ) (see Figure 2). ATP and ADP

binding stabilizes the structure of TAP which is indicative for an induced conformational rearrangement (Van Endert 1999).

The NBD of TAP1 binds ATP much more efficiently than that of TAP2 (Russ *et al.* 1995; Alberts *et al.* 2001). It is proposed that the NBDs are functionally distinct, and that their nonequivalence is an intrinsic property rather than an effect induced by their different membrane domain fusion partners (Daumke and Knittler 2001). There is also evidence for functionally distinct NBDs in other ABC transporters (e.g., in CFTR and SUR1) (Matsuo *et al.* 1999, 2000; Aleksandrov *et al.* 2002).

ATP hydrolysis strictly depends on prior peptide binding to the MSDs, whereas peptide transport requires ATP hydrolysis at the NBDs, indicating a tight functional coupling of the structural domains of TAP (Gorbulev *et al.* 2001).

Peptide binding and specificity

The TAP complex seems to contain a single peptide binding site, which is formed by residues of the MSDs of both subunits (Figure 2) (Nijenhuis and Hämmerling 1996). Peptide photo-cross-linking studies were used to map the peptide binding region of TAP to the cytosolic loops between core-transmembrane helix (TM) 4 and TM5 and a carboxy-terminal stretch of approximately 15 residues after TM6 (Nijenhuis and Hämmerling 1996). In addition, using a series of human TAP1 deletion mutants expressed in TAP1$^{-/-}$ mice, Ritz et al. (2001) were able to confirm that the amino acids 366–405 located in the putative cytosolic loop following core-TM4 are essential for peptide transport activity of the TAP complex.

The binding site can accommodate a large variety of peptides. Using combinatorial peptide libraries in a peptide binding and competition assay Uebel et al. (1997) revealed the peptide binding motif of human TAP. It was shown that peptides of 8–16 residues with specific amino acids at the first three positions and at the carboxy-terminus (basic or hydrophobic) are preferentially bound, but peptides with up to 40 residues can be transported by TAP (reviewed in Uebel and Tampé, 1999) (Figure 3). Peptides are fixed to TAP via backbone contacts and the free amino- and carboxy-termini. The T-cell receptor (TCR), on the other hand, contacts mainly residues 5–8 of an MHC-I-associated nonapeptide where TAP is promiscuous, and thus TAP does not restrict the pool of peptides available for presentation by the MHC I–TCR complex (Garboczi et al. 1996). Indicative of coevolution of the proteins, the selection principles and size distribution of peptides transported by TAP overlap with those of peptides generated by the proteasome and peptides bound by MHC I. Peptides which are longer than suitable for MHC I undergo amino-terminal trimming in the ER (Komlosh et al. 2001).

Peptide translocation cycle and loading of MHC class I molecules

The NBDs play a crucial role in the overall TAP function because they are required for energizing peptide translocation and synchronization of peptide transport, loading and

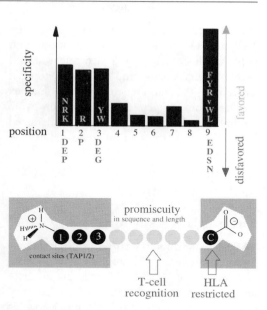

Figure 3 Peptide binding specificity of TAP. Screening a combinatorial nonapeptide library revealed that the three amino-terminal as well as the carboxy-terminal amino acids contribute to the overall peptide affinity. Favored and disfavored amino acids are indicated (upper panel). A model of the peptide binding pocket is shown in the lower panel. Peptides are anchored by their free amino- and carboxy-termini as well as by the three amino-terminal and the carboxy-terminal amino acids. TAP shows promiscuity for the overall length as well as for the sequence of the mid-part of the peptide, which is accessible for recognition by the T-cell receptor.

dissociation of MHC I molecules from the TAP complex (Knittler et al. 1999). The peptide translocation cycle has not been elucidated in detail, but based on the currently available data, Alberts et al. (2001) developed the following working model. Step 1 represents the ground state with ATP bound to TAP1 and ADP bound to TAP2. Peptides can bind to TAP in this state and thereby (step 2) stimulate ATP hydrolysis at TAP1 and ADP release at TAP2. Then ATP is bound to TAP2 (step 3) and hydrolyzed (step 4), whereas ADP produced from the previous step is released from TAP1. Binding of new ATP to TAP1 leads to recovery of the ground state of the TAP complex. MHC I molecules loaded with peptides during this cycle are released and empty MHC I can bind to TAP (step 5). Currently, the number of ATP molecules required for one peptide translocation cycle has not been determined, but has been deduced from other ABC transporters to be one or two.

DYSFUNCTION OF TAP AND HUMAN DISEASES

Bare lymphocyte syndrome type I

TAP plays an important role in the immunologic response. In mammalian cells lacking *TAP* genes, the MHC I molecules are unstable and are degraded in the cytosol. Thus delivery of peptides into the ER by TAP, which enables peptide binding by MHC I, is necessary for MHC I, stability and function. Mutations in either *TAP1* or *TAP2* which lead to deficiency in functional TAP complex are often the cause of the bare lymphocyte syndrome (BLS) type I, the only known inherited trait connected with TAP.

BLS is a rare autosomal recessive disorder first described by Touraine *et al.* (1978). Three types of BLS can be distinguished: BLS type I, II, and III patients have MHC class I, II, and combined MHC I and II deficiency, respectively (Bernaerts *et al.* 2001).

BLS type II and III patients suffer from a complete lack of cellular and humoral immune responses to foreign antigens. The symptoms resemble those of severe combined immunodeficiency (SCID) patients with recurrent bronchopulmonary infections, general high susceptibility to viral, bacterial, and fungal infections, and chronic diarrhea. BLS type II and III patients usually die within the first 3–4 years of life (Klein *et al.* 1993).

BLS type I patients can be divided into three subgroups (reviewed by Gadola *et al.* 2000). Patients from the first subgroup – like BLS type II patients – suffer from severe recurrent bacterial or parasitic infections in early childhood and die within the first 3 years of life

from infectious complications. A strongly decreased MHC I surface expression and lack of antibody production are characteristic. Gene loci responsible for these symptoms were not identified. Patients of the second subgroup have no symptoms, but MHC I surface expression is downregulated as can be determined by fluorescence-activated cell sorter (FACS) analysis on peripheral blood mononuclear cells (Teisserenc *et al.* 1997). The third subgroup contains patients with the TAP deficiency syndrome. In these patients, downregulation of MHC I surface expression is due to mutations in either *TAP1* or *TAP2* which usually lead to a premature translation stop (de la Salle *et al.* 1994, 1999; Tesserenc *et al.* 1997 Furukawa *et al.* 1999). Typical symptoms are recurrent bacterial infections and necrotizing granulomatous skin lesions, often on the legs and in the mid-face. Only about a dozen cases have been described. The patients, who survive into adulthood, do not suffer from severe viral infections; thus their cell-mediated immune response seems to work at least to some extent.

Recurrent bacterial infections of the upper respiratory tract are usually the first symptoms occurring in the childhood of TAP-deficient BLS type I patients. From the early adulthood on, the lower respiratory tract and the skin are also effected in most patients. The clinical manifestations found in TAP-deficiency patients are listed in the Box: Pathophysiology of Bare Lymphocyte Syndrome Type I.

Pathophysiology of bare lymphocyte syndrome I

In most TAP-deficient patients, an expansion in NK and $\gamma\delta$ T cells was found in the peripheral blood lymphocytes. The reactivity of NK lymphocytes from healthy individuals to autologous cells is negatively controlled by MHC-I-binding inhibitory NK receptors (KIRs). Activation and cytotoxicity of NK cells is enhanced by IFN-α. In one TAP2-deficient patient, treatment with IFN-α led to a progression of the granulomatous lesions. Therefore, in TAP-deficient patients, the threshold of

PATHOPHYSIOLOGY OF BARE LYMPHOCYTE SYNDROME TYPE I (BLS I)

PATHOGENESIS

- Impaired positive selection of T cells
- Lack of MHC-I-dependent antigen presentation

MOLECULAR IMMUNOLOGY

- Autosomal recessive

LABORATORY DIAGNOSIS

- Decreased number of CD8$^+$ T cells
- Normal humoral immune response
- Lack of MHC I surface expression

COMMON CLINICAL MANIFESTATIONS

Ear/nose/throat

Chronic sinusitis, nasal disease (discharge, polyps, septum ulcers), postnasal drip syndrome, otitis media, mastoiditis, erosion/ destruction of facial tissue around the nose.

Lungs

Chronic spastic bronchitis, recurrent bacterial pneumonia, bronchiectasis.

Skin

Necrotizing granulomatous skin lesions, leukokytoclastic vasculitis.

Nervous system

Cerebral abscess, encephalomyelitis.

Gastrointestinal tract

Chronic gastritis, pseudomembranous colitis.

Other organs

Nonerosive symmetrical polyarthritis, retinal vasculitis.

Treatment

Only symptomatic treatment; future: bone marrow transplantation?

inhibition provided by inhibitory NK receptors may not be reached, leading to autoimmune lesions (Moins-Teisserenc *et al.* 1999). Autoreactive NK cells and γδ T cells were found in several patients, in some cases also in the skin lesions (Moins-Teisserenc *et al.* 1999), and these cells might be involved in the development of the skin lesions since they are capable of promoting inflammatory responses leading to granuloma formation. The NK and γδ T cells may also account for the lack of severe viral infections, because they can recognize virally infected cells in a TAP-independent manner (Tay *et al.* 1998). Resting TAP-deficient NK lymphocytes are unable to lyse MHC I-deficient cells in vitro. NK cells of TAP-deficient patients could participate in immune defense, at least through antibody-dependent cellular cytotoxicity. TAP-deficient NK cells stimulated by interleukins (IL-2) are cytotoxic to autologous B lymphocytes and thus may be involved in autoimmune processes (Zimmer *et al.* 1998).

Activated autologous NK cells can kill skin fibroblasts. Inflammatory cytokines such as interferons protect normal, but not TAP2-deficient, fibroblasts because the cytokines increase the levels of cell-surface MHC I molecules. This suggests a role of MHC–I/KIR interactions in the control of immune responses in normal individuals. The absence of this regulation in MHC-deficient patients could lead to autoimmune reaction (de la Salle *et al.* 1999).

Today, the treatment of BLS I patients is focused on the individual symptoms. The respiratory infections are treated with antibiotics to prevent bronchiectasis. Surgery for chronic sinusitis should be avoided, since this seems to promote nasal disease, leading to bronchial infection (Gadola *et al.* 2000). Skin ulcers are carefully cleansed to decrease bacterial infections. No immunosuppressive or immunomodulating therapy should be used, since both have been found to worsen the symptoms in two patients (Gadola *et al.* 2000). Bone marrow transplantation was successfully applied in several patients suffering from

BLS type II (Klein 1993), but none of the BLS type I patients has yet received a transplant. Although it might be beneficial, bone marrow transplantation may also induce a severe graft-versus-host reaction due to the presence of the donor's NK cells.

Viral infections and virus persistence

During a million years of coevolution, viruses have evolved elaborated strategies to evade the host's immune response, leading to chronic or latent infections (reviewed by Ploegh 1998). Several human pathogenic viruses are known to induce immune suppression, which is often associated with opportunistic secondary viral or bacterial infections (e.g., human cytomegalovirus, human immunodeficiency virus, Epstein–Barr virus, and the measles virus). These combined multiple infections represent a serious clinical problem. The immune suppression caused by viruses may also facilitate tumor development.

Blocking of antigen presentation as one of many strategies to escape from the host's immune system is used by several DNA viruses. Every step of the antigen processing and presentation pathway is a potential target, either on the DNA or at the protein level. Several members of the herpesvirus family (herpes simplex virus, human cytomegalovirus, Epstein–Barr virus), as well as from the adenovirus family, inhibit antigen presentation on the level of TAP. Also, inhibition of TAP has been proposed for Kaposi's sarcoma associated herpesvirus (Brander *et al.* 2000) and human papilloma virus (Vambutas *et al.* 2000, 2001), but clear evidence is still lacking.

Herpes simplex virus

The herpes simplex virus (HSV) belongs to the α subfamily of herpes viruses. HSV type I infects epithelia, leading to blisters and orofacial lesions ("cold sore") or sometimes to lesions on the fingers ("whitlows"). HSV-2, which is commonly referred to as genital herpes, produces lesions on the genitals, urethra,

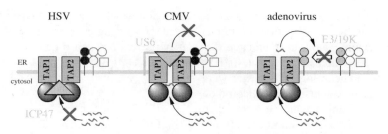

Figure 4 Viral factors affecting TAP function. Viral escape mechanism of herpes simplex virus (HSV) protein ICP47, human cytomegalovirus (HCMV) protein US6, and adenovirus E3 19K (E3/19K) protein blocking TAP function by different mechanisms. ICP47 binds TAP from the cytosolic face, thereby inhibiting peptide binding and transport. US6 acts on TAP from the ER luminal face. By occupying parts of the ATP binding site, peptide binding on the cytosolic face is inhibited. E3/19K can act on either TAP or MHC I. Binding to TAP prevents the association of TAP and MHC I, leading to inhibition of peptide loading onto MHC I and subsequent surface presentation.

and bladder. Both HSV types invade the innervating sensory neurons.

In most cases, the immune system can fight the epithelial infection. In addition, nucleoside analogs like aciclovir, ganciclovir, and related compounds are used in standard therapy for HSV as well as for other herpesviruses (i.e., cytomegalovirus and Epstein–Barr virus). These compounds stop the viral replication by inhibiting the virus-specific DNA polymerase. While propagation of activated viruses is stopped in this way, latent virions are not effected. Latent virions escape the attacks of the immune system because they reside in neurons, which have a small number of surface MHC I molecules. Furthermore, the virus expresses only a few proteins in the latent state so that hardly any viral peptide is presented to the immune system. Different stress situations, like bacterial infections or immune suppression, can lead to the reactivation of the virus.

TAP is the target of the immediate-early gene product ICP47 (infected cell peptide 47) of HSV type 1 and 2. ICP47 is a cytosolic protein, which inhibits the TAP-dependent peptide translocation into the ER (Hill *et al.* 1995; Früh *et al.* 1995). Because it binds at least in part to the peptide-binding site on the cytosolic side of TAP, it blocks peptide binding to TAP, thereby blocking peptide transport (Figure 4). ICP47 does not act as a simple competitor to the peptides but probably forms multivalent contacts with TAP by an induced fit mechanism (Pfänder *et al.* 1999). ATP- and ADP-binding are not effected by ICP47. The inhibition of TAP by ICP47 is not absolute, possibly because it has a similar affinity to TAP as presentable peptides ($K_{D, ICP47} \approx 50\,nM$, $K_{D, antigenic\ peptides} \approx 10$–$500\,nM$) (Ahn *et al.* 1996; Tomazin *et al.* 1996; Uebel *et al.* 1997). High amounts of TAP within the cell enable the presentation of MHC I peptides, leading to triggering of the CTL-mediated immune response.

Human cytomegalovirus

The human cytomegalovirus (HCMV) is a β-herpesvirus. Seventy to hundred percent of all humans are carriers of the virus, which was named for the observation that infected cells are enlarged compared with uninfected cells. The intranuclear inclusions of HCMV-infected cells are also characteristic. The primary infection is usually asymptomatic or mild, but can lead to a complex disease in immunocompromised individuals such as patients receiving immunosuppressive therapy, AIDS patients, embryos, and neonates. In these individuals,

HCMV can cause retinitis, pneumonitis, enterocolitis, esophagitis, and hepatitis. In infants infected with cytomegalovirus-associated disease, "cytomegalic inclusion disease" (CID), loss of hearing and neurological damages were observed. In AIDS patients, the most common manifestation is retinitis which occurs when $CD4^+$ T cells decrease in number below a threshold of about 1×10^8 cells/L. Following primary infection, HCMV can establish a lifelong persistence in a latent state without causing any disease. In the active state, HCMV causes MHC I and MHC II downregulation in dendritic cells, while infected macrophages can escape MHC I downregulation (Hengel et al. 2000). Reduction of MHC I surface expression interferes with the activation of virus-specific CTLs, but makes the infected cells more susceptible to NK-cell-mediated lysis. Following maturation, dendritic cells acquire resistance to the virus.

HCMV can escape the host immune response by inhibiting surface expression of MHC I (reviewed by Wiertz et al. 1997). Several virus gene products are involved in this process, and they act at different points in the antigen processing pathway. The US3 protein retains class I molecules in the ER (Jones et al. 1996), while US2 and US11 mistarget MHC I molecules to the cytosol where they are degraded by the proteasome (Wiertz et al. 1997). TAP is the target of the ER-resident US6 glycoprotein (Ahn et al. 1997; Hengel et al. 1997; Lehner et al. 1997). By binding to the ER-luminal part of TAP, US6 specifically blocks the binding of ATP, but not of ADP, at the cytosolic side of TAP, thereby inhibiting the ATP-driven peptide transport into the ER (Figure 4) (Ahn et al. 1997; Hewitt et al. 2001; Kyritsis et al. 2001). Neither peptide binding nor the association of TAP with tapasin, calreticulin, and MHC I are affected by US6. It seems that binding of US6 to TAP stabilizes TAP in a conformation which is unable to bind ATP and thus prevents peptide translocation (Hewitt et al. 2001; Kyritsis et al. 2001). US6 probably acts in concert with another HCMV-encoded protein, UL18, an MHC I heavy-chain

homolog which is targeted to the plasma membrane. By acting as a surrogate MHC I, UL18 may prevent NK cell lysis (Farrell et al. 1997; Reyburn et al. 1997).

Epstein–Barr virus

The Epstein–Barr virus (EBV) is a member of the γ subfamily of the herpesviridae. Like HCMV, EBV is a ubiquitous virus, infecting about 90% of all adult humans. EBV causes infectious mononucleosis, an infection of the B lymphocytes. In addition, EBV can contribute to the development of malignant diseases, for example, Hodgkin's disease, Burkitt's lymphoma, nasopharyngeal carcinoma, and post-transplant lymphoproliferatative diseases (Zhang and Pagano 2001). The malignant transformations are due to the ability of EBV to immortalize B lymphocytes, which are one of its target cells. EBV infection leads to a T-cell response, but EBV can survive the attack by establishing a type-I-like latency state in which the latent membrane protein 1 (LMP-1) is not expressed.

During the acute phase of an EBV infection, the expression of LMP-1 indirectly induces expression of *TAP2*. Activation of the *TAP2* promoter is dependent on an intact interferon-stimulated response element sequence. LMP-1 is thought to activate the interferon regulatory factor-7 (IRF-7), which then mediates the activation of the *TAP2* gene. LMP-1 seems to be required for immortilization of infected primary B lymphocytes. Thus survival of both host and virus is possible for many years. While *TAP2* expression is induced by EBV, TAP1 is downregulated, and this disequilibrium of TAP1 and TAP2 leads to the formation of only a few functional TAP complexes and therefore disturbance of peptide presentation and immune reaction.

In addition, EBV applies other strategies to avoid immune recognition. For example, it encodes an interleukin 10 homolog (BCRF1) which results in the downregulation of TAP1 (Zeidler et al. 1997) as well as inhibition of CTLs and NK-cell activity and the upregulation of B-cell activity. The CTL-mediated response is thereby disrupted, and the target cells for virus replication can proliferate (Shen *et al.* 2001).

Adenovirus

Adenoviruses cause mild infections, usually of the upper respiratory tract, in immunocompetent children. In contrast to this, adenoviral infections are a severe problem in immuno-compromised patients. They can be associated with a number of homology group-specific clinical manifestations. Adenovirus of homology group E cause conjunctivitis and pharyngoconjunctival fever. This virus inhibits MHC I surface expression because of the association of MHC I with the E3 region of the ER-resident E19 protein (Andersson *et al.* 1985; Burgert and Kvist 1985, 1987). E19 of adenovirus type E3 (E3/19K) can bind to both MHC I and TAP but, unlike tapasin, not simultaneously. Binding to TAP inhibits interaction with tapasin and thereby also the MHC I–TAP association, so that MHC I expression and antigen presentation are decreased (Figure 4) (Bennett *et al.* 1999).

Tumor development

Several ABC transporters are associated with cancer in humans. Resistance to anticancer drugs is often due to expression of the ABC transporter MDR1 (ABCB1) or members of the MRP subfamily (ABCC) (see Chapter 18) (Chen *et al.* 1986; Kruh *et al.* 2001). TAP can also be involved in tumor development since impairment of processing and presentation of tumor-associated antigens provides an important mechanism for immune escape of malignant cells.

Many tumor cells have lost the ability to present MHC-I-restricted antigens. MHC-I surface expression may be either downregulated or completely lost due to β2-m gene mutations. In addition, selective loss of MHC I allospecificities has been described (Marincola *et al.* 1997; Garrido *et al.* 1997; Seliger *et al.*

2000). The loss of MHC I surface presentation would normally lead to an increased susceptibility to lysis by NK cells and thereby to elimination of the transformed cells. Despite the lack of MHC I at the cell surface, many tumors are "ignored" by the immune system (Johnsen *et al.* 1999). The reasons are not fully understood, but may involve MHC I surrogates like the HCMV UL18 protein (Farrell *et al.* 1997).

Several tumors have been found to have reduced amounts of TAP complexes, which is caused by either a mutation in one of the *TAP* genes or malfunction of one or more regulatory mechanisms of *TAP* expression (Restifo *et al.* 1993; Singal *et al.* 1996; Kageshita *et al.* 1999; Seliger *et al.* 2002b). Mutations in the TAP genes have only been found in two cell lines. One is a small-cell lung carcinoma cell line containing a point mutation near the ATP-binding site of TAP1, which results in a transport-deficient TAP protein (Chen *et al.* 1996). The second is a melanoma cell line with a one base pair deletion in the *TAP1* gene, which causes a premature translation stop. This latter cell line also has another defect, a two-nucleotide insertion in an MHC I gene (Seliger *et al.* 2001).

The more frequently observed reason for reduced MHC I surface expression lies in reduction of expression of the *TAP* genes. Several human tumor cell lines show reduced TAP expression, for example, Hodgkin/Burkitt's lymphoma cells (often associated with EBV infection; see above), small-cell lung carcinoma cells (Singal *et al.* 1996), and cervical carcinomas, which are usually associated with papilloma virus infections (Cromme *et al.* 1994), as well as in breast cancers (Kaklamanis *et al.* 1995; Vitale *et al.* 1998) and lung carcinomas (Korkolopoulou *et al.* 1996). Since TAP downregulation is correlated with the metastatic potential of primary melanoma lesions, determination of *TAP* expression can be used as a prognostic indicator (Kageshita *et al.* 1999; Kamarashev *et al.* 2001). In concordance with this, reduced TAP expression has also been found with significantly higher frequency in metastases than in primary lesions of breast cancer (Kaklamanis *et al.* 1995; Seliger *et al.* 1997) and melanomas (Kageshita *et al.* 1999). The reduced MHC I surface expression may represent a strategy used by tumors to evade CTL-mediated recognition and elimination.

Impaired *TAP* expression could be overcome in small-cell lung carcinoma cell culture models by transfection of the *TAP1* gene (Singal *et al.* 1996; Alimonti *et al.* 2000) or transfection of the *TAP1, TAP2,* and *MHC I* genes in human melanoma cell lines (White *et al.* 1997). Additionally, defects in *TAP* gene regulation can often be corrected by application of the transcription activator IFN-γ (Maeurer *et al.* 1996; White *et al.* 1998; Corrias *et al.* 2001).

SUMMARY

The TAP transporter complex plays in important role in antigen presentation by the MHC I pathway. Therefore loss of TAP function due to mutations, downregulation of expression, or inhibition by viral proteins leads to severe deficiency in the cytotoxic T-lymphocyte-dependent immune response. Though we have learned much about the function of TAP during the last decade, many questions still remain unanswered. For example, it will be necessary to understand how TAP can be substituted in TAP-deficient cells. While some viruses specifically inhibit TAP to escape immune response, bare lymphocyte syndrome I patients are not especially susceptible to viral infections. The regulation between TAP-dependent and TAP-independent pathways for immune recognition and the mechanism of TAP-independent antigen presentation itself are poorly understood. Therefore the development of efficient immunotherapies for malignancies caused by lack of TAP function requires further insight into the interactions between all proteins and signaling molecules of the MHC-I-dependent antigen presentation pathway.

REFERENCES

Ahn, K., Meyer, T.H., Uebel, S., Sempe, P., Djaballah, H., Yang, Y., *et al.* (1996). Molecular Mechanism and Species Specificity of TAP Inhibition by Herpes Simplex Virus ICP47. *EMBO J.*, **15**, 3247–3255.

Ahn, K., Gruhler, A., Galocha, B., Jones, T.R., Wiertz, E.J., Ploegh, H.L., *et al.* (1997). The ER-Luminal Domain of the HCMV Glycoprotein US6 Inhibits Peptide Translocation by TAP. *Immunity*, **6**, 613–621.

Alberts, P., Daumke, O., Deverson, E.V., Howard, J.C., and Knittler, M.R. (2001). Distinct Functional Properties of the TAP Subunits Coordinate the Nucleotide-Dependent Transport Cycle. *Curr. Biol.*, **11**, 242–251.

Aleksandrov, L., Aleksandrov, A.A., Chang, X.B., and Riordan, J.R. (2002). The First Nucleotide Binding Domain of Cystic Fibrosis Transmembrane Conductance Regulator is a Site of Stable Nucleotide Interaction, whereas the Second is a Site of Rapid Turnover. *J. Biol. Chem.*, **277**, 15419–15425.

Alimonti, J., Zhang, Q.J., Gabathuler, R., Reid, G., Chen, S.S., and Jefferies, W.A. (2000). TAP Expression Provides a General Method for Improving the Recognition of Malignant Cells In Vivo. *Nat. Biotechnol.*, **18**, 515–520.

Andersson, M., Paabo, S., Nilsson, T., and Peterson, P.A. (1985). Impaired Intracellular Transport of Class I MHC Antigens as a Possible Means for Adenoviruses to Evade Immune Surveillance. *Cell*, **43**, 215–222.

Antoniou, A.N., Ford, S., Alphey, M., Osborne, A., Elliott, T., and Powis, S.J. (2002). The Oxidoreductase ERp57 Efficiently Reduces Partially Folded in Preference to Fully Folded MHC Class I Molecules. *EMBO J.*, **21**, 2655–2663.

Arons, E., Kunin, V., Schechter, C., and Ehrlich, R. (2001). Organization and Functional Analysis of the Mouse Transporter Associated with Antigen Processing 2 Promoter. *J. Immunol.*, **166**, 3942–3951.

Arora, S., Lapinski, P.E., and Raghavan, M. (2001). Use of Chimeric Proteins to Investigate the Role of Transporter Associated with Antigen Processing (TAP) Structural Domains in Peptide Binding and Translocation. *Proc. Natl Acad. Sci USA*, **98**, 7241–7246.

Baumeister, W., Walz, J., Zühl, F., and Seemuller, E. (1998). The Proteasome: Paradigm of a Self-Compartmentalizing Protease. *Cell*, **92**, 367–380.

Belich, M.P., Glynne, R.J., Senger, G., Sheer, D., and Trowsdale, J. (1994). Proteasome Components with Reciprocal Expression to that of the MHC-Encoded LMP Proteins. *Curr. Biol.*, **4**, 769–776.

Beninga, J., Rock, K.L., and Goldberg, A.L. (1998). Interferon-Gamma can Stimulate Post-Proteasomal Trimming of the N Terminus of an Antigenic Peptide by Inducing Leucine Aminopeptidase. *J. Biol. Chem.*, **273**, 18734–18742.

Bennett, E.M., Bennink, J.R., Yewdell, J.W., and Brodsky, F.M. (1999). Cutting Edge: Adenovirus E19 has Two Mechanisms for Affecting Class I MHC Expression. *J. Immunol.*, **162**, 5049–5052.

Bernaerts, A., Vandevenne, J.E., Lambert, J., de Clerck, L.S., and de Schepper, A.M. (2001). Bare Lymphocyte Syndrome: Imaging Findings in an Adult. *Eur. Radiol.*, **11**, 815–818.

Brander, C., Suscovich, T., Lee, Y., Nguyen, P.T., O'Connor, P., Seebach, J., *et al.* (2000). Impaired CTL Recognition of Cells Latently Infected with Kaposi's Sarcoma-Associated Herpes Virus. *J. Immunol.*, **165**, 2077–2083.

Burgert, H.G., and Kvist, S. (1985). An Adenovirus Type 2 Glycoprotein Blocks Cell Surface Expression of Human Histocompatibility Class I Antigens. *Cell*, **41**, 987–997.

Burgert, H.G. and Kvist, S. (1987). The E3/19K Protein of Adenovirus Type 2 Binds to the Domains of Histocompatibility Antigens Required for CTL Recognition. *EMBO J.*, **6**, 2019–2026.

Chen, C.J., Chin, J.E., Ueda, K., Clark, D.P., Pastan, I., Gottesman, M.M., *et al.* (1986). Internal Duplication and Homology with Bacterial Transport Proteins in the mdr1 (P-glycoprotein) Gene from Multidrug-Resistant Human Cells. *Cell*, **47**, 381–389.

Chen, H.L., Gabrilovich, D., Tampe, R., Girgis, K.R., Nadaf, S., and Carbone, D.P. (1996). A Functionally Defective Allele of TAP1 Results in Loss of MHC Class I Antigen Presentation in a Human Lung Cancer. *Nat. Genet.*, **13**, 210–213.

Colonna, M., Bresnahan, M., Bahram, S., Strominger, J.L., and Spies, T. (1992). Allelic Variants of the Human Putative Peptide Transporter Involved in Antigen Processing. *Proc. Natl Acad. Sci. USA*, **89**, 3932–3936.

Corrias, M.V., Occhino, M., Croce, M., De Ambrosis, A., Pistillo, M.P., Bocca, P., *et al.*, (2001). Lack of HLA-Class I Antigens in Human Neuroblastoma Cells: Analysis of its Relationship to TAP and Tapasin Expression. *Tissue Antigens*, **57**, 110–117.

Cramer, L.A., and Klemsz, M.J. (1997). Altered Kinetics of Tap-1 Gene Expression in Macrophages Following Stimulation with both IFN-Gamma and LPS. *Cell Immunol.*, **178**, 53–61.

Cromme, F.V., Airey, J., Heemels, M.T., Ploegh, H.L., Keating, P.J., Stern, P.L., *et al.* (1994). Loss of Transporter Protein, Encoded by the TAP-1 Gene, is Highly Correlated with Loss of HLA Expression in Cervical Carcinomas. *J. Exp. Med.*, **179**, 335–340.

Daniel, S., Caillat-Zucman, S., Hammer, J., Bach, J.F., and van Endert, P.M. (1997). Absence of Functional Relevance of Human Transporter Associated with Antigen Processing Polymorphism for Peptide Selection. *J. Immunol.*, **159**, 2350–2357.

Daumke, O., and Knittler, M.R. (2001). Functional Asymmetry of the ATP-Binding-Cassettes of the ABC Transporter TAP is Determined by Intrinsic Properties of the Nucleotide Binding Domains. *Eur. J. Biochem.*, **268**, 4776–4786.

de la Salle, H., Hanau, D., Fricker, D., Urlacher, A., Kelly, A., Salamero, J., *et al.* (1999). Homozygous Human TAP Peptide Transporter Mutation in HLA Class I Deficiency. *Science*, **265**, 237–241.

de la Salle, H., Zimmer, J., Fricker, D., Angenieux, C., Cazenave, J.P., Okubo, M., *et al.* (1999). HLA Class I Deficiencies due to Mutations in Subunit 1 of the Peptide Transporter TAP1. *J. Clin. Invest.*, **103**, R9-R13.

Diedrich, G., Bangia, N., Pan, M., and Cresswell, P. (2001). A Role for Calnexin in the assembly of the MHC Class I Loading Complex in the Endoplasmic Reticulum. *J. Immunol.*, **166**, 1703–1709.

Driscoll, J., Brown, M.G., Finley, D., and Monaco, J.J. (1993). MHC-Linked LMP Gene Products Specifically Alter Peptidase Activities of the Proteasome. *Nature*, **365**, 262–264.

Ehring, B., Meyer, T.H., Eckerskorn, C., Lottspeich, F., and Tampé, R. (1996). Effects of Major-Histocompatibility-Complex-Encoded Subunits on the Peptidase and Proteolytic Activities of Human 20S Proteasomes. Cleavage of Proteins and Antigenic Peptides. *Eur. J. Biochem.*, **235**, 404–415.

Epperson, D.E., Arnold, D., Spies, T., Cresswell, P., Pober, J.S., and Johnson, D.R. (1992). Cytokines Increase Transporter in Antigen Processing-1 Expression More Rapidly than HLA Class I Expression in Endothelial Cells. *J. Immunol.*, **149**, 3297–3301.

Farrell, H.E., Vally, H., Lynch, D.M., Fleming, P., Shellam, G.R., Scalzo, A.A., *et al.* (1997). Inhibition of Natural Killer Cells by a Cytomegalovirus MHC Class I Homologue In Vivo. *Nature*, **386**, 510–514.

Früh, K., Ahn, K., Djaballah, H., Sempé, P., van Endert, P.M., Tampé, R., Peterson, P.A., and Yang, Y. (1995). A viral inhibitor of peptide transporters for antigen presentation. Nature, **375**, 415–418.

Früh, K., and Yang, Y. (1999). Antigen Presentation by MHC Class I and its Regulation by Interferon Gamma. *Curr. Opin. Immunol.*, **11**, 76–81.

Furukawa, H., Murata, S., Yabe, T., Shimbara, N., Keicho, N., Kashiwase, K., *et al.* (1999). Splice Acceptor site Mutation of the Transporter Associated with Antigen Processing-1 Gene in Human Bare Lymphocyte Syndrome. *J. Clin. Invest.*, **103**, 755–758.

Gadola, S.D., Moins-Teisserenc, H.T., Trowsdale, J., Gross, W.L., and Cerundolo, V. (2000). TAP Deficiency Syndrome. *Clin. Exp. Immunol.*, **121**, 173–178.

Gao, B., Adhikari, R., Howarth, M., Nakamura, K., Gold, M.C., Hill, A.B., *et al.* (2002). Assembly and Antigen-Presenting Function of MHC Class I Molecules in Cells Lacking the ER Chaperone Calreticulin. *Immunity*, **16**, 99–109.

Garboczi, D.N., Ghosh, P., Utz, U., Fan, Q.R., Biddison, W.E., and Wiley, D.C. (1996). Structure of the Complex between Human T-Cell Receptor, Viral Peptide and HLA-A2. *Nature*, **384**, 134–141.

Garrido, F., Ruiz-Cabello, F., Cabrera, T., Perez-Villar, J.J., Lopez-Botet, M., Duggan-Keen, M., et al. (1997). Implications for Immunosurveillance of Altered HLA Class I Phenotypes in Human Tumours. Immunol. Today, 18, 89–95.

Goldberg, A.L. and Rock, K.L. (1992). Proteolysis, Proteasomes and Antigen Presentation. Nature, 357, 375–379.

Gorbulev, S., Abele, R., and Tampé, R. (2001). Allosteric Crosstalk between Peptide-Binding, Transport, and ATP Hydrolysis of the ABC Transporter TAP. Proc. Natl Acad. Sci. USA, 98, 3732–3737.

Groettrup, M., Kraft, R., Kostka, S., Standera, S., Stohwasser, R., and Kloetzel, P.M. (1996). A Third Interferon-Gamma-Induced Subunit Exchange in the 20S Proteasome. Eur. J. Immunol., 26, 863–869.

Guo, Y., Yang, T., Liu, X., Lu, S., Wen, J., Durbin, J.E., et al. (2002). Cis Elements for Transporter Associated with Antigen-Processing-2 Transcription: Two New Promoters and an Essential Role of the IFN Response Factor Binding Element in IFN-Gamma-Mediated Activation of the Transcription Initiator. Int. Immunol., 14, 189–200.

Hanson, I.M. and Trowsdale, J. (1991). Colinearity of Novel Genes in the Class II Regions of the MHC in Mouse and Human. Immunogenetics, 34, 5–11.

Harty, J.T., and Bevan, M.J. (1999). Responses of CD8$^{(+)}$ T Cells to Intracellular Bacteria. Curr. Opin. Immunol., 11, 89–93.

Hengel, H., Koopmann, J.O., Flohr, T., Muranyi, W., Goulmy, E., Hämmerling, G.J., Koszinowski, U.H., and Momburg, F. (1997). A viral ER-resident glycoprotein inactivates the MHC-encoded peptide transporter. Immunity, 6, 623–632.

Hengel, H., Reusch, U., Geginat, G., Holtappels, R., Ruppert, T., Hellebrand, E., et al. (2000). Macrophages Escape Inhibition of Major Histocompatibility Complex Class I-Dependent Antigen Presentation by Cytomegalovirus. J. Virol., 74, 7861–7868.

Hewitt, E.W., Gupta, S.S., and Lehner, P.J. (2001). The Human Cytomegalovirus Gene Product US6 Inhibits ATP Binding by TAP. EMBO J., 20, 387–396.

Hill, A., Jugovic, P., York, I., Russ, G., Bennink, J., et al. (1995). Herpes Simplex Virus Turns Off the TAP to Evade Host Immunity. Nature, 375, 411–415.

Jackson, D.G. and Capra, J.D. (1993). TAP1 Alleles in Insulin-Dependent Diabetes Mellitus: A Newly Defined Centromeric Boundary of Disease Susceptibility. Proc. Natl Acad. Sci. USA, 90, 11079–11083.

Johnsen, A.K., Templeton, D.J., Sy, M., and Harding, C.V. (1999). Deficiency of Transporter for Antigen Presentation (TAP) in Tumor Cells Allows Evasion of Immune Surveillance and Increases Tumorigenesis. J. Immunol., 163, 4224–4231.

Jones, T.R., Wiertz, E.J., Sun, L., Fish, K.N., Nelson, J.A., and Ploegh, H.L. (1996). Human Cytomegalovirus US3 Impairs Transport and Maturation of Major Histocompatibility Complex Class I Heavy Chains. Proc. Natl Acad. Sci. USA, 93, 11327–11333.

Kageshita, T., Hirai, S., Ono, T., Hicklin, D.J., and Ferrone, S. (1999). Down-Regulation of HLA Class I Antigen-Processing Molecules in Malignant Melanoma: Association with Disease Progression. Am. J. Pathol., 154, 745–754.

Kaklamanis, L., Leek, R., Koukourakis, M., Gatter, K.C., and Harris, A.L. (1995). Loss of Transporter in Antigen Processing 1 Transport Protein and Major Histocompatibility Complex Class I Molecules in Metastatic versus Primary Breast Cancer. Cancer Res., 55, 5191–5194.

Kamarashev, J., Ferrone, S., Seifert, B., Boni, R., Nestle, F.O., Burg, G., et al. (2001). TAP1 Down Regulation in Primary Melanoma Lesions: An Independent Marker of Poor Prognosis. Int. J. Cancer, 95, 23–28.

Kaufman, S.H. (1993). Immunity to Intracellular Bacteria. Annu. Rev. Immunol., 11, 129–163.

Kisselev, A.F., Akopian, T.N., Woo, K.M., and Goldberg, A.L. (1999). The Sizes of Peptides Generated from Protein by Mammalian 26 and 20 S Proteasomes. Implications for Understanding the Degradative Mechanism and Antigen Presentation. J. Biol. Chem., 274, 3363–3371.

Klein, C., Lisowska-Grospierre, B., LeDeist, F., Fischer, A., and Griscelli, C. (1993). Major Histocompatibility Complex Class II Deficiency: Clinical Manifestations, Immunologic Features, and Outcome. J. Pediatr., 123, 921–928.

Kloetzel, P.M. (2001). Antigen Processing by the Proteasome. Nat. Rev. Mol. Cell. Biol., 2, 179–187.

Knittler, M.R., Alberts, P., Deverson, E.V., and Howard, J.C. (1999). Nucleotide Binding by TAP Mediates Association with Peptide and Release of Assembled MHC Class I Molecules. Curr. Biol., 9, 999–1008.

Komlosh, A., Momburg, F., Weinschenk, T., Emmerich, N., Schild, H., Nadav, E., et al. (2001). A Role for a Novel Luminal Endoplasmic Reticulum Aminopeptidase in Final Trimming of 26 S Proteasome-Generated Major Histocompatability Complex Class I Antigenic Peptides. J. Biol. Chem., 276, 30050–30056.

Konno, Y., Numaga, J., Mochizuki, M., Mitsui, H., Hirata, R., and Maeda, H. (1998). TAP Polymorphism is not Associated with Ankylosing Spondylitis and Complications with Acute Anterior Uveitis in HLA-B27-Positive Japanese. Tissue Antigens, 52, 478–483.

Korkolopoulou, P., Kaklamanis, L., Pezzella, F., Harris, A.L., and Gatter, K.C. (1996). Loss of Antigen-Presenting Molecules (MHC Class I and TAP-1) in Lung Cancer. Br. J. Cancer, 73, 148–153.

Kruh, G.D., Zeng, H., Rea, P.A., Liu, G., Chen, Z.S., Lee, K., et al. (2001). MRP Subfamily Transporters and Resistance to Anticancer Agents. J. Bioenerg. Biomembr., 33, 493–501.

Kyritsis, C., Gorbulev, S., Hutschenreiter, S., Pawlitschko, K., Abele, R., and Tampé, R. (2001).

Molecular Mechanism and Structural Aspects of Transporter Associated with Antigen Processing Inhibition by the Cytomegalovirus Protein US6. *J. Biol. Chem.*, **276**, 48031–48039.

Lankat-Buttgereit, B. and Tampé, R. (2002). The Transporter Associated with Antigen Processing: Function and Implications in Human Diseases. *Physiol. Rev.*, **82**, 187–204.

Lee, H.J., Ha, S.J., Han, H., and Kim, J.W. (2001). Distribution of HLA-A, B Alleles and Polymorphisms of TAP and LMP Genes in Korean Patients with Atopic Dermatitis. *Clin. Exp. Allergy*, **31**, 1867–1874.

Lehner, P.J., Karttunen, J.T., Wilkinson, G.W., and Cresswell, P. (1997). The Human Cytomegalovirus US6 Glycoprotein Inhibits Transporter Associated with Antigen Processing-Dependent Peptide Translocation. *Proc. Natl Acad. Sci. USA*, **94**, 6904–6909.

Ljunggren, H.G. and Karre, K. (1990). In Search of the 'Missing Self': MHC Molecules and NK Cell Recognition. *Immunol. Today*, **11**, 237–244.

Maeurer, M.J., Gollin, S.M., Martin, D., Swaney, W., Bryant, J., Castelli, C., *et al.* (1996). Tumor Escape from Immune Recognition: Lethal Recurrent Melanoma in a Patient Associated with Downregulation of the Peptide Transporter Protein TAP-1 and Loss of Expression of the Immunodominant MART-1/Melan- A Antigen. *J. Clin. Invest.*, **98**, 1633–1641.

Marincola, F.M., Shamamian, P., Alexander, R.B., Gnarra, J.R., Turetskaya, R.L., Nedospasov, S.A., *et al.* (1994). Loss of HLA Haplotype and B Locus Down-Regulation in Melanoma Cell Lines. *J. Immunol.*, **153**, 1225–1237.

Matsuo, M., Kioka, N., Amachi, T., and Ueda, K. (1999). ATP Binding Properties of the Nucleotide-Binding Folds of SUR1. *J. Biol. Chem.*, **274**, 37479–37482.

Matsuo, M., Tanabe, K., Kioka, N., Amachi, T., and Ueda, K. (2000). Different Binding Properties and Affinities for ATP and ADP among Sulfonylurea Receptor Subtypes, SUR1, SUR2A, and SUR2B. *J. Biol. Chem.*, **275**, 28757–28763.

Moins-Teisserenc, H.T., Gadola, S.D., Cella, M., Dunbar, P.R., Exley, A., Blake, N., *et al.* (1999). Association of a Syndrome resembling Wegener's Granulomatosis with Low Surface Expression of HLA Class-I Molecules. *Lancet*, **354**, 1598–1603.

Nandi, D., Jiang, H., and Monaco, J.J. (1996). Identification of MECL-1 (LMP-10) as the Third IFN-Gamma-Inducible Proteasome Subunit. *J. Immunol.*, **156**, 2361–2364.

Nijenhuis, M. and Hämmerling, G.J. (1996). Multiple Regions of the Transporter Associated with Antigen Processing (TAP) Contribute to its Peptide Binding Site. *J. Immunol.*, **157**, 5467–5477.

Obst, R., Armandola, E.A., Nijenhuis, M., Momburg, F., and Hämmerling, G.J. (1995). TAP Polymorphism does not Influence Transport of Peptide Variants in Mice and Humans. *Eur. J. Immunol.*, **25**, 2170–2176.

Ohashi, P.S. (1996). T Cell Selection and Autoimmunity: Flexibility and Tuning. *Curr. Opin. Immunol.*, **8**, 808–814.

Pamer, E. and Cresswell, P. (1998). Mechanisms of MHC Class I-Restricted Antigen Processing. *Annu. Rev. Immunol.*, **16**, 323–358.

Pfänder, R., Neumann, L., Zweckstetter, M., Seger, C., Holak, T.A., and Tampé, R. (1999). Structure of the Active Domain of the Herpes Simplex Virus Protein ICP47 in Water/Sodium Dodecyl Sulfate Solution Determined by Nuclear Magnetic Resonance Spectroscopy. *Biochemistry*, **38**, 13692–13698.

Ploegh, H.L. (1998). Viral Strategies of Immune Evasion. *Science*, **280**, 248–253.

Powis, S.H., Mockridge, I., Kelly, A., Kerr, L.A., Glynne, R., Gileadi, U., *et al.* (1992). Polymorphism in a Second ABC Transporter Gene Located within the Class II Region of the Human Major Histocompatibility Complex. *Proc. Natl Acad. Sci. USA*, **89**, 1463–1467.

Raghavan, M. (1999). Immunodeficiency due to Defective Antigen Processing: The Molecular Basis for Type 1 Bare Lymphocyte Syndrome. *J. Clin. Invest.*, **103**, 595–596.

Restifo, N.P., Esquivel, F., Kawakami, Y., Yewdell, J.W., Mule, J.J., Rosenberg, S.A., *et al.* (1993). Identification of Human Cancers Deficient in Antigen Processing. *J. Exp. Med.*, **177**, 265–272.

Reyburn, H.T., Mandelboim, O., Vales-Gomez, M., Davis, D.M., Pazmany, L., and Strominger, J.L. (1997). The Class I MHC Homologue of Human Cytomegalovirus Inhibits Attack by Natural Killer Cells. *Nature*, **386**, 514–517.

Ritz, U., Momburg, F., Pircher, H.P., Strand, D., Huber, C., and Seliger, B. (2001). Identification of Sequences in the Human Peptide Transporter Subunit TAP1 Required for Transporter Associated with Antigen Processing (TAP) Function. *Int. Immunol.*, **13**, 31–41.

Rock, K.L. and Goldberg, A.L. (1999). Degradation of Cell Proteins and the Generation of MHC Class I-Presented Peptides. *Annu. Rev. Immunol.*, **17**, 739–779.

Russ, G., Esquivel, F., Yewdell, J.W., Cresswell, P., Spies, T., and Bennink, J.R. (1995). Assembly, Intracellular Localization, and Nucleotide Binding Properties of the Human Peptide Transporters TAP1 and TAP2 Expressed by Recombinant Vaccinia Viruses. *J. Biol. Chem.*, **270**, 21312–21318.

Saulquin, X., Bodinier, M., Peyrat, M.A., Hislop, A., Scotet, E., Lang, F., *et al.* (2001). Frequent Recognition of BCRF1, a Late Lytic Cycle Protein of Epstein–Barr Virus, in the HLA-B*2705 Context: Evidence for a TAP-Independent Processing. *Eur. J. Immunol.*, **31**, 708–715.

Seliger, B., Maeurer, M.J., and Ferrone, S. (1997). TAP Off – Tumors On. *Immunol. Today*, **18**, 292–299.

Seliger, B., Maeurer, M.J., and Ferrone, S. (2000). Antigen-Processing Machinery Breakdown and Tumor Growth. *Immunol. Today*, **21**, 455–464.

Seliger, B., Bock, M., Ritz, U., and Huber, C. (2002a). High Frequency of a Non-functional TAP1/LMP2

Promoter Polymorphism in Human Tumors. *Int. J. Oncol.*, **20**, 349–353.

Seliger, B., Cabrera, T., Garrido, F., and Ferrone, S. (2002b). HLA Class I Antigen Abnormalities and Immune Escape by Malignant Cells. *Semin. Cancer Biol.*, **12**, 3–13.

Seliger, B., Ritz, U., Abele, R., Bock, M., Tampé, R., Sutter, G., *et al*. (2001). Immune Escape of Melanoma: First Evidence of Structural Alterations in Two Distinct Components of the MHC Class I Antigen Processing Pathway. *Cancer Res.*, **61**, 8647–8650.

Shen, L., Chiang, A.K., Liu, W.P., Li, G.D., Liang, R.H., and Srivastava, G. (2001). Expression of HLA Class I, Beta(2)-Microglobulin, TAP1 and IL-10 in Epstein–Barr Virus-Associated Nasal NK/T-Cell Lymphoma: Implications for Tumor Immune Escape Mechanism. *Int. J. Cancer*, **92**, 692–696.

Singal, D.P., Ye, M., Ni, J., and Snider, D.P. (1996). Markedly Decreased Expression of TAP1 and LMP2 Genes in HLA Class I-Deficient Human Tumor Cell Lines. *Immunol. Lett.*, **50**, 149–154.

Singal, D.P., Ye, M., and Bienzle, D. (1998). Transfection of TAP 1 Gene Restores HLA Class I Expression in Human Small-Cell Lung Carcinoma. *Int. J. Cancer*, **75**, 112–116.

Sprent, J. and Kishimoto, H. (2001). The Thymus and Central Tolerance. *Philos. Trans. R. Soc. London B. Biol. Sci.*, **356**, 609–616.

Sprent, J., Lo, D., Gao, E.K., and Ron, Y. (1988). T Cell Selection in the Thymus. *Immunol. Rev.*, **101**, 173–190.

Tay, C.H., Szomolanyi-Tsuda, E., and Welsh, R.M. (1998). Control of Infections by NK Cells. *Curr. Top. Microbiol. Immunol.*, **230**, 193–220.

Teisserenc, H., Schmitt, W., Blake, N., Dunbar, R., Gadola, S., Gross, W.L., *et al*. (1997). A Case of Primary Immunodeficiency due to a Defect of the Major Histocompatibility Gene Complex Class I Processing and Presentation Pathway. *Immunol. Lett.*, **57**, 183–187.

Toes, R.E., Nussbaum, A.K., Degermann, S., Schirle, M., Emmerich, N.P., Kraft, M., *et al*. (2001). Discrete Cleavage Motifs of Constitutive and Immuno-proteasomes Revealed by Quantitative Analysis of Cleavage Products. *J. Exp. Med.*, **194**, 1–12.

Tomazin, R., Hill, A.B., Jugovic, P., York, I., van Endert, P., Ploegh, H.L., *et al*. (1996). Stable Binding of the Herpes Simplex Virus ICP47 Protein to the Peptide Binding Site of TAP. *EMBO J.*, **15**, 3256–3266.

Touraine, J.L., Betuel, H., Souillet, G., and Jeune, M. (1978). Combined Immunodeficiency Disease Associated with Absence of Cell-Surface HLA-A and -B Antigens. *J. Pediatr.*, **93**, 47–51.

Townsend, A.R., Bastin, J., Gould, K., and Brownlee, G.G. (1986). Cytotoxic T Lymphocytes Recognize Influenza Haemagglutinin that Lacks a Signal Sequence. *Nature*, **324**, 575–577.

Townsend, A., Ohlen, C., Bastin, J., Ljunggren, H.G., Foster, L., and Karre, K. (1989). Association of Class I

Major Histocompatibility Heavy and Light Chains Induced by Viral Peptides. *Nature*, **340**, 443–448.

Trowsdale, J., Ragoussis, J., and Campbell, R.D. (1991). Map of the Human MHC. *Immunol. Today*, **12**, 443–446.

Uebel, S. and Tampé, R. (1999). Specificity of the Proteasome and the TAP Transporter. *Curr. Opin. Immunol.*, **11**, 203–208.

Uebel, S., Kraas, W., Kienle, S., Wiesmüller, K.H., Jung, G., and Tampé, R. (1997). Recognition Principle of the TAP Transporter Disclosed by Combinatorial Peptide Libraries. *Proc. Natl Acad. Sci. USA*, **94**, 8976–8981.

Vambutas, A., Bonagura, V.R., and Steinberg, B.M. (2000). Altered Expression of TAP-1 and Major Histocompatibility Complex Class I in Laryngeal Papillomatosis: Correlation of TAP-1 with Disease. *Clin. Diagn. Lab. Immunol.*, **7**, 79–85.

Vambutas, A., DeVoti, J., Pinn, W., Steinberg, B.M., and Bonagura, V.R. (2001). Interaction of Human Papillomavirus Type 11 E7 Protein with TAP-1 Results in the Reduction of ATP-Dependent Peptide Transport. *Clin. Immunol.*, **101**, 94–99.

van Endert, P.M. (1999). Role of Nucleotides and Peptide Substrate for Stability and Functional State of the Human ABC Family Transporters Associated with Antigen Processing. *J. Biol. Chem.*, **274**, 14632–14638.

Vejbaesya, S., Luangtrakool, P., Luangtrakool, K., Sermduangprateep, C., and Parivisutt, L. (2000). Analysis of TAP and HLA-DM Polymorphism in Thai Rheumatoid Arthritis. *Hum. Immunol.*, **61**, 309–313.

Vitale, M., Rezzani, R., Rodella, L., Zauli, G., Grigolato, P., Cadei, M., *et al*. (1998). HLA Class I Antigen and Transporter Associated with Antigen Processing (TAP1 and TAP2) Down-Regulation in High-Grade Primary Breast Carcinoma Lesions. *Cancer Res.*, **58**, 737–742.

White, C.A., Thomson, S.A., Cooper, L., van Endert, P.M., Tampé, R., Coupar, B., *et al*. (1998). Constitutive Transduction of Peptide Transporter and HLA Genes Restores Antigen Processing Function and Cytotoxic T Cell-Mediated Immune Recognition of Human Melanoma Cells. *Int. J. Cancer*, **75**, 590–595.

Wiertz, E., Hill, A., Tortorella, D., and Ploegh, H. (1997). Cytomegaloviruses use Multiple Mechanisms to Elude the Host Immune Response. *Immunol. Lett.*, **57**, 213–216.

Wilson, C.M., Farmery, M.R., and Bulleid, N.J. (2000). Pivotal Role of Calnexin and Mannose Trimming in Regulating the Endoplasmic Reticulum-Associated Degradation of Major Histocompatibility Complex Class I Heavy Chain. *J. Biol. Chem.*, **275**, 21224–21232.

Wright, K.L., White, L.C., Kelly, A., Beck, S., Trowsdale, J., and Ting, J.P. (1995). Coordinate Regulation of the Human TAP1 and LMP2 Genes from a Shared Bidirectional Promoter. *J. Exp. Med.*, **181**, 1459–1471.

Yan, G., Shi, L., and Faustman, D. (1999). Novel Splicing of the Human MHC-Encoded Peptide Transporter Confers Unique Properties. *J. Immunol.*, **162**, 852–859.

York, I.A. and Rock, K.L. (1996). Antigen Processing and Presentation by the Class I Major Histocompatibility Complex. *Annu. Rev. Immunol.*, **14**, 369–396.

Zeidler, R., Eissner, G., Meissner, P., Uebel, S., Tampé, R., Lazis, S., *et al.* (1997). Downregulation of TAP1 in B Lymphocytes by Cellular and Epstein–Barr Virus-Encoded Interleukin-10. *Blood*, **90**, 2390–2397.

Zhang, L. and Pagano, J.S. (2001). Interferon Regulatory Factor 7 Mediates Activation of Tap-2 by Epstein–Barr Virus Latent Membrane Protein 1. *J. Virol.*, **75**, 341–350.

Zhu, K., Wang, J., Zhu, J., Jiang, J., Shou, J., and Chen, X. (1999). p53 Induces TAP1 and Enhances the Transport of MHC Class I Peptides. *Oncogene*, **18**, 7740–7747.

Zimmer, J., Donato, L., Hanau, D., Cazenave, J.P., Tongio, M.M., Moretta, A., *et al.* (1998). Activity and Phenotype of Natural Killer Cells in Peptide Transporter (TAP)-Deficient Patients (Type I Bare Lymphocyte Syndrome). *J. Exp. Med.*, **187**, 117–122.

TRANSPORTERS
INVOLVED IN
SIGNAL TRANSDUCTION

22

STEFAN BRÖER*

Introduction

SIGNALING BETWEEN CELLS

Two classes of molecules, hormones and neurotransmitters, are used to transmit most signals between cells. Both classes are released from one type of cell and bind to receptors on another cell, thereby transmitting a signal. The signal can be an ion flow across the cell membrane or the initiation of a metabolic signal transduction cascade that results in the generation of a second messenger within the target cell. Transporters are involved in both types of signaling events, either to remove neurotransmitters or to remove second messengers after the signal has been transmitted. In this part we will focus on the important role of transporters in the removal of neurotransmitters, particularly of biogenic amines, and in the removal of the second messenger Ca^{2+}.

NEUROTRANSMISSION

Neurons are comprised of three anatomically distinct parts: dendrites, the cell body, and the axon. Each neuron has multiple branched dendrites that originate from the cell body. Each neuron has only one axon that originates from the cell body opposite to the side where the dendrites originate. Dendrites and axons are the signal input site and the signal output site

of neurons, respectively. Axons can be highly branched and may extend for long distances throughout the brain and the spinal cord. Signals are transmitted between neurons by synaptic contacts. Most synapses use a chemical transmitter for transmission of an action potential from one neuron to the next. The basic design of a chemical synapse is similar in all parts of the brain. Each synapse has a presynaptic element, which is the terminus of an axon, and a postsynaptic element, a dendritic spine (Figure 1). Each synapse uses a specific neurotransmitter for signal transduction. The fast processing in the brain, that is, the processing of information coming from the senses, analysis of that information, memory and exertion of movement, etc., relies on glutamate as a signaling molecule (to excite the postsynaptic neuron) and γ-aminobutyric acid (GABA) or glycine to inhibit propagation of excitatory signals (inhibitory transmitters).

Neurotransmitters are stored inside the presynaptic axon terminal in synaptic vesicles, some of which fuse with the plasma membrane when an incoming action potential depolarizes the presynaptic region. The energy to accumulate neurotransmitters inside synaptic vesicles is provided by vacuolar ATPases (*ATP6 family*) (Moriyama *et al.* 1992) which pump protons into the vesicle (Figure 1). The proton electrochemical gradient is subsequently used to accumulate neurotransmitters inside the synaptic vesicles by vesicular neurotransmitter transporters. Each neurotransmitter has a specific vesicular transporter (Eiden 2000). Thus

* School of Biochemistry and Molecular Biology, Faculty of Science, Australian National University, Canberra ACT 0200, Australia

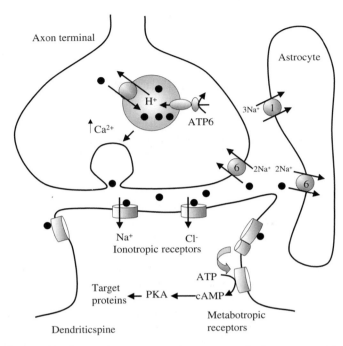

Figure 1   Basic design of a chemical synapse. In the axon terminal neurotransmitters (●) are loaded into synaptic vesicles by vesicular neurotransmitter transporters. The driving force for this process is provided by vacuolar ATPases (ATP6) that pump protons into the vesicles. An increase in the intracellular Ca^{2+} concentration triggers the fusion of the synaptic vesicle with the plasma membrane. Neurotransmitter is released into the synaptic cleft and binds to receptors on the postsynaptic dendritic spine. Binding to ionotropic receptors elicits ion fluxes that either depolarize or hyperpolarize the postsynaptic neuron. Binding to metabotropic receptors initiates signal transduction cascades that modify the responsiveness of the postsynaptic cell to neurotransmitters. Neurotransmitters are removed from the synaptic cleft by neurotransmitter transporters (SLC1 and SLC6 family) that reside on either presynaptic neurons or neighboring astrocytes.

glutamatergic synapses have vesicular glutamate transporters (*SLC17*) (Takamori *et al.* 2000) and GABAergic neurons have a vesicular GABA transporter (*SLC32*) (McIntire *et al.* 1997). After being released into the synaptic cleft, neurotransmitters bind to receptors on the postsynaptic membrane. Glutamate, GABA, and glycine receptors are ion channels that open in response to neurotransmitter binding. Excitatory neurotransmitters open Na^+ channels which depolarize the postsynaptic cell; inhibitory neurotransmitters open chloride channels which hyperpolarize the postsynaptic membrane. The postsynaptic signals, being depolarizations or hyperpolarizations, are summed up at the axon hillock. If a depolarization threshold is reached, a new action potential is generated. The transmission of a signal across a synapse and subsequent generation of the signal on the postsynaptic neuron occurs in a few milliseconds.

Neurotransmitters are finally removed from the synaptic cleft by uptake into neurons or astrocytes. Both cell types express a number of neurotransmitter transporters. A hallmark of these transporters is the complex transport mechanism which involves the cotransport of two or three Na^+ ions, ensuring that substrates can be accumulated inside the cell leaving only traces of neurotransmitter in the synaptic cleft. Glutamate transporters (*SLC1 family*) couple the uptake of glutamate to a cotransport of $3Na^+$ ions and $1H^+$ ion (Kanai 1997) and the exit of $1K^+$ ion. Other neurotransmitters, such as GABA, glycine, norepinephrine, serotonin, and dopamine are taken up by members of the *SLC6 family* (Nelson 1998). Most members of this family mediate the transport of neurotransmitters together with $2Na^+$ ions and $1Cl^-$ ion, although the Na^+ cotransport stoichiometry can vary from one to three. Decreased transporter activity will increase the half-life of neurotransmitters in the extracellular space and thus enhance the action of neurotransmitters.

The brain has a limited metabolic capacity. Instead of breakdown and resynthesis, neurotransmitters are reused after release. Surprisingly, recycling of neurotransmitters may involve different cell types. Glutamate, for example, is recycled by the glutamate–glutamine cycle which involves uptake of glutamate into astrocytes (*SLC1 family*), conversion into glutamine, release of glutamine from astrocytes (*SLC38 and SLC7 families*), and uptake of glutamine into neurons (*SLC 38*), where it is converted back into glutamate (Bröer and Brookes 2001).

NEUROTRANSMITTERS OF THE MODULATORY SYSTEM

In addition to the fast circuits that use glutamate, GABA, or glycine, there are diffuse modulatory systems that regulate the general activity of the brain. For example, broad regions of the brain are less active during sleep and come back to normal activity after waking up. Moreover, motivation to start things and initiate actions differs widely with the prevalent mood. Thus the modulatory system does not change the principal response of a neuron to the incoming action potentials from other neurons but modifies the amplitude of the response. Neuronal cell bodies of the diffuse modulatory system originate from a limited number of nuclei in the brainstem. Although limited in number, the axons of these neurons are highly branched and may have more than 100,000 synaptic contact sites along the axon (called varicosities) throughout the brain. Neurons of the diffuse modulatory system use four neurotransmitters, namely norepinephrine, serotonin (5-hydroxytryptamine, 5HT), dopamine, and acetylcholine. In this section we will only focus on the former three, the biogenic amines, as much more is known about the function of these neurotransmitters compared with the role of acetylcholine in the central nervous system.

Synapses that release biogenic amines are anatomically different from fast synapses that use glutamate or GABA. Fast synapses are anatomically confined. Care is taken that the neurotransmitter is rapidly removed and does not spill over to other synapses. Synapses using biogenic amines (aminergic synapses), by contrast, are less confined and anatomically structured. Release is diffuse and spillover of the neurotransmitter to neighboring cells is part of its function as a modulatory agent. Another difference is observed in the postsynaptic response to these neurotransmitters. The conduction of an action potential at glutamatergic or GABAergic synapses relies on the opening of ion channels (ionotropic receptors). Neurotransmitters of the modulatory system, in contrast, bind to metabotropic G-protein coupled receptors that trigger synthesis or release of second messengers, such as cAMP and Ca^{2+}. The response of these receptors is much slower than that of the ionotropic receptors but widespread because of the spillover and diffusion of the neurotransmitter to neighboring cells. As a result, modulatory neurotransmitters could equally well be considered as hormones.

Glutamatergic synapses also display metabotropic glutamate receptors in the membrane. In an analogous manner to the modulatory system, they initiate modifying signals rather than directly participating in the propagation of action potentials. Thus it is conceivable that a limited number of neurons, having highly

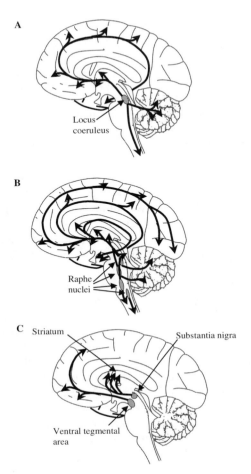

Figure 2 Innervation of the cortex by noradrenergic, dopaminergic, and serotonergic neurons.
(A) The noradrenergic system. Cell bodies of noradrenergic neurons are located in the brainstem in
the locus ceruleus. Their axons reach out into
almost every part of the brain and have multiple
branches and varicosities. A single neuron may
have as many as 250,000 synapses. (B) The serotonergic system. Cell bodies of serotonergic neurons, are clustered in the raphe nuclei of the
brainstem. Similar to noradrenergic neurons, they
project intensively into all areas of the brain.
(C) Dopaminergic system. Dopmainergic neurons
can be found throughout the brain, but the majority are clustered in two areas, the substantia nigra
and the ventral tegmental area. Neurons of the
substantia nigra project to the striatum; those of
the ventral tegmental area project to the frontal
cortical and limbic areas.

branched and extended axons, can influence
the overall activity of large brain areas.

The cell bodies of *noradrenergic neurons* are
located in the locus ceruleus in the brainstem
(Figure 2A). There is one locus in each half of
the brain, each containing about 12,000 neurons. Noradrenergic neurons innervate almost
every part of the brain. One neuron of the locus
ceruleus may have as much as 250,000 contact
sites to other neurons throughout the brain.
Noradrenergic neurons are involved in the modulation of attention, arousal, sleep–wake cycle,
learning and memory, anxiety, pain, mood, and
brain metabolism (Chapter 23). Locus ceruleus
neurons are activated by new unexpected non-
painful sensations, but they are almost silent
at rest.

Serotonergic neurons are located in the
raphe nuclei of the brainstem (Figure 2B).
There are several nuclei, each of them innervating a different area of the brain. Serotonergic
inputs, like noradrenergic inputs, can be found
all over the brain. Raphe neurons appear to be
intimately involved in the control of sleep–wake
cycles and the different stages of sleep. Similar
to neurons of the locus ceruleus, they are active
when people are awake and almost silent during
sleep. Serotonergic neurons are implicated in
the control of mood, aggression, and depression
(Chapter 23).

Dopaminergic neurons are located in two
areas in the brainstem, the substantia nigra
and the ventral tegmental area (Figure 2C).
Neurons of the substantia nigra project axons
to the striatum. They are involved in the control of movement, particularly in the initiation
of movement. Degeneration of dopaminergic
neurons in the substantia nigra is observed in
Parkinson's disease, which is characterized
by a general slowdown of movements, stiffness, and problems in initiating movements.
Neurons in the ventral tegmental area, on the
other hand, are more related to the diffuse
modulatory system. They innervate the frontal
lobe of the brain. Dopaminergic neurons of
this area are involved in a reward system that
assigns positive values to certain behaviors.
The reward for activation of this system is

likely to be a pleasurable sensation. Hence it is thought that the dopaminergic system plays a major role in addiction and drug abuse (Chapter 23).

A unusual impetus in the research of the mechanism of Parkinson's disease was provided by the toxicity of a compound called MPTP (*N*-methyl-4-phenyl-1,2,3,6-tetrahydropyridine), which was detected as an impurity in a designer drug. MPTP is lipid soluble and diffuses into the brain. Inside astrocytes it is converted into MPP^+ (1-methyl-4-phenylpyridinium) by the action of monoamine oxidase B. MPP^+ is a specific substrate of the dopamine transporter and therefore it is accumulated inside dopaminergic neurons. The accumulation causes MPP^+ to become toxic for mitochondrial respiration, which finally causes the death of the neurons, particularly those of the substantia nigra. As a result, the clinical symptoms of MPP^+ intoxication are strikingly similar to those of Parkinson's disease.

Neurotransmitter transporters of the diffuse regulatory system are targets for a variety of drugs, such as tricyclic antidepressants, cocaine, amphetamine, etc., which generally extend the half-life of the corresponding neurotransmitter in the extracellular space. In agreement with the role of neurotransmitter transporters in the clearance of biogenic amines, changes in the expression of these transporters may cause *mood disorders*, such as *orthostatic intolerance, obsessive–compulsive disorder*, and *serotonin spectrum disorders* as described in Chapter 23.

1993). Superoxide dismutase protects cells against the detrimental effects of oxygen radicals. However, only the presence of a mutated form of superoxide dismutase causes the disease, whereas a lack of the enzyme does not. A lack of glutamate transporters is thought to result in elevated extracellular glutamate concentrations (Howland *et al.* 2002). This, in turn, results in a prolonged opening of ionotropic glutamate receptors, causing excessive influx of Ca^{2+} ions. Long-term elevation of cytosolic Ca^{2+} is toxic and leads to cell death (see Chapter 24).

As pointed out above, transporters are not only involved in the removal neurotransmitters but are also critical for the removal of second messengers. In the retina, 11-*cis* retinal is converted into all-*trans* retinal by absorption of light. Conversion off all-*trans* retinal back into 11-*cis* retinal is a multistep process. First, all-*trans* retinal is reduced into all-*trans* retinol. This compound is transported from the rods to the retinal pigment epithelial cell, where it is converted into 11-*cis* retinol via an esterified intermediate. Finally, it is oxidized back to 11-*cis* retinal, which is transported back to the photoreceptor cells. The recycling is even more complicated because part of all-*trans* retinal reacts with phospholipids to form *N*-retinylidene-phosphatidylethanolamine. As explained in detail in Chapter 20, this compound needs to be transported by the ABCA4 protein before it can be recycled. Mutations in this protein are causative of *Stargardt disease, cone–rod dystrophy, atypical retinitis pigmentosa*, and *age-related macular degeneration*.

DISEASES OF OTHER NEUROTRANSMITTER SYSTEMS

Loss of glutamatergic neurons, particularly that of motor neurons, is observed in *amyotrophic lateral sclerosis (ALS)*. Loss of glutamate transporters has been implicated in the development of ALS (Lin *et al.* 1998). In contrast, mutations in the superoxide dismutase gene have been detected in patients suffering from a familiar form of ALS (Rosen *et al.*

SIGNAL TRANSDUCTION WITHIN CELLS

Signal transduction by neurotransmitters of the diffuse modulatory system is similar to that of hormones. Both generate second messengers that trigger subsequent reactions which modulate the activity of a cell. In agreement with this notion, norepinephrine is not only active in the central nervous system but is also a neurotransmitter/hormone of the peripheral autonomic nervous system. Noradrenergic neurons with cell bodies next to the spinal chord, for example, innervate the heart. Norepinephrine is released from these neurons during stressful situations. It binds to β receptors on cardiac myocytes, which are G-protein-coupled receptors (Kamp and Hell 2000). Binding of norepinephrine to β receptors recruits membrane-associated heterotrimeric G proteins (α, β, γ subunit) to the receptor (Figure 3). The α subunit of G proteins binds GDP, when it is not bound to receptors. Binding to the receptor causes cytosolic GTP to displace GDP. This in turn causes dissociation of the βγ heterodimer from the α subunit and exposes the active site of the α subunit. GTP–GDP exchange and splitting of the G-protein heterotrimer also causes the α subunit to dissociate from the receptor. Although the α subunit remains associated with the membrane because of a lipid anchor, it can now move independently in the plane of the membrane and activate a number of effector proteins, most notably adenylate cyclase and a number of ion channels. The activated adenylate cyclase synthesizes cAMP from ATP. cAMP acts as a second messenger and activates a number of other proteins, such as protein kinase A. Thus one G protein can activate several molecules of adenylate cyclase, which in turn generate many more cAMP molecules. The cAMP molecules can further activate many protein kinase A molecules which may phosphorylate ion channels or other proteins. Thus second-messenger cascades act as molecular amplifiers. The phosphorylation of the final target protein modulates the cellular activity. In the case of cardiac myocytes, binding of epinephrine/norepinephrine to β receptors finally results in the phosphorylation of plasma membrane Ca^{2+} channels by protein kinase A, thereby increasing their open probability (Figure 3) (Kamp and Hell 2000). This has a dual function. On the one hand, influx of Ca^{2+} depolarizes the plasma membrane of cardiac myocytes and triggers further Ca^{2+} release from intracellular stores. On the other hand, it causes contraction of the muscle

fibers. An increased open probability of plasma membrane Ca^{2+} channels, as a result, causes the heart to contract more strongly. Subsequently, Ca^{2+} has to be removed quickly from the cytosol to allow muscle relaxation after each contraction (see below and Chapter 24).

G-protein-coupled receptors not only activate adenylate cyclase but may instead inhibit this enzyme. G proteins are accordingly referred to as G_s or G_i to reflect stimulatory or inhibitory action, respectively. The binding of acetylcholine to the muscarinic M_2 receptor on cardiac myocytes, for example, inhibits adenylate cyclase and therefore reduces Ca^{2+} influx.

Another important class of G proteins (G_q) activates phospholipase C. Phospholipase C hydrolyzes phosphatidylinositol-(4,5)bisphosphate (PIP_2) to generate diacylglycerol (DAG) and the soluble second-messenger inositol-(1,4,5)trisphosphate (IP_3) (see Chapter 24, Figure 1). IP_3 is an agonist of IP_3 receptors, which are ligand-gated Ca^{2+} channels located on the endoplasmic reticulum (ER). The extracellular space together with the ER or the sarcoplasmic reticulum (SR) in muscle constitute the two major Ca^{2+} stores. Thus intracellular Ca^{2+} can increase as a result of the opening of either plasma membrane Ca^{2+} channels or Ca^{2+} channels in the membrane of the ER or SR. Elevated Ca^{2+} usually triggers or initiates events, such as fusion of vesicles with the plasma membrane, muscle contraction, or fertilization. The fusion of synaptic vesicles with the plasma membrane, for example, is triggered by a rise of cytosolic Ca^{2+} caused by the opening of plasma membrane Ca^{2+} channels. Contraction of skeletal muscle, on the other hand, is largely caused by the opening of the ryanodine channel on the SR. Cardiac muscle requires opening of both plasma membrane calcium channels and ryanodine channels to increase intracellular Ca^{2+}.

Analogous to the function of neurotransmitter transporters, increased cytosolic Ca^{2+} levels are rapidly brought back to resting concentrations by the action of Ca^{2+}-ATPases (ATP2) and Na^{2+}/Ca^{2+} exchangers (SLC8). Ca^{2+}-ATPases reside in the plasma membrane,

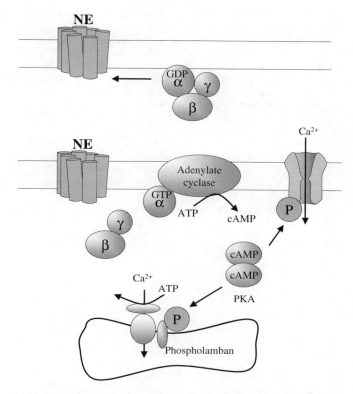

Figure 3 Signaling mediated by G-protein-coupled-receptors. Norepinephrine binds to β receptors on target cells such as cardiac myocytes. Binding of norepinephrine recruits heterotrimeric G proteins to the receptor. This causes replacement of GDP by GTP on the α subunit and dissociation of the complex. The α subunit activates adenylate cyclase, which results in the production of the second messenger cAMP. cAMP in turn activates protein kinase A (PKA) which may phosphorylate different target proteins such as phospholamban or Ca^{2+} channels.

the membrane of the ER, and the membranes of the vesicles of the secretory pathway, and are referred to as plasma membrane calcium ATPase (PMCA), Sarcoplasmic/endoplasmic reticulum calcium ATPase (SERCA), and secretory pathway calcium ATPase (SPCA), respectively. Several isoforms are known for each type of pump.

As outlined in Chapter 24, activation of adrenergic β receptors on cardiac myocytes, apart from activating plasma membrane Ca^{2+} channels, also results in the phosphorylation of phospholamban by protein kinase A (Figure 3). Phospholamban is one of the major regulators of the sarcoplasmic Ca^{2+}-ATPase in cardiac

muscle, thereby enhancing its capacity to remove Ca^{2+}.

Mutations of the SERCA1 Ca^{2+} ATPase result in muscle stiffness and cramping (*Brody disease*). Surprisingly, mutations of SERCA2 and SPCA1 cause blistering disorders of the skin (*Darier* disease and *Hailey–Hailey disease*), suggesting a role of Ca^{2+} in long-term maintenance of cell-to-cell contacts.

REFERENCES

Broer, S. and Brookes, N. (2001). Transfer of Glutamine Between Astrocytes and Neurons. *J. Neurochem.*, **77**, 705–719.

Eiden, L.E. (2000). The Vesicular Neurotransmitter Transporters: Current Perspectives and Future Prospects. *FASEB J.*, **14**, 2396–2400.

Howland, D.S., Liu, J., She, Y., Goad, B., Maragakis, N.J., Kim, B., *et al.* (2002). Focal Loss of the Glutamate Transporter EAAT2 in a Transgenic Rat Model of SOD1 Mutant-Mediated Amyotrophic Lateral Sclerosis (ALS). *Proc. Natl Acad. Sci. USA*, **99**, 1604–1609.

Kamp, T.J. and Hell, J.W. (2000). Regulation of Cardiac L-Type Calcium Channels by Protein Kinase A and Protein Kinase C. *Circ. Res.*, **87**, 1095–1102.

Kanai, Y. (1997). Family of Neutral and Acidic Amino Acid Transporters: Molecular Biology, Physiology, and Medical Implications. *Curr. Opin. Cell Biol.*, **9**, 565–572.

Lin, C.L., Bristol, L.A., Jin, L., Dykes-Hoberg, M., Crawford, T., Clawson, L. *et al.* (1998). Aberrant RNA Processing in a Neurodegenerative Disease: The Cause for Absent EAAT2, a Glutamate Transporter, in Amyotrophic Lateral Sclerosis. *Neuron.*, **20**, 589–602.

McIntire, S.L., Reimer, R.J., Schuske, K., Edwards, R.H., and Jorgensen, E.M. (1997). Identification and Characterization of the Vesicular GABA Transporter. *Nature*, **389**, 870–876.

Moriyama, Y., Maeda, M., and Futai, M. (1992). The Role of V-ATPase in Neuronal and Endocrine Systems. *J. Exp. Biol.*, **172**, 171–178.

Nelson, N. (1998). The Family of Na^+/Cl^- Neurotransmitter Transporters. *J. Neurochem.*, **71**, 1785–1803.

Rosen, D.R., Siddique, T., Patterson, D., Figlewicz, D.A., Sapp, P., Hentati, A., *et al.* (1993). Mutations in Cu/Zn Superoxide Dismutase Gene are Associated with Familial Amyotrophic Lateral Sclerosis. *Nature*, **362**, 59–62.

Takamori, S., Rhee, J.S., Rosenmund, C., and Jahn, R. (2000). Identification of a Vesicular Glutamate Transporter that Defines a Glutamatergic Phenotype in Neurons. *Nature*, **407**, 189–194.

KLAUS PETER LESCH* AND DENNIS L. MURPHY**

Molecular genetics of transporters for norepinephrine, dopamine, and serotonin in behavioral traits and complex diseases

INTRODUCTION

Sodium/chloride-dependent neurotransmitter cotransporters, which constitute a gene superfamily (SLC6), are crucial for limiting neurotransmitter activity (Hertting and Axelrod 1961; Iversen 1971). At the plasma membrane, substrate influx is directly coupled to transmembrane ion gradients that provide the energy for transport against a concentration gradient. On the basis of their remarkable amino acid identity and their properties as targets of tri- and heterocyclic antidepressants, as well as of psychostimulants including amphetamine, cocaine, and their analogs, the carriers for the monoamines norepinephrine (NE), dopamine (DA), and serotonin (5-hydroxytryptamine, 5HT) comprise a distinct subfamily (Lesch et al. 1996; Amara and Sonders

1998). Monoamine transporters are located in the plasma membrane of presynaptic neuronal cells and mediate the removal of NE, DA, and 5HT from the synaptic cleft, thereby terminating their action at pre- and postsynaptic receptors and fine tuning the spatiotemporal characteristics of neuronal communication.

Although some variations exist, the structure of monoamine transporter proteins is consistent with a model of 12 transmembrane domains (TMDs), intracellular amino and carboxyl termini, and an extended extracellular loop with several glycosylation sites between TMD3 and TMD4 [for a review see (Torres et al. 2003)]. This loop is most divergent among family members and therefore may participate in substrate recognition and inhibitor binding. Monoamine transporters are electrogenic and possess several unexpected ion-channel-like properties.

New insights into neurotransmitter transporter diversity provide the means for novel approaches to studying neurotransmitter uptake processes at the molecular level. Recent research has focused not only on molecular mechanisms of substrate

*Department of Psychiatry and Psychotherapy, University of Würzburg, Füchsleinstrasse 15, 97080 Würzburg, Germany

**Laboratory of Clinical Science, National Institute of Mental Health, NIH Clinical Center, 10-3D4J, MSC 1264, Bethesda, MD 20892-1264, USA

translocation and inhibitor binding but also on regulation of transporter gene expression, post-translational modification, cellular and subcellular localizations, and short- and long-term regulation of transporter functions. Crucial information is also derived from the analysis of genomic regulatory elements as well as from modeling monoamine-transporter-related traits or diseases and novel therapeutic strategies in genetically engineered animals. Several pertinent reviews have comprehensively discussed the molecular biology, regulation of function, and phenotypic consequences of targeted gene inactivation (Zahniser and Doolen 2001; Gainetdinov *et al.* 2001; Hahn and Blakely 2001), and this chapter will cover the molecular and clinical genetics of the three monoamine transporters with special emphasis on behavioral, neuropsychiatric, and other complex diseases.

NOREPINEPHRINE TRANSPORTER

Norepinephrine (NE), an essential neurotransmitter in both the central and peripheral nervous systems, is synthesized by dopamine-β-hydroxylase through the oxidation of DA to NE. By directing the magnitude and duration of postsynaptic receptor-mediated signaling, the NE transporter (NET) plays a pivotal role in norepinephrine neurotransmission. The NET is a target for both noradrenergic antidepressants (e.g., desipramine, reboxetine) and psychostimulants (e.g., methylphenidate) as well as drugs of abuse (e.g., amphetamines, cocaine).

The cDNA sequence of the human NET predicts a protein of 617 amino acids with 12 highly hydrophobic TMDs and significant amino acid identity with other sodium/chloride-dependent monoamine transporters (Figure 1) (Pacholczyk *et al.* 1991). The human NET gene (*SLC6A2*) has been mapped to chromosome

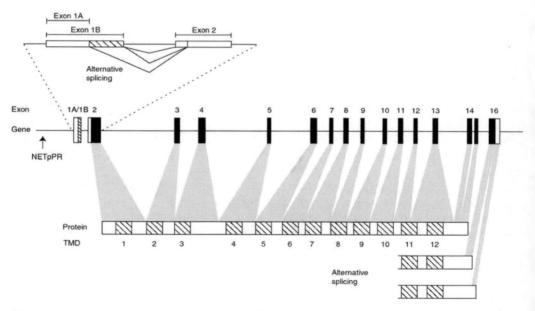

Figure 1 Genomic organization and corresponding protein segments of the human norepinephrine transporter (NET). Positions of exons are shown as black and white boxes for the coding and noncoding regions, respectively, and transmembrane domains (TMD) depicted as hatched areas are indicated. The alternatively spliced exon 1 and the variable splice acceptor of intron 1 as well as the alternative splicing of the NET protein's C-terminus are indicated. The location of the NET promoter polymorphic region (NETpPR) in the 5'-flanking region of the NET gene is also shown.

ORTHOSTATIC INTOLERANCE

Orthostatic intolerance (OI) is an autonomic syndrome characterized by lightheadedness, fatigue, altered mentation, syncope, and postural tachycardia. Biochemical features may include plasma norepinephrine (NE) concentration that is disproportionately high in relation to sympathetic outflow, decreased NE clearance with standing, resistance to the NE-releasing effect of tyramine, and increased sensitivity to adrenergic agonists. A subset of OI patients show pathophysiologic features that are associated with an Ala457Pro variant of the NE transporter gene (NET, *SLC6A2*) located on chromosome 16q12.2. α-Methyldopa, β-adrenoceptor antagonists (β-blockers) and clonidine, a partial agonist of the α_2-adrenoceptor that acts centrally to reduce sympathetic outflow and lower blood pressure, have been effective in the treatment of this condition.

16q12.2 and is composed of 15 exons spanning approximately 45 kb (Bruss *et al.* 1993; Porzgen *et al.* 1995). Northern blot analysis revealed 5.8 kb and 3.6 kb mRNA species due to alternatively terminated transcripts of the same gene. In the brain the NET is mainly expressed in the locus ceruleus, although NET transcripts have also been detected in other brain areas as well as in peripheral sympathetic ganglia, adrenal medulla, vas deferens, lung, and the placental syncytiotrophoblast (Ramamoorthy *et al.* 1995; Wakade *et al.* 1996). NET function is acutely modulated by post-translational modification including phosphorylation through protein kinase C (PKC) and p38 MAP kinase (Apparsundaram *et al.* 1998, 2001).

The human NET protein displays three splicing variants resulting in different carboxy terminals, designated NET, NET C-t var1, and NET C-t var2 (Figure 1) (Kitayama *et al.* 2001). Functional characterization of these isoforms revealed a significant increase in substrate uptake and inhibitor binding of NET C-t var2, but not NET C-t var1, compared with hNET, which was related to their different cellular localization. Kinetic and pharmacological analyses of NE uptake also showed different characteristics between NET and NET C-t var2. Differential isoform expression may modulate efficiency of NE transport at noradrenergic synapses and in other tissues. Moreover, determinants within the C-terminus of the NET have been shown to control transporter trafficking, stability, and activity (Baumann and Blakely 2002). Two residues

located in TMD6 and TMD7 (Phe316 and Val356, respectively) of NET seem to play an important role in tricyclic antidepressant interaction, and a critical region in TMD8 (Gly400Leu) is likely to be involved in the tertiary structure allowing the high-affinity binding of tricyclic antidepressants (Roubert *et al.* 2001).

Functional mapping of the NET gene promoter demonstrated constitutive activity in NET-expressing cells. A negative attenuating element and several positive elements contained within 4 kb of the 5'-flanking sequence are required to confer its cell-selective expression of NET (Meyer *et al.* 1998). In the 5'-untranslated region (5'-UTR) alternative splicing of exon 1A/B and alternate usage of the splice acceptor of intron 1 result in multiple mRNA species likely to contribute to differential regulation of gene expression in humans (Figure 1). An E-box motif residing in the exon–intron 1 junction was shown to control both transcriptional activation and splicing of the NET gene (Kim *et al.* 2001). A domain in the proximal promoter, containing a homeodomain-binding core motif, interacts with multiple transcription factors, including HoxA5 and Phox2 proteins, and critically participates in the regulation of noradrenergic cell type-selective gene transcription (Kim *et al.* 2002). Recently, length variation of an AAGG repeat island, the NET gene promoter polymorphic region (NETpPR), located approximately 4.2 kb upstream of the transcription start site and resulting in loss or gain of a putative Elk-1

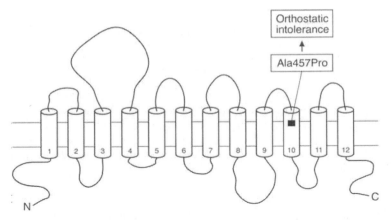

Figure 2 Structure of the human norepinephrine transporter (NET) protein characterized by 12 transmembrane domains (TMDs) and intracellular N- and C-termini. The position of the Ala457Pro substitution, which causes orthostatic intolerance, is indicated by a black box in TMD10.

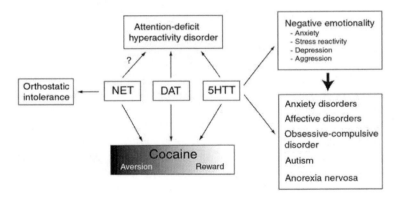

Figure 3 Norepinephrine (NET), dopamine (DAT), and serotonin transporter (5HTT/SERT) in behavioral traits and complex diseases.

transcription factor site was demonstrated to be associated with anorexia nervosa of the restrictive subtype (Urwin *et al.* 2002). Since long-term weight-restored patients with anorexia nervosa display lower norepinephrine concentration than controls, allelic variation in NET expression may lead to differential NET function and thus increase the risk of developing anorexia nervosa.

The NET tightly limits synaptic NE concentrations in the heart and an intrinsic NET dysfunction has been implicated in orthostatic dysregulation. An alanine-to-proline substitution in a highly conserved region of TMD9 of the NET, which causes an almost complete loss of NET function, was recently identified in a subset of patients presenting with orthostatic intolerance (Figures 2 and 3) (Shannon *et al.* 2000). Orthostatic intolerance is characterized by protracted postural tachycardia (>30/min) without hypotension and frequently associated with chronic fatigue, headache, dizziness, and syncope (Robertson 1999; see Box: Orthostatic Intolerance). The

Ala457Pro variant was associated with a 50-fold higher K_m indicating lower affinity for NE and a partial failure of NET protein to mature and reach the cell surface. Pharmacological comparison with the wild-type NET demonstrated a twofold lower affinity for the NET inhibitor nisoxetine but unchanged affinity for the antidepressant desipramine and a fivefold higher affinity for cocaine for the Ala457Pro variant (Paczkowski et al. 2002). Several noncoding and coding single-nucelotide polymorphisms (SNPs) were identified in the NET gene, but none showed association with depression, bipolar disorder, schizophrenia, and Tourette's syndrome (for review see Hahn and Blakely 2002). While some of these variants with the potential to influence uptake function may also cause changes in substrate affinity or trafficking abnormality, and thus loss of function similar to the Ala457Pro variant, others could lead to a gain of function that will facilitate the delineation of functional extent of transport activity.

DOPAMINE TRANSPORTER

As with NET, a single gene (*SLC6A3*), located at 5p15.3, encodes the dopamine (DA) transporter (DAT) (Frazer et al. 1999). In brain DAT expression is limited exclusively to a subset of central nervous system (CNS) dopaminergic neurons in the substantia nigra and ventral tegmental area. Peripheral locations include the stomach, pancreas, and kidney. A wide spectrum of neuropsychiatric disorders, including Parkinson's disease, attention-deficit hyperactivity disorder (ADHD), drug abuse, Tourette's syndrome, and possibly affective disorders and schizophrenia, are thought to involve dopaminergic systems and the DAT (Torres et al. 2003). The DAT is an important target for therapeutics (e.g., methylphenidate, buproprion) and illicit drugs (methamphetamine, cocaine), and serves as the point of entry for DA-specific neurotoxins [e.g., 1-methyl-4-phenylpyridine ion (MPP$^+$)] (Lesch et al.

1996). Single-photon emission tomography and positron emission tomography studies of DAT binding provide an in vivo measure of dopaminergic cell integrity and the efficacy of therapeutic intervention in neurodegenerative disease (Verhoeff 1999; Laakso and Hietala 2000; Tatsch 2001).

Abnormalities in DAT function have long been suggested for a variety of neurodegenerative disorders. Post-mortem studies indicate that [^{3}H]cocaine and [^{3}H]GBR-12935 binding are decreased in Parkinson's disease. Hemisphere-to-hemisphere differences in the binding of [^{11}C]nomifensine in the striatum and accumulation of [^{18}F]fluorodopa in the putamen were confirmed by positron emission tomography studies; loss of striatal [^{123}I]β-CIT binding was shown with single-photon emission tomography (Innis et al. 1993; Boja et al. 1994). Neurodegeneration of the nigrostriatal dopaminergic system induced by MPP$^+$, the neurotoxic metabolite of MPTP, has implicated the DAT in the etiology of some toxicologic forms of Parkinson's disease. Since MPP$^+$ enters neurons through the DAT, a dysfunctional transport process may contribute to an increased susceptibility to exogenous MPP$^+$-like neurotoxins. This vulnerability to neurotoxins may be further aggravated by an impaired capacity of the brain vesicular monoamine transporter 2 (VMAT2), which plays a central role in the sequestration of cytoplasmic toxins and thus in the limitation of mitochondrial damage. With aging, even modest reductions of vesicular uptake would lead to accumulation of MPP$^+$-like toxins in the cytoplasmic pool and cause neuronal cell death. In this regard, it is of interest to note that in humans there is evidence for progression of MPP$^+$-induced dopaminergic lesions, thus indicating that transient exposure to a toxin may cause a more rapid protracted decline in nigrostriatal dopaminergic function than in normal aging which is similar to Parkinson's disease (Vingerhoets et al. 1994). While studies of VMAT2 expression in Parkinson's disease yielded mixed results, an age-related drastic and sudden-onset reduction of DAT mRNA, which contrasts with a more gradual decline in DAT protein, has been described in post-mortem substantia nigra (Bannon et al. 1992; Bannon and Whitty 1997). Functional and structural analyses of the DAT (and VMAT2) using in vivo and in vitro strategies, such as neuroimaging with single-photon emission tomography and positron emission tomography and expression profiling in cultured DA cells, are currently enhancing our understanding of the DAT's role in the pathogenesis and progression of Parkinson's disease.

An interaction between α-Synuclein and DAT has been suggested as a second mechanism which might contribute to the pathogenesis of Parkinson's disease. α-Synuclein is found in Lewy bodies, and mutations in α-Synuclein result in an autosomal dominant rare familial form of Parkinons's disease (Polymeropoulos et al. 1997; Kruger et al. 1998). α-Synuclein complexes to the carboxyl-terminal tail of DAT, resulting in accelerated cellular DA uptake and an enhancement of DA-induced apoptosis (Lee et al. 2001). An enhanced MPP^+-induced or rotenone-induced toxicity was observed in HEK-293 cells stably coexpressing α-Synuclein and DAT (Lehmensiek et al. 2002).

Finally, alterations in DAT expression in neurodegenerative processes may represent adaptive responses to dysfunction of dopaminergic cells or associated neural circuits. In Parkinson's disease, which is characterized by extensive loss of dopaminergic neurons, DAT gene expression is decreased in surviving dopaminergic cells (Harrington et al. 1996; Joyce et al. 1997). A reduced capacity of the DAT to recapture extracellular DA may represent a compensatory mechanism to enhance dopaminergic neurotransmission in the brain of Parkinson's disease patients. A better understanding of DAT gene regulation may provide novel therapeutic strategies for the treatment of Parkinson's disease and other DA-linked neurodegenerative disorders.

The human DAT gene spans 60 kb and consists of 15 exons (Bannon et al. 2001). There is no evidence for splice variants or the use of multiple polyadenylation sites. The 3'-UTR

adjacent to the polyadenylation site contains a 40 bp repeat polymorphism. Nine- and ten-repeat alleles are the most common alleles in the 3'-UTR of the human DAT gene, with a range from 3 to 13 repeat units (Vandenbergh et al. 1992). The GC-rich sequence of the repeat polymorphism is likely to give rise to the formation of DNA secondary structure that has the potential to regulate DAT gene transcription and/or mRNA stability. Fuke et al. (2001) have demonstrated that repeat-length variation differentially modulates transcriptional activity of the CMV promoter driving a luciferase reporter gene-DAT exon 15-fusion construct in cell lines. Luciferase expression of constructs containing the 10-repeat allele was significantly higher than that of constructs with seven or nine repeats. While the effect of DAT repeat-length variation on mRNA concentration/stability, inhibitor binding, and rate of specific 5HT uptake remains to be investigated in vitro, the relationship between in vivo binding and the genotype of DAT repeat polymorphisms has been examined using $[^{123}I]β$-CIT single-photon emission tomography, although with inconclusive results. Jacobsen et al. (2000) found that individuals with 10-repeat alleles had lower DAT binding than 9-repeat subjects, whereas other investigators reported that 10-repeat subjects displayed higher DAT binding than heterozygous 9/10-repeat subjects (Heinz et al. 2000).

The sequence of the DAT gene transcript predicts a hydrophobic 620 amino acid protein, and several features of the deduced amino acid sequence and secondary structure are shared by other members of the monoamine transporter gene family (for review see Uhl and Johnson 1994). The amino- and carboxy-termini reside in the cytoplasm. The DAT contains an extended extracellular loop with several glycosylation sites as well as intracellular sites for serine and threonine phosphorylation between transmembrane helices 3 and 4. Differential glycosylation of the DAT occurs during postnatal development and aging (Patel et al. 1995). While the glycosylation status of a transporter protein appears to be relevant to folding stability, transport, and possibly ligand recognition, post-translational modification such as phosphorylation of the amino-terminal tail is linked to acute hormone-dependent regulation of transport activity. Studies utilizing chimeric monoamine transporters have revealed that the amino terminal transmembrane domains including an aspartate residue in the first transmembrane segment appear to be involved in the uptake mechanism and ion dependence, a cluster of serine residues in TMD6–TMD8 define antagonist and neurotoxin binding, and the remaining carboxy-terminal region is likely to determine stereoselectivity and high-affinity sites for the respective substrate (Kitayama *et al.* 1992; Gros *et al.* 1994). Moreover, cysteines in the DAT's second extracellular loop may provide sulfide residues crucial to full transporter expression, possibly through interference with membrane insertion or formation of quarternary structures. Moreover, it contains a potential degenerate leucine zipper motif and thus could oligomerize to a tetrameric assembly of identical units.

Characterization of the 5′-flanking regulatory region of *SLC6A3* has allowed elucidation of similarities in regulation of gene expression between each member of this gene family (Donovan *et al.* 1995). Sequences upstream of the transcriptional start sites contain GC-rich regions with multiple Sp1 sites but no canonical TATA and CAAT motifs. However, a single transcription start site, the lack of an initiator motif, and a conserved TATA-like sequence (TAAGA) located at 32 bp relative to the transcription start site are consistent with a TATA-containing promoter. These sequence elements, as well as specific upstream and intronic silencers, are likely to control DAT's dopaminergic neuron-selective expression.

ADHD is a genetically complex (polygenic) disorder likely due to the effects of genes modulating dopaminergic, serotonergic, and noradrenergic neurotransmitter pathways (Thapar *et al.* 1999). The efficacy of DAT ligands (e.g., methylphenidate and amphetamine) in the treatment of ADHD as well as DAT knockout mice's behavioral and biochemical hyperdopaminergic functions and failure to respond to psychostimulants indicates a role of the DAT in the pathophysiology and treatment response of this disorder. Several, but not all, population- and family-based association analyses have shown an increased frequency of the 10-copy repeat polymorphisms, suggesting that allelic variation in DAT function may be a modest risk factor in (Figure 3) (Waldman *et al.* 1988; Cook *et al.* 1995; Gill *et al.* 1997; Daly *et al.* 1999). Winsberg and Comings (1999) suggested that homozygosity of the 10-copy repeat is associated with poor response of ADHD symptomatology to methylphenidate. The significance of the repeat polymorphism and other coding and noncoding variations (e.g., SNPs) in the DAT gene as a risk factor in a variety of neuropsychiatric disorders, such as bipolar disorder, schizophrenia, Tourette's syndrome, and Parkinson's disease, as well as drug abuse including alcoholism, requires further clarification (for review see Vandenbergh *et al.* 2000).

The preferential action of cocaine and its analogs on DA uptake and release via the DAT, in addition to its interaction with NET and serotonin transporter (5HTT/SERT), induces multiple neurochemical and behavioral effects that are thought to represent the molecular basis of euphoria/reward, sensitization, and addictive behavior. Recent work indicates that cocaine reward/reinforcement, elicited by cocaine place preferences, is similar in both mice with a targeted inactivation and wild-type mice (Sora *et al.* 1998, 2001). NET and DAT knockout mice even display enhanced cocaine reward. However, deletions of both the DAT and 5HTT/SERT genes in double-knockout mice completely eliminated cocaine reward, thus defining the minimal set of gene products necessary for the euphoric and addictive effects of cocaine. In contrast, double knockouts of both 5HTT/SERT and NET not only failed to reduce cocaine reward but strongly enhanced place preference, further supporting the notion of an involvement of multiple transporters in the rewarding and aversive actions of psychostimulants and drugs of abuse (Figure 3) (Uhl *et al.* 2002).

SEROTONIN TRANSPORTER

Mood, cognition, and motor functions as well as circadian and neuroendocrine rhythms including food intake, sleep, and reproductive activity are modulated by the brainstem raphe serotonin (5HT) system. While 5HT controls a highly complex system of neural communication mediated by multiple pre- and post-synaptic 5HT receptor subtypes, high-affinity 5HT transport into the presynaptic neuron is mediated by a single protein, the 5HT transporter (5HTT/SERT), which is regarded as the initial site of action of antidepressant drugs and several neurotoxic compounds. Tricyclic antidepressants, such as prototypical imipramine, and the selective 5HT uptake inhibitors (paroxetine, citalopram, and sertraline) occupy several pharmacologically distinct sites overlapping at least partially the substrate binding site. These agents are widely used in the treatment of depression, anxiety, and impulse control disorders, as well as substance abuse including alcoholism.

Following systematic attempts to characterize genetically driven variation in 5HT uptake function, the 5HTT/SERT has assumed importance as a piece in the mosaic-like texture of personality traits such as anxiety and impulsivity.

The contribution of variability of the 5HTT/SERT gene-linked polymorphic region to individual phenotypic differences in temperament, personality, and behavior has been explored in several independent population/family genetic studies (for review see Lesch 1997; Lesch *et al.* 2002). The findings show that the 5HTT/SERT gene-linked polymorphic region influences traits of negative emotionality-related anxiety and aggression. Nevertheless, several efforts to detect associations between the 5HTT/SERT gene-linked polymorphic region and personality traits have been complicated by the use of small sample sizes, heterogeneous subject populations, ethnic and sociocultural characteristics, and differing methods of personality assessment. The studies indicate that the 5HTT/SERT gene-linked polymorphic region (Figure 4) has a moderate influence on these behavioral predispositions which corresponds to less than 4% of the total variance and approximately 8% of the genetic variance, based on estimates from twin studies using these and related measures which have consistently demonstrated that genetic factors contribute 40–60% of the variance in personality traits. This is consistent with the view that the influence of a single common polymorphism on continuously distributed traits is likely to be

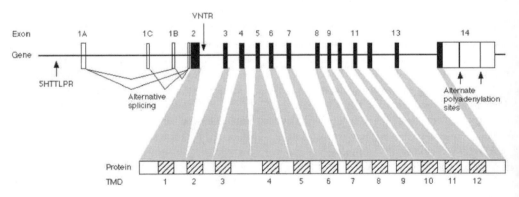

Figure 4 Genomic organization and corresponding protein segments of the human serotonin transporter (5HTT/SERT). Positions of exons shown as black and white boxes for the coding and noncoding regions, respectively, and TMDs are depicted as hatched areas. Alternate polyadenylation site usage resulting in multiple mRNA species is indicated. The location of the 5HTT/SERT gene-linked polymorphic region (5HTTLPR) in the 5'-flanking region and the VNTR in intron 2 is also shown.

small in humans. In addition to the exploration of the impact of allelic variation in 5HTT/SERT expression on anxiety, depression, and aggression-related personality traits, a role of the 5HTT/SERT gene-linked polymorphic region has been suggested in a variety of diseases such as depression, bipolar disorder, anxiety disorders, eating disorders, substance abuse, schizophrenia, and neurodegenerative disorders (for review see Lesch 2001). Based on a consistent overrepresentation of the high 5HTT/SERT activity l-allele, several studies implicated allelic variation of 5HTT/SERT function in the pathophysiology of ADHD (Figure 3) (Gainetdinov *et al.* 1999; Auerbach *et al.* 2001; Manor *et al.* 2001; Seeger *et al.* 2001; Kent *et al.* 2002; Retz *et al.* 2002; Zoroglu *et al.* 2002), obsessive–compulsive disorder (OCD), and autism.

A modulatory effect of the 5HTT/SERT gene-linked polymorphic region on cortical activity provided the first evidence that genotype–phenotype correlations may be accessible by functional imaging of the brain. Recently, Hariri *et al.* (2002) reported that individuals with one or two copies of the short variant of the 5HTT/SERT gene-linked polymorphic region exhibit greater amygdala neuronal activity, as assessed by functional magnetic resonance imaging (fMRI), in response to fearful stimuli compared with individuals homozygous for the long allele. These findings confirm that genetically driven variation of serotonergic function contributes to the response of brain regions underlying human emotional behavior and indicate that differential excitability of the amygdala to emotional stimuli may contribute to increased fear- and anxiety-related responses.

Most variants that change the structure of 5HTT/SERT protein transport activity are rare and their potential to alter 5HT activity remains to be determined (Lesch *et al.* 1995; Altemus *et al.* 1996; Di Bella *et al.* 1996; Glatt *et al.* 2001). Most of these variants have yet to be explored with respect to a functional effect on transport activity or association with a behavioral phenotype or disorder. Nevertheless, two nonsynonymous SNPs that change the coding sequence of the 5HTT/SERT were found to segregate with complex serotonergic dysfunction-related phenotypes, including OCD and other 5HT spectrum disorders (SSDs), or were associated with severe depression (see Box: Obsessive–Compulsive Disorder and Serotonin Spectrum Disorders) usually begins

---

OBSESSIVE–COMPULSIVE DISORDER AND SEROTONIN SPECTRUM DISORDERS

Serotonergic dysregulation and altered serotonin (5HT) transporter (5HTT/SERT) function have been implicated in anxiety and affective disorders, and 5HT-selective reuptake inhibitors (SSRIs) are widely used in the treatment of depression, anxiety disorders, anorexia nervosa, and substance abuse as well as obsessive–compulsive disorder (OCD). 5HT spectrum disorders (SSDs) are characterized by changes in mood, cognition, motor activation, circadian rhythms including sleep and food intake, reproduction, pain, and immune response. OCD is characterized by recurrent and intrusive thoughts and/or repetitive and ritualized behaviors. Obsessions and compulsions are experienced by the individual as irrational and significantly restrict occupational, social, and interpersonal actions. A rare Ile425Val variant of the 5HTT/SERT gene (*SLC6A4*) located on chromosome 17q11.2 was detected in two affected individuals and their family members with OCD and related disorders. Affected individuals and family members carrying the Ile425Val variant met diagnostic criteria for other SSDs including Asperger's syndrome, social phobia, anorexia nervosa, tic disorder, depression, and alcohol abuse/dependence. SSRIs that enhance central serotonergic function have been highly effective in the treatment of this OC symptomatology.

in adolescence or early adulthood and is characterized by either obsessions or compulsions that cause marked distress and are time consuming or significantly interfere with an individual's normal routine or functioning. Obsessions are recurrent and persistent ideas, thoughts, impulses, or images that an individual attempts to ignore or suppress, whereas compulsions are repetitive, purposeful, and intentional behaviors performed in response to an obsession to neutralize or prevent discomfort, worry, and anxiety, or some dreaded event. OCD often co-occurs with other disorders such as anxiety, depression, anorexia nervosa, schizophrenia, and Tourette's syndrome

A missense mutation resulting in a conserved Ile425Val substitution in the 5HTT/SERT gene was detected in two affected individuals and their family members with OCD and related disorders (Figure 5) (Ozaki et al. 2003). Six of seven family members with the variant had OCD or obsessive–compulsive personality disorder. In addition, the affected individuals and their immediate family members carrying the Ile425Val variant met diagnostic criteria for other disorders including Asperger's syndrome, social phobia, anorexia nervosa,

tic disorder, depression, and alcohol abuse/dependence. The evolutionarily conserved Ile425Val substitution is located in TMD8 and may modify the (α-helical secondary structure of the 5HTT/SERT protein and consequently transport function (Figure 6.B). Expression studies of the mutant 5HTT/SERT cDNA in human cells demonstrated a gain of function via constitutive activation of 5HT transport in a nitric-oxide-stimulated pathway resulting in a twofold increase in 5HT uptake. Taken together, these findings strongly indicate that gain-of-function mutations associated with coding sequence variants may contribute to the expression of psychopathology related to serotonergic dysfunction in some families.

Interestingly, two brothers from one of the families suffering from OCD and Asperger's syndrome and carrying the Ile425Val variant also had the l/l genotype of the 5HTT/SERT gene-linked polymorphic region, which was previously found to be asociated with or preferentially transmitted in both OCD (Figure 3) (McDougle et al. 1998; Bengel et al. 1999) and autism (Klauck et al. 1997; Yirmiya et al. 2001). Moreover, a conservative Leu255Met substitution located in TMD4 was detected in

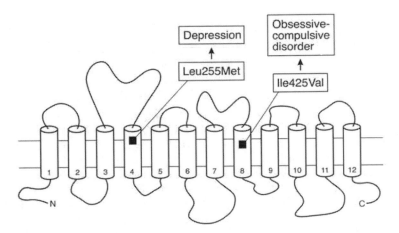

Figure 5 Structure of the human serotonin transporter (5HTT/SERT) protein characterized by 12 TMDs and intracellular N- and C-termini. The positions of the Leu255Met substitution (associated with depression) and Ile425Val variant (segregating with OCD) are indicated by black boxes in TMD4 and TMD8, respectively.

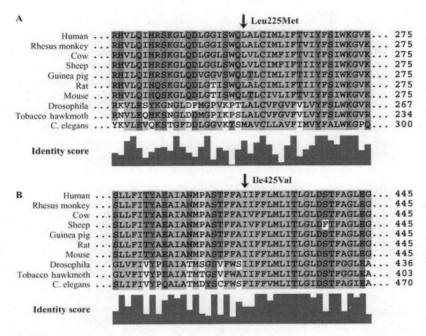

Figure 6 Alignment of the 5HTT/SERT protein segments containing (A) the highly conserved Leu255Met substitution and (B) the Ile425Val variant.

a patient with delusional depression, who was also found to carry an s/s genotype of the polymorphic region, which is implicated in anxiety- and depression-related traits (Figure 5) (Di Bella *et al.* 1996). Possibly, low 5HTT/SERT gene expression together with the Leu255Met variant, a residue which is highly conserved among various species, could additively perturb 5HTT/SERT function or regulation (Figure 6A). These two examples of co-occurence and possible cooperativity of allelic variation in gene expression and protein structure might represent a "double hit" with functional consequences in the same gain- or loss-of-function direction for both of these 5HTT/SERT gene variations.

Although this notion requires further functional analysis, such as alteration of substrate or ion transport or response to drugs that alter serotonergic signaling, advances in 5HTT/SERT gene knockout studies are also changing views of the relevance of adaptive 5HT uptake function and homeostasis in the developing human brain as well as molecular processes underlying anxiety- and aggression-related traits (Bengel *et al.* 1998; Murphy *et al.* 2001; Persico *et al.* 2001; Salichon *et al.* 2001; Holmes *et al.* 2002), as well as SSDs including depression, bipolar disorder, autism, and anorexia nervosa (Lesch 2001). Nevertheless, despite growing evidence for a potential role of the 5HTT/SERT in the integration of synaptic connections in the rodent, nonhuman primate, and human brain during critical periods of development, adult life, and old age, knowledge of the molecular mechanisms involved in these fine-tuning processes remains fragmentary and requires major research efforts in the years to come.

SUMMARY

Transporters for the monoamines norepinephrine (NE), dopamine (DA), and serotonin (5HT) comprise a distinct subgroup of the extended family of neurotransporters. NE, DA, and 5HT transporters (NET, DAT, and 5HTT/SERT, respectively) are sodium/chloride-dependent high-affinity transport proteins located in the plasma membrane of presynaptic neurons that mediate the clearance and recycling of the monoamine neurotransmitters/modulators from the synaptic cleft and thereby tightly control monoamine homeostasis. In the past, pharmacological approaches have accumulated persuasive proof that behavioral traits related to NET, DAT, and 5HTT/SERT activity, such as autonomic function, locomotion, cognition, and emotional responses, are influenced by genetic factors. However, the genetic component of behavioral traits related to NET, DAT, and 5HTT/SERT function, particularly including anxiety, depression, aggression, and euphoria/reward, is highly complex, polygenic, and epistatic. Elucidation of monoamine transporter gene regulation, intracellular trafficking, and post-translational modification as well as short- /long-term regulation provide the means of studying the impact of monoamine transporter gene variants on behavioral traits and complex disorders including orthostatic intolerance (OI), obsessive–compulsive disorder (OCD), and other 5-HT spectrum disorders (SSDs) at the molecular level. Orthostatic intolerance is an autonomic syndrome characterized by lightheadedness, fatigue, altered mentation, syncope, and postural tachycardia. A subset of patients show pathophysiologic features that are associated with a Ala457Pro variant of the NET gene located on chromosome 16q12.2. OCD is characterized by recurrent and intrusive thoughts and/or repetitive and ritualized behaviors. A rare Ile425Val variant of the 5HTT/SERT gene mapped at 17q11.2 was detected in two affected individuals and their family members with OCD and related disorders. Affected individuals and family members carrying the Ile425Val variant met diagnostic criteria for other SSDs including Asperger's syndrome, social phobia, anorexia nervosa, tic disorder, depression, and alcohol abuse/dependence. Modeling monoamine-transporter-related traits in genetically modified mice has eventually entered the mainstream of research aimed at the identification of molecular mechanisms underlying the pathophysiology and pharmacologic therapy of neuropsychiatric behavioral disorders, such as attention deficit hyperactivity disorder and drug abuse, and other medical conditions.

REFERENCES

Altemus, M., Murphy, D.L., Greenberg, B., and Lesch, K.P. (1996). Intact Coding Region of the Serotonin Transporter in Obsessive–Compulsive Disorder. *Am. J. Med. Genet.*, **67**, 104–109.

Amara, S.G. and Sonders, M.S. (1998). Neurotransmitter Transporters as Molecular Targets for Addictive Drugs. *Drug Alcohol Depend.*, **51**, 87–96.

Apparsundaram, S., Galli, A., DeFelice, L.J., Hartzell, H.C., and Blakely, R.D. (1998). Acute Regulation of Norepinephrine Transport. I. Protein Kinase-C Linked Muscarinic Receptors Influence Transport Capacity and Transporter Density in SK–N–Sh Cells. *J. Pharmacol. Exp. Ther.*, **287**, 733–743.

Apparsundaram, S., Sung, U., Price, R.D., and Blakely, R.D. (2001). Trafficking-Dependent and Independent Pathways of Neurotransmitter Transporter Regulation Differentially Involving p38 Mitogen-Activated Protein Kinase Revealed in Studies of Insulin Modulation of Norepinephrine Transport in SK–N–SH Cells. *J. Pharmacol. Exp. Ther.*, **299**, 666–677.

Auerbach, J.G., Benjamin, J., Faroy, M., Geller, V., and Ebstein, R. (2001). DRD4 Related to Infant Attention and Information Processing: A Developmental Link to ADHD. *Psychiatr. Genet.*, **11**, 31–35.

Bannon, M.J. and Whitty, C.J. (1997). Age-Related and Regional Differences in Dopamine Transporter mRNA Expression in Human Midbrain. *Neurology*, **48**, 969–977.

Bannon, M.J., Poosch, M.S., Xia, Y., Goebel, D.J., Cassin, B., and Kapatos, G. (1992). Dopamine Transporter mRNA Content in Human Substantia Nigra Decreases Precipitously with Age. *Proc. Natl Acad. Sci. USA*, **89**, 7095–7099.

Bannon, M.J., Michelhaugh, S.K., Wang, J., and Sacchetti, P. (2001). The Human Dopamine Transporter Gene: Gene Organization, Transcriptional Regulation, and Potential Involvement in Neuropsychiatric Disorders. *Eur. Neuropsychopharmacol.*, **11**, 449–445.

Bauman, P.A. and Blakely, R.D. (2002). Determinants within the C-Terminus of the Human Norepinephrine Transporter Dictate Transporter Trafficking, Stability, and Activity. *Arch. Biochem. Biophys.*, **404**, 80–91.

Bengel, D., Murphy, D.L., Andrews, A.M., Wichems, C.H., Feltner, D., Heils, A., et al. (1998). Altered Brain Serotonin Homeostasis and Locomotor Insensitivity to 3,4-Methylenedioxymethamphetamine ("Ecstasy") in Serotonin Transporter-Deficient Mice. Mol. Pharmacol. 53, 649–655.

Bengel, D., Greenberg, B., Cora-Locatelli, G., Altemus, M., Heils, A., Li, Q., et al. (1999). Association of the Serotonin Transporter Promoter Regulatory Region Polymorphism and Obsessive–Compulsive Disorder. Mol. Psychiatry, 4, 463–466.

Boja, J.W., Vaughan, R., Patel, A., Shaya, E.K., and Kuhar, M.J. (1994). The Dopamine Transporter. In H.B. Niznik (Ed.), Dopamine Receptors and Transporters, New York: Marcel Dekker, pp. 611–644.

Bruss, M., Kunz, J., Lingen, B., and Bonisch, H. (1993). Chromosomal Mapping for the Tryclic Antidepressant-Sensitive Noradrenaline Transporter. Hum. Genet., 91, 278–280.

Cook, E.H., Jr., Stein, M.A., Krasowski, M.D., Cox, N.J., Olkon, D.M., Kieffer, J.E., et al. (1995). Association of Attention-Deficit Disorder and the Dopamine Transporter Gene. Am. J. Hum. Genet., 56, 993–998.

Daly, G., Hawi, Z., Fitzgerald, M., and Gill, M. (1999). Mapping Susceptibility Loci in Attention Deficit Hyperactivity Disorder: Preferential Transmission of Parental Alleles at DAT1, DBH and DRD5 to Affected Children. Mol. Psychiatry, 4, 192–196.

Di Bella, D., Catalano, M., Balling, U., Smeraldi, E., and Lesch, K.P. (1996). Systematic Screening for Mutations in the Coding Region of the 5-HTT Gene Using PCR and DGGE. Am. J. Med. Genet., 67, 541–545.

Donovan, D.M., Vandenbergh, D.J., Perry, M.P., Bird, G.S., Ingersoll, R., Nanthakumar, E., et al. (1995). Human and Mouse Dopamine Transporter Genes: Conservation of 5′-Flanking Sequence Elements and Gene Structures. Brain Res. Mol. Brain Res., 30, 327–335.

Frazer, A., Gerhardt, G.A., and Daws, L.C. (1999). New Views of Biogenic Amine Transporter Function: Implications for Neuropsychopharmacology. Int. J. Neuropsychopharmacol., 2, 305–320.

Fuke, S., Suo, S., Takahashi, N., Koike, H., Sasagawa, N., and Ishiura, S. (2001). The VNTR Polymorphism of the Human Dopamine Transporter (DAT1) Gene Affects Gene Expression. Pharmacogenom. J., 1, 152–156.

Gainetdinov, R.R., Wetsel, W.C., Jones, S.R., Levin, E.D., Jaber, M., et al. (1999). Role of Serotonin in the Paradoxical Calming Effect of Psychostimulants on Hyperactivity. Science, 283, 397–401.

Gainetdinov, R.R., Sotnikova, T.D., and Caron, M.G. (2002). Monoamine Transporter Pharmacology and Mutant Mice. Trends Pharmacol., 23, 367–373.

Gill, M., Daly, G., Heron, S., Hawi, Z., and Fitzgerald, M. (1997). Confirmation of Association between Attention Deficit Hyperactivity Disorder and a Dopamine Transporter Polymorphism. Mol. Psychiatry, 2, 311–313.

Giros, B., Wang, Y.M., Suter, S., Mcleskey, S.B., Pifl, C., and Caron, M.G. (1994). Delineation of Discrete Domains for Substrate, Cocaine and Tricyclic Antidepressant Interactions Using Chimeric Dopamine–Norepinephrine Transporters. J. Biol. Chem., 295, 15985–15988.

Glatt, C.E., DeYoung, J.A., Delgado, S., Service, S.K., Giacomini, K.M., Edwards, R.H., et al. (2001). Screening a Large Reference Sample to Identify Very Low Frequency Sequence Variants: Comparisons between Two Genes. Nat. Genet., 27, 435–438.

Hahn, M.K. and Blakely, R.D. (2002). Gene Organization and Polymorphisms of Monoamine Transporters. Relationship to Psychiatric and other Complex Diseases. In M.E.A. Reith (ed.), Neurotransmitter Transporters: Structure, Function, and Regulation, 2nd edn, Totowa, NJ: Humana Press, pp. 111–169.

Hariri, A.R., Mattay, V.S., Tessitore, A., Kolachana, B., Fera, F., Goldman, D., et al. (2002). Serotonin Transporter Genetic Variation and the Response of the Human Amygdala. Science, 297, 400–403.

Harrington, K.A., Augood, S.J., Kingsbury, A.E., Foster, O.J., and Emson, P.C. (1996). Dopamine Transporter (Dat) and Synaptic Vesicle Amine Transporter (VMAT2) Gene Expression in the Substantia Nigra of Control and Parkinson's Disease. Brain Res. Mol. Brain Res., 36, 885–897.

Heinz, A., Goldman, D., Jones, D.W., Palmour, R., Hommer, D., Gorey, J.G., et al. (2000). Genotype Influences In Vivo Dopamine Transporter Availability in Human Striatum. Neuropsychopharmacology, 22, 133–139.

Hertting, G. and Axelrod, J. (1961). Fate of Tritiated Noradrenaline at the Sympathetic Nerve Endings. Nature, 192, 172–173.

Holmes, A., Murphy, D.L., and Crawley, J.N. (2002). Reduced Aggression in Mice Lacking the Serotonin Transporter. Psychopharmacology (Berlin), 161, 160–167.

Innis, R.B., Seibyl, J.P., Scanley, B.E., Laruelle, M., Abidargham, A., Wallace, E., et al. (1993). Single-Photon Emission Computed Tomographic Imaging Demonstrates Loss of Striatal Dopamine Transporters in Parkinson's Disease. Proc. Natl Acad. Sci. USA, 90, 11965–11969.

Iversen, L.L. (1971). Role of Transporter Uptake Mechanisms in Synaptic Neurotransmission. Br. J. Pharmacol., 41, 571–591.

Jacobsen, L.K., Staley, J.K., Zoghbi, S.S., Seibyl, J.P., Kosten, T.R., Innis, R.B., et al. (2000). Prediction of Dopamine Transporter Binding Availability by Genotype: A Preliminary Report. Am. J. Psychiatry, 157, 1700–1703.

Joyce, J.N., Smutzer, G., Whitty, C.J., Myers, A., and Bannon, M.J. (1997). Differential Modification of Dopamine Transporter and Tyrosine Hydroxylase mRNAs in Midbrain of Subjects with Parkinson's,

Alzheimer's with Parkinson's, and Alzheimer's Disease. *Mov. Disord.*, **12**, 885–897.

Kent, L., Doerry, U., Hardy, E., Parmar, R., Gingell, K., Hawi, Z., et al. (2002). Evidence that Variation at the Serotonin Transporter Gene Influences Susceptibility to Attention Deficit Hyperactivity Disorder (ADHD): Analysis and Pooled Analysis. *Mol. Psychiatry*, **7**, 908–912.

Kim, C.H., Ardayfio, P., and Kim, K.S. (2001). An E-Box Motif Residing in the Exon/Intron 1 Junction Regulates Both Transcriptional Activation and Splicing of the Human Norepinephrine Transporter Gene. *J. Biol. Chem.*, **276**, 24797–24805.

Kim, C.H., Hwang, D.Y., Park, J.J., and Kim, K.S. (2002). A Proximal Promoter Domain Containing a Homeodomain-Binding Core Motif Interacts with Multiple Transcription Factors, Including HoxA5 and Phox2 Proteins, and Critically Regulates Cell Type-Specific Transcription of the Human Norepinephrine Transporter Gene. *J. Neurosci.*, **22**, 2579–2589.

Kitayama, S., Shimuda, S., Xu, H., Markham, L., Donovan, D.M., and Uhl, G.R. (1992). Dopamine Transporter Site-Directed Mutations Differentially Alter Substrate Transport and Cocaine Binding. *Proc. Natl Acad. Sci. USA*, **89**, 7782–7785.

Kitayama, S., Morita, K., and Dohi, T. (2001). Functional Characterization of the Splicing Variants of Human Norepinephrine Transporter. *Neurosci. Lett.*, **312**, 108–112.

Klauck, S.M., Poustka, F., Benner, A., Lesch, K.P., and Poustka, A. (1997). Serotonin Transporter (5-HTT) Gene Variants Associated with Autism? *Hum. Mol. Genet.*, **6**, 2233–2238.

Kruger, R., Kuhn, W., Muller, T., Woitalla, D., Graeber, M., Kosel, S., et al. (1998). Ala30Pro Mutation in the Gene Encoding Alpha-Synuclein in Parkinson's Disease. *Nat. Genet.*, **18**, 106–108.

Laakso, A. and Hietala, J. (2000). PET Studies of Brain Monoamine Transporters. *Curr. Pharm. Des.*, **6**, 1611–1623.

Lee, F.J., Liu, F., Pristupa, Z.B., and Niznik, H.B. (2001). Direct Binding and Functional Coupling of Alpha-Synuclein to the Dopamine Transporters Accelerates Dopamine-Induced Apoptosis. *FASEB J.*, **15**, 916–926.

Lehmensiek, V., Tan, E.M., Schwarz, J., and Storch, A. (2002). Expression of Mutant Alpha-Synucleins Enhances Dopamine Transporter-Mediated MPP$^+$ Toxicity In Vitro. *NeuroReport*, **13**, 1279–1283.

Lesch, K.P. (1997). Molecular Biology, Pharmacology, and Genetics of the Serotonin Transporter: Psychological and Clinical Implications. In H.G. Baumgarten and M. Göthert (Eds), *Serotonergic Neurons and 5-HT Receptors in the CNS*, Berlin: Springer, pp. 671–705.

Lesch, K.P. (2001). Serotonin Transporter: From Genomics and Knockouts to Behavioral Traits and Psychiatric Disorders. In M. Briley and F. Sulser (eds), *Molecular Genetics of Mental Disorders*, London: Martin Dunitz, pp. 221–267.

Lesch, K.P., Gross, J., Wolozin, B.L., Murphy, D.L., and Riederer, P. (1995). Primary Structure of the Serotonin Transporter in Unipolar Depression and Bipolar Disorder. *Biol. Psychiatry*, **37**, 215–223.

Lesch, K.P., Heils, A., and Riederer, P. (1996). The Role of Neurotransporters in Exitotoxicity, Neuronal Cell Death and other Neurodegenerative diseases. *J. Mol. Med.*, **74**, 365–378.

Lesch, K.P., Greenberg, B.D., Higley, J.D., and Murphy, D.L. (2002). Serotonin Transporter, Personality, and Behavior: Toward Dissection of Gene–Gene and Gene–Environment Interaction. In J. Benjamin, R. Ebstein, and R.H. Belmaker (eds), *Molecular Genetics and the Human Personality*, Washington, DC: American Psychiatric Press, pp. 109–135.

Manor, I., Eisenberg, J., Tyano, S., Sever, Y., Cohen, H., Ebstein, R.P., et al. (2001). Family-Based Association Study of the Serotonin Transporter Promoter Region Polymorphism (5-HTTLPR) in Attention Deficit Hyperactivity Disorder. *Am. J. Med. Genet.*, **105D**, **91–95**.

McDougle, C.J., Epperson, C.N., Price, L.H., and Gelernter, J. (1998). Evidence for Linkage Disequilibrium between Serotonin Transporter Gene (*SLC6A4*) and Obsessive–Compulsive Disorder. *Mol. Psychiatry*, **3**, 270–273.

Meyer, J., Wiedemann, P., Okladnova, O., Bruss, M., Staab, T., Stober, G., et al. (1998). Cloning and Functional Characterization of the Human Norepinephrine Transporter Gene Promoter. *J Neural Transm.*, **105**, 1341–1350.

Murphy, D.L., Li, Q., Engel, S., Wichems, C., Andrews, A., Lesch, K.P., et al. (2001). Genetic Perspectives on the Serotonin Transporter. *Brain Res. Bull.*, **56**, 487–494.

Ozaki, N., Goldman, D., Kaye, W.H., Plotnicov, K., Greenberg, B.D., and Murphy, D.L. (2003). Functional Missense Mutation in the Serotonin Transporter Gene Associated with Obsessive–Compulsive Disorder and Related Neuropsychiatric Disorders. *Mol. Psychiatry*, in press.

Pacholczyk, T., Blakely, R.D., and Amara, G.S. (1991). Expression Cloning of a Cocaine- and Antidepressant-Sensitive Human Noradrenaline Transporter. *Nature*, **350**, 350–354.

Paczowski, F.A., Bonisch, H., and Bryan-Lluka, L.J. (2002). Pharmacalogical Properties of the Naturally Occurring Ala457 Pro Variant of the Human Norepinephrine Transporter. *Pharmacogenetics*, **12**, 165–173.

Patel, A.P., Cerruti, C., Vaughan, R.A., and Kuhar, M.J. (1995). Developmentally Regulated Glycosylation of Dopamine Transporter. *Brain Res. Dev. Brain Res.*, **83**, 53–58.

Persico, A.M., Revay, R.S., Mössner, R., Conciatori, M., Marino, R., Baldi, A., et al. (2001). Barrel Pattern Formation in Somatosensory Cortical Layer IV

Requires Serotonin Uptake by Thalamocortical Endings, while Vesicular Monoamine Release is Necessary for Development of Supragranular Layers. *J. Neurosci.*, **21**, 6862–6873.

Polymeropoulos, M.H., Lavedan, C., Leroy, E., Ide, S.E., Dehejia, A., Dutra, A., *et al.* (1997). Mutation in the Alpha-Synuclein Gene Identified in Families with Parkinson's Disease. *Science*, **276**, 2045–2047.

Porzgen, P., Bonisch, H., and Bruss, M. (1995). Molecular Cloning and Organization of the Coding Region of the Human Norepinephrine Transporter Gene. *Biochem. Biophys. Res. Commun.*, **215**, 1145–1150.

Ramamoorthy, J.D., Ramamoorthy, S., Papapetropoulos, A., Catravas, J.D., Leibach, F.H., and Ganapathy, V. (1995). Cyclic AMP-Independent Up-Regulation of the Human Serotonin Transporter by Stauroporine in Choriocarcinoma Cells. *J. Biol. Chem.*, **270**, 17189–17195.

Retz, W., Thome, J., Blocher, D., Baader, M., and Rosler, M. (2002). Association of Attention Deficit Hyperactivity Disorder-Related Psychopathology and Personality Traits with the Serotonin Transporter Promoter Region Polymorphism. *Neurosci. Lett.*, **319**, 133–136.

Robertson, D. (1999). The Epidemic of Orthostatic Tachycardia and Orthostatic Intolerance. *Am. J. Med. Sci.*, **317**, 75–77.

Roubert, C., Cox, P.J., Bruss, M., Hamon, M., Bonisch, H., and Giros, B. (2001). Determination of Residues in the Norepinephrine Transporter that are Critical for Tricyclic Antidepressant Affinity. *J. Biol. Chem.*, **276**, 8254–8260.

Salichon, N., Gaspar, P., Upton, A.L., Picaud, S., Hanoun, N., Hamon, M., *et al* . (2001). Excessive Activation of Serotonin (5-HT) 1B Receptors Disrupts the Formation of Sensory Maps in Monoamine Oxidase A and 5-HT Transporter Knock-Out Mice. *J. Neurosci.*, **21**, 884–896.

Seeger, G., Schloss, P., and Schmidt, M.H. (2001). Marker Gene Polymorphisms in Hyperkinetic Disorder – Predictors of Clinical Response to Treatment with Methylphenidate? *Neurosci. Lett.*, **313**, 45–48.

Shannon, J.R., Flattem, N.L., Jordan, J., Jacob, G., Black, B.K., Biaggioni, I., *et al.* (2000). Orthostatic Intolerance and Tachycardia Associated with Norepinephrine-Transporter Deficiency. *N. Engl. J. Med.*, **342**, 541–549.

Sora, I., Wichems, C., Takahashi, N., Li, X.F., Zeng, Z., Revay, R., *et al.* (1998). Cocaine Reward Models: Conditioned Place Preference can be Established in Dopamine- and in Serotonin-Transporter Knockout Mice. *Proc. Natl Acad. Sci. USA*, **95**, 7699–7704.

Sora, I., Hall, F.S., Andrews, A.M., Itokawa, M., Li, F.X., Wei, H.B., *et al.* (2001). Molecular Mechanisms of Cocaine Reward: Combined Dopamine and Serotonin Knockouts Eleiminate Cocaine Place Preference. *Proc. Natl Acad. Sci. USA*, **98**, 5300–5305.

Tatsch, K. (2001). Imaging of the Dopaminergic System in Parkinsonism with SPET. *Nucl. Med. Commun.*, **22**, 819–827.

Thapar, A., Holmes, J., Poulton, K., and Harrington, R. (1999). Genetic Basis of Attention Deficit and Hyperactivity. *Br. J. Psychiatry*, **174**, 105–111.

Torres, G.E., Gainetdinov, R.R., and Caron, M.G. (2003). Plasma Membrane Monoamine Transporters: Structure, Regulation and Function. *Nat. Rev. Neurosci.*, **4**, 13–25.

Uhl, G.R. and Johnson, P.S. (1994). Neurotransmitter Transports: Three Important Gene Families for Neuronal Function. *J. Exp. Biol.*, **196**, 229–236.

Uhl, G.R., Hall, F.S., and Sora, I. (2002). Cocaine, Reward, Movement and Monoamine Transporters. *Mol. Psychiatry*, **7**, 21–26.

Urwin, R.E., Bennetts, B., Wilcken, B., Lampropoulos, B., Beumont, P., Clarke, S., *et al.* (2002). Anorexia Nervosa (Restrictive Type) Is Associated with a Polymorphism in the Novel Norepinephrine Transporter Gene Promoter Polymorphic Region. *Mol. Psychiatry*, **7**, 652–657.

Vandenbergh, D.J., Persico, A.M., Hawkins, A.L., Griffin, C.A., Li, X., Jabs, E.W., *et al.* (1992). Human Dopamine Transporter Gene (DAT1) Maps to Chromosome 5p15.3 and Displays a VNTR. *Genomics*, **14**, 1104–1106.

Vandenbergh, D.J., Thompson, M.D., Cook, E.H., Bendahhou, E., Nguyen, T., Krasowski, M.D., *et al.* (2000). Human Dopamine Transporter Gene: Coding Region Conservation Among Normal, Tourette's Disorder, Alcohol Dependence and Attention-Deficit Hyperactivity Disorder Populations. *Mol. Psychiatry*, **5**, 283–292.

Verhoeff, N.P. (1999). Radiotracer Imaging of Dopaminergic Transmission in Neuropsychiatric Disorders. *Psychopharmacology (Berlin)*, **147**, 217–249.

Vingerhoets, F.J.G., Snow, B.J., Tetrud, J.W., Langston, J.W., Schulzer, M., and Calne, D.B. (1994). Positron Emission Tomographic Evidence for Possession of Human MPTP-Induced Dopaminergic Lesions. *Ann. Neurol.*, **36**, 765–770.

Wakade, A.R., Wakade, T.D., Poosch, M., and Bannon, M.J. (1996). Noradrenaline Transport and Transporter mRNA of Rat Chromaffin Cells Are Controlled by Dexamethasone and Nerve Growth Factor. *J. Physiol.*, **494**, 67–75.

Waldman, I.D., Rowe, D.C., Abramowitz, A., Kozel, S.T., Mohr, J.H., Sherman, S.L., *et al.* (1998). Association and Linkage of the Dopamine Transporter Gene and Attention-Deficit Hyperactivity Disorder in Children: Heterogeneity Owing to Diagnostic Subtype and Severity. *Am. J. Hum. Genet.*, **63**, 1767–1776.

Winsberg, B.G. and Comings, D.E. (1999). Association of the Dopamine Transporter Gene (DAT1) with Poor

Methylphenidate Response. *J. Am. Acad. Child Adolesc. Psychiatry*, **38**, 1474–1477.

Yirmiya, N., Pilowsky, T., Nemanov, L., Arbelle, S., Feinsilver, T., Fried, I., *et al.* (2001). Evidence for an Association with the Serotonin Transporter Promoter Region Polymorphism and Autism. *Am. J. Med. Genet.*, **105**, 381–386.

Zahniser, N.R. and Doolen, S. (2001). Chronic and Acute Regulation of Na^+/Cl^--Dependent Neurotransmitter Transporters: Drugs, Substrates, presynaptic Receptors, and Signaling Systems. *Pharmacol. Ther.*, **92**, 21–55.

Zoroglu, S.S., Erdal, M.E., Alasehirli, B., Erdal, N., Sivasli, E., Tutkun, H., *et al.* (2002). Significance of Serotonin Transporter Gene 5-HTTLPR and Variable Number of Tandem Repeat Polymorphisms in Attention Deficit Hyperactivity Disorder. *Neuropsychobiology*, **45**, 176–181.

RAJINI RAO* AND GIUSEPPE INESI**

Ca^{2+}-ATPase Genes and Related Diseases

CELLULAR CA^{2+} AND THE ROLE OF CA^{2+}-TRANSPORTING ATPASES

Transient and localized elevations in cytosolic calcium act as signals for diverse cellular events including muscle contraction, secretion, cell cycle control, fertilization, transcriptional activation, and the regulation of complex metabolic pathways. Calcium release channels mediate the rapid and downhill flow of calcium from extracellular and intracellular stores to trigger these signaling events (Berridge *et al.* 2000). However, prolonged elevation of cytosolic Ca^{2+} is unequivocally toxic and can lead to apoptosis. Therefore a critical prerequisite for all calcium signaling events is the return to submicromolar cytoplasmic calcium levels by an array of Ca^{2+}-transporting ATPases located at the plasma membrane and various endomembranes of the cell (Figure 1). Proteins that modulate the activity of these transporters, as well as Ca^{2+}-binding proteins that buffer cellular Ca^{2+}, are also important in the control of cellular calcium. Clearly, dysfunction of any component of the Ca^{2+} signaling and homeostatic machinery can potentially lead to disease. This chapter will focus primarily on

dysfunction of Ca^{2+}-pumping ATPases, with a discussion on the role of their regulatory proteins, phospholamban, and sarcolipin.

The Ca^{2+}-ATPases belong to the superfamily of P-type cation pumps, so called because they undergo transient phosphorylation from ATP during the catalytic cycle. Members of this family include other clinically relevant ion pumps, such as the Cu^{2+}-ATPases defective in Menkes and Wilson disease (see Chapter 5), the ubiquitous Na$^+$, K$^+$-ATPase that is the target of cardiac glycosides used in the treatment of failing hearts (Akera and Brody 1977), and the gastric H$^+$, K$^+$-ATPase that is the target of the anti-ulcer drug omeprazole (Sachs 1997).

MOLECULAR DIVERSITY AMONG CA^{2+}-ATPASES

Overview

In recent years, it has become clear that there are three phylogenetically distinct subtypes of Ca^{2+}-ATPases, each with unique non-overlapping subcellular distributions, distinct biochemical properties, and consequently distinct physiological roles. The Ca^{2+}-ATPases have been named after the membranes in which they reside: hence, the *s*arco/*e*ndoplasmic *r*eticulum *C*a^{2+}-*A*TPase (SERCA), the *p*lasma *m*embrane *C*a^{2+}-*A*TPase (PMCA), and the Golgi/*s*ecretory *p*athway *C*a^{2+}-*A*TPase (SPCA).

*Johns Hopkins University School of Medicine, 725 N. Wolfe Street, Baltimore, MD 21205, USA

** University of Maryland, School of Medicine, 108 N. Greene Street, Baltimore, MD 21201-1503, USA

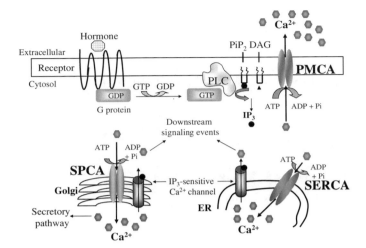

Figure 1   Cellular role of Ca^{2+}-ATPases. Calcium from extracellular or intracellular stores enters the cytoplasm through Ca^{2+} channels, which open in response to one of several possible signaling events such as the G-protein-mediated cascade shown here. For example, the binding of a hormone to a 7-transmembrane helix receptor triggers production of the secondary messenger IP$_3$, which binds to and activates ligand-gated receptor channels on the ER and Golgi membranes (Berridge *et al.* 2000). The removal of cytosolic Ca^{2+} by a family of ATP-driven Ca^{2+} pumps (PMCA, SPCA, and SERCA) that localize to distinct membrane compartments terminates the calcium signal.

The SERCA pumps are responsible for sequestering Ca^{2+} into the extensive reticular network of the cell that serves as a large agonist-sensitive reservoir. In the specialized sarcoplasmic reticulum of skeletal muscle, this pump is expressed at impressively high densities, nearing 50% of total membrane protein, which has facilitated its purification, extensive characterization, and, most recently, structural determination by X-ray crystallography (reviewed in Hussain and Inesi 1999; East 2000; MacLennan 2000). A distinguishing feature of the SERCAs is their inhibition by nanomolar concentrations of the plant sesquiterpene, thapsigargin (Sagara and Inesi 1991). The extrusion of cytosolic Ca^{2+} from the cell is accomplished by the PMCA pumps that occur ubiquitously in the plasma membranes of all mammalian cells. Although Ca^{2+} extrusion at the cell membrane is assisted by a high-capacity Na$^+$/Ca^{2+} exchanger in most cell types, PMCA has a higher affinity for Ca^{2+} ($K_M = 0.1\,\mu$M K_M) so that the steady-state resting Ca^{2+} concentrations are set by the pump (reviewed by Zylinska and Soszynski 2000).

In some cell types, notably erythrocytes and auditory hair cells, the PMCAs are solely responsible for Ca^{2+} extrusion. PMCAs are characterized by an extended C-terminal tail with an autoinhibitory domain. Binding of Ca^{2+}-calmodulin to this domain relieves inhibition, directly correlating pump activity with cytosolic Ca^{2+} concentrations (Penniston and Enyedi 1998).

The third type of Ca^{2+} pump, and the most recent to be discovered, are the SPCAs that constitute a distinct subtype from the SERCAs and PMCAs (Rudolph *et al.* 1989; Hu *et al.* 2000; Ton *et al.* 2002). The SPCAs sequester Ca^{2+} into the compartments of the Golgi apparatus and support the normal functions of the secretory pathway, including protein sorting and processing. There is emerging evidence that the SPCA pumps have uniquely high affinity for Mn^{2+} ions, with an apparent K_M of about 20 nM (Mandal *et al.* 2000). It appears that Mn^{2+} transport activity of the SPCA is important for delivery of lumenal Mn^{2+} to Golgi enzymes, such as mannosyl transferases, and also in cellular Mn^{2+} detoxification

(Bolton *et al.* 2002). While it is clear that Golgi Ca^{2+} can reach millimolar levels (Chandra *et al.* 1994), the storage capacity of the Golgi is small relative to that of the endoplasmic reticulum, and therefore the contribution of this organelle as an agonist-sensitive store remains to be established.

Isoforms and splice variants

The number and diversity of Ca^{2+}-ATPases in mammalian species are increased by the occurrence of isoforms and splice variants with tissue-specific distribution, function, and mode of regulation. Thus there are three genes, *ATP2A1*, *ATP2A2*, and *ATP2A3*, encoding the SERCA1, SERCA2, and SERCA3 enzymes, respectively. Differential splicing of the primary transcripts yields two or three splice variants for each (SERCA1a/b, SERCA2a/b, and SERCA3a/b/c) (East 2000). SERCA1a is the best known of these isoforms, being expressed in fast-twitch skeletal muscle where its high turnover makes it ideally suited to excitation–contraction coupling. SERCA2a is found in cardiac and slow-twitch muscle, where it is noteworthy for regulation by phospholamban, while SERCA2b is the nonmuscle ubiquitously expressed isoform (Baba-Aissa *et al.* 1998). The function of SERCA3 is less well defined, with these isoforms having specialized distribution in the cells of hematopoietic origin, Purkinje neurons, and certain epithelial/endothelial cells (Wuytack *et al.* 1994), where they are coexpressed with the "housekeeping" SERCA2b isoform. Their unusually high K_M for Ca^{2+} has led to speculation that they may function in a high-Ca^{2+} environment, with the reservation that activities were measured in COS cells where a specific activator may have been missing (Dode *et al.* 1998).

The complexity of the PMCA family is even higher than that of SERCA, with four non-allelic genes designated *ATP2B1*, *ATP2B2*, *ATP2B3*, and *ATP2B4* in humans encoding four isoforms: PMCA1, PMCA2, PMCA3, and PMCA4. These genes are alternatively spliced,

resulting in at least 20 distinct enzymes (Strehler and Zacharias 2001). Of these, PMCA1 and PMCA4 are ubiquitously expressed in most adult cells and therefore are regarded as the "housekeeping" isoforms, whereas PMCA2 and PMCA3 are largely restricted to the excitable cells of the nervous system and muscle. Notably, PMCA2 is found in cerebellar Purkinje neurons and cochlear hair cells, and is massively upregulated in lactating mammary glands, where its transcript was up to 60 times more abundant than all other Ca^{2+}-ATPases and approached that of actin (Reinhardt and Horst 1999). Although PMCA3 has a particularly restricted distribution in the choroid plexus of the brain, its transcript was widely detected early in the embryonic tissues of the nervous system, limb, and lung, suggesting a role in organ development (Zacharias and Kappen 1999). Alternative splicing of the PMCA genes occurs at sites clearly associated with regulatory regions, and thus directly impacts on ATPase function. Rearrangements of the C-terminus at splice "site C" affect binding and regulation by calmodulin, phosphorylation sites for protein kinases A and C, and binding to PDZ domains. Similarly, alternative splicing at "site A" in the first intracellular loop of the pump affects a downstream sequence known to confer activation by acidic phospholipids, including phosphoinositides.

Much less is known about the molecular biology of SPCA genes. The *ATP2C1* gene was cloned recently from human (Hu *et al.* 2000; Sudbrak *et al.* 2000) and encodes a protein, SPCA1, that is recognizably related to the rat clone RS10-31 identified by Shull and colleagues (Gunteski-Hamblin *et al.* 1992) and to the founding member of this family, yeast PMR1 (Rudolph *et al.* 1989). It has a widespread tissue distribution consistent with a "housekeeping" role in the cell. Interestingly, the rat homolog, clone RS10, is the only Ca^{2+}-ATPase subtype to be induced *before* parturition and is also highly expressed in lactating mammary glands like PMCA2, suggesting a prominent role in Ca^{2+} secretion (Reinhardt and Horst 1999). A second cDNA, KIAA0703, encoding a protein with 64% identity to SPCA1 has been isolated from brain and found to map to chromosome 16q24.1 on the human genome. This uncharacterized gene has been tentatively named *ATP2C2* (protein SPCA2) by us, although the tissue distribution and physiological role of this isoform remains to be determined. Splice variants are likely for both genes, and may be inferred from the cDNA sequences deposited in databases that suggest variations at the N- and C-termini. *ATP2C1a* and *ATP2C1b* encode polypeptides of 919 and 888 amino acids, differing at the C-termini beyond amino acid 877 (Hu *et al.* 2000).

STRUCTURAL AND MECHANISTIC FEATURES OF Ca^{2+}-ATPASES

Catalysis and transport mechanism

All three Ca^{2+}-ATPase subtypes undergo transient phosphorylation during the course of the catalytic cycle. The cycle of events leading to transmembrane transport against a concentration gradient has been best described for skeletal muscle SERCA1 (reviewed by East 2000), and is summarized briefly here. A fundamental feature of the transport mechanism is the alternation of the enzyme between two major conformational states, known as E_1 and E_2. The E_1 conformation of the pump is obtained upon high affinity (between 0.1 and 2 µM, depending on the particular isoform) and cooperative binding of two Ca^{2+} in exchange for two H^+ from the cytoplasmic side of the membrane. The Ca^{2+}-activated enzyme then binds ATP, and the γ-phosphate of ATP is transferred to an aspartyl residue of the Ca^{2+}-ATPase to form a covalent aspartyl–phosphate intermediate. The K_{eq} for this reaction is nearly 1, so that the free energy of the ATP γ-phosphate remains conserved in the phosphorylated intermediate. Following the release of ADP, a rate-limiting conformational change reorients bound Ca^{2+} toward the lumen or extracellular space. The so-called "high-energy" $E_1{\sim}P$ intermediate converts to the low-energy E_2–P

intermediate, and the affinity for Ca^{2+} falls by three orders of magnitude. As a result, Ca^{2+} is released at the opposite side of the membrane and vectorial translocation is accomplished. The cycle is then completed by hydrolytic cleavage of the phosphoenzyme intermediate, resulting in production of inorganic phosphate and countertransport of $2Ca^{2+}$ and $2H^+$ per cycle. The overall mechanism of the pump is based on utilization of ATP free energy to disrupt high-affinity Ca^{2+} binding. The catalytic cycles of the PMCA and SPCA pumps are believed to be essentially like that of SERCA, although only one Ca^{2+} is transported per reaction cycle.

Structure

There is a large body of experimental work on the structure of the Ca^{2+}-ATPases, culminating in the recent determination of crystallographic structure of the SERCA1 pump in both major conformations (reviewed by East 2000). A single catalytic subunit (about 100–120 kDa) is embedded in the membrane by 10 hydrophobic helices, with large cytosolic domains that contribute to the hydrolysis of ATP and conformational coupling to transport. The cation binding domain resides within the membrane and is composed of the core helices M4, M5, M6, and M8 (see Chapter 1). Extensive mutagenic analyses of SERCA1, expressed heterologously in cultured cells, led to the identification of the side chains contributing to Ca^{2+} sites I (Asn 768 and Glu 771 in M5, Thr 799 and Asp 800 in M6, and Glu 908 in M8) and site II (Asn 796 and Asp 800 in M6) (Zhang *et al.* 2000). These were subsequently observed in the crystal structure of the Ca^{2+}-bound E_1 conformation of SERCA1, along with the unexpected contribution of main-chain carbonyls that are made available by partial unwinding of helices M4 and M6 (Toyoshima *et al.* 2000). The two Ca^{2+} ions lie side by side about halfway within the membrane bilayer. Biochemical and amino acid sequence analyses point to the absence of Ca^{2+} site I in the PMCA and SPCA pumps. The cytoplasmic segments of the pump consist of the actuator (A) domain that is made up of a predominantly β-strand structure between M2 and M3, the phosphorylation (P) domain, and the nucleotide binding (N) domain, both extending between M4 and M5. In the thapsigargin-bound E$_2$-like structure, the three cytoplasmic domains gather to form a single headpiece and are accompanied by large rearrangements of several membrane helices (see Chapter 1) which result in a drastic reduction in affinity for Ca^{2+} (Toyoshima and Nomura 2002). The elegant crystal structures offer tremendous insight into the molecular mechanism of Ca^{2+} transport and will clearly serve as a reference for interpretation of future studies.

REGULATORY PROTEINS OF THE CA^{2+}-ATPASES

An interesting feature of Ca^{2+}-ATPases is their regulation by smaller proteins that associate with the enzyme and are in turn responsive to additional regulation such as that by kinase-assisted phosphorylation. These proteins provide a linkage between ATPase and signaling mechanisms for optimal response of Ca^{2+} transport as required for homeostasis and specific cell functions.

Phospholamban and sarcolipin

The presence of a 22 kDa protein component undergoing phosphorylation in cardiac sarcoplasmic reticulum (SR) was reported by Tada et al. (1975). It was later demonstrated that the 22 kDa protein is a pentamer of identical subunits undergoing reversible association (Wegener and Jones 1984). Phospholamban undergoes concerted and additive phosphorylation catalyzed by calmodulin-dependent and cAMP-dependent kinases. Phosphorylation produces a proportional stimulation of Ca^{2+} transport and ATPase activity in cardiac SR (Lepeuch et al. 1979; Kranias 1985). The increase in ATPase activity is mostly observed at low Ca^{2+} concentrations, as phospholamban

phosphorylation appears to lower the Ca^{2+} concentration required for ATPase activation (Tada et al. 1983). In fact, interaction of the ATPase with nonphosphorylated phospholamban inhibits ATPase activity by displacing the Ca^{2+} concentration required for catalytic activation to a higher range. Reversal of this effect is produced by phospholamban phosphorylation (Inui et al. 1986). It is clear that phospholamban phosphorylation by the cAMP-dependent kinase links ATPase regulation to the adrenergic signaling mechanism of cardiac and slow skeletal muscle, and is not observed in fast skeletal muscle (Kirchberger and Tada 1976).

Amino acid sequencing (Fujii et al. 1987) and cDNA cloning (Fujii et al. 1997) demonstrated that the phospholamban monomer is a 6 kDa protein composed of 52 amino acids, residing in a prevalently rod-shaped helical structure (Simmerman et al. 1989). A smaller NH$_2$-terminal domain protrudes from the cytosolic side of the SR membrane and includes the serine and threonine residues undergoing phosphorylation (Simmerman et al. 1986). Although phospholamban is assembled prevalently as a pentamer in the lipid bilayer, its tendency to polymerize is enhanced by phosphorylation (Cornea et al. 1997). On the other hand, ATPase inhibition is enhanced under conditions favoring dissociation of the polymer, indicating that interaction of the ATPase with phospholamban monomers is involved in the inhibitory effect (Autry et al. 1997; Kimura et al. 1997). Cross-linking experiments have demonstrated an interaction site for phospholamban on the cytosolic headpiece of the ATPase, spanning over the P and N domain (James et al. 1989). In agreement with this finding, mutational studies (Toyofuku et al. 1994a, b) have identified the cytosolic ATPase site with the peptide segment intervening between Leu397 and Val402, and suggested that the phospholamban sequence between Glu2 and Ile18 is required for this interaction. Phospholamban binding results in helical stabilization and restricted amplitude of motions within the nucleotide binding domain of the ATPase (Negash *et al.* 1999). These changes are likely to interfere with rate-limiting Ca^{2+}-dependent conformational changes required for ATPase activation (Cantilina *et al.* 1993). The COOH-terminal domain of phospholamban and the transmembrane helix M6 of the ATPase provide an additional site of interaction that may, by itself, produce the inhibitory effect (Kimura *et al.* 1996; Asahi *et al.* 1999). Such interactions are strong in the SERCA1 and SERCA2 isoforms of the ATPase, but rather weak in the SERCA3 isoform, due to poor sequence homology in the corresponding segments (Toyofuku *et al.* 1993). High resolution structural methods for the interaction of phospholamban with SERCA were recently published (Toyoshima *et al.* 2003; Chen *et al.* 2003).

It has been convincingly demonstrated that regulation of the SR ATPase by phospholamban is reflected on cardiac contractile parameters in vivo. In fact, exposure of cardiac myocytes to antibodies against phospholamban (Sham *et al.* 1991) and phospholamban gene knockout in whole animals (Luo *et al.* 1994; Koss and Kranias 1996) produce faster rates of contraction and relaxation, as well as diminished response to β-adrenergic stimulation. On the other hand, overexpression of phospholamban in transgenic mice is associated with reduced Ca^{2+} kinetics and contractile parameters (Kadambi *et al.* 1996). Evidence has been presented suggesting that unbalanced interaction of phospholamban with the Ca^{2+}-ATPase is a critical Ca^{2+} cycling defect in dilated cardiomyopathy (Minamisawa *et al.* 1999).

It was recently reported that sarcolipin, a 31 amino acid protein, is a component of the SR membrane and a member of the same gene family as phospholamban (Odermatt *et al.* 1997, 1998; Hellstern *et al.* 2001). Although phospholamban is prevalently expressed in cardiac muscle and sarcolipin in fast skeletal muscle coexpression of the two proteins in cardiac muscle has been reported, at least in some animal species (Gayan-Ramirez *et al.* 2000). Sarcolipin may have a regulatory function on the SR ATPase, alternative or additional to that of phospholamban (Odermatt *et al.* 1998).

Calmodulin

While the SR ATPase can be inhibited by non-phosphorylated phospholamban, the PMCA is auto-inhibited by interaction of its own peptide segments. The "inhibitory" peptide is a segment of the ATPase extended carboxyl-terminus, while the "receptor" segment lies between the ATP binding site and the aspartate residue involved in formation of the phosphorylated enzyme intermediate. In turn, the PMCA auto-inhibition is relieved by calmodulin in the presence of Ca^{2+}. Calmodulin is the most common member of a large family of Ca^{2+} binding proteins, referred to as the EF-hand family (Kawasaki *et al.* 1998). These proteins, including troponin C as well as calmodulin, play a key role in Ca^{2+} signaling inasmuch as they are able to bind and dissociate Ca^{2+} at concentrations compatible with the high and low levels produced by cytosolic Ca^{2+} transients related to membrane excitation. In turn, Ca^{2+} binding triggers conformational changes that are transmitted to specific "partner" enzymes, resulting in catalytic activation.

Crystallographic structural analysis (Babu *et al.* 1985) indicates that the calmodulin molecule consists of two globular lobes connected by a long exposed α-helix. Each lobe binds two calcium ions through helix–loop–helix domains similar to those of other calcium-binding proteins. The long helix between the lobes can then be involved in interactions of calmodulin with drugs and various proteins. The NMR solution structure is similar to the X-ray crystal structure, but indicates considerable backbone plasticity within the long domains of calmodulin which is likely to be relevant to its ability to bind a wide range of targets (Chou *et al.* 2001). In fact, the mechanism of signaling is based on conformational changes triggered by Ca^{2+} binding at the globular lobes, resulting in collapse of the calmodulin elongated shape for recognition and binding to specific peptide segments of target enzymes (Ikura *et al.* 1992; Meador *et al.* 1992).

The calmodulin binding domain of the PMCA has been identified (James *et al.* 1989, 1996; Falchetto *et al.* 1991) with the same peptide segment producing auto-inhibition by interacting with the "receptor" segment intervening between the nucleotide and the phosphorylation sites. The inhibition would then be relieved by Ca-calmodulin interference with "inhibitory" and "receptor" domain interaction. It is of interest that a similar relief of inhibition, in the absence of calmodulin, can be obtained by proteolytic C-terminal truncation or phosphorylation by protein kinase C (Hofmann *et al.* 1994). Furthermore, binding of acidic phospholipids to a PMCA peptide segment in the second cytosolic extramembranous domain can also relieve inhibition (Niggli *et al.* 1981), consistent with a conformational mechanism for occurrence and relief of inhibition.

INSIGHTS FROM MODEL ORGANISMS

Studies in model organisms have led to invaluable advances in the study of Ca^{2+} transporters, from the initial cloning of new genes to the characterization of transport properties and the phenotypes of null mutants. The contributions of two model systems, yeast and mouse, are of particular significance in the understanding of Ca^{2+}-ATPases and will be discussed briefly.

Yeast

The search for new P-ATPases in *Saccharomyces cerevisiae* resulted in the discovery of *PMR1*, encoding a putative Ca^{2+}-ATPase with about 30% identity with the SERCA pumps (Rudolph *et al.* 1989). The localization of this pump to the medial-Golgi (Antebi and Fink 1992), rather than to the endoplasmic reticulum, and the cloning of a closely homologous cDNA from rat (Gunteski-Hamblin *et al.* 1992), provided the first indications for a potentially novel Ca^{2+} pump subtype. This was confirmed by biochemical characterization of yeast PMR1, which showed that the Golgi pump differed from the SERCA and PMCA pumps in inhibitor sensitivity and affinity for Ca^{2+} (Sorin *et al.* 1997), and in a uniquely high affinity for Mn^{2+} (Mandal *et al.* 2000). The similarity between PMR1 and certain bacterial Ca^{2+}-ATPases, and the absence of a SERCA homolog in yeast, led to the idea that the Golgi/secretory pathway pumps may have arisen early in evolution, possibly predating the SERCA pumps (Sorin *et al.* 1997). Since then, numerous PMR1 homolog have been identified in other organisms, including other fungi, worm (*Caenorhabditis elegans*), fruitfly (*Drosophila*), and other vertebrates (bovine, human), although, curiously, not in the genomes of any plants. The phenotypes of *pmr1* null mutants in yeast are consistent with the symptoms of SPCA deficiency in human. These include defective proteolytic processing and underglycosylation of secreted proteins, and hypersecretion of heterologously expressed proteins (Rudolph *et al.* 1989; Durr *et al.* 1998). Addition of excess Ca^{2+} to the growth medium corrected the sorting and processing defects, but not the glycosylation defect, which was shown to be specific for Mn^{2+}. Interestingly, the *pmr1* null mutant showed a large compensatory increase in the expression of a PMCA homolog, PMC1, which localizes to the yeast vacuolar/lysosomal membrane (Marchi *et al.* 1999). Induction of *PMC1* was shown to occur via activation of calcineurin, the Ca^{2+}- and calmodulin-activated protein phosphatase, which in turn activated the transcription factor TCN1 (Matheos *et al.* 1997). These studies demonstrate how elucidation of Ca^{2+} transport and homeostasis in a simple model organism can shed light on more complex signaling events and on the compensatory changes in gene expression seen upon gene knockouts in vertebrates.

Mouse

Transgenic approaches have been used to knock out one or both copies of individual Ca^{2+}-ATPase genes or to manipulate their overexpression. The study of these mouse models has provided much insight into the physiological roles of Ca^{2+} pumps, and in understanding the basis for disease in human (reviewed by Shull (2000)). Homozygous null mutants in SERCA2 were nonviable, consistent with the ubiquitous presence of the gene product in all cells, particularly in cardiac tissue. The heterozygous animals appeared healthy, but close examination revealed that cardiac muscle contractility and relaxation rates were impaired, although the impairment was not severe enough to lead to symptoms of cardiac disease (Periasamy et al. 1999). These results were consistent with experiments showing that transgenic overexpression of SERCA genes in heart increased both contractility and relaxation rates (He et al. 1997). The blistering disorder of skin found in humans with SERCA2 mutations (Darier disease) did not present in the mice, possibly correlating with the late onset of disease in human. However, older mice heterozygous for SERCA2 deletion developed squamous cell carcinoma, providing an interesting link between calcium homeostasis and cancer (Liu et al. 2001). The pattern of expression and properties of SERCA3 suggested that this pump may be important in embryonic development or may play some critical specialized role in cells. However, SERCA3-deficient mice appeared fully normal and only showed impairment of endothelial relaxation in smooth muscle of aorta and trachea, with no disease phenotype (Liu et al. 1997). Thus the specific contribution of this isoform remains to be determined.

While there have been no inherited diseases linked to the plasma membrane Ca^{2+}-ATPases in human, the phenotypes of PMCA2 null mice have been very informative. Although normal at birth, null mutants grew slowly, displayed severe balance deficit, and were deaf (Kozel et al. 1998). Histological analysis of the inner ear revealed loss of sensory hair cells and other auditory structures. In addition, the calcium carbonate crystals, or otoconia, that are required for detection of balance and motion were absent from endolymph. Similar phenotypes were observed in *deafwaddler* and

wriggle mouse sagami mouse lines which carry spontaneously generated mutations in PMCA2 (Street et al. 1998; Takahashi and Kitamura 1999). Thus PMCA2 plays a critical role in hearing and balance, and may potentially play similar roles in human.

Ca^{2+}-ATPase MUTATIONS AND HUMAN DISEASE

To date, mutations in three Ca^{2+}-ATPase genes have been linked to human disease. *Brody disease* is an inherited skeletal muscle disorder resulting from autosomal recessive mutations in *ATP2A1*, the gene encoding fast-twitch skeletal muscle SERCA1 (Figure 2). *Darier disease* and *Hailey–Hailey disease* are, surprisingly, both blistering disorders of the skin resulting from dominantly inherited mutations in one allele of *ATP2A2* (SERCA2) and *ATP2C1* (SPCA1), respectively (Figure 2). Although inherited disorders linked to PMCA genes are yet to be identified, based on the phenotypes of mouse PMCA knockouts, it would be surprising if there are none.

Brody disease

The *ATP2A1* gene accounts for 99% of the SERCA1 isoform found in fast-twitch skeletal muscle (type 2 fibers) (Wu et al. 1995), where muscle relaxation is initiated by the rapid transport of cytosolic calcium into the lumen of the sarcoplasmic reticulum. Brody disease is a rare disorder (1 in 100,000) characterized by a lifelong history of exercise- and cold-induced impairment of muscle relaxation, particularly in the arms, legs, and eyelids. As a result of sustained contractions, patients develop stiffened muscles and largely painless cramping. Early studies identified reductions in Ca^{2+} uptake and Ca^{2+}-ATPase activities in the sarcoplasmic reticulum of Brody patients, consistent with defects in the pump. A systematic sequencing of the *ATP2A1* gene in Brody patients led to the identification of mutations in six of ten families examined

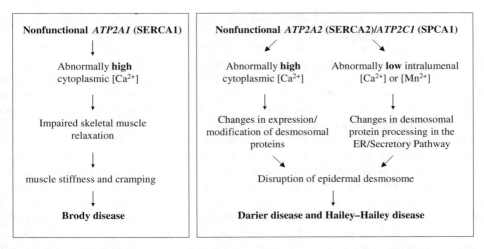

Figure 2 Pathophysiology of diseases resulting from mutations in Ca^{2+}-ATPase genes. Loss-of-function mutations in *ATP2A1*, encoding the SERCA1 pump expressed in skeletal muscle, lead to Brody disease, inherited as a recessive trait. Loss-of-function mutations in *ATP2A2* or *ATP2C1*, encoding the ubiquitously expressed SERCA2 or SPCA1 pumps, respectively, lead to distinct skin blistering disorders known as Darier disease and Hailey–Hailey disease, inherited as dominant traits, as discussed in the text.

(Odermatt *et al.* 1996, 1997, 2000). All but one mutation results in truncations of SERCA1 and the deletion of essential functional domains, very likely leading to complete loss of Ca^{2+} pumping activity. The single missense mutation results in substitution of Pro789 with Leu near the lumenal interface of transmembrane helix M6. Heterologous expression of this mutant in HEK-293 cells resulted in normal levels of protein expression, but with at least 20-fold reduction in Ca^{2+} affinity which would effectively abolish Ca^{2+} transport activity at physiological Ca^{2+} levels. Formation of the phosphoenzyme intermediate was normal, indicating that catalytic activity in the mutant enzyme was not compromised. Interestingly, in an earlier study using alanine scanning mutagenesis, substitution of Pro789 with Ala was not found to alter either affinity for calcium or V_{max} (Rice and MacLennan 1996). Thus it is likely that the bulky Leu residue leads to steric hindrance, disturbing the position of M6 or the conformational flexibility of the lumenal M5–M6 loop. Another apparent missense mutation found in a Brody patient, Arg819 to Cys, had normal activity when tested and was

concluded to represent a rare polymorphism (Odermatt *et al.* 2000). This case underlines the importance of testing the functional effect of mutations in order to confirm the molecular basis of disease. In fact, Brody disease is clearly heterogenous in genetic origin, with mutations in at least one other gene leading to the recessive form of the disease, as well as in another unknown gene responsible for a dominant mode of inheritance. It is noteworthy that no Brody disease mutations have been found to date in the *SLN* gene encoding sarcolipin, which might in theory result in superinhibition of the SERCA pump (Odermatt *et al.* 1997). Finally, the fact that SERCA1 function is not essential for muscle function indicates that there must be significant compensatory upregulation of other SERCA isoforms, or other Ca^{2+}-ATPase subtypes (SPCA and PMCA), to clear cytosolic Ca^{2+} and to refill the stores in Brody patients.

Darier disease

Darier–White or Darier disease, also known as keratosis follicularis, is a disfiguring skin disorder characterized by warty papules and plaques covering most of the body, particularly in sebarrhoeic regions. Inheritance is autosomal dominant and is linked to mutations in the *ATP2A2* gene encoding SERCA2 (Sakuntabhai *et al.* 1999). The disease occurs with an estimated prevalence of 1 in 55,000 and has a late onset, usually presenting between 11 and 20 years of age. The *ATP2A2* gene gives rise to two splice variants, which have the first 993 amino acids in common. SERCA2a has four additional amino acids and predominates in cardiac and slow-twitch skeletal muscle, whereas SERCA2b has a 49 amino acid C-terminal extension, likely to form an eleventh transmembrane domain (Campbell *et al.* 1992). SERCA2b has been reported to have a higher affinity for Ca^{2+} but a lower turnover than SERCA2a. Although the Darier mutations are expected to alter expression and function of both splice variants, it is the SERCA2b splice variant that is exclusively expressed in various epidermal structures of adult skin (Ruiz-Perez *et al.* 1999).

At the ultrastructural level, lesions show loss of desmosomal contacts, perinuclear aggregation of keratin filaments, and cytoplasmic vacuolation, indicative of a loss of structure and function of desmosomes (Burge and Garrod 1991). Desmosomes consist of desmosomal cadherins that interact via their extracellular domains, and plaque proteins which link the desmosomes to the cytoplasmic keratin–intermediate filament network. In cell culture models, assembly of desmosomes is initiated by a rise in extracellular calcium (Watt *et al.* 1984). A role for SERCA has been demonstrated by studies with thapsigargin, which blocks sequestration of calcium in the endoplasmic reticulum. Treatment of the Madin–Darby canine kidney epithelial cell model with low concentrations of thapsigargin perturbed the formation of tight junctions and desmosomes, and delayed the sorting of desmosomal proteins (Stuart *et al.* 1996). Interestingly, the low concentrations of thapsigargin used (100 nM) did not alter cytoplasmic calcium dynamics temporarily associated with

the formation of junctions, strongly suggesting that it is the perturbation of lumenal rather than cytoplasmic Ca^{2+} concentrations that disrupt junction biogenesis, at least in this model.

Darier disease mutations range from premature terminations, in-frame deletions, and insertions to defined missense mutations. Most of the mutations occur at evolutionarily conserved sites that are likely to disrupt essential functional domains. Based on the vast body of data from site-directed mutational analysis of SERCA1 and related Ca^{2+}-ATPases, it seems reasonably certain that Darier disease mutations lead to loss rather than gain of function. Further, it is generally accepted that oligomerization is not necessary for pump function. Therefore it appears that dominant inheritance of the disease is due to haploinsufficiency rather than dominant-negative effects. No patients were found with mutations in both alleles of the gene, strongly implying that such mutations would be lethal, as suggested by nonviability of the SERCA2 homozygous knockout in mouse. Remarkably, specific cutaneous manifestations of the disease may correlate with specific missense mutations. For example, the mutation N767S in transmembrane segment 5 was indentified in four unrelated families where the disease presented with an acral hemorrhaging (Ruiz-Perez *et al.* 1999). This mutation affects a conserved residue that contributes to, but is not essential for, Ca^{2+} binding (Zhang *et al.* 2000). N767S is predicted to cause partial loss of function based on mutational analysis of the equivalent residue in SERCA1. Substitution with Ala led to reduction in Ca^{2+} binding affinity by 0.8 pCa units, while the bulkier Ile led to complete loss of transport (Rice and MacLennan 1996). It is interesting to note that this position is normally occupied by a Ser in the related SPCA and PMCA pumps. Another mutation, K683E, at a widely conserved motif common to P-ATPases correlated with a "classical" phenotype of the disease (Ruiz-Perez *et al.* 1999; Ringpfeil *et al.* 2001).

Darier disease has also been associated with various neuropsychiatric features, including mental handicap, schizophrenia, bipolar disorder, and epilepsy. Since SERCA2 expression is widespread in brain (Baba-Aissa et al. 1998), it is tempting to attribute these disorders to the *ATP2A2* mutations. However, the connection is sporadic and a direct association of neuropsychiatric symptoms with Darier disease has been problematic, indicating the involvement of other genetic and environmental factors (Ruiz-Perez et al. 1999).

Hailey–Hailey disease

First described by the Hailey brothers in 1939 as familial benign chronic pemphigus, this disease has been mapped to dominantly inherited mutations in *ATP2C1* encoding the secretory pathway/Golgi Ca^{2+}- and Mn^{2+}-ATPase (Hu et al. 2000; Sudbrak et al. 2000). Missense mutations target functionally important regions of the pump, including transmembrane spans M4, M5, and M6, and the catalytic cytoplasmic domain. Like Darier disease, Hailey–Hailey disease is a defect in keratinocyte adhesion in which acantholysis is more severe so that the keratinocytes dissociate spontaneously upon culture (De Dobbeleer et al. 1989). Although there is some overlap, the clinical symptoms of Darier disease and Hailey–Hailey disease can be distinguished from one another, with reports of pure acantholysis in Hailey–Hailey disease and acantholysis with dyskeratinization in Darier disease. This indicates that the two pumps play distinct roles in cell adhesion, possibly because they control ion levels at different and functionally distinct stages in the secretory pathway. Inadequate concentrations of Ca^{2+} and Mn^{2+} in the Golgi are known to impair post-translational modifications of proteins, including protein processing and O-linked glycosylation (Oda 1992; Kaufman et al. 1994). Addition of membrane-permeant cation chelators have been shown to inhibit both anterograde and retrograde traffic from the Golgi (Chen et al. 2002). Thus haploinsufficiency of SPCA may alter critical concentrations of divalent cations in the Golgi lumen and impair the cell-surface delivery and functionality of desmosomal proteins. Hu et al. (2000) have also demonstrated impairment of cytoplasmic Ca^{2+} signaling in Hailey–Hailey keratinocytes compared with normal cells: resting Ca^{2+} concentrations were higher but the response to increases in extracellular Ca^{2+} were lower in Hailey–Hailey cells. Together, Darier disease and Hailey–Hailey disease emphasize the importance of calcium in the induction of cell–cell junctions and the differentiation of keratinocytes to form the final cornified layer (Figure 2). Although there is not yet a mouse model for SPCA1 null mutants, the absence of mutations in both alleles of the affected gene in Hailey–Hailey patients (as in Darier patients) suggests that homozygous mutations may be lethal.

CARDIAC FUNCTION AND CARDIAC DISEASE

Ca^{2+} signaling is a general biological function that is widely distributed to many organisms and cell types. On the other hand, the role of Ca^{2+} in cardiac muscle has been studied in greatest detail, since the initial observation of Ca^{2+} requirement for contraction of isolated hearts (Ringer 1883). It is now well known that Ca^{2+} is involved in several physiological and pathological aspects of cardiac function, including electrical potentials, excitation contraction coupling, regulation of contractile tension, cell proliferation, and apoptosis (Misquitta et al. 1999). Furthermore, Ca^{2+}-dependent transcriptional activation may contribute to the development of hypertrophy (Sugden and Clerk 1998). Finally, it is clear that active transport is a very important determinant of cardiac function. For instance, the improvement of cardiac performance by digitalis has been attributed to primary inhibition of the Na^{2+}/K$^+$ pump and secondary reduction of Ca^{2+} export by the Na$^+$/Ca^{2+} exchanger. Therefore it is worth considering the relevance of Ca^{2+} transport ATPases to cardiac diseases, in the light of their large statistics and impact on public health.

In ventricular muscle, the most important contributor to beat-to-beat regulation of cytosolic Ca^{2+} signaling is, in addition to the release channel, the SR ATPase. The importance of the SR ATPase can be clearly demonstrated (Figure 3) in isolated myocytes subjected to stimuli at progressively shorter intervals. Thereby, the time allowed to the SR ATPase to refill the intracellular stores becomes limiting, and the amount of Ca^{2+} released by the next stimulus is consequently reduced. On the other hand, if the SR ATPase is overexpressed by transfection of an exogenous gene, the myocytes become more tolerant and retain full Ca^{2+} signals at higher frequency of stimulation (Figure 3). In fact, it has been shown by direct measurements that the rate of Ca^{2+} transport by the SR ATPase is accelerated by SERCA overexpression in cultured myocytes (Giordano et al. 1997; Hajjar et al. 1997) and transgenic animals (He et al. 1997; Loukianov et al. 1998). It is then apparent that inadequate function of the SR ATPase, due to either reduced copy number or defective regulation, would result in inadequte relaxation and contractile activation. Relevant to these findings is the observation that SERCA mRNA and protein levels decrease in end-stage heart failure (Gwathmey et al. 1987; Mercadier et al. 1990; Takahashi et al. 1992; Hasenfuss et al. 1994). It was also reported that unbalanced interaction of phospholamban with the Ca^{2+} ATPase is a critical Ca^{2+} cycling defect in dilated cardiomyopathy (Minamisawa et al. 1999). Defective performance of the SR ATPase may play an important role in the pathogenesis of heart failure (Morgan et al. 1990; Hasenfuss et al. 1994, 1997; Schmidt et al. 1998). At present, medical concern with defective Ca^{2+} homeostasis is directed to correction of passive fluxes by the wide use of Ca^{2+}-channel blockers. It is likely that transcription, expression, and/or regulation of the SR ATPase will be targets of therapeutic intervention in the near future.

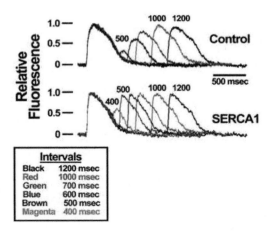

Figure 3 Exogenous SERCA gene expression improves the ability of the sarcoplasmic reticulum to "refill with Ca^{2+}" when challenged by progressively shorter diastolic times. Cytosolic Ca^{2+} transients were measured in cardiac myocytes loaded with fluo-4 and were triggered with a two-pulse protocol after pacing at 1.0 Hz. The last paced transient and the extrasystolic transient were recorded while varying the extrasystolic interval. Each record was normalized to its conditioning transient and superimposed as a series: upper series, control myocyte; lower series, myocyte overexpressing SERCA1. Average of 16–20 experiments. (Reproduced from Cavagni et al. 2000.)

THERMOGENESIS

It was recently demonstrated that ATP consumption by the Ca^{2+}-ATPase may result in heat production, likely to be an important factor in physiological and pathological thermogenesis (deMeis 2003). Heat production is related to slippage of the pump, and a reduced stoichiometric ratio of transported Ca^{2+} and ATP consumed (Yu and Inesi 1995).

SUMMARY

Active transport of cytoplasmic Ca^{2+} across the cell and organellar membranes is accomplished by a family of ubiquitously distributed Ca^{2+}-ATPases that are members of the superfamily of P-type cation pumps. Three distinct subtypes of Ca^{2+}-ATPases localize to the plasma membrane, endoplasmic reticulum, and Golgi membranes, where they play important roles in diverse Ca^{2+} signaling events. Isoforms and splice variants of these pumps show tissue-specific distribution or are associated with distinct physiological functions, such as cardiac muscle contraction, keratinocyte adhesion, and hearing. Ca^{2+}-transporting ATPases maintain submicromolar concentrations of cytoplasmic Ca^{2+}, refill intracellular stores, and regulate extracellular Ca^{2+} levels. The multiplicity of Ca^{2+}-ATPase subtypes and their variants ensures that Ca^{2+} signaling is tightly controlled and adaptable to changes in the expression and activity of the transporters. However, despite the redundancy in the number of transporters, specific diseases result from mutations in Ca^{2+}-ATPase genes. Abnormally high Ca^{2+} in the cytosol or abnormally low Ca^{2+} in lumenal/extracellular spaces can lead to tissue-specific defects relating to muscle relaxation, as in Brody disease, or in blistering of the skin, as is seen in both Darier disease and Hailey–Hailey disease.

REFERENCES

Akera, T. and Brody, T.M. (1997). The role of Na$^+$, K$^+$-ATPase in the Inotropic Action of Digitalis. *Pharmacol. Rev.*, **29**, 187–220.

Antebi, A. and Fink, G.R. (1992). The Yeast Ca^{2+}-ATPase Homologue, PMR1, is Required for Normal Golgi Function and Localizes in a Novel Golgi-Like Distribution. *Mol. Biol. Cell.*, **3**, 633–654.

Asahi, M., Kimura, Y., Kurzydlowski, K., Tada, M., and MacLennan, D.H. (1999). Transmembrane Helix M6 in Sarco(endo)plasmic Reticulum Ca^{2+}-ATPase Forms a Functional Interaction Site with Phospholamban. Evidence for Physical Interactions at other Sites. *J. Biol. Chem.*, **274**, 32855–32862.

Autry, J.M. and Jones, L.R. (1997). Functional Co-expression of the Canine Cardiac Ca^{2+} Pump and Phospholamban in *Spodoptera frugiperda* (Sf21) Cells Reveals New Insights on ATPase Regulation. *J. Biol. Chem.*, **272**, 15872–15880.

Baba-Aissa, F., Raeymaekers, L., Wuytack, F., Dode, L., and Casteels, R. (1998). Distribution and Isoform Diversity of the Organellar Ca^{2+} Pumps in the Brain. *Mol. Chem. Neuropathol.*, **33**, 199–208.

Babu, Y.S., Sack, J.S., Greenhough, T.J., Bugg, C.E., Means, A.R., and Cook, W.J. (1985). Three-Dimensional Structure of Calmodulin. *Nature*, **315**, 37–40.

Berridge, M.J., Lipp, P., and Bootman, M.D. (2000). The Versatility and Universality of Calcium Signaling. *Nature Rev. Mol. Cell Biol.*, **1**, 11–21.

Bolton, E.C., Mildvan, A.S., and Boeke, J.D. (2002). Inhibition of Reverse Transcription In Vivo by Elevated Manganese Ion Concentration. *Mol. Cell.*, **9**, 879–889.

Burge, S.M. and Garrod, D.R. (1999). An Immunohistological Study of Desmosomes in Darier's Disease and Hailey–Hailey Disease. *Br. J. Dermatol.*, **124**, 242–251.

Campbell, A.M., Kessler, P.D., and Fambrough, D.M. (1992). The Alternative Carboxyl Termini of Avian Cardiac and Brain Sarcoplasmic Reticulum/Endoplasmic Reticulum Ca^{2+}-ATPases are on Opposite Sides of the Membrane. *J. Biol. Chem.*, **267**, 9321–9325.

Cantilina, T., Sagara, Y., Inesi, G., and Jones, L.R. (1993). Comparative Studies of Cardiac and Skeletal Sarcoplasmic Reticulum ATPases. Effect of a Phospholamban Antibody on Enzyme Activation by Ca^{2+}. *J. Biol. Chem.*, **268**, 17018–17025.

Cavagna, M., O'Donnell, J.M., Sumbilla, C., Inesi, G., and Klein, M.G. (2000). Exogenous Ca^{2+}-ATPase Isoform Effects on Ca^{2+} Transients of Embryonic Chicken and Neonatal Rat Cardiac Myocytes. *J. Physiol.*, **528**, 53–63.

Chandra, S., Fewtrell, C., Millard, P.J., Sandison, D.R., Webb, W.W., and Morrison, G.H. (1994). Imaging of Total Intracellular Calcium and Calcium Influx and Efflux in Individual Resting and Stimulated Tumor Mast Cells using Ion Microscopy. *J. Biol. Chem.*, **269**, 15186–15194.

Chen, J.L., Ahluwalia, J.P., and Stamnes, M. (2002). Selective Effects of Calcium Chelators on Anterograde and Retrograde Protein Transport in the Cell. *J. Biol. Chem.*, **277**, 35682–35687.

Chen, Z., Stokes, D.L., Rice, W.J., Jones, L.R. (2003). Spatial and dynamic interactions between phospholamban and the canine cardiac Ca^{2+} pump revealed with use of heterobifunctional cross-linking agents. *J. Biol. Chem.* Sep 12 [Epub ahead of print].

Chou, J.J., Li, S., Klee, C.B., and Bax, A. (2001). Solution Structure of Ca^{2+}-Calmodulin Reveals Flexible Hand-Like Properties of its Domains. *Nat. Struct. Biol.*, **8**, 990–997.

Cornea, R.L., Jones, L.R., Autry, J.M., and Thomas, D.D. (1997). Mutation and Phosphorylation Change the Oligomeric Structure of Phospholamban in Lipid Bilayers. *Biochemistry*, **36**, 2960–2967.

De Dobbeleer, G., De Graef, C., M'Poudi, E., Gourdain, J.M., and Heenen, M. (1989). Reproduction of the Characteristic Morphologic Changes of Familial Benign Chronic Pemphigus in Cultures of Lesional Keratinocytes onto Dead Deepidermized Dermis. *J. Am. Acad. Dermatol.*, **21**, 961–965.

De Meis, L. (2003). Brown Adipose Tissue Ca^{2+}-ATPase: UNCOUPLED ATP HYDROLYSIS AND THERMOGENIC ACTIVITY. *J. Biol. Chem.*, **278**(43), 41856–61.

Dode, L., De Greef, C., Mountain, I., Attard, M., Town, M.M., Casteels, R., *et al.* (1998). Structure of the Human Sarco/Endoplasmic Reticulum Ca^{2+}-ATPase 3

Gene-Promoter Analysis and Alternative Splicing of the Serca3 Pre-mrna. *J. Biol. Chem.*, **273**, 13982–13994.

Durr, G., Strayle, J., Plemper, R., Elbs, S., Klee, S.K., Catty, P., *et al.* (1998). The Medial-Golgi Ion Pump Pmr1 Supplies the Yeast Secretory Pathway with Ca^{2+} and Mn^{2+} Required for Glycosylation, Sorting, and Endoplasmic Reticulum-Associated Protein Degradation. *Mol. Biol. Cell.*, **9**, 1149–1162.

East, J.M. (2000). Sarco(endo)plasmic Reticulum Calcium Pumps: Recent Advances in our Understanding of Structure/Function and Biology. *Mol. Membr. Biol.*, **17**, 189–200.

Falchetto, R., Vorherr, T., Brunner, J., and Carafoli, E. (1991). The Plasma Membrane Ca^{2+} Pump Contains a Site that Interacts with its Calmodulin-Binding Domain. *J. Biol. Chem.*, **266**, 2930–2936.

Fujii, J., Kadoma, M., Tada, M., Toda, H., and Sakiyama, F. (1986). Characterization of Structural Unit of Phospholamban by Amino Acid Sequencing and Electrophoretic Analysis. *Biochem. Biophys. Res. Commun.*, **138**, 1044–1050.

Fujii, J., Ueno, A., Kitano, K., Tanaka, S., Kadoma, M., and Tada, M. (1987). Complete Complementary DNA-Derived Amino Acid Sequence of Canine Cardiac Phospholamban. *J. Clin. Invest.*, **79**, 301–304.

Gayan-Ramirez, G., Vanzeir, L., Wuytack, F., and Decramer, M. (2000). Corticosteroids Decrease mRNA Levels of SERCA Pumps, whereas they Increase Sarcolipin mRNA in the Rat Diaphragm. *J. Physiol.*, **524**, 387–397.

Giordano, F.J., He, H., McDonough, P., Meyer, M., Sayen, M.R., and Dillmann, W.H. (1997). Adenovirus-Mediated Gene Transfer Reconstitutes Depressed Sarcoplasmic Reticulum Ca^{2+}-ATPase Levels and Shortens Prolonged Cardiac Myocyte Ca^{2+} Transients. *Circulation*, **96**, 400–403.

Gunteski-Hamblin, A.M., Clarke, D.M., and Shull, G.E. (1992). Molecular Cloning and Tissue Distribution of Alternatively Spliced mRNAs Encoding Possible Mammalian Homologues of the Yeast Secretory Pathway Calcium Pump. *Biochemistry*, **31**, 7600–7608.

Gwathmey, J.K., Copelas, L., MacKinnon, R., Schoen, F.J., Feldman, M.D., and Grossman, W. (1987). Abnormal Intracellular Calcium Handling in Myocardium from Patients with End-Stage Heart Failure. *Circ. Res.*, **61**, 70–76.

Hajjar, R.J., Kang, J.X., Gwathmey, J.K., and Rosenzweig, A. (1997). Physiological Effects of Adenoviral Gene Transfer of Sarcoplasmic Reticulum Calcium ATPase in Isolated Rat Myocytes. *Circulation*, **95**, 423–429.

Hasenfuss, G., Reinecke, H., Studer, R., Meyer, M., Pieske, B., Holtz, J., *et al.* (1994). Relation between Myocardial Function and Expression of Sarcoplasmic Reticulum Ca^{2+}-ATPase in Failing and Nonfailing Human Myocardium. *Circ. Res.*, **75**, 434–442.

Hasenfuss, G., Meyer, M., Schillinger, W., Preuss, M., Pieske, B., and Just, H. (1997). Calcium Handling

Proteins in the Failing Human Heart. *Basic Res. Cardiol.*, **92**, 87–93.

He, H., Giordano, F.J., Hilal-Dandan, R., *et al.* (1997). Overexpression of the Rat Sarcoplasmic Reticulum Ca^{2+} ATPase Gene in the Heart of Transgenic Mice Accelerates Calcium Transients and Cardiac Relaxation. *J. Clin. Invest.*, **100**, 380–389.

Hellstern, S., Pegoraro, S., Karim, C.B., *et al.* (2001). Sarcolipin, the Shorter Homologue of Phospholamban, Forms Oligomeric Structures in Detergent Micelles and in Liposomes. *J. Biol. Chem.*, **276**, 30845–30852.

Hofmann, F., Anagli, J., Carafoli, E., and Vorherr, T. (1994). Phosphorylation of the Calmodulin Binding Domain of the Plasma Membrane Ca^{2+} Pump by Protein Kinase C Reduces its Interaction with Calmodulin and with its Pump Receptor Site. *J. Biol. Chem.*, **269**, 24298–24303.

Hu, Z., Bonifas, J.M., Beech, J., Bench, G., Shigihara, T., Ogawa, H., *et al.* (2000). Mutations in ATP2C1, Encoding a Calcium Pump, Cause Hailey–Hailey Disease. *Nat. Genet.*, **24**, 61–65.

Hussain, A. and Inesi, G. (1999). Involvement of Sarco/Endoplasmic Reticulum Ca^{2+} ATPases in Cell Function and the Cellular Consequences of their Inhibition. *J. Membr. Biol.*, **172**, 91–99.

Ikura, M., Clore, G.M., Gronenborn, A.M., Zhu, G., Klee, C.B., and Bax, A. (1992). Solution Structure of a Calmodulin-Target Peptide Complex by Multidimensional NMR. *Science*, **256**, 632–638.

Inui, M., Chamberlain, B.K., Saito, A., and Fleischer, S. (1986). The Nature of the Modulation of Ca^{2+} Transport as Studied by Reconstitution of Cardiac Sarcoplasmic Reticulum. *J. Biol. Chem.*, **261**, 1794–1800.

James, P., Maeda, M., Fischer, R., *et al.* (1989). Identification and Primary Structure of a Calmodulin Binding Domain of the Ca^{2+} Pump of Human Erythrocytes. *J. Biol. Chem.*, **263**, 2905–2910.

James, P., Inui, M., Tada, M., Chiesi, M., and Carafoli, E. (1988). Nature and Site of Phospholamban Regulation of the Ca^{2+} Pump of Sarcoplasmic Reticulum. *Nature*, **342**, 90–92.

Kadambi, V.J., Ponniah, S., Harrer, J.M., *et al.* (1996). Cardiac-Specific Overexpression of Phospholamban Alters Calcium Kinetics and Resultant Cardiomyocyte Mechanics in Transgenic Mice. *J. Clin. Invest.*, **97**, 533–539.

Kaufman, R., Swaroop, M., and Murtha-Riel, P. (1994). Depletion of Manganese within the Secretory Pathway Inhibits O-linked Glycosylation in Mammalian Cells. *Biochemistry*, **33**, 9813–9819.

Kawasaki, H., Nakayama, S., and Kretsinger, R.H. (1998). Classification and Evolution of EF-Hand Proteins. *Biometals*, **11**, 277–295.

Kranias, E.G. (1985). Regulation of Ca^{2+} Transport by Cyclic 3', 5'-AMP-Dependent and Calcium-Calmodulin-Dependent Phosphorylation of Cardiac Sarcoplasmic Reticulum. *Biochim. Biophys. Acta*, **844**, 193–199.

Kimura, Y., Kurzydlowski, K., Tada, M., and MacLennan, D.H. (1996). Phospholamban Regulates the Ca^{2+}-ATPase through Intramembrane Interactions. J. Biol. Chem., 271, 21726–21731.

Kimura, Y., Kurzydlowski, K., Tada, M., and MacLennan, D.H. (1997). Phospholamban Inhibitory Function is Activated by Depolymerization. J. Biol. Chem., 272, 15061–15064.

Kirchberger, M.A. and Tada, M. (1976). Effects of Adenosine 3':5'-Monophosphate-Dependent Protein Kinase on Sarcoplasmic Reticulum Isolated from Cardiac and Slow and Fast Contracting Skeletal Muscles. J. Biol. Chem., 251, 725–729.

Koss, K.L. and Kranias, E.G. (1996). Phospholamban: A Prominent Regulator of Myocardial Contractility. Circ. Res., 79, 1059–1063.

Kozel, P.J., Friedman, R.A., Erway, L.C., Yamoah, E.N., Liu, L.H., Riddle, T., et al. (1998). Balance and Hearing Deficits in Mice with a Null Mutation in the Gene Encoding Plasma Membrane Ca^{2+}-ATPase Isoform 2. J. Biol. Chem., 273, 18693–18696.

Lepeuch, C.J., Haiech, J., and Demaille, J.G. (1979). Concerted Regulation of Cardiac Sarcoplasmic Reticulum Calcium Transport by Cyclic Adenosine Monophosphate-Dependent and Calcium-Calmodulin-Dependent Phosphorylations. Biochemistry, 18, 5150–5157.

Liu, L.H., Paul, R.J., Sutliff, R.L., Miller, M.L., Lorenz, J.N., Pun, R.Y., et al. (1997). Defective Endothelium-Dependent Relaxation of Vascular Smooth Muscle and Endothelial Cell Ca^{2+} Signaling in Mice Lacking Sarco(Endo)Plasmic Reticulum Ca^{2+}-ATPase Isoform 3. J. Biol. Chem., 272, 30538–30545.

Liu, L.H., Boivin, G.P., Prasad, V., Periasamy, M., and Shull, G.E. (2001). Squamous Cell Tumors in Mice Heterozygous for a Null Allele of ATP2A2, Encoding the Sarco(endo)plasmic Reticulum Ca^{2+}-ATPase Isoform 2 Ca^{2+} pump. J Biol Chem, 276, 26737–26740.

Loukianov, E., Ji, Y., Grupp, I.L., et al. (1998). Enhanced Myocardial Contractility and Increased Ca^{2+} Transport Function in Transgenic Hearts Expressing the Fast-Twitch Skeletal Muscle Sarcoplasmic Reticulum Ca^{2+}-ATPase. Circ. Res., 83, 889–897.

Luo, W., Grupp, I.L., Harrer, J., et al. (1994). Targeted Ablation of the Phospholamban Gene is Associated with Markedly Enhanced Myocardial Contractility and Loss of Beta-Agonist Stimulation. Circ. Res., 75, 401–409.

MacLennan, D.H. (2000). Ca^{2+} Signalling and Muscle Disease. Eur. J. Biochem., 267, 5291–5297.

Mandal, D., Woolf, T.B., and Rao, R. (2000). Manganese Selectivity of Pmr1, the Yeast Secretory Pathway Ion Pump, is Defined by Residue Gln783 in Transmembrane Segment 6. Residue Asp778 is Essential for Cation Transport. J. Biol. Chem., 275, 23933–23938.

Marchi, V., Sorin, A., Wei, Y., and Rao, R. (1997). Induction of Vacuolar Ca^{2+}-ATPase and H$^+$/Ca^{2+} Exchange Activity in Yeast Mutants Lacking Pmr1, the Golgi Ca^{2+}-ATPase. FEBS Lett., 454, 181–186.

Matheos, D.P., Kingsbury, T.J., Ahsan, U.S., and Cunningham, K.W. (1997). Tcn1p/Crz1p, a Calcineurin-Dependent Transcription Factor that Differentially Regulates Gene Expression in Saccharomyces cerevisiae. Genes Dev., 11, 3445–3458.

Meador, W.E., Means, A.R., and Quiocho F.A. (1992). Target Enzyme Recognition by Calmodulin: 2.4 Å Structure of a Calmodulin–Peptide Complex. Science, 257, 1251–1255.

Mercadier, J.J., Lompre, A.M., Duc, P., et al. (1990). Altered Sarcoplasmic Reticulum Ca^{2+}-ATPase Gene Expression in the Human Ventricle During End-Stage Heart Failure. J. Clin. Invest., 85, 305–309.

Minamisawa, S., Hoshijima, M., Chu, G., et al. (1999). Chronic Phospholamban–Sarcoplasmic Reticulum Calcium ATPase Interaction Is the Critical Calcium Cycling Defect in Dilated Cardiomyopathy. Cell, 99, 313–322.

Misquitta, C.M., Mack, D.P., and Grover, A.K. (1999). Sarco/Endoplasmic Reticulum Ca^{2+} (SERCA)-Pumps: Link to Heart Beats and Calcium Waves. Cell Calcium, 25, 277–290.

Morgan, J.P., Erny, R.E., Allen, P.D., Grossman, W., and Gwathmey, J.K. (1990). Abnormal Intracellular Calcium Handling, a Major Cause of Systolic and Diastolic Dysfunction in Ventricular Myocardium from Patients with Heart Failure. Circulation, 81, III21–III32.

Negash, S., Huang, S., and Squier, T.C. (1999). Rearrangement of Domain Elements of the Ca^{2+}-ATPase in Cardiac Sarcoplasmic Reticulum Membranes upon Phospholamban Phosphorylation. Biochemistry, 38, 8150–8158.

Niggli, V., Adunyah, E.S., Penniston, J.T., and Carafoli, E. (1981). Purified (Ca^{2+}-Mg^{2+})-ATPase of the Erythrocyte Membrane. Reconstitution and Effect of Calmodulin and Phospholipids. J. Biol. Chem., 256, 395–401.

Oda, K. (1992). Calcium Depletion Blocks Proteolytic Cleavages of Plasma Protein Precursors which Occur at the Golgi and/or Trans-Golgi Network. Possible Involvement of Ca^{2+}-Dependent Golgi Endoproteases. J. Biol. Chem., 267, 17465–17471.

Odermatt, A., Taschner, P.E., Khanna, V.K., Busch, H.F., Karpati, G., Jablecki, C.K., et al. (1996). Mutations in the Gene Encoding SERCA1, the Fast-Twitch Skeletal Muscle Sarcoplasmic Reticulum Ca^{2+} ATPase, are Associated with Brody Disease. Nat Genet, 14, 191–194.

Odermatt, A., Taschner, P.E., Scherer, S.W., et al. (1997). Characterization of the Gene Encoding Human Sarcolipin (SLN), a Proteolipid Associated with SERCA1: Absence of Structural Mutations in Five Patients with Brody Disease. Genomics, 45, 541–553.

Odermatt, A., Becker, S., Khanna, V.K., et al. (1998). Sarcolipin Regulates the Activity of SERCA1, the Fast-

Twitch Skeletal Muscle Sarcoplasmic Reticulum Ca^{2+}-ATPase. *J. Biol. Chem.*, **273**, 12360–12369.

Odermatt, A., Barton, K., Khanna, V.K., Mathieu, J., Escolar, D., Kuntzer, T., *et al.* (2000). The Mutation of Pro789 to Leu Reduces the Activity of the Fast-Twitch Skeletal Muscle Sarco(endo)plasmic Reticulum Ca^{2+} ATPase (SERCA1). *Hum. Genet.*, **106**, 482–491.

Penniston, J.T. and Enyedi, A. (1998). Modulation of the Plasma Membrane Ca^{2+} Pump. *J. Membr Biol.*, **165**, 101–109.

Periasamy, M., Reed, T.D., Liu, L.H., Ji, Y., Loukianov, E., Paul, R.J., *et al.* (1999). Impaired Cardiac Performance in Heterozygous Mice with a Null Mutation in the Sarco(endo)plasmic Reticulum Ca^{2+}-ATPase Isoform 2 (SERCA2) Gene. *J. Biol. Chem.*, **274**, 2556–2562.

Reinhardt, T.A. and Horst, R.L. (1999). Ca^{2+}-ATPases and their Expression in the Mammary Gland of Pregnant and Lactating Rats. *Am. J. Physiol.*, **276**, C796–C802.

Rice, W.J. and MacLennan, D.H. (1996). Scanning Mutagenesis Reveals a Similar Pattern of Mutation Sensitivity in Transmembrane Sequences M4, M5, and M6, but not in M8, of the Ca^{2+}-ATPase of Sarcoplasmic Reticulum (SERCA1a). *J. Biol. Chem.*, **271**, 31412–31419.

Ringer, S. (1883). A Further Contribution Regarding the Influence of Different Constituents of the Blood on the Contraction of the Heart. *J. Physiol. (London)*, **4**, 29–43.

Ringpfeil, F., Raus, A., DiGiovanna, J.J., Korge, B., Harth, W., Mazzanti, C., *et al.* (2001). Darier Disease – Novel Mutations in *ATP2A2* and Genotype–Phenotype Correlation. *Exp. Dermatol.*, **10**, 19–27.

Rudolph, H.K., Antebi, A., Fink, G.R., Buckley, C.M., Dorman, T.E., LeVitre, J., *et al.* (1989). The Yeast Secretory Pathway is Perturbed by Mutations in PMR1, a Member of a Ca^{2+} ATPase Family. *Cell*, **58**, 133–145.

Ruiz-Perez, V.L., Carter, S.A., Healy, E., Todd, C., Rees, J.L., Steijlen, P.M., *et al.* (1999). ATP2A2 Mutations in Darier's Disease: Variant Cutaneous Phenotypes are Associated with Missense Mutations, but Neuropsychiatric Features Are Independent of Mutation Class. *Hum. Mol. Genet.*, **8**, 1621–630.

Sachs, G. (1997). Proton Pump Inhibitors and Acid-Related Diseases. *Pharmacotherapy*, **17**, 22–37.

Sagara, Y. and Inesi, G. (1991). Inhibition of the Sarcoplasmic Reticulum Ca^{2+} Transport ATPase by Thapsigargin at Subnanomolar Concentrations. *J. Biol. Chem.*, **266**, 13503–13506.

Sakuntabhai, A., Ruiz-Perez, V., Carter, S., Jacobsen, N., Burge, S., Monk, S., *et al.* (1999). Mutations in ATP2A2, Encoding a Ca^{2+} Pump, Cause Darier Disease. *Nat. Genet.*, **21**, 271–277.

Schmidt, U., Hajjar, R.J., Helm, P.A., Kim, C.S., Doye, A.A., and Gwathmey, J.K. (1998). Contribution of Abnormal Sarcoplasmic Reticulum ATPase Activity to Systolic and Diastolic Dysfunction in Human Heart Failure. *J. Mol. Cell. Cardiol.*, **30**, 1929–1937.

Sham, J.S., Jones, L.R., and Morad, M. (1991). Phospholamban Mediates the Beta-Adrenergic-Enhanced Ca^{2+} Uptake in Mammalian Ventricular Myocytes. *Am. J. Physiol.*, **261**, H1344–H1349.

Shull, G.E. (2000). Gene Knockout Studies of Ca^{2+}-Transporting ATPases. *Eur. J. Biochem.*, **267**, 5284–5290.

Simmerman, H.K., Collins, J.H., Theibert, J.L., Wegener, A.D., and Jones, L.R. (1989). Sequence Analysis of Phospholamban. Identification of Phosphorylation Sites and Two Major Structural Domains. *J. Biol. Chem.*, **261**, 13333–13341.

Simmerman, H.K., Lovelace, D.E., and Jones, L.R. (1989). Secondary Structure of Detergent-Solubilized Phospholamban, a Phosphorylatable, Oligomeric Protein of Cardiac Sarcoplasmic Reticulum. *Biochim. Biophys. Acta*, **997**, 322–329.

Sorin, A., Rosas, G., and Rao, R. (1997). PMR1, a Ca^{2+}-ATPase in Yeast Golgi, has Properties Distinct from Sarco/Endoplasmic Reticulum and Plasma Membrane Calcium Pumps. *J. Biol. Chem.*, **272**, 9895–9901.

Street, V.A., McKee-Johnson, J.W., Fonseca, R.C., Tempel, B.L., and Noben-Trauth, K. (1998). Mutations in a Plasma Membrane Ca^{2+}-ATPase Gene Cause Deafness in *deafwaddler* Mice. *Nat. Genet.*, **19**, 390–394.

Strehler, E.E. and Zacharias, D.A. (2001). Role of Alternative Splicing in Generating Isoform Diversity Among Plasma Membrane Calcium Pumps. *Physiol. Rev.*, **81**, 21–50.

Stuart, R.O., Sun, A., Bush, K.T., and Nigam, S.K. (1996). Dependence of Epithelial Intercellular Junction Biogenesis on Thapsigargin-Sensitive Intracellular Calcium Stores. *J. Biol. Chem.*, **271**, 13636–13641.

Sudbrak, R., Brown, J., Dobson-Stone, C., Carter, S., Ramser, J., White, J., *et al.* (2000). Hailey–Hailey Disease is Caused by Mutations in ATP2C1 Encoding a Novel Ca^{2+} Pump. *Hum. Mol. Genet.*, **9**, 1131–1140.

Sugden, P.H. and Clerk, A. (1988). Cellular Mechanisms of Cardiac Hypertrophy. *J. Mol. Med.*, **76**, 725–746.

Tada, M., Kirchberger, M.A., and Katz, A.M. (1975). Phosphorylation of a 22,000-dalton Component of the Cardiac Sarcoplasmic Reticulum by Adenosine 3′:5′-Monophosphate-Dependent Protein Kinase. *J. Biol. Chem.*, **250**, 2640–2647.

Tada, M., Inui, M., Yamada, M., *et al.* (1983). Effects of Phospholamban Phosphorylation Catalyzed by Adenosine 3′:5′-Monophosphate *J. Mol. Cell. Cardiol.*, **15**, 335–346.

Takahashi, T., Allen, P.D., and Izumo, S. (1992). Expression of A-, B-, and C-Type Natriuretic Peptide Genes in Failing and Developing Human Ventricles. Correlation with Expression of the Ca^{2+}-ATPase Gene. *Circ. Res.*, **71**, 9–17.

Tatulian, S.A., Chen, B., and Li, J. (2002). The Inhibitory Action of Phospholamban Involves Stabilization of Alpha-Helices within the Ca^{2+}-ATPase. *Biochemistry*, **41**, 741–751.

Ton, V.K., Mandal, D., Vahadji, C., and Rao, R. (2002). Functional Expression in Yeast of the Human Secretory Pathway Ca^{2+}, Mn^{2+}-ATPase Defective in Hailey–Hailey Disease. *J. Biol. Chem.*, **277**, 6422–6427.

Toyofuku, T., Kurzydlowski, K., Tada, M., and MacLennan, D.H. (1993). Identification of Regions in the Ca^{2+}-ATPase of Sarcoplasmic Reticulum that Affect Functional Association with Phospholamban. *J. Biol. Chem.*, **268**, 2809–2815.

Toyofuku, T., Kurzydlowski, K., Tada, M., and MacLennan, D.H. (1994a). Amino Acids Lys-Asp-Asp-Lys-Pro-Val402 in the Ca^{2+}-ATPase of Cardiac Sarcoplasmic Reticulum are Critical for Functional Association with Phospholamban. *J. Biol. Chem.*, **269**, 22929–22932.

Toyofuku, T., Kurzydlowski, K., Tada, M., and MacLennan, D.H. (1994b). Amino Acids Glu2 to Ile18 in the Cytoplasmic Domain of Phospholamban are Essential for Functional Association with the Ca^{2+}-ATPase of Sarcoplasmic Reticulum. *J. Biol. Chem.*, **269**, 3088–3094.

Toyoshima, C., Asahi, M., Sugita, Y., Khanna, R., Tsuda, T., MacLennan, D.H. (2003). Modelling of the inhibitory interaction of phospholamban with the Ca^{2+} ATPase. *Proc. Natl Acad. Sci. USA.* Jan 21; **100**(2), 467–72. Epub 2003 Jan 13.

Toyoshima, C. and Nomura, H. (2002). Structural Changes in the Calcium Pump Accompanying the Dissociation of Calcium. *Nature*, **418**, 605–611.

Toyoshima, C., Nakasako, M., Nomura, H., and Ogawa, H. (2000). Crystal Structure of the Calcium Pump of Sarcoplasmic Reticulum at 2.6 Å Resolution. *Nature*, **405**, 647–655.

Watt, F.M., Mattey, D.L., and Garrod, D.R. (1984). Calcium-Induced Reorganization of Desmosomal Components in Cultured Human Keratinocytes. *J. Cell. Biol.*, **99**, 2211–2215.

Wegener, A.D. and Jones, L.R. (1984). Phosphorylation-Induced Mobility Shift in Phospholamban in Sodium Dodecyl Sulfate–Polyacrylamide Gels. Evidence for a Protein Structure Consisting of Multiple Identical Phosphorylatable Subunits. *J. Biol. Chem.*, **259**, 1834–1841.

Wu, K.D., Lee, W.S., Wey, J., Bungard, D., and Lytton, J. (1995). Localization and Quantification of Endoplasmic Reticulum Ca^{2+}-ATPase Isoform Transcripts. *Am. J. Physiol.*, **269**, C775–C784.

Wuytack, F., Papp, B., Verboomen, H., Raeymaekers, L., Dode, L., Bobe, R., *et al.* (1994). A Sarco/Endoplasmic Reticulum Ca^{2+}-ATPase 3-Type Ca^{2+} Pump Is Expressed in Platelets, in Lymphoid Cells, and in Mast Cells. *J. Biol. Chem.*, **269**, 1410–1416.

Yu, X., Inesi, G. (1995). Variable stoichiometric efficiency of Ca2+ and Sr2+ transport by the sarcoplasmic reticulum ATPase. *J. Boil. Chem.*, **270**(9), 4361–7.

Zacharias, D.A. and Kappen, C. (1999). Developmental Expression of the Four Plasma Membrane Calcium ATPase (Pmca) Genes in the Mouse. *Biochim. Biophys. Acta*, **1428**, 397–405.

Zhang, Z., Lewis, D., Strock, C., Inesi, G., Nakasako, M., Nomura, H., *et al.* (2000). Detailed Characterization of the Cooperative Mechanism of Ca^{2+} Binding and Catalytic Activation in the Ca^{2+} Transport (SERCA) ATPase. *Biochemistry*, **39**, 8758–8767.

Zylinska, L. and Soszynski, M. (2000). Plasma Membrane Ca^{2+}-ATPase in Excitable and Nonexcitable Cells. *Acta Biochim. Pol.*, **47**, 529–539.

Index